AF411344

Mathematical Modeling and Control of Bioprocesses

Mathematical Modeling and Control of Bioprocesses

Editors

Philippe Bogaerts
Alain Vande Wouwer

MDPI • Basel • Beijing • Wuhan • Barcelona • Belgrade • Manchester • Tokyo • Cluj • Tianjin

Editors
Philippe Bogaerts
Université Libre de Bruxelles
Belgium

Alain Vande Wouwer
University of Mons
Belgium

Editorial Office
MDPI
St. Alban-Anlage 66
4052 Basel, Switzerland

This is a reprint of articles from the Special Issue published online in the open access journal *Processes* (ISSN 2227-9717) (available at: https://www.mdpi.com/journal/processes/special_issues/Model_Control_Bioprocesses).

For citation purposes, cite each article independently as indicated on the article page online and as indicated below:

LastName, A.A.; LastName, B.B.; LastName, C.C. Article Title. *Journal Name* **Year**, *Volume Number*, Page Range.

ISBN 978-3-0365-7140-9 (Hbk)
ISBN 978-3-0365-7141-6 (PDF)

Cover image courtesy of Philippe Bogaerts

Contents

About the Editors

Philippe Bogaerts

Philippe Bogaerts (full professor at the Université libre de Bruxelles) received his master's degree in Chemical Engineering (1992), in Control Engineering (1993) and his Ph.D. degree in Applied Sciences (1999) from the Faculty of Applied Sciences at Université libre de Bruxelles (Belgium). He was the president of the Brussels Bioengineering School from 2008 to 2012. Since 2013, he has worked as a full professor at Brussels School of Engineering, Université libre de Bruxelles (Belgium), where he has been the head of the BioControl, BioInfo & BioMatter (3BIO) lab and of the Biosystems Modeling and Control (3BIO-BioControl) research unit since 2003. His research topics include mathematical modeling, parameter identification, non-linear state estimation, model-based optimization and process control of biochemical processes. He is the co-author of more than 100 journal and conference papers on these topics.

Alain Vande Wouwer

Alain Vande Wouwer (full professor at the University of Mons) received his master's degree in Electrical Engineering from the Faculté Polytechnique de Mons, Belgium, in 1988, and his Ph.D. degree in Applied Sciences from the same university (together with a European doctorate degree in collaboration with Stuttgart Universität, Germany) in 1994. In 1994, he underwent a postdoctoral research stay in the department of mechanical engineering at Laval University, Quebec. Since 2009, he has been a full professor in the department of Electrical Engineering of the University of Mons, Belgium, and is currently the head of the Systems, Estimation, Control, and Optimization group (SECO). He is the co-author of two books on the numerical analysis of partial differential equations, and of more than 250 conference and journal articles on various aspects of process modeling and control. His research interests include non-linear dynamics, parameter and state estimation, dynamic optimization, and process control, with applications to biochemical processes and unmanned aerial vehicles.

Preface to "Mathematical Modeling and Control of Bioprocesses"

Over the years, we have been involved in several invited sessions at international conferences, as well as Special Issues dedicated to topics revolving around mathematical modeling and the analysis of biological systems, as well as bioprocess optimization and control. All this would not have been possible without the participation of a great circle of colleagues and friends who have contributed high-quality work to these events. The present book is no exception to this fruitful synergy and communication, and we would like to express our gratitude to the contributors to this collective publication. The array of subjects considered in the several chapters reflects the vibrant research in the area of biosystem modeling and control. Mathematical modeling and analysis have always been central in unveiling behavioral aspects of biosystems and today, more than ever, to forge strategies to improve the quality, productivity, and sustainability of bioprocesses.

Philippe Bogaerts and Alain Vande Wouwer
Editors

Editorial

Special Issue: Mathematical Modeling and Control of Bioprocesses

Philippe Bogaerts [1],* and Alain Vande Wouwer [2],*

[1] Brussels School of Engineering, Université Libre de Bruxelles, 3BIO-BioControl, Av. F.-D. Roosevelt 50, CP 165-61, B-1050 Brussels, Belgium
[2] Faculté Polytechnique de Mons, Université de Mons, Systems, Estimation, Control and Optimization (SECO), B-7000 Mons, Belgium
* Correspondence: philippe.bogaerts@ulb.be (P.B.); alain.vandewouwer@umons.ac.be (A.V.W.)

This Special Issue (SI) of Processes on Mathematical Modeling and Control of Bioprocesses (MMCB) contains papers focusing, on the one hand, on mathematical modeling of biological processes at different scales ranging from microscopic to macroscopic levels and, on the other hand, on model-based estimation, optimization and control of these processes.

1. Mathematical Modeling of Biological Processes at Microscopic Scale

At the microscopic scale, metabolic networks are often used to model cell behavior in different culture conditions. Three papers of the SI on MMCB are focused on metabolic network modeling. One of the major issues with models based on metabolic networks is their underdetermination, in the sense that the number of unknown intracellular fluxes to be determined is usually higher than the number of available equations corresponding to the mass balances and the available measurements. Bogaerts and Vande Wouwer [1] review various methods to tackle this underdetermination, among which are flux pathway analysis, flux balance analysis, flux variability analysis and sampling of the flux solution space. One of these methods, namely Dynamic Flux Balance Analysis, is used by Shen and Budman [2] to infer metabolite concentrations for which hardware measurements are not available. A variable structure system, describing different regions of the state space, is introduced and a set membership-based approach is used to estimate the unmeasured concentrations from few available measurements. To completely circumvent the abovementioned problem of system underdetermination, Bastin et al. [3] use relatively simple macroscopic models that allow obtaining a unique flux distribution to describe VERO cell behavior in different bioreactor culture conditions, e.g., exponential growth phase or substrate-limited growth phase. Such models could be used to develop feedback control strategies in fed-batch or perfused bioreactors.

2. Mathematical Modeling of Biological Processes at Macroscopic Scale

At a more macroscopic scale, four papers of the SI on MMCB focus on dynamic models for describing different processes, i.e., production of biopolymers [4], production of biopesticides [5], biodegradation [6] and biorefinery of second-generation biomass [7]. Duvigneau et al. [4] propose a new kinetic model for describing the production of the bio-copolymer poly(3-hydroxybutyrate-co-3-hydroxyvalerate) (PHBV), which has a broad range of applications and is easier to process than the classical homopolymer poly(3-hydroxybutyrate). Given the coupling of CO_2 online measurements in the exhaust gas to biomass production, the model allows for predicting the composition and current yield of PHBV along the process. Monroy et al. [5] propose and compare different kinetic models for three different strains of *Bacillus thuringiensis* ssp *Kurstaki*, a microorganism used for the production of biopesticides. The main goal of the model is to estimate the total protein productivity, yield and titer. Based on several experimental datasets, a dynamical model is finally selected with the Akaike information criterion. Dimitrova and Zlateva [6]

Citation: Bogaerts, P.; Vande Wouwer, A. Special Issue: Mathematical Modeling and Control of Bioprocesses. *Processes* 2022, 10, 1372. https://doi.org/10.3390/pr10071372

Received: 4 July 2022
Accepted: 12 July 2022
Published: 14 July 2022

propose a mathematical model for the biodegradation of a phenol and cresol mixture in a continuously stirred bioreactor, with a specific growth rate that involves sum kinetics with interaction parameters (SKIP) and inhibition effects. The global stabilizability of the model dynamics towards equilibrium points is analyzed and illustrated with numerical examples. Sbarciog et al. [7] propose a biorefinery model that integrates several processes, among which are steam refining, anaerobic digestion, ammonia stripping and composting. The overall goal is the model-based optimization of the process. The authors illustrate with simulation results the potential to efficiently produce oligosaccharides, lignin, fibers, biogas, fertilizer and compost from real collected biowaste.

3. Model-Based Estimation, Optimization and Control of Biological Processes

This SI on MMBC also focuses attention on recent developments in monitoring and optimization of biological systems, where mathematical models can of course play an important role, as reflected by four articles.

Sokač et al. [8] review the application of mathematical modeling and optimization for enhancing composting, which is a complex process whose efficiency is influenced by temperature, pH, moisture content, C/N ratio, particle size, nutrient content and oxygen supply. In [9], Djema et al. derive optimal control strategies based on the Pontryagin maximum principle in order to ensure the domination of the strain of interest in cultures of microalgae in the chemostat described by the Droop model. Sari [10] considers one-step and two-step models of anaerobic digestion together with a large class of growth functions encountered in applications, and studies their operating diagrams as a function of the dilution rate and inlet concentration, so as to define the best operating conditions for biogas production. In [11], Wallocha and Popp present an off-gas-based software sensor for real-time biomass estimation in continuous single-use bioreactors with CHO cell lines. They discuss the interest of considering viable cell volume concentration (instead of the density) as cell size or volume can have a direct impact on oxygen demand.

Finally, two articles are devoted to controller design and implementation, addressing important practical aspects of bioprocess control strategy deployment.

Colin-Robles et al. [12] consider the maximization of the hydrogen production rate in a microbial electrolysis cell. Using the golden section search optimization algorithm coupled with a robust super-twisting controller, the cell is brought to the maximum of a static performance map. The proposed optimization strategy is embedded in an FPGA throughout different digital architectures that are executed in parallel without hardware sharing. In [13], Butkus et al. propose a gain scheduling approach, which is based on controller input/output signals only and does not require additional online measurements of cultivation process variables for the adaptation of controller parameters. The approach is used to control dissolved oxygen concentration in a bioreactor operated in fed-batch mode.

Author Contributions: Both authors contributed equally. All authors have read and agreed to the published version of the manuscript.

Funding: This research received no external funding.

Conflicts of Interest: The authors declare no conflict of interest.

References

1. Bogaerts, P.; Vande Wouwer, A. How to Tackle Underdeterminacy in Metabolic Flux Analysis? A Tutorial and Critical Review. *Processes* **2021**, *9*, 1577. [CrossRef]
2. Shen, X.; Budman, H. Set Membership Estimation with Dynamic Flux Balance Models. *Processes* **2021**, *9*, 1762. [CrossRef]
3. Bastin, G.; Chotteau, V.; Wouwer, A.V. Metabolic Flux Analysis of VERO Cells under Various Culture Conditions. *Processes* **2021**, *9*, 2097. [CrossRef]
4. Duvigneau, S.; Dürr, R.; Behrens, J.; Kienle, A. Advanced Kinetic Modeling of Bio-co-polymer Poly(3-hydroxybutyrate-co-3-hydroxyvalerate) Production Using Fructose and Propionate as Carbon Sources. *Processes* **2021**, *9*, 1260. [CrossRef]

5. Monroy, T.S.; Abdelmalek, N.; Rouis, S.; Kallassy, M.; Saad, J.; Abboud, J.; Cescut, J.; Bensaid, N.; Fillaudeau, L.; Aceves-Lara, C.A. Dynamic Model for Biomass and Proteins Production by Three Bacillus Thuringiensis ssp Kurstaki Strains. *Processes* **2021**, *9*, 2147. [CrossRef]
6. Dimitrova, N.; Zlateva, P. Global Stability Analysis of a Bioreactor Model for Phenol and Cresol Mixture Degradation. *Processes* **2021**, *9*, 124. [CrossRef]
7. Sbarciog, M.; De Buck, V.; Akkermans, S.; Bhonsale, S.; Polanska, M.; Van Impe, J.F.M. Design, Implementation and Simulation of a Small-Scale Biorefinery Model. *Processes* **2022**, *10*, 829. [CrossRef]
8. Sokač, T.; Valinger, D.; Benković, M.; Jurina, T.; Kljusurić, J.G.; Redovniković, I.R.; Tušek, A.J. Application of Optimization and Modeling for the Composting Process Enhancement. *Processes* **2022**, *10*, 229. [CrossRef]
9. Djema, W.; Bayen, T.; Bernard, O. Optimal Darwinian Selection of Microorganisms with Internal Storage. *Processes* **2022**, *10*, 461. [CrossRef]
10. Sari, T. Best Operating Conditions for Biogas Production in Some Simple Anaerobic Digestion Models. *Processes* **2022**, *10*, 258. [CrossRef]
11. Wallocha, T.; Popp, O. Off-Gas-Based Soft Sensor for Real-Time Monitoring of Biomass and Metabolism in Chinese Hamster Ovary Cell Continuous Processes in Single-Use Bioreactors. *Processes* **2021**, *9*, 2073. [CrossRef]
12. Colin-Robles, J.; Torres-Zúñiga, I.; Ibarra-Manzano, M.A.; Alcaraz-González, V. FPGA-Based Implementation of an Optimization Algorithm to Maximize the Productivity of a Microbial Electrolysis Cell. *Processes* **2021**, *9*, 1111. [CrossRef]
13. Butkus, M.; Levišauskas, D.; Galvanauskas, V. Simple Gain-Scheduled Control System for Dissolved Oxygen Control in Bioreactors. *Processes* **2021**, *9*, 1493. [CrossRef]

processes

Article

Metabolic Flux Analysis of VERO Cells under Various Culture Conditions

Georges Bastin [1,*], Véronique Chotteau [2] and Alain Vande Wouwer [3]

[1] Department of Mathematical Engineering, ICTEAM, UCLouvain, 1348 Louvain-la-Neuve, Belgium
[2] Cell Technology Group, Division of Industrial Biotechnology, KTH Royal Institute of Technology, 100 44 Stockholm, Sweden; veronique.chotteau@biotech.kth.se
[3] Systems, Estimation, Control and Optimization (SECO), University of Mons, 7000 Mons, Belgium; alain.vandewouwer@umons.ac.be
* Correspondence: georges.bastin@uclouvain.be

Abstract: Although the culture of VERO cells in bioreactors is an important industrial bioprocess for the production of viruses and vaccines, surprisingly few reports on the analysis of the flux distribution in the cell metabolism have been published. In this study, an attempt is made to fill this gap by providing an analysis of relatively simple metabolic networks, which are constructed to describe the cell behavior in different culture conditions, e.g., the exponential growth phase (availability of glucose and glutamine), cell growth without glutamine, and cell growth without glucose and glutamine. The metabolic networks are kept as simple as possible in order to avoid underdeterminacy linked to the lack of extracellular measurements, and a unique flux distribution is computed in each case based on a mild assumption that the macromolecular composition of the cell is known. The result of this computation provides some insight into the metabolic changes triggered by the culture conditions, which could support the design of feedback control strategies in fed batch or perfusion bioreactors where the lactate concentration is measured online and regulated by controlling the delivery rates of glucose and, possibly, of some essential amino acids.

Keywords: metabolic flux analysis; metabolic network; VERO cells; biotechnology

Citation: Bastin, G.; Chotteau, V.; Vande Wouwer, A. Metabolic Flux Analysis of VERO Cells under Various Culture Conditions. *Processes* **2021**, *9*, 2097. https://doi.org/10.3390/pr9122097

Academic Editor: Florian M. Wurm

Received: 10 October 2021
Accepted: 15 November 2021
Published: 23 November 2021

1. Introduction

The production of biopharmaceuticals using cultures of genetically modified strains has gained tremendous importance in the drug manufacturing sector. In this context, it is important to understand and assess the influence of the culture conditions, and the impact of metabolic engineering, on the yield of the products of interest. This can be achieved through an analysis of the flux distribution inside the metabolic network of the cells or microorganisms under consideration. Various computational procedures have been proposed for that purpose, including metabolic flux analysis and flux balance analysis [1].

Even though there has been a significant number of reports of the application of these procedures to cultures of CHO cells and hybridoma cells (e.g., [2–8]), there has been surprisingly few reports focusing on the metabolism of VERO cell cultures [9]. However VERO cells are important vectors for the production of viruses (and vaccines) (e.g., [10–18]).

The objective of this study is to apply metabolic flux analysis to small metabolic networks of VERO cells, on the basis of experimental data collected in three different culture conditions. In each case, the network is designed to be fully compatible with the data while being kept as simple as possible to avoid the underdeterminacy that usually prevails when manipulating large metabolic networks. In this study, the considered metabolic networks allow keeping the underdeterminacy at a minimum, and to compute a unique solution based on the only additional mild assumption that the macromolecular (proteins, nucleic acids, membrane lipids) composition of the cell is as reported in the literature [19] (p. 113, Table 7.1).

The paper is organized as follows. The next section presents the experimental data, cell densities, and metabolite concentrations, collected in two batch cultures of VERO cells. From these data, depending on the availability of glucose and glutamine, three types of culture growth are distinguished. The metabolic network and its analysis are then detailed and discussed in Section 3 for the exponential growth phase with glucose and glutamine as carbon and nitrogen sources, respectively. Next, in Section 4, we consider the case where glutamine is replaced by glutamate as the source of nitrogen. Finally, the case where lactate becomes the source of carbon instead of glucose is addressed in Section 5. Final conclusions are presented in the last section.

2. Experimental Data

In this paper, we use data from two batch cultures of VERO cells, labeled (a1) and (a2), which were simultaneously carried out over a period of eight days in parallel spinner-flasks, using the same culture medium. In particular, the two cultures were seeded from the same pool of cells. The only difference between the two cultures lies in the initial concentration of glucose. A full description of the materials and methods of these experiments can be found in [20] (Section 3).

The experiments were performed in spinner-flasks (paddle impeller type). The culture volume was 250–270 mL. VERO cells (passage 136–146) were grown adherently on Cytodex 1 microcarriers (3.5 g/L). The spinner-flasks were inoculated with approximately 10^5 cells/mL, which corresponded to eight cells per microcarrier on average. The basic culture medium was M199 supplemented at inoculation with fetal calf serum (10% v/v) and antibiotic (neomycin sulfate 5% v/v). The culture was magnetically stirred at 45 RPM. The oxygen supply was provided by transfer via the head space. The atmosphere of the head space was renewed twice a day.

The culture medium was sampled (2 mL) twice a day for analysis (except on day 5, where there was only one sample). Cell counting was done with a hemacytometer using crystal violet staining. Glucose and lactate concentrations were determined with a Yellow–Springer analyzer. Amino acids and ammonia were determined with the HPLC method.

The data collected during these cultures are presented in various figures hereafter. The time evolution of cell densities (counting) is shown in Figure 1. We can readily observe two different successive phases in both cultures. The growth begins with a classical exponential phase during the first four days. Then, from the fourth to the eighth days, there was a shift to a slower quasi-linear growth. An explanation for this behavior can be found in Figure 2 where the time evolution of the concentrations of glucose and glutamine in the culture medium are shown. Indeed, it can be seen that, after the fourth day, both glucose and glutamine are depleted in culture (a1). However the growth proceeds more slowly, with lactate as the carbon source, while alanine and glutamate are alternative nitrogen sources in the central metabolism (see Figure 3). In contrast, culture (a2) is operated with an excess of glucose, so that only glutamine is depleted on the fourth day and replaced by glutamate as a source of nitrogen. These three different situations of the culture conditions are summarized in the Table 1 below. Our purpose, in this paper, is to perform a metabolic flux analysis in order to compute and compare the distributions of the intracellular metabolic fluxes in these three situations.

Table 1. Three different culture conditions.

Culture Conditions	Carbon Source	Nitrogen Source	
Exponential growth	Glucose	Glutamine	$\mu \approx 0.6\,\mathrm{day}^{-1}$
Growth without Gln	Glucose	Glutamate	$\mu \approx 0.18\,\mathrm{day}^{-1}$
Growth without Glc and Gln	Lactate	Glutamate, Alanine	$\mu \approx 0.04\,\mathrm{day}^{-1}$

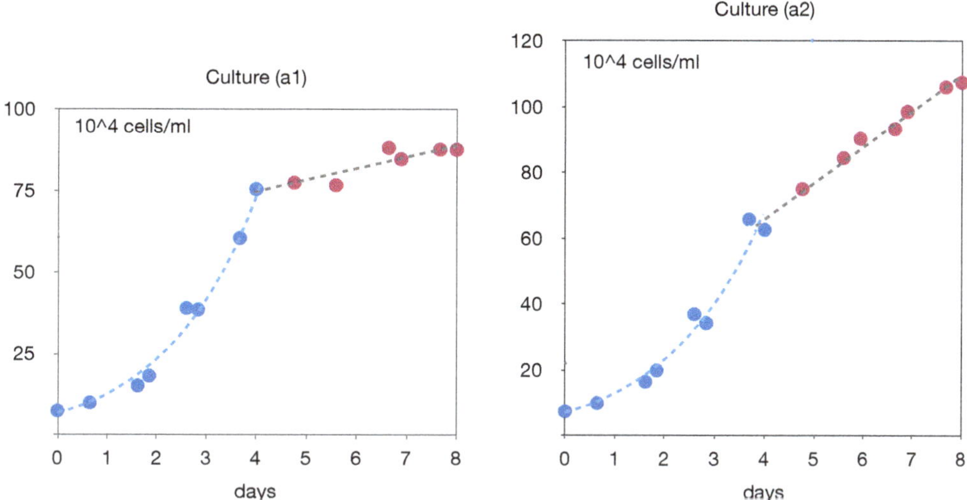

Figure 1. Time evolution of cell density in cultures a1 (**left**) and a2 (**right**).

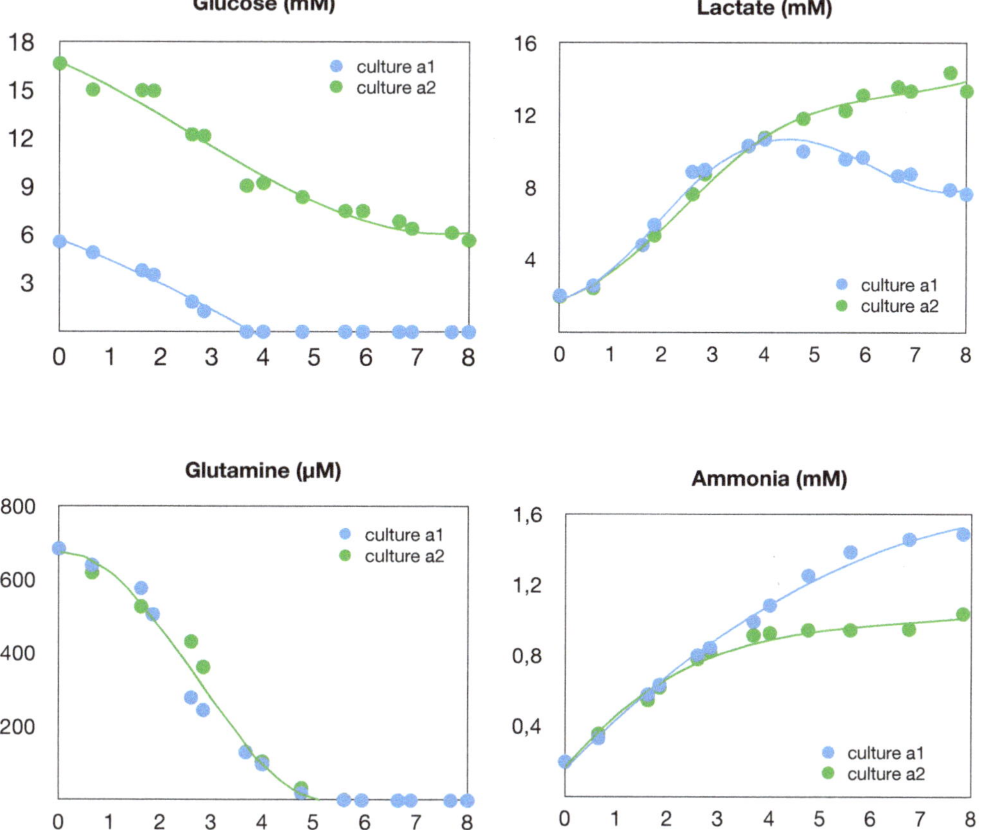

Figure 2. Time evolution of substrates and products of the central metabolism in cultures a1 and a2.

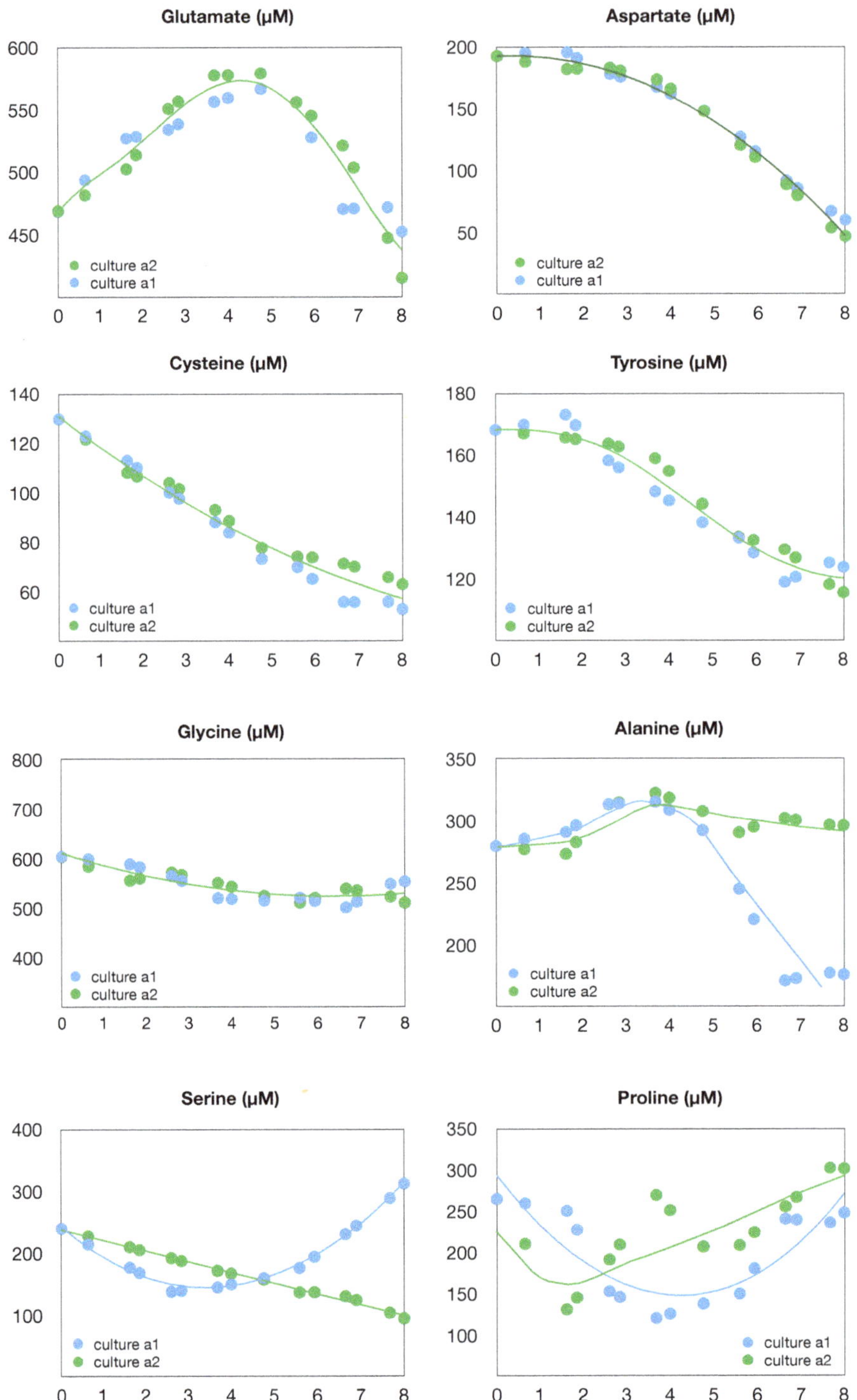

Figure 3. Time evolution of non-essential amino acids in cultures a1 and a2.

Data of Amino Acids

Concentration measurements of eighteen amino acids in the culture medium were measured with an HPLC method. In Figure 4, we present the data for nine essential amino acids, which are naturally consumed in correlation with the cell growth: arginine, histidine, isoleucine, leucine, lysine, methionine, phenylalanine, tryptophan, and valine. Data are not available for threonine. It can be seen that the shape of the consumption is quite similar for both cultures, (a1) and (a2), with only marginal deviations for leucine and isoleucine.

Moreover, in addition to glutamine in Figure 2, we present in Figure 3 the measurements for eight other non-essential amino acids: alanine, aspartate, cysteine, glutamate, glycine, proline, serine, and tyrosine. It can be seen that these measurements do not always follow the shape of the cellular growth. In particular, glutamate and alanine are accumulated in the culture medium during the exponential growth, but are significantly consumed when glutamine is depleted. We also note that the medium does not contain asparagine at the start of the culture and that asparagine data are not available throughout the culture.

In Tables 2–4, the specific uptake and/or excretion rates of all the species measured in the culture medium are given for the three considered culture conditions (Tables 2 and 4 are complementary and when a rate does not appear in one table (symbol ——) it appears in the other one). These rates are estimated from the slopes of the solid curves that fit the experimental data in Figures 2–4, at time $t = 2.90$ days for the exponential growth and at time $t = 6.64$ days for the growth without glucose and/or glutamine.

Table 2. Specific uptake rates (μM/d $\times 10^7$ cell) Glucose, Lactate and non-essential AA.

Species		Exponential Growth	Growth without Gln	Growth without Glc & Gln
Glucose	v_{Glc}	42.308	8.933	0.0
Lactate	v_{Lac}	——	——	10.218
Glutamine	v_{Gln}	4.994	0.0	0.0
Alanine	v_{Ala}	——	0.058	0.702
Aspartate	v_{Asp}	0.338	0.337	0.376
Cysteine	v_{Cys}	0.264	0.068	0.077
Glutamate	v_{Glu}	——	0.499	0.558
Glycine	v_{Gly}	0.325	——	——
Proline	v_{Pro}	0.565	——	——
Serine	v_{Ser}	0.231	0.182	——
Tyrosine	v_{Tyr}	0.205	0.063	0.071

Table 3. Specific uptake rates (μM/d $\times 10^7$ cell) for essential AA.

Species		Exponential Growth	Growth without Gln	Growth without Glc & Gln
Arginine	v_{Arg}	0.949	0.179	0.200
Histidine	v_{His}	0.195	0.045	0.051
Isoleucine	v_{Ile}	0.428	0.105	0.171
Leucine	v_{Leu}	0.821	0.200	0.459
Lysine	v_{Lys}	0.484	0.199	0.222
Methionine	v_{Met}	0.244	0.100	0.112
Phenylalanine	v_{Phe}	0.228	0.091	0.101
Tryptophan	v_{Trn}	0.091	0.037	0.042
Valine	v_{Val}	0.499	0.146	0.164

Table 4. Specific excretion rates (μM/d $\times 10^7$ cell) Lactate, NH3 and non-essential AA.

Species		Exponential Growth	Growth without Gln	Growth without Glc & Gln
Lactate	v_{Lac}	63.770	8.818	——
NH3	v_{NH3}	5.316	0.195	1.089
Alanine	v_{Ala}	0.513	——	——
Glutamate	v_{Glu}	0.659	——	——
Glycine	v_{Gly}	——	0.019	0.021
Proline	v_{Pro}	——	0.474	0.859
Serine	v_{Ser}	——	——	0.624

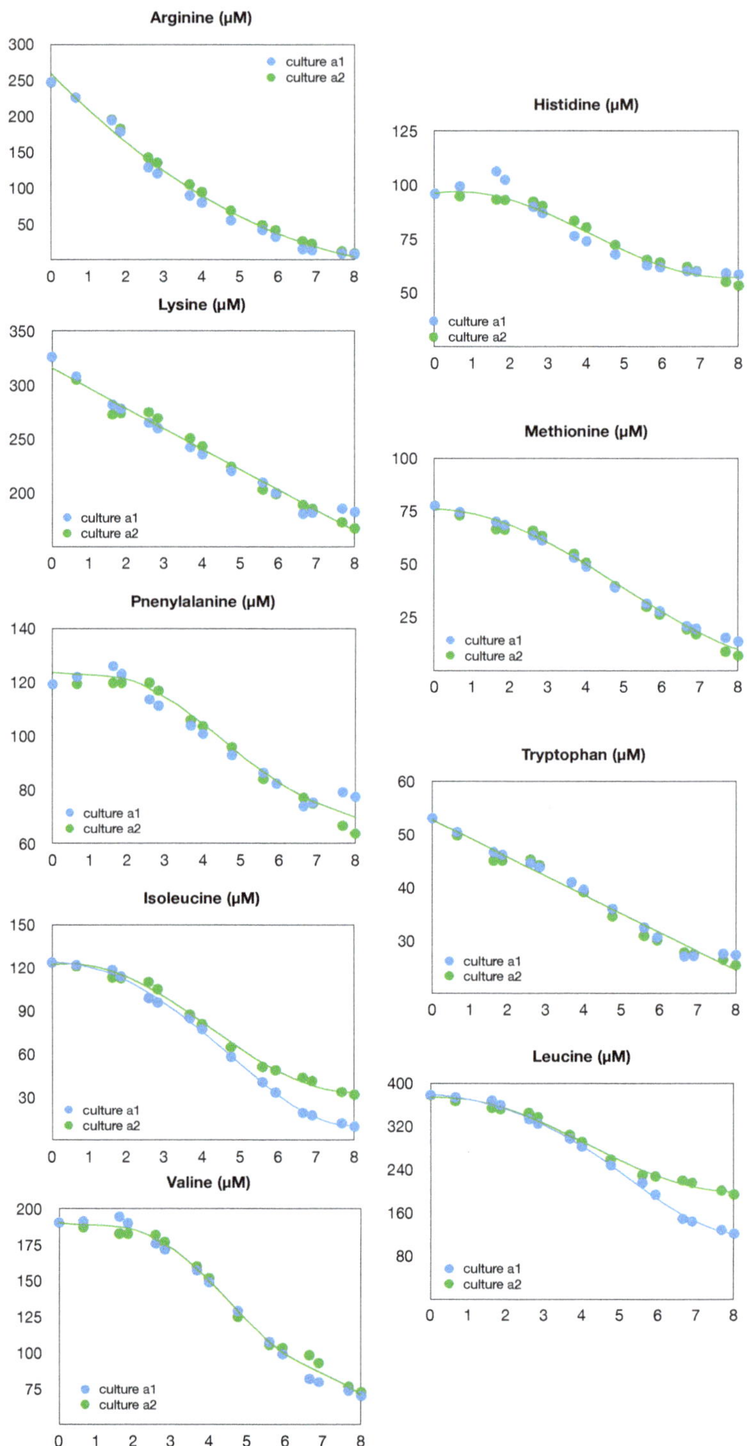

Figure 4. Time evolution of essential amino acids in cultures a1 and a2.

3. Metabolic Flux Analysis of the Exponential Growth Phase

3.1. Metabolic Network

The metabolic network considered for the exponential growth is made up of all the biochemical reactions in Figures 5–7. The main motivations behind the set-up of this network are given in the present section.

3.1.1. Central Metabolism

For the growth of mammalian cells, the central metabolism involves glycolysis, TCA cycle, and glutaminolysis, as represented in Figure 5. For simplicity, the pentose phosphate pathway is neglected.

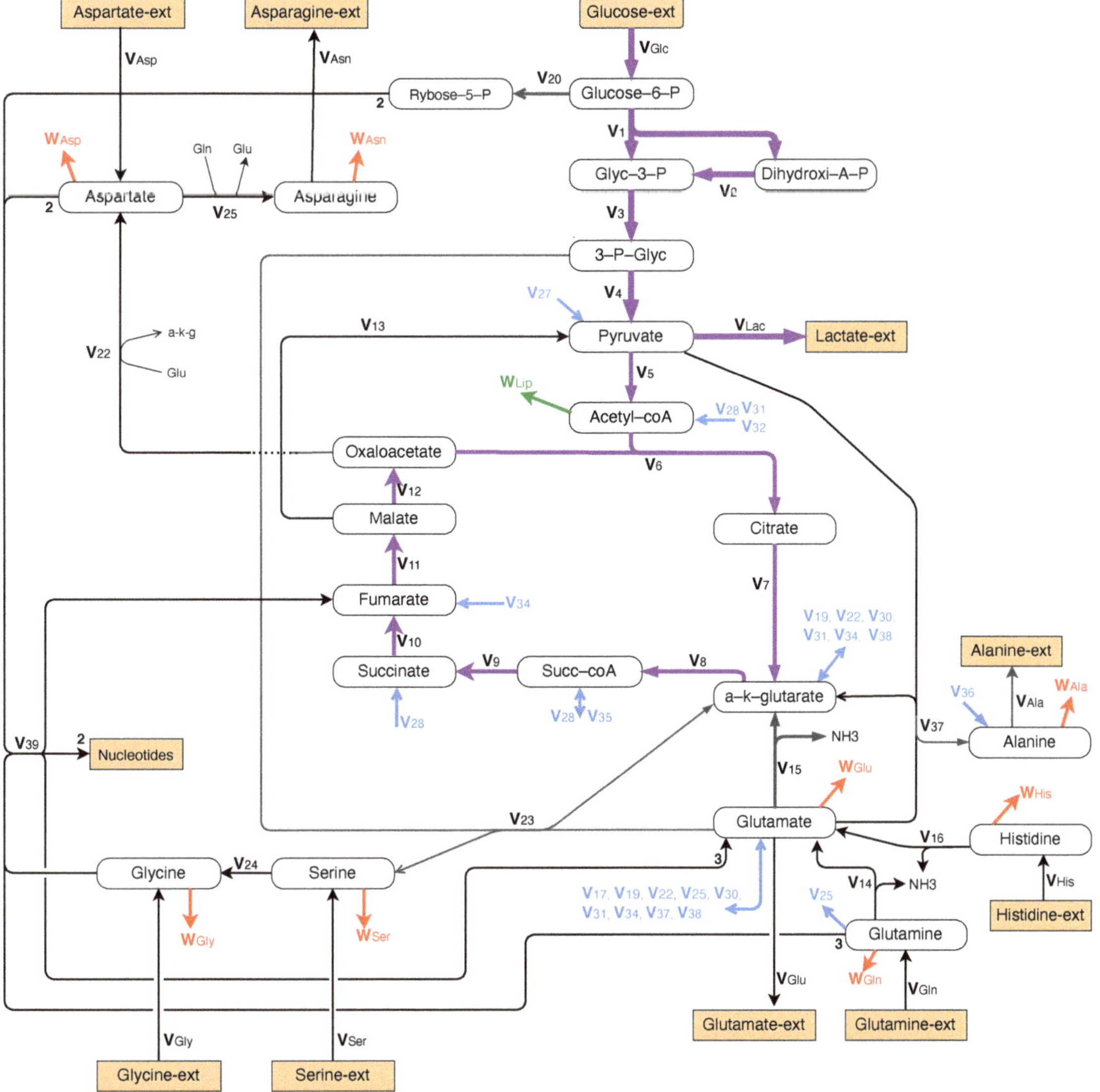

Figure 5. Exponential growth: central metabolism involving glycolysis, TCA, and glutaminolysis, nucleotide synthesis, and metabolism of alanine, asparagine, aspartate, histidine, glycine, and serine.

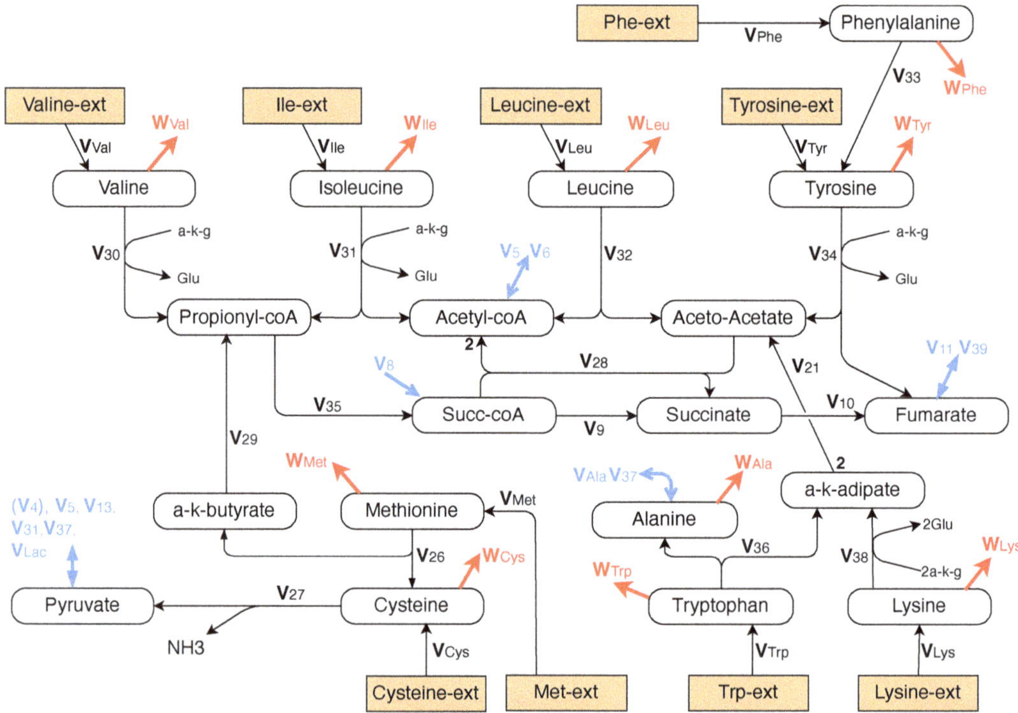

Figure 6. Metabolic network for seven essential amino acids (isoleucine, leucine, lysine, methionine, phenylalanine, tryptophan, valine) and two non-essential amino acids (cysteine, tyrosine).

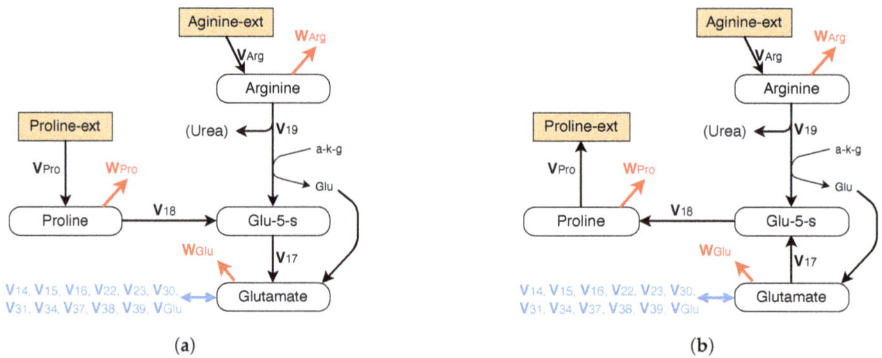

Figure 7. Metabolic network for arginine (essential) and proline (non-essential). (**a**) Exponential growth; (**b**) growth without glutamine.

3.1.2. Synthesis of Proteins

Essential amino acids cannot be synthesized and must be provided in the culture medium. Therefore, the maximum possible production rate of proteins is determined by the essential amino acid with the lowest ratio between its external uptake rate (from Table 3) and its frequency in protein composition as given in Table 5.

Table 5. Specific consumption rates of amino acids (AA) for protein production [μM/d $\times 10^7$ cell]. (Essential AA are in bold).

Amino Acid (AA)	Frequency [1] f (%)		Exponential Growth	Growth without Gln	Growth without Glc & Gln
Alanine	8.0	w_{Ala}	0.493	0.150	0.033
Arginine	5.0	w_{Arg}	0.306	0.093	0.020
Asparagine	4.3	w_{Asn}	0.266	0.081	0.018
Aspartate	5.5	w_{Asp}	0.339	0.103	0.023
Cysteine	2.5	w_{Cys}	0.151	0.046	0.010
Glutamine	4.2	w_{Gln}	0.260	0.079	0.017
Glutamate	5.9	w_{Glu}	0.366	0.111	0.024
Glycine	7.5	w_{Gly}	0.464	0.141	0.031
Histidine	2.4	w_{His}	0.150	0.045	0.010
Isoleucine	4.6	w_{Ile}	0.286	0.087	0.019
Leucine	8.4	w_{Leu}	0.515	0.157	0.034
Lysine	7.2	w_{Lys}	0.443	0.135	0.030
Methionine	2.0	w_{Met}	0.126	0.038	0.008
Phenylalanine	3.7	w_{Phe}	0.228	0.070	0.015
Proline	4.9	w_{Pro}	0.305	0.093	0.020
Serine	7.1	w_{Ser}	0.435	0.132	0.029
Threonine	5.9	w_{Thr}	0.366	0.111	0.024
Tryptophan	1.1	w_{Trn}	0.069	0.021	0.005
Tyrosine	3.1	w_{Tyr}	0.189	0.057	0.013
Valine	6.5	w_{Val}	0.402	0.122	0.027
		$\sum w_{AA}$	6.159	1.874	0.411

[1] Average of frequencies given in [3,21,22].

In the case of exponential growth, among all essential measured amino acids, phenylalanine is the one with this lower ratio. Hence, assuming a maximization of the biomass production, we suppose that phenylalanine is exclusively used for protein production. Therefore the protein production flux from phenylalanine w_{Phe} must be equal to the external uptake rate given in Table 3, i.e., $w_{Phe} = v_{Phe} = 0.228$ μM/d $\times 10^7$ cell. On this basis, we can then compute the contribution of each amino acid to the production rate of proteins given in Table 5 with the formula:

$$w_{AA} = w_{Phe} \frac{f_{AA}}{f_{Phe}}. \tag{1}$$

where w_{AA} is a specific intracellular consumption rate of amino acid AA for protein production, f_{AA} is the frequency of amino acid AA in the protein composition (and $AA = Phe$ for phenylalanine).

3.1.3. Synthesis of Nucleotides

The synthesis of nucleotides is represented by the following standard overall biochemical reactions:

1 Ribose-5-P + 2 Glutamine + 1 Aspartate + 1 Glycine + 1 CO2 + 5 ATP

$\longrightarrow$ 2 Glutamate + 1 Fumarate + 4 ADP + 1 AMP + 1 Purine

1 Ribose-5-P + 1 Glutamine + 1 Aspartate + 2 ATP

$\longrightarrow$ 1 Glutamate + 2 ADP + 1 Pyrimidine

Furthermore, we assume that DNA and RNA are made up with, approximately, equal shares of purine and pyrimidine nucleotides. It results that, omitting the co-factors ATP,

ADP and AMP, nucleotide synthesis is represented in the network of Figure 5 by the single overall reaction:

$$2 \text{ Ribose-5-P} + 3 \text{ Glutamine} + 2 \text{ Aspartate} + 1 \text{ Glycine} + 1 \text{ CO2}$$
$$\longrightarrow 3 \text{ Glutamate} + 1 \text{ Fumarate} + 2 \text{ Nucleotides}. \quad (2)$$

3.1.4. Catabolism of Essential Amino Acids

Only catabolic pathways must be considered for essential amino acids since they cannot be synthesized in the cell. We adopt the standard catabolic reactions represented in Figures 6 and 7. It can be verified that this representation is fully consistent with the available data because we have $0 < w_{AA} \leqslant v_{AA}$ for all essential amino acids (with v_{AA} from Table 3 and w_{AA} from Table 5).

3.1.5. Metabolism of Non-Essential Amino Acids

For non-essential amino acids, both catabolic and anabolic pathways can be taken into account. From the data of Tables 2, 4 and 5, it appears that the metabolism of non-essential AA may strongly depend on the culture conditions.

In the phase of exponential growth, anabolic pathways must be considered for alanine and glutamate because, as seen in Table 4, they are excreted in the culture medium and, therefore, produced inside the cell at a level that is widely in excess, with respect to the amount needed for protein production. The metabolism of alanine and glutamate is represented in Figure 5.

Moreover, from the data of Table 2, it appears that the uptake rates of extracellular aspartate, glycine, and serine are not sufficient to reach the protein production level given in Table 5, and that an intracellular synthesis must be provided, too. The metabolism of these amino acids is also represented in Figure 5. In Figure 5, a synthesis pathway is provided for asparagine together with an excretion. This assumption will be motivated in the next section.

In contrast, again from Table 2, we see that the uptake rate of cysteine, proline, and tyrosine is large enough for protein production, and that an additional catabolic pathway is needed. The catabolic reactions are represented in Figure 6 for cysteine and tyrosine, and in Figure 7a for proline.

3.1.6. Synthesis of Lipids

Finally, in addition to nucleotides and amino acids, we consider the lipids as the last fundamental building blocks of the biomass, in order to set up a model that is consistent, in terms of mass balance with a sufficient accuracy. For simplicity, we assume however that acetyl-CoA is the only significant contributor to the molecular mass of lipids, with a rate denoted w_{Lip} as shown in Figure 5. Indeed, the other necessary precursors of membrane lipids (e.g., serine, choline, ethanolamine, or dihydroxyacetone phosphate) are used in such low proportions that they can be neglected without significant loss of accuracy.

3.2. Metabolic Flux Analysis
3.2.1. Balance Equations

The metabolic fluxes satisfy the following set of balance equations.

Internal Metabolite	Flux Balance Equation
Glucose-6-P	$v_1 + v_{20} = v_{Glc} = 42.308$
Glyceraldehyde-3-P	$v_1 + v_2 - v_3 = 0$
Dihydroxyacetone P	$v_1 - v_2 = 0$
3-Phosphoglycerate	$v_3 - v_4 - v_{23} = 0$
Pyruvate	$v_4 - v_5 + v_{13} + v_{27} - v_{37} = v_{Lac} = 63.970$
Acetyl-coA	$v_5 - v_6 + 2v_{28} + v_{31} + v_{32} - w_{Lip} = 0$

Internal Metabolite	Flux Balance Equation
Citrate	$v_6 - v_7 = 0$
α-ketoglutarate	$v_7 - v_8 + v_{15} - v_{19} + v_{22} + v_{23} - v_{30} - v_{31} - v_{34} + v_{37} - 2v_{38} = 0$
Succinyl-CoA	$v_8 - v_9 - v_{28} + v_{35} = 0$
Succinate	$v_9 - v_{10} + v_{28} = 0$
Fumarate	$v_{10} - v_{11} + v_{34} + v_{39} = 0$
Malate	$v_{11} - v_{12} - v_{13} = 0$
Oxaloacetate	$-v_6 + v_{12} - v_{22} = 0$
Glutamate-5-semialdehyde	$-v_{17} + v_{18} + v_{19} = 0$
α-ketobutyrate	$v_{26} - v_{29} = 0$
Propionyl-CoA	$v_{29} + v_{30} + v_{31} - v_{35} = 0$
α-ketoadipate	$v_{36} + v_{38} - 2v_{21} = 0$
Acetoacetate	$v_{21} - v_{28} + v_{32} + v_{34} = 0$
Ribose-5-P	$v_{20} - 2v_{39} = 0$
Arginine	$v_{19} = v_{Arg} - w_{Arg} = 0.643$
Aspartate	$-v_{22} + v_{25} + 2v_{39} = v_{Asp} - w_{Asp} = -0.001$
Asparagine	$v_{25} - v_{Asn} = w_{Asn} = 0.266$
Cysteine	$-v_{26} + v_{27} = v_{Cys} - w_{Cys} = 0.113$
Glutamate	$v_{14} - v_{15} + v_{16} + v_{17} + v_{19} - v_{22} - v_{23} + v_{25} + v_{30} + v_{31} + v_{34} - v_{37} - 2v_{38} + 3v_{39} = v_{Glu} + w_{Glu} - 1.025$
Glutamine	$v_{14} + v_{25} + 3v_{39} = v_{Gln} - w_{Gln} = 4.734$
Glycine	$-v_{24} + v_{39} = v_{Gly} - w_{Gly} = -0.139$
Histidine	$v_{16} = v_{His} - w_{His} = 0.045$
Methionine	$v_{26} = v_{Met} - w_{Met} = 0.118$
Phenylalanine	$v_{33} = v_{Phe} - w_{Phe} = 0$
Proline	$v_{18} = v_{Pro} - w_{Pro} = 0.260$
Serine	$-v_{23} + v_{24} = v_{Ser} - w_{Ser} = -0.204$
Tyrosine	$-v_{33} + v_{34} = v_{Tyr} - w_{Tyr} = 0.016$
Valine	$v_{30} = v_{Val} - w_{Val} = 0.097$
Isoleucine	$v_{31} = v_{Ile} - w_{Ile} = 0.142$
Leucine	$v_{32} = v_{Leu} - w_{Leu} = 0.306$
Lysine	$v_{38} = v_{Lys} - w_{Lys} = 0.041$
Tryptophan	$v_{36} = v_{Trp} - w_{Trp} = 0.022$
Alanine	$v_{36} + v_{37} = v_{Ala} + w_{Ala} = 1.006$
NH3	$v_{14} + v_{15} + v_{16} + v_{27} = v_{NH3} = 5.3$

3.2.2. Computation of Metabolic Fluxes

The above 39×41 system of linear equations is underdetermined. One of the reasons for this indeterminacy is that asparagine data are not available and the rate v_{Asn} of asparagine transfer between the cell and the external culture medium is unknown. In order to get a unique solution, we introduce the additional constraint that the macromolecular (proteins, nucleic acids, membrane lipids) composition is as reported, e.g., in [19] (p. 113, Table 7.1) for mammalian cells. From this reference, the mass of proteins is roughly 12-fold larger than the mass of nucleic acids in mammalian cells. Furthermore, we know that the average mass of a nucleotide is about three-fold larger than the average mass of an amino acid. Using molar units as we do in this paper, it follows that the sum $\sum w_{AA}$ (see Table 5) of amino acid rates for protein production must be approximately 36-fold larger than the nucleotide production rate and, consequently, 72-fold larger that the rate v_{39} of reaction (2), i.e.,

$$\frac{\sum w_{AA}}{v_{39}} \approx 72. \tag{3}$$

Similarly, the mass of protein is roughly 3.5-fold larger than the mass of membrane lipids, while the average mass of a phospholipid is about 7-fold larger than the average mass of an amino acid. Then, since the production of one mole of phospholipids consumes about 18 moles of acetyl-CoA, we deduce that we have approximately:

$$\frac{\sum w_{AA}}{w_{Lip}} \approx \frac{3.5 \times 7}{18} = 1.36, \tag{4}$$

where w_{Lip} denotes the consumption rate of acetyl-CoA for lipid synthesis (see Figure 5).

Under these additional constraints (3) and (4), the system of equations is determined and has the following solution:

$v_1 = v_2 = 42.136$	$v_{13} = 3.198$	$v_{23} = 0.429$	$v_{33} = 0.0$
$v_3 = 84.272$	$v_{14} = 2.731$	$v_{24} = 0.225$	$v_{34} = 0.016$
$v_4 = 83.843$	$v_{15} = 2.306$	$v_{25} = 1.745$	$v_{35} = 0.357$
$v_5 = 22.318$	$v_{16} = 0.045$	$v_{26} = 0.118$	$v_{36} = 0.022$
$v_6 = v_7 = 18.973$	$v_{17} = 0.903$	$v_{27} = 0.231$	$v_{37} = 0.984$
$v_8 = 23.630$	$v_{18} = 0.260$	$v_{28} = 0.354$	$v_{38} = 0.041$
$v_9 = 23.633$	$v_{19} = 0.643$	$v_{29} = 0.118$	$v_{39} = 0.086$
$v_{10} = 23.987$	$v_{20} = 0.172$	$v_{30} = 0.097$	
$v_{11} = 24.089$	$v_{21} = 0.032$	$v_{31} = 0.142$	$v_{Asn} = 1.479$
$v_{12} = 20.891$	$v_{22} = 1.918$	$v_{32} = 0.306$	$w_{Lip} = 4.52$

These results can be summed up as follows:

Assuming that

(a) The essential amino acids are not produced inside the cell;
(b) The biomass production is maximal;
(c) The production rates of proteins, nucleic acids, and membrane lipids are in the same proportions as the respective mass fractions of these macromolecules inside the cells;

Then

(a) A metabolic flux analysis based on the considered metabolic network allows computing the entire intracellular flux distribution from the measured extracellular uptake and excretion rates of Tables 2–4.
(b) Closing the overall flux balance necessarily implies that asparagine (which is not measured) is significantly excreted with a rate of about $1.5 \ \mu M/d \times 10^7$ cells. This is quite natural because it is well known that "mammalian cell culture metabolism is characterized by a high glucose and glutamine uptake combined with a high rate of lactate and non-essential amino acid secretion" [3]. From our results, it appears that glutamate, alanine, and especially asparagine, are the main excreted non-essential amino acids for this culture of VERO cells during the exponential growth. It is clearly the reason why a medium without asparagine may be used at the start of the culture without problem.

4. Metabolic Flux Analysis for the Growth without Glutamine

4.1. Metabolic Network

4.1.1. Central Metabolism and Nucleotide Synthesis

As explained in Section 1, we now consider a case where there is no glutamine in the culture medium while glucose is in excess and not limiting. Glutamate (and aspartate to a lesser extent) become the main nitrogen sources. Obviously, in that case, glutamine must be synthesized inside the cells. The central metabolism is therefore slightly modified as represented in Figure 8 with a pathway for the synthesis of glutamine from glutamate.

Moreover, as shown in Figure 8, the pathway to nucleotide synthesis is assumed to be identical to that of exponential growth (see Section 3.1.3).

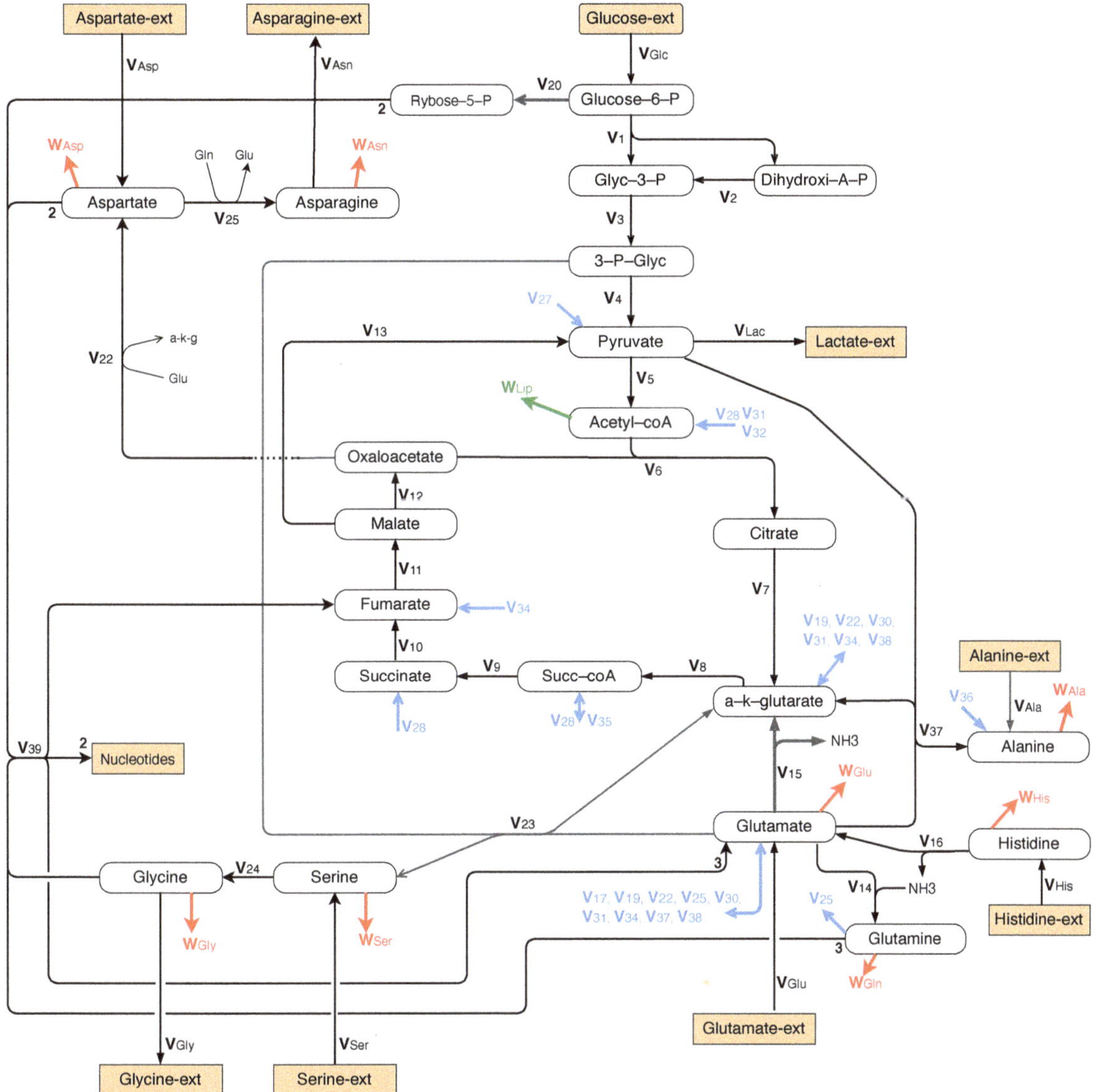

Figure 8. Growth without glutamine: central metabolism involving glycolysis, TCA and glutamine synthesis, nucleotide synthesis and metabolism of alanine, asparagine, aspartate, histidine, glycine, and serine.

4.1.2. Synthesis of Proteins

In this phase of growth without glutamine, histidine turns out to be the essential amino acid with the lowest ratio between its external uptake rate (from Table 3) and its frequency in protein composition (from Table 5). If we assume (as in the previous section) that the cell growth is maximized and that histidine is exclusively used for protein formation, then the total protein production rate can be estimated as

$$\left(\sum w_{AA}\right)_{wg} \approx \frac{v_{His}}{f_{His}} = \frac{0.045}{0.024} = 1.87 \, \frac{\mu M}{d \times 10^7 \, cells} \tag{5}$$

where the values of v_{His} and f_{His} are taken from Tables 3 and 5, respectively. We can then compute the contribution of each amino acid to the production rate of proteins. The result is given in Table 5.

From the available experimental data, the assumption of cell growth maximization could however be questionable when glutamine is depleted. The reason is that, as it can be seen in Figure 4, the net uptake rate of essential amino acids is similar in magnitude to the exponential growth while the specific cell growth rate is much smaller. An alternative natural assumption is to suppose that the rate of protein production is proportional to the rate of cell growth. Under this assumption, we have another manner to compute a plausible estimate of the protein production rate, as follows:

$$\left(\sum w_{AA}\right)_{wg} \approx \frac{\mu_{wg}}{\mu_{eg}} \left(\sum w_{AA}\right)_{eg} = \frac{0.18}{0.6} \times 6.159 = 1.85 \, \frac{\mu M}{d \times 10^7 \text{ cells}}. \tag{6}$$

In Equations (5) and (6), the subscripts '*wg*' and '*eg*' refer to the growth without glutamine and to the exponential growth, respectively. The values of the growth rates μ_{eg} and μ_{wg} are taken from Table 1.

It is remarkable that these two estimations of the protein production rate are almost equal, although they are obtained under totally different assumptions. In our viewpoint, this certainly gives a strong consistency to the validity of our experimental data and to the relevance of the assumption of growth maximization, which appears to be very plausible, not only for the exponential growth with non-limiting glucose and glutamine resources, but also in the case of growth with glutamine limitation. It will be seen, in the next section, that the situation is very different for the culture without glucose.

4.1.3. Metabolism of Amino Acids

For the catabolism of essential amino acids, we adopt the same standard catabolic reactions represented in Figures 6 and 7.

Moreover anabolic pathways must be provided for glycine and proline, which are excreted into the extracellular medium under the current conditions (see Table 2). The metabolic pathways are given in Figure 7b for proline and in Figure 8 for glycine. Note that the difference between Figure 7a,b lies in the inversion of fluxes v_{17}, v_{18} and v_{Pro}.

Finally, catabolic pathways are used for aspartate, cysteine, serine, and tyrosine because the uptake rate from the culture medium is larger than their contribution to the flux in protein production given in Table 5. The metabolism of these amino acids is represented in Figure 6 for cysteine and tyrosine, and in Figure 8 for aspartate and serine.

4.2. Metabolic Flux Analysis

In this case of growth without glutamine, the metabolic fluxes satisfy the following set of balance equations.

Internal Metabolite	Flux Balance Equation
Glucose-6-P	$v_1 + v_{20} = v_{Glc} = 8.933$
Glyceraldehyde-3-P	$v_1 + v_2 - v_3 = 0$
Dihydroxyacetone P	$v_1 - v_2 = 0$
3-Phosphoglycerate	$v_3 - v_4 - v_{23} = 0$
Pyruvate	$v_4 - v_5 + v_{13} + v_{27} - v_{37} = v_{Lac} = 8.818$
Acetyl-CoA	$v_5 - v_6 + 2v_{28} + v_{31} + v_{32} - w_{Lip} = 0$
Citrate	$v_6 - v_7 = 0$
α-ketoglutarate	$v_7 - v_8 + v_{15} - v_{19} + v_{22} + v_{23} - v_{30} - v_{31} - v_{34} + v_{37} - 2v_{38} = 0$
Succinyl-CoA	$v_8 - v_9 - v_{28} + v_{35} = 0$
Succinate	$v_9 - v_{10} + v_{28} = 0$
Fumarate	$v_{10} - v_{11} + v_{25} + v_{34} = 0$

Internal Metabolite	Flux Balance Equation
Malate	$v_{11} - v_{12} - v_{13} = 0$
Oxaloacetate	$-v_4 - v_6 + v_{12} + v_{22} = 0$
Glutamate-5-semialdehyde	$v_{17} - v_{18} + v_{19} = 0$
α-ketobutyrate	$v_{26} - v_{29} = 0$
Propionyl-CoA	$v_{29} + v_{30} + v_{31} - v_{35} = 0$
α-ketoadipate	$v_{36} + v_{38} - 2v_{21} = 0$
Acetoacetate	$v_{21} - v_{28} + v_{32} + v_{34} = 0$
Ribose-5-P	$v_{20} - 2v_{39} = 0$
Arginine	$v_{19} = v_{Arg} - w_{Arg} = 0.086$
Aspartate	$-v_{22} + v_{25} + 2v_{39} = v_{Asp} - w_{Asp} = 0.234$
Asparagine	$v_{25} - v_{Asn} = w_{Asn} = 0.081$
Cysteine	$-v_{26} + v_{27} = v_{Cys} - w_{Cys} = 0.022$
Glutamate	$-v_{14} - v_{15} + v_{16} + v_{17} + v_{19} - v_{22} - v_{23} + v_{25} + v_{30} + v_{31} + v_{34} - v_{37} + 2v_{38} + 3v_{39} = -v_{Glu} + w_{Glu} = -0.388$
Glutamine	$v_{14} - v_{25} - 3v_{39} = w_{Gln} = 0.079$
Glycine	$v_{24} - v_{39} = v_{Gly} + w_{Gly} = 0.160$
Histidine	$v_{16} = v_{His} - w_{His} = 0.0$
Methionine	$v_{26} = v_{Met} - w_{Met} = 0.062$
Phenylalanine	$v_{33} = v_{Phe} - w_{Phe} = 0.021$
Proline	$v_{18} = v_{Pro} + w_{Pro} = 0.567$
Serine	$-v_{23} + v_{24} = v_{Ser} - w_{Ser} = 0.050$
Tyrosine	$-v_{33} + v_{34} = v_{Tyr} - w_{Tyr} = 0.006$
Valine	$v_{30} = v_{Val} - w_{Val} = 0.024$
Isoleucine	$v_{31} = v_{Ile} - w_{Ile} = 0.018$
Leucine	$v_{32} = v_{Leu} - w_{Leu} = 0.043$
Lysine	$v_{38} = v_{Lys} - w_{Lys} = 0.064$
Tryptophan	$v_{36} = v_{Trp} - w_{Trp} = 0.016$
Alanine	$v_{36} + v_{37} = w_{Ala} - v_{Ala} = 0.092$
NH3	$-v_{14} + v_{15} + v_{16} + v_{27} = v_{NH3} = 0.195$

Under conditions (3) and (4), this system of linear equations is determined and has the following solution.

$v_1 = v_2 = 8.881$	$v_{13} = 0.741$	$v_{23} = 0.136$	$v_{33} = 0.021$
$v_3 = 17.762$	$v_{14} = 0.544$	$v_{24} = 0.186$	$v_{34} = 0.027$
$v_4 = 17.626$	$v_{15} = 0.655$	$v_{25} = 0.545$	$v_{35} = 0.104$
$v_5 = 9.557$	$v_{16} = 0.0$	$v_{26} = 0.062$	$v_{36} = 0.016$
$v_6 = v_7 = 8.478$	$v_{17} = 0.481$	$v_{27} = 0.084$	$v_{37} = 0.076$
$v_8 = 9.426$	$v_{18} = 0.567$	$v_{28} = 0.110$	$v_{38} = 0.064$
$v_9 = 9.420$	$v_{19} = 0.086$	$v_{29} = 0.062$	$v_{39} = 0.026$
$v_{10} = 9.530$	$v_{20} = 0.052$	$v_{30} = 0.024$	
$v_{11} = 9.583$	$v_{21} = 0.040$	$v_{31} = 0.018$	$v_{Asn} = 0.465$
$v_{12} = 8.842$	$v_{22} = 0.363$	$v_{32} = 0.043$	$w_{Lip} = 1.36$

We can conclude, as above, that the flux balance analysis based on the considered metabolic network allows computing the entire intracellular flux distribution from the measured extracellular uptake and excretion rates of Tables 2–4.

In this case, closing the overall flux balance requires that the non-essential amino acids that are significantly excreted be asparagine (0.47 μM/d $\times$ 10^7 cells) and proline (0.47 μM/d $\times$ 10^7 cells).

5. Metabolic Flux Analysis for the Growth without Glucose and Glutamine

5.1. Metabolic Network

5.1.1. Central Metabolism and Nucleotide Synthesis

We now consider the case where there is neither glucose nor glutamine in the culture, and where lactate, glutamate, and alanine are the main carbon and nitrogen sources for the central metabolism. This is represented by a metabolic network, shown in Figure 9, which involves a gluconeogenesis pathway for the synthesis of glucose-6-phosphate and a pathway for the synthesis of glutamine from alanine and glutamate. Moreover, the nucleotide synthesis pathway remains unchanged.

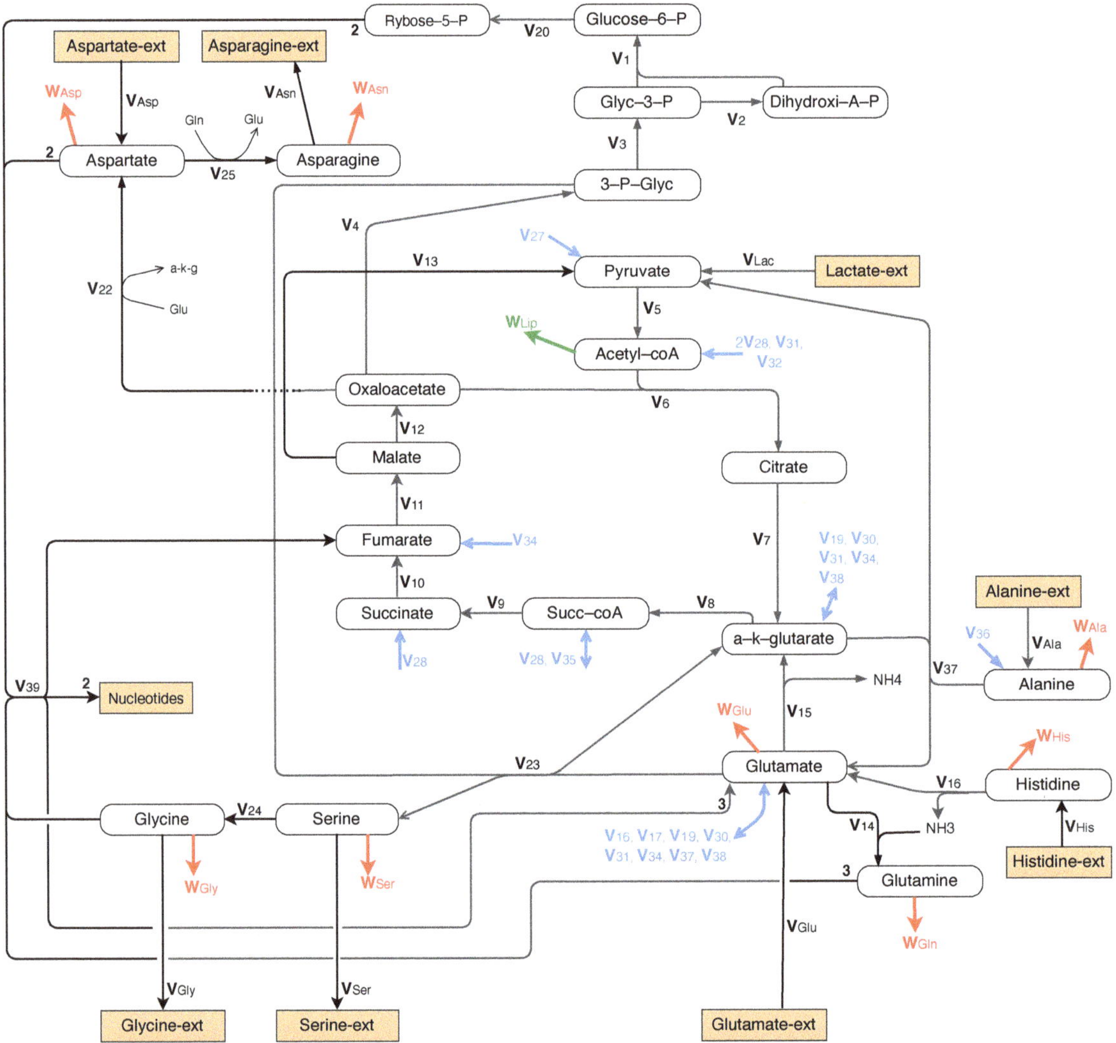

Figure 9. Growth without glucose and glutamine: central metabolism involving glyconeogenesis, TCA, and glutamine synthesis, nucleotide synthesis and metabolism of alanine, asparagine, aspartate, histidine, glycine, and serine.

5.1.2. *Synthesis of Proteins*

In this phase of growth without glucose and glutamine, histidine is the essential amino acid with the lowest ratio between its external uptake rate (from Table 3) and its frequency in protein composition (from Table 5). From the assumption that the protein production rate is proportional to the cell growth rate, it appears however that maximization of cell growth is not applicable. Indeed, using equations of the form (5) and (6), we have here:

$$\left(\sum w_{AA}\right)_{wgg} \approx \frac{\mu_{wgg}}{\mu_{eg}}\left(\sum w_{AA}\right)_{eg} = \frac{0.04}{0.6}\times 6.159 = 0.41\ \frac{\mu\mathrm{M}}{\mathrm{d}\times 10^7\mathrm{cells}} \tag{7}$$

$$\ll \frac{v_{His}}{f_{His}} = \frac{0.051}{0.024} = 2.12\ \frac{\mu\mathrm{M}}{\mathrm{d}\times 10^7\mathrm{cells}}. \tag{8}$$

In Equation (7), the subscript '*wgg*' refers to growth without glucose and glutamine. It follows clearly from (7) and (8) that the actual protein production rate must be much smaller than the maximal rate that could be reached from the measured amino acid uptakes. Hence, the contributions of the amino acids are computed, in Table 5, by using the value $\sum w_{AA} = 0.41$ from Equation (7).

5.1.3. Metabolism of Amino Acids

For the catabolism of essential amino acids, we adopt the same standard catabolic reactions represented in Figures 6 and 7.

Moreover, anabolic pathways are provided for glycine, serine, and proline, which are excreted into the extracellular medium under the current conditions (see Table 2). The metabolic pathways for glycine and serine are given in Figure 9. The metabolism of proline is represented in Figure 7b.

Furthermore, catabolic pathways are used for aspartate, cysteine, and tyrosine, because the uptake rate from the culture medium is larger that their contribution to the flux in protein production. The metabolism of these amino acids is represented in Figure 9 for aspartate and in Figure 6 for cysteine and tyrosine.

5.2. *Metabolic Flux Analysis*

In this case of growth without glucose and without glutamine, the metabolic fluxes satisfy the following set of balance equations.

Internal Metabolite	Flux Balance Equation
Glucose-6-P	$v_1 - v_{20} = 0$
Glyceraldehyde-3-P	$-v_1 - v_2 + v_3 = 0$
Dihydroxyacetone P	$-v_1 + v_2 = 0$
3-Phosphoglycerate	$-v_3 + v_4 - v_{23} = 0$
Pyruvate	$v_5 + v_{13} - v_{27} - v_{37} = v_{Lac} = 10.218$
Acetyl-coA	$v_5 - v_6 + 2v_{28} + v_{31} + v_{32} = 0$
Citrate	$v_6 - v_7 = 0$
α-ketoglutarate	$v_7 - v_8 + v_{15} - v_{19} + v_{22} + v_{23} - v_{30} - v_{31} - v_{34} + v_{37} - 2v_{38} = 0$
Succinyl-CoA	$v_8 - v_9 - v_{28} + v_{35} = 0$
Succinate	$v_9 - v_{10} + v_{28} = 0$
Fumarate	$v_{10} - v_{11} + v_{34} + v_{39} = 0$
Malate	$v_{11} - v_{12} = 0$
Oxaloacetate	$-v_4 - v_6 + v_{12} + v_{13} - v_{22} = 0$
Glutamate-5-semialdehyde	$v_{17} - v_{18} + v_{19} = 0$
α-ketobutyrate	$v_{26} - v_{29} = 0$
Propionyl-CoA	$v_{29} + v_{30} + v_{31} - v_{35} = 0$
α-ketoadipate	$-2v_{21} + v_{36} + v_{38} = 0$
Acetoacetate	$v_{21} - v_{28} + v_{32} + v_{34} = 0$
Ribose-5-P	$v_{20} - 2v_{39} = 0$

Internal Metabolite	Flux Balance Equation
Arginine	$v_{19} = v_{Arg} - w_{Arg} = 0.180$
Asparagine	$v_{25} - v_{Asn} = w_{Asn} = 0.018$
Aspartate	$-v_{22} + v_{25} + 2v_{39} = v_{Asp} - w_{Asp} = 0.353$
Cysteine	$-v_{26} + v_{27} = v_{Cys} - w_{Cys} = 0.067$
Glutamate	$v_{14} + v_{15} - v_{16} - v_{17} - v_{19} + v_{22} + v_{23} - v_{25} - v_{30} - v_{31} - v_{34} + v_{37} - 2v_{38} - 3v_{39} = v_{Glu} - w_{Glu} = 0.534$
Glutamine	$v_{14} - v_{25} - 3v_{39} = w_{Gln} = 0.017$
Glycine	$v_{24} - v_{39} = v_{Gly} + w_{Gly} = 0.052$
Histidine	$v_{16} = v_{His} - w_{His} = 0.041$
Methionine	$v_{26} = v_{Met} - w_{Met} = 0.104$
Phenylalanine	$v_{33} = v_{Phe} - w_{Phe} = 0.086$
Proline	$v_{18} = v_{Pro} + w_{Pro} = 0.879$
Serine	$v_{23} - v_{24} = v_{Ser} + w_{Ser} = 0.653$
Tyrosine	$-v_{33} + v_{34} = v_{Tyr} - w_{Tyr} = 0.058$
Valine	$v_{30} = v_{Val} - w_{Val} = 0.137$
Isoleucine	$v_{31} = v_{Ile} - w_{Ile} = 0.152$
Leucine	$v_{32} = v_{Leu} - w_{Leu} = 0.425$
Lysine	$v_{38} = v_{Lys} - w_{Lys} = 0.192$
Tryptophan	$v_{36} = v_{Trp} - w_{Trp} = 0.037$
Alanine	$-v_{36} + v_{37} = v_{Ala} - w_{Ala} = 0.669$
NH3	$-v_{14} + v_{15} + v_{16} + v_{25} + v_{27} = v_{NH3} = 1.089$

Under conditions (3) and (4), this system of equations is determined and has the following solution.

$v_1 = v_2 = 0.012$	$v_{13} = 0.567$	$v_{23} = 0.711$	$v_{33} = 0.086$
$v_3 = 0.024$	$v_{14} = 0.874$	$v_{24} = 0.058$	$v_{34} = 0.144$
$v_4 = 0.735$	$v_{15} = 1.751$	$v_{25} = 0.839$	$v_{35} = 0.393$
$v_5 = 11.662$	$v_{16} = 0.041$	$v_{26} = 0.104$	$v_{36} = 0.037$
$v_6 = v_7 = 13.306$	$v_{17} = 0.699$	$v_{27} = 0.171$	$v_{37} = 0.706$
$v_8 = 14.563$	$v_{18} = 0.879$	$v_{28} = 0.684$	$v_{38} = 0.192$
$v_9 = 14.272$	$v_{19} = 0.180$	$v_{29} = 0.104$	$v_{39} = 0.006$
$v_{10} = 17.956$	$v_{20} = 0.012$	$v_{30} = 0.137$	
$v_{11} = 15.106$	$v_{21} = 0.114$	$v_{31} = 0.152$	$v_{Asn} = 0.821$
$v_{12} = 14.539$	$v_{22} = 0.498$	$v_{32} = 0.425$	$w_{Lip} = 0.3$

Here, the excreted amino acids are proline (0.86 μM/d $\times$ 10^7 cells), asparagine (0.82 μM/d $\times$ 10^7 cells), and serine (0.62 μM/d $\times$ 10^7 cells).

6. Final Remarks and Conclusions

In this paper, we applied metabolic flux analysis to investigate the behavior of VERO cells in three different culture conditions.

As long as glucose is not limiting, this analysis supports the validity of a maximum growth hypothesis in which the amino acids are primarily used as building blocks for the formation of the biomass, even in case of glutamine deprivation. In the latter case, glutamine is replaced by glutamate as the nitrogen source and it can be observed that the biomass yield is even slightly higher (while the productivity is lower).

When glucose is exhausted, the cell growth does not stop, but continues at a smaller rate with the consumption of lactate as an alternative source of carbon, while using only a small part (about 20%) of available amino acids for biomass synthesis. As represented in the network of Figure 9, lactate is reintroduced into the cell, transformed into pyruvate, and integrated in the TCA cycle in order to provide a part of the required energy, which is no longer given by glycolysis. The rest of the energy is provided by the degradation of that part of amino acids, which are not used as building blocks for the biomass synthesis.

This analysis provides a metabolic foundation for the design of feedback control strategies in fed batch or perfusion bioreactors where the lactate concentration is measured online and regulated by controlling the delivery rates of glucose and, possibly, of some essential amino acids. Applications of such control strategies to CHO cells are discussed, e.g., in [23,24], while, to our knowledge, applications to VERO cell cultures have not been reported in the literature.

Let us finally mention that, in our metabolic analysis, one amino acid, namely *threonine*, was omitted from the model because experimental measurements are (unfortunately) missing. Obviously, this is equivalent to implicitly assume that, in the considered experiments, threonine, which is an essential amino acid, is consumed at the rate required for protein synthesis as given in Table 5. We can however confirm that this assumption is quite plausible on the basis of threonine data, which were obtained for the same cell line grown in similar conditions, but with a slightly different fetal serum (bovine instead of calf), as reported in [20], Chapter 3. This means that including threonine catabolism in the model, if actual data were available, should not significantly alter our results.

Author Contributions: Conceptualization, G.B., V.C. and A.V.W.; methodology, G.B., V.C. and A.V.W.; data curation and visualization, G.B. and V.C.; software, G.B.; investigation, G.B., V.C. and A.V.W.; writing—original draft preparation, G.B.; writing—review and editing, G.B. and A.V.W. All authors have read and agreed to the published version of the manuscript.

Funding: This research received no external funding.

Data Availability Statement: All the data used in this study are available in the graphs and tables of this article.

Conflicts of Interest: The authors declare no conflict of interest.

References

1. Antoniewicz, M.R. Methods and advances in metabolic flux analysis: A mini-review. *J. Ind. Microbiol. Biotechnol.* **2015**, *42*, 317–325. [CrossRef] [PubMed]
2. Bonarius, H.P.J.; Hatzimanikatis, V.; Meesters, K.P.H.; de Gooijer, C.D.; Schmid, G.; Tramper, J. Metabolic flux analysis of hybridoma cells in different culture media using mass balances. *Biotechnol. Bioeng.* **1996**, *50*, 299–318. [CrossRef]
3. Quek, L.E.; Dietmair, S.; Krömer, J.O.; Nielsen, L.K. Metabolic flux analysis in mammalian cell culture. *Metab. Eng.* **2010**, *12*, 161–171. [CrossRef] [PubMed]
4. Zamorano, F.; Wouwer, A.V.; Bastin, G. A detailed metabolic flux analysis of an underdetermined network of CHO cells. *J. Biotechnol.* **2010**, *150*, 497–508. [CrossRef] [PubMed]
5. Niklas, J.; Heinzle, E. Metabolic flux analysis in systems biology of mammalian cells. *Adv. Biochem. Eng. Biotechnol.* **2011**, *127*, 109–132.
6. Fernandes de Sousa, S.; Bastin, G.; Jolicoeur, M.; Vande Wouwer, A. Dynamic metabolic flux analysis using a convex analysis approach: Application to hybridoma cell cultures in perfusion. *Biotechnol. Bioeng.* **2016**, *113*, 1102–1112. [CrossRef] [PubMed]
7. Hagrot, E.; Oddsdóttird, H.Æ.; Mäkinena, M.; Forsgrend, A.; Chotteau, V. Novel column generation-based optimization approach for poly-pathway kinetic model applied to CHO cell culture. *Metab. Eng. Commun.* **2018**, *8*, e00083. [CrossRef]
8. Chen, Y.; McConnell, B.O.; Dhara, V.G.; Naik, H.M.; Li, C.T.; Antoniewicz, M.R.; Betenbaugh, M.J. An unconventional uptake rate objective function approach enhances applicability of genome-scale models for mammalian cells. *Npj Syst. Biol. Appl.* **2019**, *5*, 25. [CrossRef] [PubMed]
9. Petiot, E.; Guedon, E.; Blanchard, F.; Gény, C.; Pinton, H.; Marc, A. Kinetic characterization of Vero cell metabolism in a serum-free batch culture process. *Biotechnol. Bioeng.* **2010**, *107*, 143–153. [CrossRef] [PubMed]
10. Montagnon, B.; Fanget, B.; Vincent-Falquet, J. Industrial-scale production of inactivated poliovirus vaccine prepared by culture of Vero cells on microcarrier. *Rev. Infect. Dis.* **1984**, *6*, S341–S344. [CrossRef] [PubMed]
11. Srivastava, A.K.; Putnak, J.R.; Lee, S.H.; Hong, S.P.; Moon, S.B.; Barvir, D.A.; Zhao, B.; Olson, R.A.; Kim, S.O.; Yoo, W.D.; et al. A purified inactivated Japanese encephalitis virus vaccine made in Vero cells. *Vaccine* **2001**, *19*, 4557–4565. [CrossRef]
12. Trabelsi, K.; Rourou, S.; Loukil, H.; Majoul, S.; Kallel, H. Optimization of virus yield as a strategy to improve rabies vaccine production by Vero cells in a bioreactor. *J. Biotechnol.* **2006**, *121*, 261–271. [CrossRef]
13. Paillet, C.; Forno, G.; Kratje, R.; Etcheverrigaray, M. Suspension-Vero cell cultures as a platform for viral vaccine production. *Vaccine* **2009**, *27*, 6464–6467. [CrossRef] [PubMed]
14. Montomoli, E.; Khadang, B.; Piccirella, S.; Trombetta, C.; Mennitto, E.; Manini, I.; Stanzani, V.; Lapini, G. Cell culture-derived influenza vaccines from Vero cells: A new horizon for vaccine production. *Expert Rev. Vaccines* **2012**, *11*, 587–594. [CrossRef] [PubMed]

15. Orr-Burks, N.; Murray, J.; Wu, W.; Kirkwood, C.D.; Todd, K.V.; Jones, L.; Bakre, A.; Wang, H.; Jiang, B.; Tripp, R.A. Gene-edited vero cells as rotavirus vaccine substrates. *Vaccine X* **2019**, *3*, 100045. [CrossRef] [PubMed]

16. Pato, T.P.; Souza, M.C.; Mattos, D.A.; Caride, E.; Ferreira, D.F.; Gaspar, L.P.; Freire, M.S.; Castilho, L.R. Purification of yellow fever virus produced in Vero cells for inactivated vaccine manufacture. *Vaccine* **2019**, *37*, 3214–3220. [CrossRef] [PubMed]

17. Rourou, S.; Zakkour, M.B.; Kallel, H. Adaptation of Vero cells to suspension growth for rabies virus production in different serum free media. *Vaccine* **2019**, *37*, 6987–6995. [CrossRef] [PubMed]

18. Kiesslich, S.; Losa, J.P.V.C.; Gélinas, J.F.; Kamen, A.A. Serum-free production of rVSV-ZEBOV in Vero cells: Microcarrier bioreactor versus scale-X™ hydro fixed-bed. *J. Biotechnol.* **2020**, *310*, 32–39. [CrossRef] [PubMed]

19. Palsson, B.Ø. *Systems Biology: Simulation of Dynamic Network States*; Cambridge University Press: Cambridge, UK, 2011.

20. Chotteau, V. A General Modelling Methodology for Animal Cell Cultures. Ph.D. Thesis, Faculty of Engineering, Louvain University, Louvain-la-Neuve, Belgium, 1995.

21. Gross, L.J. Observed Frequency of Amino Acids in Vertebrates. 1999. Available online: http://www.tiem.utk.edu/~gross/bioed/webmodules/aminoacid.htm (accessed on 14 November 2021).

22. Provost, A. Metabolic Design of Dynamic Bioreaction Models. Ph.D. Thesis, Faculty of Engineering, Louvain University, Louvain-la-Neuve, Belgium, 2006.

23. Gagnon, M.; Hiller, G.; Luan, Y.T.; Kittredge, A.; DeFelice, J.; Drapeau, D. High-end pH-controlled delivery of glucose effectively suppresses lactate accumulation in CHO fed-batch cultures. *Biotechnol. Bioeng.* **2011**, *108*, 1328–1337. [CrossRef] [PubMed]

24. Hartley, F.; Walker, T.; Chung, V.; Morten, K. Mechanisms driving the lactate switch in Chinese hamster ovary cells. *Biotechnol. Bioeng.* **2018**, *115*, 1890–1903. [CrossRef] [PubMed]

Article

Set Membership Estimation with Dynamic Flux Balance Models

Xin Shen and Hector Budman *

Department of Chemical Engineering, University of Waterloo, 200 University Ave W, Waterloo, ON N2L3G1, Canada; xin.shen@uwaterloo.ca
* Correspondence: hbudman@uwaterloo.ca

Abstract: Dynamic flux balance models (DFBM) are used in this study to infer metabolite concentrations that are difficult to measure online. The concentrations are estimated based on few available measurements. To account for uncertainty in initial conditions the DFBM is converted into a variable structure system based on a multiparametric linear programming (mpLP) where different regions of the state space are described by correspondingly different state space models. Using this variable structure system, a special set membership-based estimation approach is proposed to estimate unmeasured concentrations from few available measurements. For unobservable concentrations, upper and lower bounds are estimated. The proposed set membership estimation was applied to batch fermentation of *E. coli* based on DFBM.

Keywords: set membership estimation; dynamic flux balance model; multiparametric programming; observability; variable structure system

1. Introduction

The increasing demand of bio-pharmaceutical products requires continuous improvement in monitoring and control strategies for the fermentation processes. Model-based control and optimization strategies are crucial to boost productivity. Unlike traditional unstructured biochemical models, dynamic flux balance models (DFBM) have gained increasing attention since they contain more detailed information about the distribution of metabolic fluxes [1,2]. The strength of DFBM relies on their use of stoichiometric information about the cell metabolic network. The use of this information often results in models that require a smaller number of parameters as compared to another type of modelling approaches and thus are less prone to over-fitting. However, regardless of the choice of model, monitoring and control of industrial fermentation processes remains challenging because feedback control strategies require many states to be measured online. In reality, most states cannot be measured online either due to the expense of measuring equipment and its maintenance or the lack of online measurement devices [3–5]. Some states, including concentration of amino acids, metals, vitamins, ATP and precursors have great effect on the fermentation process but are either difficult or impossible to measure online.

To address the lack of online measurements, soft sensors have been proposed. Soft sensors are algorithms that estimate the values of the states based on few available online measurements. Data-driven soft sensors are currently very popular, driven by the interest in the artificial intelligence research area. Reported data-driven soft sensors are generally based on artificial neural networks, support vector machines, partial least squares, and fuzzy inference [4]. However, despite their popularity, the main drawback of data-driven soft sensors is their limited applicability to the region of data used for model training and the scarcity of data available for calibration [6]. Moreover, the lack of mechanistic information of these black box models introduces concerns about the safety and reliability of the controllers designed based on these models [6].

Another category of soft sensors are state observers based on mechanistic models such as a Luenberger observer, Kalman filter, and particle filter. These state observers estimate

Citation: Shen, X.; Budman, H. Set Membership Estimation with Dynamic Flux Balance Models. *Processes* **2021**, *9*, 1762. https://doi.org/10.3390/pr9101762

Academic Editors: Philippe Bogaerts and Alain Vande Wouwer

Received: 25 August 2021
Accepted: 26 September 2021
Published: 1 October 2021

Publisher's Note: MDPI stays neutral with regard to jurisdictional claims in published maps and institutional affiliations.

the values of some states based on convergence of state prediction errors and provided that sufficient measurements are available [7]. A key prerequisite of theses state observer designs is that some observability condition is satisfied with respect to the estimated states. It will be shown later in the manuscript that, unless enough states of a DFBM model are measured online, it is difficult to satisfy full observability for all the states.

In the absence of observability of some states, instead of estimating their specific values it is possible to estimate intervals (ranges) of values based on a priori known range of initial conditions, i.e., range of values at time $= 0$. This type of problem is referred to in the literature as an initial values problem with parameter uncertainty or set-valued ODE integration. The parameter here refers to either uncertain initial states or some model parameter such as a kinetic constant. In the past several decades, different methods have been proposed to find tight bounds containing the reachable sets, including interval analysis [8], Taylor models with different remainder bounds [9], set-base parameter estimation [10], and different relaxation methods [11]. Due to the uncertainty amplification effect, interval analysis can diverge quickly and only suits a small part of the system. Set-based parameter estimation is computationally expensive because the parameter space need to be validated in a piecewise manner and each validation test requires the solution of a semi-definite programming problem. Taylor models can be used to find tight and nonconvex bounds of reachable sets but cannot be easily formulated to take measurements into consideration. To find the reachable sets compatible with available measurements, different relaxation methods and domain reduction are required which are computationally expensive.

When measurements are available, the trajectories that are not compatible with those measurements should be removed from the reachable sets. Estimation algorithms that efficiently deal with reachable sets subject to measurements including interval observers and set membership estimation algorithms [12,13]. An interval observer is usually composed of two classical observers (framers) which estimate the lower and upper bounds of states, respectively. However, sufficient measurements and fulfillment of observability are still required to build the two classical observers [14,15]. Most interval observers exploit the order-preserving properties of cooperative systems to estimate bounds of states [16]. Set membership estimation is an alternative method for estimating the uncertainty of a set of states that has been applied to linear systems [17]. The propagation of uncertainty along time is performed by a series of affine mapping operations over sets. Different shapes of sets have been used to contain the uncertainty, including zonotopes [18], parallelotopes [17] and ellipsoids [19].

In this research, a set membership estimation approach is proposed for nonlinear systems described by DFBM models. The DFBM is converted into a variable structure system composed of several continuous systems in different region of state space by multiparametric linear programming. To address the lack of measurements an Extended Kalman Filter (EKF) is used to estimate nominal values of some states which are important for determining metabolic fluxes. Then, a set membership estimation algorithm is applied for DFBM to estimate bounds of all states. A detector is proposed to detect the switch between different subsystems.

The paper is organized as follows. Section 2.1 introduces background of DFBM. Section 2.2 describes the use of multiparametric linear programming to convert the DFBM into a variable structure system composed of subsystems. Section 2.3 describes the EKF used to estimate some states which are important for determining metabolic fluxes. Section 2.4 presents the main ideas of set propagation and error compensation for calculation of states' bounds. Section 2.5 presents the algorithm for detecting the switch between different subsystems. Section 3 provides the application of the proposed techniques to the batch fermentation of *E. coli*. Section 4 presents a Discussion of the results followed by Conclusions.

2. Materials and Methods

2.1. Dynamic Flux Balance Models

Dynamic flux balance models (DFBM) are structured genome-based metabolic models developed from flux balance models. The key assumption of DFBM is that the cells act as agents distributing resources through metabolic reaction networks to boost a biological objective, e.g., growth rate [1]. Accordingly, the DFBM is formulated as an optimization problem. In the literature [20], both dynamic and static optimization approaches are reported. In the dynamic approach, the nonlinear programming problem is solved over a relatively large time period which is computationally expensive and thus less convenient for uncertainty propagation. In this investigation, static optimization approach is adopted for its simplicity. DFBM is interpreted as a local linear programming problem to maximize a biological objective. In terms of dynamics of intracellular metabolites, there are two type of DFBM models in the literature. One type of DFBM differentiates intracellular and extracellular environments and assumes that the intracellular metabolic reactions are fast enough such as it can be assumed at quasi-steady state [2,21]. Accordingly, only the extracellular metabolites and the biomass are described by dynamic state equations. It has been argued that the intracellular metabolite concentrations are not constant and may change over time [22]. Accordingly, there is a second type of DFBM, used in the current study, which does not differentiate between intracellular and extracellular compartments and the dynamics of all the metabolites are considered [20,23]. The governing equations of DFBM are based on discretized mass balances for all metabolites and these are defined by Equations (1a)–(1d).

$$x_{k+1} = Bx_k + \Delta t x_{bio,k} Av_k + h \tag{1a}$$

$$y_k = Cx_k + r_k \tag{1b}$$

$$x_0 \in \mathcal{P}_0 \tag{1c}$$

$$r_k \sim TN(0, \Sigma, l, u) \quad k = 0, 1, 2 \cdots \tag{1d}$$

where x_k is a vector of n_x state variables at the time step k. The state vector x includes concentrations of metabolites and biomass x_{bio}. y is a vector of n_y measured variables. $B \in \mathbb{R}^{n_x} \times \mathbb{R}^{n_x}$ is a constant diagonal matrix with diagonal elements $b_j, j = 1, \cdots, n_x$. Δt is a constant discrete time step size. $A \in \mathbb{R}^{n_x} \times \mathbb{R}^{n_{rct}}$ is a stoichiometry coefficient matrix, where n_{rct} is the number of reactions considered in the metabolic network. $v \in \mathbb{R}^{n_{rct}}$ is the metabolic flux vector and its calculation is discussed below. $h \in \mathbb{R}^{n_x}$ is a constant vector. The initial state x_0 is assumed to be bounded by a finite polyhedron $\mathcal{P}_0$ as Equation (1c). The underlying assumption is that in practice the initial concentrations of the culture medium components are known to be within specific ranges of values $\mathcal{P}_0$. This assumption is based on the fact that some variation in media formulation occurs due to human factor and variability in raw materials. Hence, this research focuses on the initial uncertainty and we assume all parameters in the state equations to be known accurately. In other words, the method proposed in this research cannot deal with model structure uncertainty like uncertainty in matrix A. However, the method can be extended to deal indirectly with uncertainty in parameters θ defined in the following paragraphs.

$r_k \in \mathbb{R}^{n_y}$ are measurement noise vectors of which the elements follow the truncated multivariate normal distribution (TN) [24,25]. The probability density function p for $TN(\mu, \Sigma, l, u)$ are defined as per Equation (2).

$$p(x, \mu, \Sigma, l, u) = \frac{exp\{-\frac{1}{2}(x - \mu)^T \Sigma^{-1}(x - \mu)\}}{\int_l^u exp\{-\frac{1}{2}(x - \mu)^T \Sigma^{-1}(x - \mu)\}} \tag{2}$$

For r_k, the mean vector of TN is $0 \in \mathbb{R}^{n_y}$; the covariance is $\Sigma \in \mathbb{R}^{n_y} \times \mathbb{R}^{n_y}$; the corresponding variance vector is $\sigma^2 \in \mathbb{R}^{n_y}$; the lower bound and upper bound are $l \in \mathbb{R}^{n_y}$ and $u \in \mathbb{R}^{n_y}$, respectively. $|\cdot|$ indicates the absolute value of a vector. It is assumed that $|l| \leq 3\sigma$ and $|u| \leq 3\sigma$, which indicate that the absolute values of the lower bound and upper bound, respectively, are within the range of 3σ. For simplicity, the current study assumes the process noise to be zero. Process noise could be included as an additional state but this is beyond the scope of the current work.

Following the assumption that the cell allocates resources optimally, the metabolic flux v vector at each time step is obtained by solving a linear programming (LP) problem, defined by Equations (3a) and (3b).

$$\max_{v_k} \quad c^T v_k \tag{3a}$$

$$\text{subject to} \quad Gv_k \leq F\theta_k(x_k) + z \tag{3b}$$

where $c \in \mathbb{R}^{n_{rct}}$, $F \in \mathbb{R}^{n_G} \times \mathbb{R}^{n_\theta}$, $z \in \mathbb{R}^{n_G}$, $G \in \mathbb{R}^{n_G} \times \mathbb{R}^{n_{rct}}$, $\theta \in \Theta \subseteq \mathbb{R}^{n_\theta}$. n_G is the number of linear constraints. The parameter vector θ is a nonlinear vector-valued function of states x. n_θ denotes the number of elements in the parameter vector θ. Usually, each element θ is only function of two states at most and one of these two states is biomass concentration. Θ denotes the parameter space where the optimal solution of the LP resides. Equation (3a) denotes the objective of the LP that cells are optimizing where the most commonly used objective is the biomass production rate, i.e., growth rate. Thus, cells try to maximize growth rate by allocating limited resources. The LHS (left hand-side) in Equation (3b) describes either the rate of change of metabolite concentrations or the change of metabolite concentrations over a discretization time step Δt. Matrices G are constant matrices containing the information of the stoichiometry of reactions. RHS in Equation (3b) is a function of x_k, denoting the metabolic reaction bounds for each step. The matrix F is a matrix of which the elements are the part of the right hand side of the constraints that are functions of states at the previous time interval. z is a vector containing constant values such as constant uptake rate limits. Therefore, linear constraints of flux v in Equation (3b) are reaction rate limits or bounds on available resources (nutrients). Numerical examples of these matrices and vectors are shown for the *E. coli* model in the results section.

2.2. Multiparametric Linear Programming for DFBM

2.2.1. Multiparametric Linear Programming

While set -based methods are available for uncertainty propagation for linear state space equations, these methods are not directly applicable to DFBM. The reason is that the fluxes used in the state equations are obtained from an LP and thus the problem is nonlinear due to the nonlinear function $\theta(x)$ and the occurrence of different sets of active constraints. To tackle the dependency of the state equations on the LP, the concept of multiparametric linear programming (mpLP) is used to convert the DFBM into a variable structure system which is composed of subsystems. Multiparametric linear programming divides the parameter space (Θ) into different regions corresponding to different sets of active constraints and generates explicit expressions for calculating optimal solutions (v) for each region [26–28].

Let assume a given optimal solution v of the LP (Equation (3)) where subscript $\mathcal{A}$ and $\mathcal{I}$ denote indices of active and inactive constraints, respectively. Using this notation Equation (3b) is decomposed into two parts, equalities $G_{\mathcal{A}}v_k = F_{\mathcal{A}}\theta_k(x_k) + z_{\mathcal{A}}$ and inequalities $G_{\mathcal{I}}v_k \leq F_{\mathcal{I}}\theta_k(x_k) + z_{\mathcal{I}}$. Without loss of generality, let us assume that $G_{\mathcal{A}}$ is linear independent (linear redundant rows can always been removed by Gaussian elimination). Let $H = G_{\mathcal{A}}^{-1}F_{\mathcal{A}}$ and $g = G_{\mathcal{A}}^{-1}z_{\mathcal{A}}$, then the optimal solution can be obtained by Equation (4).

$$v_k = H\theta_k(x_k) + g \tag{4}$$

Substituting Equation (4) into the inequality constraints results in Equation (5).

$$(G_{\mathcal{I}}H - F_{\mathcal{I}})\theta_k(x_k) < z_{\mathcal{I}} - G_{\mathcal{I}}g \tag{5}$$

Equation (5) defines a polyhedral region of θ where the existence of the optimal solution is ensured by Equation (4). The region defined by Equation (5) is referred to as a critical region in the multiparametric programming literature. Different critical regions are defined by different combinations of $\mathcal{A}$ and $\mathcal{I}$. Then, the entire parameter space Θ can be decomposed into connected critical regions denoted by $\{\Theta^i\}$, $i = 1, \cdots, n_\Theta$. n_Θ denotes the total number of critical regions in Θ. In practice, critical regions that are very small are ignored and assumed to be covered by the adjacent critical region. Correspondingly, superscript i is used to denotes the i-th critical region. Assume for a specific $\theta \in \Theta^i$, the optimal flux v vector can be calculated analytically by $v_k^i = H^i\theta_k + g^i$ thus bypassing the need for solving the LP. Following the literature and our previous studies, for a given θ, multiple optimal solutions can coexist [29,30]. In other words, multiple Equation (4) can coexist which results in different ways to divide the parameter space Θ. When such multiplicity issue occurs it results in different time trajectories. For simplicity, multiplicity is not addressed in the current study and it is addressed in a separate work by different methods from the one presented here.

By substituting the optimizer equation $v_k^i = H^i\theta_k + g^i$ into Equation (1a), we obtained a set of governing state equations as per Equations (6a)–(6e). Since different θ_k are within different critical regions as Equation (6b), each critical region corresponds to different state equations Equation (6a). Thus the set $\{\Theta^i\}$ defines a family of state space models and this family is referred to as a variable structure system. A variable structure system is a piecewise continuous system composed of subsystems where each subsystem corresponds to a different region of the state space. Furthermore, the region of the state space corresponding to a specific subsystem is referred to as a critical region. Each subsystem is described by a different set of state equations. Accordingly, the state equations need to be changed as soon as the states enter into a new critical region. Here, the superscript i denotes the *i-th* subsystem corresponding to a critical region Θ^i. Equations (6c)–(6e) remain the same form as Equations (1b)–(1d).

$$x_{k+1} = Bx_k + \Delta t x_{bio,k} A(H^i\theta_k(x_k) + g^i) + h \tag{6a}$$

$$\theta_k(x_k) \in \Theta^i \quad i = 1, \cdots, n_\Theta \tag{6b}$$

$$y_k = Cx_k + r_k \tag{6c}$$

$$x_0 \in \mathcal{P}_0 \tag{6d}$$

$$r_k \sim TN(0, \Sigma, l, u) \quad k = 0, 1, 2 \cdots \tag{6e}$$

2.2.2. Reaction Rate Estimability

To further simplify the system described by Equations (6a)–(6e) it is possible to exploit the sparseness of the H matrix. For instance, to take advantage of zero columns of H, Equation (4) can be re-written as shown in Equation (7). For conciseness, the subscript k is omitted here because Equation (7) applies for all time steps.

$$v^i = H^i\theta(x) + g^i = \begin{bmatrix} H_N^i & H_Z^i \end{bmatrix} \begin{bmatrix} \theta_N^i(x_N^i) \\ \theta_Z^i(x) \end{bmatrix} + g^i = H_N^i\theta_N^i(x_N^i) + g^i \tag{7}$$

In Equation (7) N and Z denote the indices of the nonzero and zero columns of the H matrix, respectively. Because H_Z is a submatrix containing the zero columns of H, the flux v is only a function of parameters $\theta_N(x_N)$ according to Equation (7). Moreover, while the parameters θ are a function of states x (see Equations Equations (1a)–(1d) and (3)), only some elements of x actually determine the entire flux vector v. The vector x_N contains, according to Equation (7), the states that determine the flux vector. Notice that for different critical regions flux-determining vector x_N contains different states. Therefore, Equation (6a) can be simplified into Equation (8).

$$x_{k+1} = Bx_k + \Delta t x_{bio,k} A(H_N^i\theta_N^i(x_{N,k}^i) + g^i) + h \tag{8}$$

The biological interpretation of the flux-determining state vector x_N is that only some resources are limiting the growth of cells, either because they are limited or because the activity of enzymes in the related reactions (fluxes) is limiting. As the fermentation progresses, the states transit into new critical regions from old critical regions. Different critical regions can be interpreted as different metabolic stages where x_N are different. Similar interpretations have been reported in [26] in the context of steady state flux balance analysis.

In Equation (8), the term $\Delta t x_{bio,k} A(H_N^i \theta_N^i(x_{N,k}^i)$ denotes the change of metabolite concentrations contributed by metabolic reactions. Therefore, the reaction rates are $x_{bio,k} A(H_N^i \theta_N^i(x_{N,k}^i)$. It is noted that this nonlinear reaction rate term is not only a function of the flux-determining states vector x_N but also of biomass concentration x_{bio}, because the fluxes are defined per unit biomass, i.e., more biomass demands more nutrients to satisfy the requirement of the growth. Once the states that determine the reaction rates, i.e., the states x_N together with the value of x_{bio}, can be estimated, the estimation problem can be simplified greatly. Since in some cases x_N contains x_{bio} but in some cases it does not, we define a reaction-rate-determining state vector x_M in Equation (9). Hence, the reaction-rate-determining state vector x_M always contains the flux-determining states x_N and the biomass state x_{bio} without any redundancy.

$$x_M = \begin{cases} x_N, & \text{if } x_N \text{ contains the biomass state } x_{bio}. \\ \begin{bmatrix} x_N \\ x_{bio} \end{bmatrix}, & \text{otherwise.} \end{cases} \tag{9}$$

The vector x_M for critical region Θ^i is denoted by x_M^i. We define reaction rate estimability as the ability to determine the reaction rates $x_{bio,k} A(H_N^i \theta_N^i(x_{N,k}^i)$ in the metabolic networks which is needed for the calculation of Equation (8). Following the above, once reaction-rate-determining state vector x_M at time step k can be estimated, the dynamic evolution of the culture at step $k+1$ as per Equation (8) can be predicted. In addition, it should be noticed that it is not necessary to measure all the reaction-rate-determining states for reaction rate estimability and instead some states can be estimated by an observer from available measurements. However, if an observer is used to estimate x_M^i, some particular combination of measurements is necessary for observability of x_M^i. Considering different measurement combinations Ω_1^i, Ω_2^i... for the critical region Θ^i, only some combinations provide full observability of x_M^i. Let $\Omega_{\mathcal{O}}^i$ be defined as a family of sets of measurements, which contains all measurement combinations that fulfill observability of x_M^i.

Although many different critical regions and corresponding combinations of measurements could be considered, in practice the possibilities will be limited because industrial fermentations usually operate in a narrow range of operating conditions. Thus, the dynamic trajectories of states only pass through a limited set of critical regions. Assume for $\forall x_0 \in \mathcal{P}_0$, the set of critical regions that the trajectories traverse are Γ. Then, the minimum set of measurements required for the reaction rate estimability of the critical region set Γ is Ω_Γ as per Equations (10a)–(10c).

$$\Omega_\Gamma = \min_j \left| \bigcup_i \Omega_j^i \right| \tag{10a}$$

$$\text{subject to} \qquad i \in \Gamma \tag{10b}$$

$$\Omega_j^i \in \Omega_{\mathcal{O}}^i \tag{10c}$$

where $|\cdot|$ is the cardinality of a finite countable set, i.e., the number of elements of a set. In Equation (10c), $\Omega_j^i \in \Omega_{\mathcal{O}}^i$ indicates that the measurement combination Ω_j^i can fulfill the observability of reaction-rate-determining states x_M^i of critical region Θ^i. If all states in set Ω_Γ are measured, the reaction rate term of any trajectory starting from $\mathcal{P}_0$ can be estimated by the observer. In other words, although x_M^i in different critical regions may be different, requiring different measurements for observability, x_M^i is always observable if the chosen set of measurements satisfy Equation (10c).

2.3. Extended Kalman Filter (EKF)

Using the minimum required set of measurements, Ω_Γ is defined in Equation (10c), x_M can be estimated by an observer. x_M corresponds to the observable subspace of the governing equation (Equations (1a)–(1d)) for each critical region. The state equation of the observable subspace for critical region Θ^i is given by Equations (11a)–(11c).

$$x^i_{M,k+1} = f^i(x^i_{M,k}) = Bx^i_{M,k} + \Delta t x_{bio,k} A_M (H^i_N \theta^i_N (x^i_{N,k}) + g^i) + h_M \tag{11a}$$

$$y_k = C^i_M x^i_{M,k} + r_k \tag{11b}$$

$$r_k \sim TN(0, \Sigma, l, u) \quad k = 0, 1, 2 \cdots \tag{11c}$$

where $x^i_{N,k}$ and $x^i_{M,k}$ are the flux-determining state vector and the reaction-rate-determining state vector for critical region Θ^i, respectively; A_M is the stoichiometry submatrix corresponding to x_M. Similarly h_M is a sub-vector of h corresponding to x_M. It should be noticed that, for different critical regions, x_M involves different states. Accordingly, each critical region requires the use of a different EKF. In addition, it should be noticed that the C^i_M matrices are different for each critical region but the measured variables Ω_Γ are the same since the same sensors are used for the entire fermentation.

To estimate x^i_M, an standard EKF is used due to its effective and simple structure [31]. The estimate $\hat{x}^i_{M,k}$ and covariance P^i_k of x^i_M for critical region Θ^i are described by Equation (12a) and Equation (12b), respectively.

$$\hat{x}^i_{M,k} = f^i(\hat{x}^i_{M,k-1}) + K_k(y_k - C^i_M f^i(\hat{x}^i_{M,k-1})) \tag{12a}$$

$$P^{i\,-1}_k = \Phi^i_{k-1} P^i_{k-1} \Phi^{i\,T}_{k-1} + C^{i\,T}_M (\Sigma\Sigma^T)^{-1} C^i_M \tag{12b}$$

where

$$K_k = \Phi^i_{k-1} P^i_{k-1} \Phi^{i\,T}_{k-1} C^{i\,T}_M (C^i_M \Phi^i_{k-1} P_{k-1} \Phi^{i\,T}_{k-1} C^{i\,T}_M + \Sigma\Sigma^T)^{-1} \tag{13a}$$

$$\Phi^i_k = \frac{\partial f^i}{\partial x^i_M}(\hat{x}^i_{M,k}) \tag{13b}$$

The measurement noise is assumed to be a truncated multivariate normal distribution as Equation (11c). This assumption is needed for estimating finite bounds as explained in the following section. Recall in Equation (2) that $|l| \leq 3\sigma$ and $|u| \leq 3\sigma$, the lower and upper bounds are located within the range of 3σ. The covariance matrix P_k is always overestimated to ensure boundedness. Although the EKF resulting from this assumption is sub-optimal, it is still sufficient to estimate x^i_M.

2.4. Set Propagation and Error Compensation

Since the minimum set of measurements defined by Equations (10a)–(10c) can only ensure the observability of x_M, the estimation of other states needs different estimation strategies. The idea is to exploit the a priori knowledge of the initial ranges of initial conditions to estimate all states. Instead of predicting specific values of states, set membership estimation (SME) approach is used to predict sets containing all possible states by a series of set operations. These set operations usually include linear mapping, projection, translation, Minkowski addition, intersection, union, and outer approximation. In this research, all sets and multiparametric linear programming operations are performed with the Multi-Parametric Toolbox 3.0 (https://www.mpt3.org/ accessed on 15 July 2021) [32] and MATLAB R2018a. The *E. coli* example can be found online (https://github.com/SetMembershipEstimationDFBM/E.coliExample, accessed on 25 September 2021). For DFBM, SME propagates the initial set $\mathcal{P}_0$ by affine mapping as Equation (14). Affine mapping involves two operations: linear mapping of the previous set and translation.

$$\hat{\mathcal{X}}_{k+1} \approx \underbrace{B\hat{\mathcal{X}}_k}_{\text{linear mapping}} + \underbrace{\Delta t \hat{x}_{bio,k} A (H_N^i \theta_N^i (\hat{x}_{N,k}^i) + g^i) + h}_{\text{translation}} \tag{14}$$

where $\hat{\mathcal{X}}_k$ represents the set of states at time step k and $\hat{\mathcal{X}}_0 = \mathcal{P}_0$, i.e., the set of initial conditions assumed to be known. In Equation (14), the translation term is approximated by using the estimate $\hat{x}_{M,k}^i$ obtained by the EKF. In the application of EKF, the estimate $\hat{x}_{M,k}^i$ needs several time steps to converge to the true flux-determining states $x_{M,k}^i$. Thus the SME described by Equation (14) may underestimate bounds while the EKF is converging. To mitigate this problem, a correction is implemented to compensate for the estimate error as described below. Since no extra information is available, the compensation of the estimate error is based on the worst case scenario.

The error in the estimate incurred by the observer for critical region Θ^i is $e_M^i = x_{M,k}^i - \hat{x}_{M,k}^i$. Since $x_{M,k}^i$ always contains biomass $x_{bio,k}^i$ and $x_{N,k}^i$, the corresponding estimate errors are defined as $e_{N,k}^i = x_{N,k}^i - \hat{x}_{N,k}^i$ and $e_{bio}^i = x_{bio,k}^i - \hat{x}_{bio,k}^i$. Let us assume that the function θ is first-order differentiable and define Jacobian matrix ψ_k^i.

$$\psi_k^i = \frac{\partial \theta_N^i}{\partial x_N^i}(\hat{x}_{N,k}^i) \tag{15}$$

Substituting the estimate error e_k^i, e_{bio}^i and Jacobian matrix ψ_k^i into Equation (8), a corrected state equation that accounts for the estimate error is obtained as Equations (16a) and (16b). Equations (16a) and (16b) uses a first order approximation to account for the state deviation ϵ_k^i caused by the estimate error $e_{M,k}^i$, while the EKF is converging. The error compensation based on linearization provides satisfactory bounds because the error between estimate and measured is small and decreases quickly due to the convergence of EKF.

$$x_{k+1} = Bx_k + \Delta t \hat{x}_{bio,k} A (H_N^i \psi_k^i \hat{x}_{N,k}^i + g^i) + h + \epsilon_k^i \tag{16a}$$

$$\epsilon_k^i = D_k e_{N,k}^i + e_{bio,k} M_k e_{N,k}^i + L_k e_{bio,k} \tag{16b}$$

where

$$D_k = \hat{x}_{bio,k} \Delta t A H_N^i \psi_k^i + h \tag{17a}$$

$$M_k = \Delta t A H_N^i \psi_k^i \tag{17b}$$

$$L_k = \Delta t A (H_N^i \theta_{N,k}^i (\hat{x}_{N,k}^i) + g^i) \tag{17c}$$

In this work, the noise was assumed to follow a truncated multivariate Gaussian distribution. The corresponding standard multivariate Gaussian distribution of noise contains the truncated one. As illustrated in Figure 1, when an EKF is used to estimate the states, the distribution of states with a standard Gaussian noise should similarly contain the one with the truncated Gaussian noise, which is the true distribution of states. Moreover, the distribution of states by standard EKF is also a multivariate Gaussian distribution. For Gaussian distribution, 99.7% of the samples are within the interval of 6 standard deviations from both sides of the mean for each state. Thus, an interval set based on 6 standard deviations can contain the distribution by standard EKF and eventually contain the true distribution of states as in Figure 1. Since P_k^i is the covariance of a standard EKF, the diagonal elements of matrix P_k^i are the variances for each state. Therefore, diagonal elements of P_k^i can be used to define the interval set to bound the error ϵ_k^i.

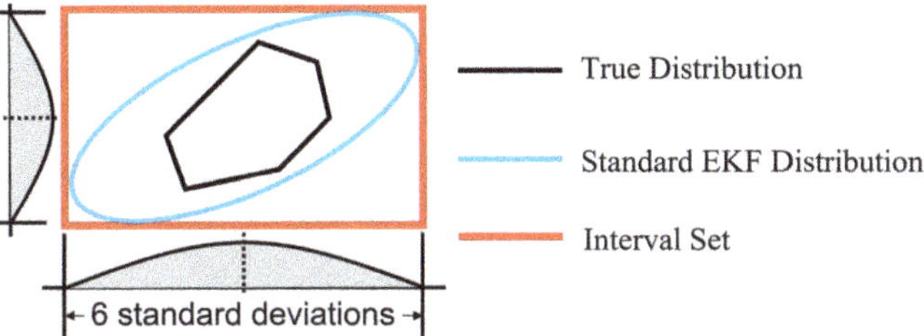

Figure 1. Illustration of the interval set containing the distribution of states.

To formulate an error compensation operation scheme several set operations are introduced first as follows. The n-dimensional interval set is $\mathcal{S}(p,q)$ with lower bound p and upper bound q as $\mathcal{S}(p,q) = \{x \in \mathbb{R}^n : p \leq x \leq q\}$. The outer approximation operation $\mathcal{Q}(\cdot)$ of a bounded set $\mathcal{W}$ is denoted by $\mathcal{Q}(\mathcal{W})$, which involves the mapping of the set $\mathcal{W}$ to a new interval set. If the infimum and supremum are denoted by $\inf(\cdot)$ and $\sup(\cdot)$, respectively, the outer approximation of the set $\mathcal{W}$ is $\mathcal{Q}(\mathcal{W}) = \mathcal{S}(\inf(\mathcal{W}), \sup(\mathcal{W}))$. The operator $\oplus$ is the Minkowski addition of two sets. For example, for two sets α and β, $\alpha \oplus \beta = \{a + b : \forall a \in \alpha, \forall b \in \beta\}$.

Notice that the diagonal elements of P_k^i are the variances of each state. Then, if the standard deviation of $e_{N,k}^i$ is $\eta_{N,k}^i$ and of $e_{bio,k}^i$ is $\eta_{bio,k}^i$, two interval sets $\mathcal{E}_{N,k}$ and $\mathcal{E}_{bio,k}$ can be defined to bound $\eta_{N,k}^i$ and $\eta_{bio,k}^i$, respectively, based on the choice of 3 standard deviation ranges, as $e_{N,k}^i \in \mathcal{E}_{N,k} = \mathcal{S}(-3\eta_{N,k}^i, 3\eta_{N,k}^i)$ and $e_{bio,k}^i \in \mathcal{E}_{bio,k} = \mathcal{S}(-3\eta_{bio,k}^i, 3\eta_{bio,k}^i)$. In Equation (16b), since $|e_{bio,k}^i| < 3\eta_{bio,k}^i$, we have $e_{bio,k} M_k e_{N,k}^i \in 3\eta_{bio,k}^i M_k \mathcal{E}_{N,k}$. Similarly, the other two terms in Equation (16b) can be bounded as $D_k e_{N,k}^i \in D_k \mathcal{E}_{N,k}$ and $L_k e_{bio,k} \in L_k \mathcal{E}_{bio,k}$, respectively. Therefore, the state deviation e_k^i term can be contained within the interval set $\mathcal{E}_{e,k}$ according to Equation (18).

$$e_k^i \in \mathcal{E}_{e,k} = \mathcal{Q}((D_k + 3\eta_{bio,k}^i M_k)\mathcal{E}_{N,k}) \oplus \mathcal{Q}(L_k \mathcal{E}_{bio,k}) \tag{18}$$

where the sets $D_k \mathcal{E}_{N,k}$ and $3\eta_{bio,k}^i M_k \mathcal{E}_{N,k}$ occurring in Equation (18) are combined together. On the other hand, $L_k \mathcal{E}_{bio,k}$ originates from a different set $\mathcal{E}_{bio,k}$ and thus Minkowski addition must be used to add the different sets. However, linear mapping of interval sets can lead to irregular convex sets. In computational geometry, traditional algorithms that perform Minkowski addition for two convex irregular high-dimensional polytopes are computationally expensive [33]. On the other hand, Minkowski addition of two interval sets is computationally efficient because intervals are axis-aligned. Thus, the operator $\mathcal{Q}(\cdot)$ that converts the irregular set to the axis-aligned set is applied to speed up the computation of the Minkowski addition.

Following the above, the set of states $\hat{\mathcal{X}}_{k+1}$ is bounded by the prior estimate set $\mathcal{P}_{k+1}^-$ according to Equations (19a) and (19b).

$$\mathcal{P}_{k+1}^- = \mathcal{Q}\{ \underbrace{B\mathcal{P}_k^+}_{\text{linear mapping}} + \underbrace{\Delta t \hat{x}_{bio,k} A(H_N^i \theta_N^i(\hat{x}_{N,k}^i) + g^i) + h}_{\text{translation}}\} \oplus \mathcal{E}_{e,k} \tag{19a}$$

$$\hat{\mathcal{X}}_{k+1} \subset \mathcal{P}_{k+1}^- \tag{19b}$$

where the set of the posterior estimates is $\mathcal{P}_k^+$. $B\mathcal{P}_k^+$ denotes the scaling of the set $\mathcal{P}_k^+$ by the diagonal matrix B. Then the set $B\mathcal{P}_k^+$ is translated by the vector in the big curly brackets. To compensate for the deviation during the convergence of EKF, the interval set $\mathcal{E}_{e,k}$ is added by Minkowski addition.

Considering the truncated measurement noise, $r_k = y_k - Cx_k$ is bounded by the lower l and upper bounds u; let us define a set $\mathcal{M}_k = \{x_k \in \mathbb{R}^{n_x} : l < y_k - Cx_k < u\}$.

Then, the posterior estimate set $\mathcal{P}_{k+1}^+$ is given by Equations (20a)–(20c). In this study, it is assumed that $\mathcal{P}_k^+$ and $\mathcal{P}_{k+1}^-$ are much smaller than the volumes of the critical regions.

$$\mathcal{P}_{k+1}^+ = \mathcal{P}_{k+1}^- \bigcap \mathcal{M}_{k+1} \tag{20a}$$

$$\hat{\mathcal{X}}_{k+1} \subset \mathcal{P}_{k+1}^+ \tag{20b}$$

$$\mathcal{P}_0^+ = \mathcal{P}_0 \tag{20c}$$

Figure 2 illustrates the set propagation using intervals for an example involving two states, e.g., glucose and biomass concentrations. The initial set $\mathcal{P}_0$ contains all possible initial values of glucose and biomass. Then $\mathcal{P}_1^+$ is generated through set operations by computational geometry algorithms. Since an interval set is used, it is computationally efficient to project the set $\mathcal{P}_1^+$ onto the biomass and glucose axes to obtain the corresponding lower bounds l_{glc}, l_{bio} and upper bounds u_{glc}, u_{bio} as shown in the figure for the set $\mathcal{P}_1^+$.

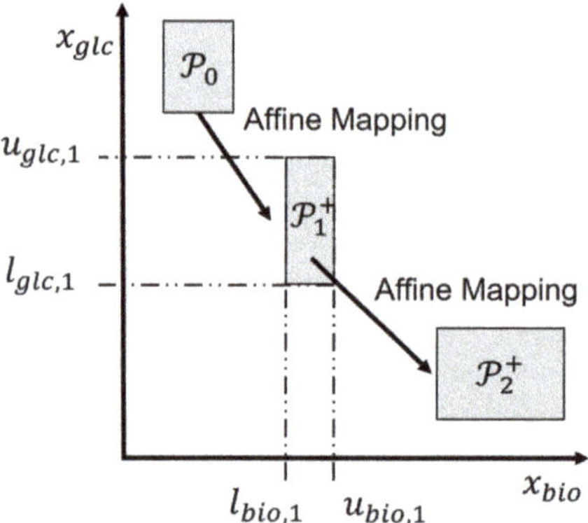

Figure 2. Illustration of set propagation of SME by set operations.

2.5. Detecting the Transition between Critical Regions

The proposed use of multiparametric programming converts the DFBM into a variable structure system composed of subsystems where each critical region corresponds to a subsystem. Along a given time trajectory the states may transit from one critical region to another. When the states estimated by the EKF leave a critical region Θ^i to enter another critical region Θ^j, the estimate $\hat{x}_{M,k}$ and the covariance P_k must be reinitialized because x_M for different critical regions may be different, even though the measured states are the same. Moreover, a criterion is required to detect whether the states are entering into a new critical region.

When the system is traversing from one critical region to another, it needs to cross a boundary between the critical regions. Over time the states may cross over several boundaries along their trajectories and these crossings must be detected. Two neighboring critical regions share a boundary where an active constraint will become inactive or vice versa. The activation of a constraint may require the change of constraints related to $\hat{x}_{N,k}$. For a given constraint, θ is usually only function of two states at most because of commonly used Michaelis—Menten kinetics [34] or constraints to prevent the depletion of nutrients [23] and one of these two states is biomass. So two special cases should be considered as follows when system switches from one critical region to the next:

Case i—x_N^i of the old critical region Θ^i have one more observable state than the x_N^j of the new critical region Θ^j. For this case, the switch between critical regions is determined by Equation (21). Equation (21) calculates the norm of the difference between the flux estimates obtained with Equation (7) in the two neighboring regions. Notice that the flux estimate of Θ^j is based on estimate $\hat{x}_{N,k}^i$ of the old critical region. The value of $\gamma(i,j,k)$ is used to detect the occurrence of a switch. If the system is exactly at the boundary of these two critical regions, the flux equation Equation (7) for these two critical regions should result in the same flux value and $\gamma(i,j,k)$ will be zero. A schematic example is shown in Figure 3. Polygons in different colors represent different critical regions in the parameter space Θ. As the state evolves with time, the corresponding θ changes along the dash line in the parameter space Θ. As the θ approaches the boundary between the critical region Θ^1 and Θ^2, $\gamma(i,j,k)$ approaches zero. Correspondingly, a value of $\gamma(i,j,k)$ smaller than a user specified tolerance indicates a switch between critical regions, thus requiring reinitialization of the EKF as follows: $\hat{x}^j_{N,k}$ is set equal to $\hat{x}^i_{N,k}$ and P^j_k is set equal to P^i_k.

$$\gamma(i,j,k) = \left\| \hat{v}^i_k - \hat{v}^j_k \right\| = \left\| H^i_N \theta^i_N(\hat{x}^i_{N,k}) + g^i - (H^j_N \theta^j_N(\hat{x}^i_{N,k}) - g^j) \right\| \tag{21}$$

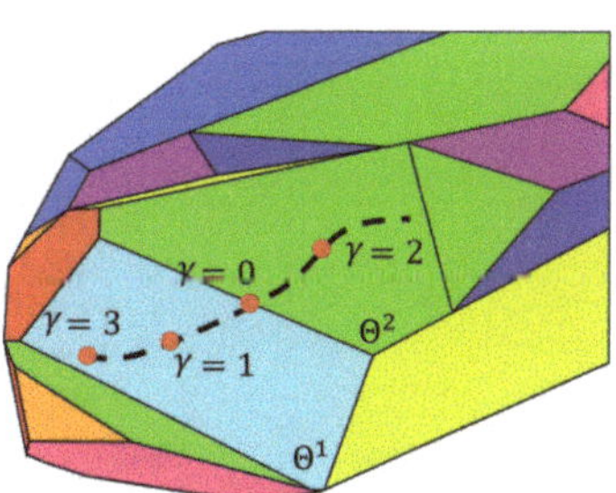

Figure 3. Illustration of detecting a critical region switch.

Case ii—x^j_N of the new critical region Θ^i has one more observable state than the x^j_N of the old critical region Θ^i. To reinitialize the EKF, $\hat{x}^j_{N,k}$ and P^j_k can be set to the old values except for the new observable state that is not observable in the old critical region, and thus it needs to be estimated for calculating $\gamma(i,j,k)$. By projecting the set $\mathcal{P}^+_k$, the lower $l_{un,k}$ and upper bounds $u_{un,k}$ can be calculated. Since no extra information is available, the mean value of the upper bound and lower bound is used as the nominal value of the unobservable state as per Equation (22).

$$\hat{x}^i_{un,k} = \frac{1}{2}(u_{un,k} + l_{un,k}) \tag{22}$$

Equation (23) is used to calculate $\gamma(i,j,k)$. The flux estimate for the new critical region Θ^j is based on the nominal values of the unobservable state $\hat{x}^i_{un,k}$ combined with $\hat{x}^i_{N,k}$ of the old critical region.

$$\gamma(i,j,k) = \left\| \hat{v}^i_k - \hat{v}^j_k \right\| = \left\| H^i_N \theta^i_N(\hat{x}^i_{N,k}) + g^i - (H^j_N \theta^j_N(\hat{x}^i_{un,k}, \hat{x}^i_{N,k}) - g^j) \right\| \tag{23}$$

To reinitialize the EKF, the estimate and covariance are used together with the estimation of the new state that is added in the new critical region. Assuming the states are close enough to the boundary between the critical regions, then Equation (24) holds.

$$\left\| H^i_N \theta^i_N(\hat{x}^i_{N,k}) + g^i - (H^j_N \theta^j_N(\hat{x}^j_{N,k}, \hat{x}^j_{un,0}) - g^j) \right\| = 0 \tag{24}$$

The initial estimate of new observable state $\hat{x}^j_{un,0}$ in the new region can be calculated by solving the Equation (24). Since the new state is between the upper bound and lower bound by SME, the half length between $u_{un,k}$ and $l_{un,k}$ is the worst possible deviation. Then, using a 3 standard deviation range, the initial variance $\eta^2_{un,k}$ can be estimated according to Equation (25) and all other covariance terms related to the new state are assumed to be zero.

$$\eta_{un,k} = \frac{1}{3} \cdot \frac{1}{2}(u_{un,k} - l_{un,k}) \tag{25}$$

Bounds of states estimated by the SME are rigorously guaranteed in each critical region separately but subject to accurate tuning of the tolerance that is used to switch between the subsystems. The tolerance of $\gamma(i, j, k)$ is the only user specified parameter in this research. If the tolerance is too large or small, the EKF may switch the subsystem too early or too late. Accordingly, if the wrong state equations are used in estimation, the bounds on the states may be violated. To avoid such a situation, exhaustive simulations that are initialized with $\mathcal{P}_0$ are conducted to find the tolerance used to switch between critical regions. As an alternative, an overestimated covariance can also be used to reinitialize the EKF when the state enters a new critical region to avoid bound violations.

3. Results

3.1. DFBM Model of E. coli

A DFBM model of *E. coli* reported in [20] is used to illustrate the proposed methodology. The DFBM in batch operation includes four states, glucose concentration x_{glc}, oxygen concentration x_{oxy}, acetate concentration x_{ace}, and biomass concentration x_{bio} as in Equations (26a)–(26e). Thus, the state vector is $x = \begin{bmatrix} x_{glc} & x_{oxy} & x_{ace} & x_{bio} \end{bmatrix}^T$. The substrates are glucose, oxygen, acetate.

$$x_{glc,k+1} = x_{glc,k} + \Delta t x_{bio,k} A_{glc} v_k \tag{26a}$$

$$x_{oxy,k+1} = (1 - k_L a \Delta t) x_{oxy,k} + \Delta t x_{bio,k} A_{oxy} v_k + 0.21 k_L a \Delta t \tag{26b}$$

$$x_{ace,k+1} = x_{ace,k} + \Delta t x_{bio,k} A_{ace} v_k \tag{26c}$$

$$x_{bio,k+1} = x_{bio,k} + \Delta t x_{bio,k} A_{bio} v_k \tag{26d}$$

$$x_0 \in \mathcal{P}_0 = \mathcal{S}\left(\begin{bmatrix} 0.38 & 0.1995 & 0.19 & 0.00095 \end{bmatrix}^T, \begin{bmatrix} 0.42 & 0.2205 & 0.21 & 0.00105 \end{bmatrix}^T\right) \tag{26e}$$

where $k_L a = 4\ h^{-1}$ is the oxygen mass transfer coefficient. The initial state vector x_0 is defined by the interval set $\mathcal{P}_0$ according to Equation (26e). The matrix A contains the stoichiometric coefficients corresponding to four reactions according to Equation (27). Each column of this matrix corresponds to one reaction and each row correspond to one component.

$$A = \begin{bmatrix} A_{glc} \\ A_{oxy} \\ A_{ace} \\ A_{bio} \end{bmatrix} = \begin{bmatrix} 0 & -9.46 & -9.84 & -19.23 \\ -35 & -12.92 & -12.73 & 0 \\ -39.43 & 0 & 1.24 & 12.12 \\ 1 & 1 & 1 & 1 \end{bmatrix} \tag{27}$$

The flux vector v_k is obtained by solving the following linear programming problem as Equations (28a)–(28g):

$$\max_{v_k} \quad A_{bio} v_k \tag{28a}$$

$$\text{subject to} \quad -A_{oxy} v_k \leq \text{OUR}_{max} \tag{28b}$$

$$A_{ace} v_k \leq 100 \tag{28c}$$

$$-\Delta t A_{glc} v_k \leq \frac{x_{glc,k}}{x_{bio,k}} = \theta_{1,k} \tag{28d}$$

$$-\Delta t A_{oxy} v_k \leq \frac{(1 - k_L a \Delta t) x_{oxy,k} + 0.21 k_L a \Delta t}{x_{bio,k}} = \theta_{2,k} \tag{28e}$$

$$-\Delta t A_{ace} v_k \leq \frac{x_{ace,k}}{x_{bio,k}} = \theta_{3,k} \tag{28f}$$

$$-A_{glc}v_k \leq \frac{GUR_{max}x_{glc,k}}{K_m + x_{glc,k}} = \theta_{4,k} \tag{28g}$$

where $OUR_{max} = 12\ \text{mM}/(\text{g-dw·h})$ is the maximum oxygen uptake rate and g-dw is grams of dry weight of biomass; $GUR_{max} = 6.5\ \text{mM}/(\text{g-dw·h})$ denotes the maximum glucose uptake rate. Equation (28a) describes that the objective of the cells is to maximize the biomass growth rate. Equation (28b) indicates that the oxygen consumption rate is limited by a maximum uptake limit. Equation (28c) indicates that the acetate generation rate is bounded by $100\ \text{mM}/(\text{g-dw·h})$. Equation (28g) indicates that the glucose consumption rate is bounded by an upper limit. All the other constraints are positivity constraints to prevent depletion of metabolites. To express these constraints in Equations (28a)–(28g) compactly, the constraints in (28a)–(28g) can be expressed in the form of Equation (3):

$$Gv_k \leq F\theta_k(x_k) + z \tag{29a}$$

$$G = \begin{bmatrix} -A_{oxy} \\ A_{ace} \\ -\Delta t A_{glc} \\ -\Delta t A_{oxy} \\ -\Delta t A_{ace} \\ -A_{glc} \end{bmatrix} \tag{29b}$$

$$F = \begin{bmatrix} 0 & 0 & 0 & 0 \\ 0 & 0 & 0 & 0 \\ 1 & 0 & 0 & 0 \\ 0 & 1 & 0 & 0 \\ 0 & 0 & 1 & 0 \\ 0 & 0 & 0 & 1 \end{bmatrix} \tag{29c}$$

$$z = \begin{bmatrix} OUR_{max} \\ 100 \\ 0 \\ 0 \\ 0 \\ 0 \end{bmatrix} \tag{29d}$$

3.2. Determination of Minimum Measurements

Due to the assumption that the initial state is contained in an interval, the problem in Equations (28a)–(28g) can be formulated as a multiparameteric linear programming (mpLP) problem. The vector θ is composed of four parameters which are nonlinear functions of states. Using the Multi-Parametric Toolbox 3.0, it can be found that the entire parameter space Θ can be decomposed into a maximum of 24 critical regions. For each critical region, the mpLP solver calculates the constraints that form the boundaries of the region and the equations that generate the optimal solutions. In order to reduce the computational effort, extensive simulations are conducted with randomly chosen initial values in set $\mathcal{P}_0$ to identify which critical regions are relevant for the problem. It is found from these simulations that, for the chosen range of initial conditions, the states only traverse through two neighboring critical regions Θ^1 and Θ^2 assuming small critical regions are ignored. According to the results of the mpLP solver, the two critical regions can be defined as Equations (30a) and (30b). Critical regions Θ^1 and Θ^2 share a boundary defined in Equation (30c). Since θ is a function of x, the critical regions are next to each other in the state space.

$$\Theta^1 : \begin{bmatrix} -0.9988 & 0 & 0 & 0.0499 & 0 \\ 0 & -1 & 0 & 0 & 0 \\ 0 & 0 & -0.9971 & -0.0767 & 0 \\ 0 & 0 & 0 & -0.0033 & -1 \\ 0 & 0 & 0 & 1 & 0 \\ 0 & 0 & 0 & -1 & 0 \end{bmatrix} \theta(x) \le \begin{bmatrix} 0 \\ -0.6 \\ -0.6740 \\ 0.0171 \\ 8.7864 \\ 0 \end{bmatrix} \tag{30a}$$

$$\Theta^2 : \begin{bmatrix} -0.9988 & 0 & 0 & 0.0499 & 0 \\ 0 & -0.7469 & 0.6630 & 0.0510 & 0 \\ 0 & 0 & -0.0254 & -0.0053 & -0.9997 \\ 0 & 0 & -1 & 0 & 0 \\ 0 & 0 & 0.9971 & 0.0767 & 0 \\ 0 & 0 & 0 & -1 & 0 \end{bmatrix} \theta(x) \le \begin{bmatrix} 0 \\ 0 \\ 0 \\ 0 \\ 0.6740 \\ 0 \end{bmatrix} \tag{30b}$$

$$\Theta^1 \bigcap \Theta^2 : \begin{bmatrix} 0 & 0 & 0.9971 & 0.0767 & 0 \end{bmatrix} \theta(x) = 0.6740 \tag{30c}$$

Accordingly, the mpLP solver also calculates the matrix H and g used in the flux equation Equation (7) for these two critical regions. By taking advantage of the sparseness of H for these two critical regions, θ_N can be determined. The equations to calculate fluxes for these two critical regions can be expressed as Equations (31a) and (31b).

$$v_k^1 = \begin{bmatrix} -0.039 & 0.1057 & 0 & 0 \end{bmatrix}^T \theta_4(x_{glc,k}) + \begin{bmatrix} 0.3429 & 0 & 0 & 0 \end{bmatrix}^T \tag{31a}$$

$$v_k^2 = \begin{bmatrix} 0.5072 & 0 & 0 & 0 \end{bmatrix}^T \theta_3(x_{ace,k}, x_{bio,k}) + \begin{bmatrix} 0 & 0.1057 & 0 & 0 \end{bmatrix}^T \theta_4(x_{glc,k}) \tag{31b}$$

where θ_N for critical region Θ^1 is θ_4 and θ_N for critical region Θ^2 is θ_3 and θ_4. By substituting the flux equation Equations (31a) and (31b) into Equations (26a)–(26e), the simplified state equations of *E. coli* model can be rewritten compactly as in Equations (32a) and (32b).

$$x_{k+1} = Bx_k + \Delta t x_{bio,k} A v_k^1(x_{glc,k}) + h \qquad \theta(x_k) \in \Theta^1 \tag{32a}$$

$$x_{k+1} = Bx_k + \Delta t x_{bio,k} A v_k^2(x_{ace,k}, x_{bio,k}, x_{glc,k}) + h \qquad \theta(x_k) \in \Theta^2 \tag{32b}$$

Following the calculations above, the original *E. coli* model is simplified into an equivalent system comprised of two subsystems of interest. Equations (32a) and (32b) describe subsystem 1 and subsystem 2, respectively. These two subsystems are continuous in the state space and they share the same boundary as per Equation (30c). Once the state crosses the boundary between the two subsystems, the governing equation is switched from Equations (32a) and (32b). Because the initial state is randomly initialized in set $\mathcal{P}_0$, $\mathcal{P}_0$ corresponds to a set in Θ^1. Thus, the state evolves within the region of subsystem 1 and gradually approximates the region of subsystem 2 governed by Equation (32b) until finally crossing the boundary given by Equation (30c). As only part of θ is known, a detector is used to detect the crossing of the boundary, thus ensuring that the switch between the regions is performed accurately.

Based on the flux equation Equations (31a) and (31b), the reaction-rate-determining states vector x_M^i for Θ^1 are biomass and glucose and for Θ^2 are biomass, acetate and glucose. Accordingly, the possible combinations of measurements needed for observing x_M^1 of Θ^1 include $\Omega_1^1 = \{Bio\}$, $\Omega_2^1 = \{Glc\}$ and $\Omega_3^1 = \{Bio, Glc\}$. Similarly, there are 7 possible combinations of measurements for observing the vector x_M^2 in Θ^2, namely $\Omega_1^2 = \{Ace\}$, $\Omega_2^2 = \{Bio\}$, $\Omega_3^2 = \{Glc\}$, $\Omega_4^2 = \{Ace, Bio\}$, $\Omega_5^2 = \{Bio, Glc\}$, $\Omega_6^2 = \{Ace, Glc\}$, and $\Omega_7^2 = \{Ace, Bio, Glc\}$. To find a combination of measurements Ω_Γ that will be suitable for both critical regions, it is necessary to perform an analysis of observability for these combinations. The Symbolic Toolbox calculation of MATLAB R2018a is used to develop an analytical equation observability rank condition and rank of Φ_k^i of the nonlinear system according to the criterion presented in [31]. Since the symbolic expressions of the rank for each critical regions for Equation (11) are very complex, it is very difficult to infer a analytical condition of observability for all possible values of the states. Instead, the rank values are calculated for different measurement combinations and rank of Φ_k^i using a Monte Carlo algorithm based on 5 million samples of Θ^1 and Θ^2, respectively. According to these Monte Carlo simulations, the only measurement required for observability of the vectors x_M^1 in Θ^1 and x_M^2 in Θ^2 is the biomass concentration, namely $\Omega_\Gamma = \{Bio\}$.

3.3. EKF for the Two Subsystems and Detection of Transition between Subsystems

Based on the aforementioned observability analysis, the biomass concentration is the only state that needs to be measured online as per Equation (33a) for implementation of the EKF. Measurement noise is assumed as a truncated normal distribution as described by Equation (33b). Since the initial $\mathcal{P}_0$ is assumed to be known, the EKF is initialized at the center of $\mathcal{P}_0$ with a variance based on 3 standard deviations and zero covariance terms. The state of the plant is initialized randomly by sampling a point within the region defined by $\mathcal{P}_0$.

$$y_k = \begin{bmatrix} 0 & 0 & 0 & 1 \end{bmatrix} x_k + r_k \tag{33a}$$

$$r_k \sim TN(0, 0.004^2, -0.0004, 0.0004) \quad k = 0, 1, 2 \cdots \tag{33b}$$

Based on the assumed $\mathcal{P}_0$, in the batch process the EKF starts in critical region Θ^1 and later it transitions into critical region Θ^2. Thus, two EKFs are required in this case study to estimate the x_M as summarized in Table 1. Based on the biomass measurement y_k, the glucose and biomass concentrations are estimated by the EKF for Θ^1 as $\hat{x}_{N,bio,k}$ and $\hat{x}_{N,glc,k}$. With the same biomass measurement, the second critical region Θ^2 has one more observable state which is the acetate concentration $\hat{x}_{N,ace,k}$.

Table 1. Observable and unobservable subpace of two subsystems of the DFBM model of *E. coli*.

	Subsystem of Θ^1	Subsystem of Θ^2
Observable Subspace (x_M)	Glc, Bio	Glc, Ace, Bio
Unobservable Subspace	Ace, Oxy	Oxy
Measurement	Bio	Bio

Since acetate and oxygen are unobservable in Θ^1, they need to be estimated by bounds. To find these bounds, SME propagates the initial set $\mathcal{P}_0$ by set operations to obtain a prior estimate set $\mathcal{P}_k^-$ as Equation (19). After obtaining the measurement of biomass, a posterior estimate set $\mathcal{P}_k^+$ as in Equation (20) is calculated by set operations. The error due to lack of convergence of the EKF is compensated by using Equation (18). By projecting $\mathcal{P}_k^+$ onto the axis of acetate and oxygen, respectively, the upper bound $u_{un,k}$ and lower bound $l_{un,k}$ of these two states are obtained.

Since Θ^2 has one more flux-determining state, acetate that is not observable from the measurement of biomass, it must be estimated as explained in Equation (22). Using the mean value of $u_{un,ace,k}$ and $l_{un,ace,k}$ the nominal values of the unobservable state $\hat{x}_{un,ace,k}$ are obtained. Using the EKF estimates of the observable flux-determining states $\hat{x}_{N,k}$ together with the nominal value of acetate $\hat{x}_{un,ace,k}$, the detection scheme explained in Section 2.5 can be implemented. Accordingly, $\gamma(i, j, k)$ is calculated from Equation (23) to determine the switch from critical region Θ^1 to critical region Θ^2. The tolerance of $\gamma(i, j, k)$ to determine the switch between the critical regions is assumed as 0.08. This tolerance is the only tuning parameter of the proposed method and it is determined by trial and error. After the switch occurs the acetate concentration is initialized by the solution of Equation (24) and the variance of acetate is initialized based on Equation (25). After the switch to critical region Θ^2, the EKF continues to generate estimates of glucose, acetate and biomass concentrations in Θ^2 and the SME approach is used to propagate the set $\mathcal{P}_k^+$ as conducted in critical region 1. Figure 4 presents the posterior estimate sets $\mathcal{P}^+$ and true plant state x at different times. Since the model is 4 dimensional, the posterior estimate sets $\mathcal{P}^+$ are projected for visualization onto two dimensional spaces: the glucose–oxygen subspace and acetate–biomass subspace. The 8 boxes denote the projected posterior estimate sets between 0 h and 7 h, and each box represents an hour. The arrows in Figure 4 indicate the direction of time evolution. The black dots denote the true plant state. Since biomass is measured, the length of the boxes along the biomass dimension is relatively smaller, as compared to the other dimensions. The switch between the critical regions occurs at around 5 h.

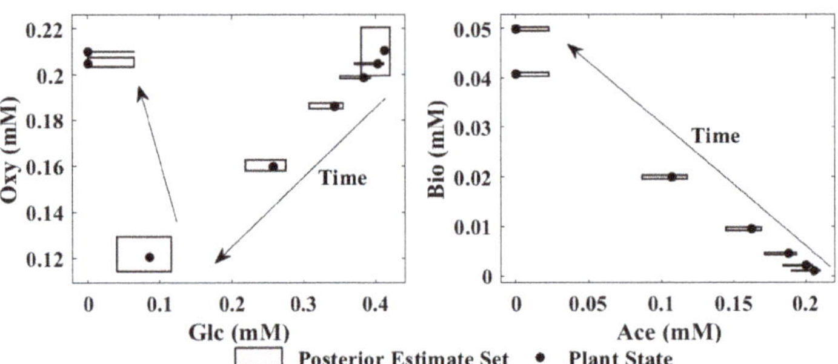

Figure 4. Posterior estimate sets projected onto the glucose–oxygen subspace and the acetate–biomass subspace at different times.

3.4. Set Membership Estimation

To verify the estimate and bounds generated by the proposed algorithm, we use a special Monte Carlo Algorithm (MCA) that takes biomass measurements into account. MCA randomly samples 100,000 different points from $\mathcal{P}_0$ and use them as initial states' values, and then calculates the corresponding trajectories with respect to time. Since, for the measurement of biomass, a truncated normal distribution measurement noise was assumed, some trajectories are not within the confidence interval of measurements. Once a trajectory is found out of the measurement range, the evolution of the trajectory is stopped and the corresponding trajectory is removed while trajectories which are still within the confidence interval of measurements are kept. Accordingly, only a part (2581) out of the trajectories starting from $\mathcal{P}_0$ are used for comparison to the bounds calculated by the proposed method. It should be noticed the fraction of trajectories kept for comparison is small because only a very narrow set of solutions are within the measurement range from the from the beginning to the end. In other words, only a small part of the samples considered in the simulation are compatible with the biomass measured trajectory that is assumed for the calculation of bounds by the set-based approach. Using parallel computation, 4 hour and 4 minutes of CPU time were required to complete all simulations. For comparison, the method proposed in this work can generate bounds with only 41 sec of CPU time without parallel computation. It should be remembered that the MCA was conducted for a specific trajectory of biomass measurements so as to enable a fair comparison with the method proposed in the current study. While it could be argued that MCA could be used to calculate bounds for all possible biomass trajectories, this will be computationally prohibitive. Thus, the proposed technique is a practical and analytical approach to the online estimation problem.

In Figure 5, the grey area denotes the trajectories randomly sampled and the two black lines represent the upper and lower bounds by SME. It is clear that the SME contains all the solutions generated by MCA, especially for the unobservable states. It can be observed that the switch from one critical region to the other occurs at approximately 5 h as shown in Figure 2. Before 5 h, the reactor has enough resources for cell growth and the limiting step is glucose uptake as Equation (31a) shows. Thus, critical region Θ^1 corresponds to the logarithmic phase of growth where the latter is driven by glucose consumption. At about 5 h, the simultaneous depletion of acetate and glucose leads to a metabolic switch from the logarithmic phase to the stationary phase. Following this metabolic switch, the culture is also acetate limited and thus acetate become a new flux-determining state. Since the oxygen feed rate is maintained constant in the model, the fact that the growth significantly decreases after the switch explains why the oxygen concentration bounces back up.

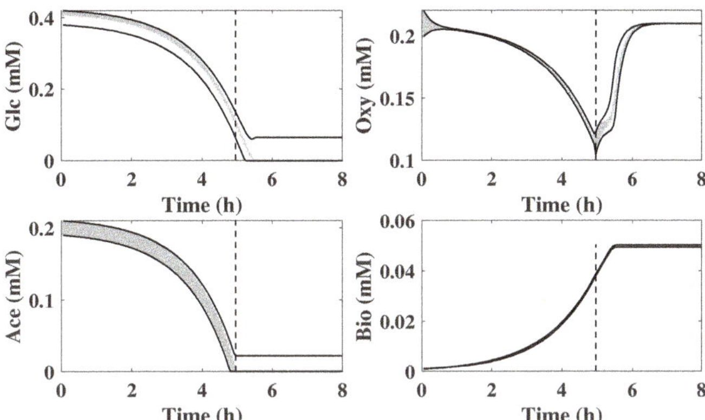

Figure 5. Comparison between MCA with bounds of 4 components estimated by SME in batch fermentation of *E. coli*.

To further verify the proposed scheme, similar MCA simulations were conducted with a larger initial uncertainty and measurement noise. In Figure 6, the bounds of 4 component concentrations estimated by SME are shown. It is clear that the simulated trajectories contained in the grey color band generated by MCA is within the bounds calculated by the proposed methodology. From comparison of Figures 5 and 6, it can be found that the SME approach copes with the larger noise and initial uncertainty by generating larger bounds.

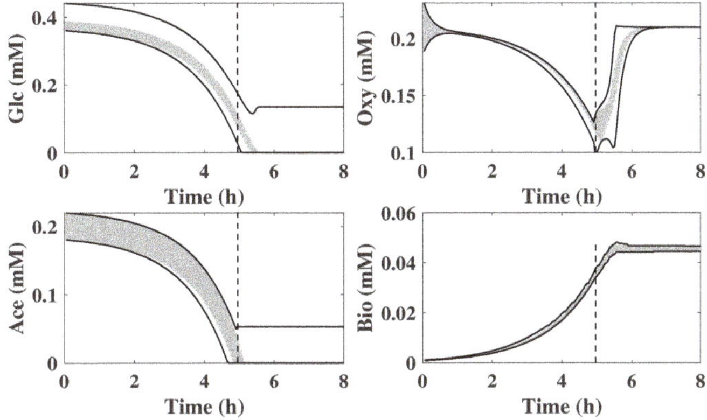

Figure 6. Comparison between MCA with bounds of 4 components estimated by SME with loud noise.

4. Discussion

DFBM models are advantageous since they contain significant detail about the cell metabolism as compared to classical unstructured models. However, due to this level of detail, DFBM contain many states thus resulting in more difficult state estimation problem. The challenge of dealing with a large number of states is further exacerbated by the fact that online measurements of metabolites are generally difficult to obtain or not available. With limited online measurements, it is often impossible to produce observability for all the states. Noticing that the diagonal matrix B in Equation (19) is a linear mapping of states, if the nonlinear term $\Delta t x_{bio,k} A v_k$ can be estimated then it is possible to estimate the other states of the DFBM.

Multiparametric LP is introduced to convert the original system into a series of piecewise continuous subsystems based on the partitioning of the parameter space into critical regions. The availability of an explicit expression for the calculation of the LP optima for each critical region significantly simplifies the solution of the problem. Although many critical regions may be mathematically possible, industrial fermentation is operated in a narrow range of initial operating conditions and as such only a few critical regions need to be considered.

Beyond their computational convenience the critical regions identified by the Multiparametric LP approach can be interpreted as corresponding changes in the cell metabolism. The relative abundance of substrates, i.e., glucose, acetate and oxygen in the *E. coli* model and their consumption towards biomass lead to the occurrence of different resources' limitations at any given time. Within some ranges of concentration, the limiting substrate remains the same corresponding to a specific metabolism strategy.

In the *E. coli* example, four reactions can synthesize the biomass from glucose, acetate and oxygen. However, since the objective is to maximize growth subject to constraints, the cell prioritizes these reactions differently at any given time due to their different efficiency for biomass synthesis. The ratio of the stoichiometry coefficients in each column of matrix *A* indicates the biomass yield of each substrate for each reaction. Reaction 1 is the only reaction that consumes acetate to synthesize biomass. The yield of acetate to biomass is $\frac{1}{39.43}$ for reaction 1, which is very low compared with reaction 2 and reaction 3. The biomass yield of reaction 2 and reaction 3 by glucose is $\frac{1}{9.46}$ and $\frac{1}{9.84}$, respectively. Reaction 4 is the only reaction that do not consume oxygen to generate biomass but it is very inefficient. Because the biomass yield of these reactions are different, reaction 2 is preferred over reaction 1 and reaction 3 when glucose and oxygen are abundant. When oxygen is very low, the cells switch their metabolism from aerobic to anaerobic to generate biomass through reaction 4.

To maximize the biomass growth rate, cells take advantage of reaction 1 and 2 to consume as much acetate and glucose as possible when oxygen is sufficient. However, the glucose amounts that can be consumed by the cells is limited by the glucose uptake rate, which is θ_4. Similarly, oxygen consumption is limited by a constant oxygen uptake rate as in Equation (28b). The oxygen is consumed first with glucose in reaction 2 to synthesize biomass and the remaining oxygen is consumed for reaction 1. Multiparametric LP captures the relative priority of different reactions towards maximization of growth and identify the key limited resources. In critical region Θ^1, glucose is the key resource that determines the flux vector according to Equation (31a). As glucose and acetate are consumed by reactions 1 and 2, biomass increases exponentially and the oxygen concentration drops fast due to oxygen demands as in Figures 5 and 6. At some point the concentration of acetate becomes very low but acetate is necessary for reaction 2 to synthesize biomass. At this point, acetate becomes the key limited resource and the system enters into a new critical region Θ^2. Then in Θ^2, the metabolism is limited by the available acetate and glucose and as they deplete the growth of cells decreases and ultimately stops. Accordingly, Θ^1 corresponds to the logarithmic phase and Θ^2 to the stationary phase of growth.

The use of EKF for each subsystem is used to estimate the reaction-rate-determining states thus reducing the need for online measurements. Since biomass is highly correlated with the reaction-rate-determining states, EKF can take advantage of biomass measurement to estimate these states. Because some of these reaction-rate-determining states are common to different critical regions, only are fewer states required to be measured or estimated, which greatly reduce the demand of online measurements of concentration. In the *E. coli* example, only biomass needs to be measured. Once biomass is measured, glucose can be estimated by the EKF in critical region Θ^1 and glucose and acetate can be estimated in Θ^2.

By using the SME upper and lower bounds for all states can be generated including the unobservable ones such as acetate and oxygen in Θ^1. Using the bounds of the acetate and biomass estimates, it was possible to determine the switch from one critical region to another and to re-initialize the estimates and covariance matrix for the EKF after the switch.

This research is helpful in DFBM-based control in bio-processes when many components cannot be measured online. Using the upper and lower bounds calculated by SME of unobservable states and estimates by EKF of observable states, robust control methods can be applied to achieve optimal operation in the presence of uncertainty. The method developed can also be extended to monitor the bio-processes and differentiate between normal and abnormal operations.

5. Conclusions

This research proposed a comprehensive DFBM-based approach to estimate the metabolites concentrations with a minimal number of online measurements. The main idea is to convert the DFBM model with uncertainty in initial conditions to an explicit variable structure system that can be analyzed by multiparametric linear programming. A key finding of the proposed work is that only a subset of the states, referred to as reaction-rate-determining states, is needed to calculate the flux vector. Identification of the reaction-rate-determining states for each critical region permitted the determination of the minimum set of measurements required for full state estimation. EKFs were used to estimate the observable states and set propagation by SME was used to identify bounds of both the observable states and unobservable states.

Author Contributions: Conceptualization, X.S. and H.B.; methodology, X.S. and H.B.; software, X.S.; validation, X.S.; formal analysis, X.S. and H.B.; investigation, X.S. and H.B.; resources, H.B.; data curation, X.S.; writing—original draft preparation, X.S.; writing—review and editing, X.S. and H.B.; visualization, X.S. and H.B.; supervision, H.B.; project administration, H.B.; funding acquisition, H.B. All authors have read and agreed to the published version of the manuscript.

Funding: This research was funded by The Natural Sciences and Engineering Research Council of Canada (NSERC) of grant number RGPIN-04609-2019, Mitacs, and Sanofi Pasteur.

Conflicts of Interest: We declare that we have no conflicts of interest to this work.

Abbreviations

The following abbreviations are used in this paper:

DFBM	Dynamic Flux Balance Models
mpLP	Multiparametric Linear Programming
LP	Linear Programming
EKF	Extended Kalman Filter
SME	Set Membership Estimation
MCA	Monte Carlo Algorithm

References

1. Orth, J.D.; Thiele, I.; Palsson, B.Ø. What is flux balance analysis? *Nat. Biotechnol.* **2010**, *28*, 245–248. [CrossRef]
2. Höffner, K.; Harwood, S.M.; Barton, P.I. A reliable simulator for dynamic flux balance analysis. *Biotechnol. Bioeng.* **2013**, *110*, 792–802. [CrossRef] [PubMed]
3. Stanbury, P.F.; Whitaker, A.; Hall, S.J. *Principles of Fermentation Technology*; Elsevier: Amsterdam, The Netherlands, 2013.
4. Kadlec, P.; Gabrys, B.; Strandt, S. Data-driven soft sensors in the process industry. *Comput. Chem. Eng.* **2009**, *33*, 795–814. [CrossRef]
5. Dochain, D. State and parameter estimation in chemical and biochemical processes: A tutorial. *J. Process. Control* **2003**, *13*, 801–818. [CrossRef]
6. Haimi, H.; Mulas, M.; Corona, F.; Vahala, R. Data-derived soft-sensors for biological wastewater treatment plants: An overview. *Environ. Model. Softw.* **2013**, *47*, 88–107. [CrossRef]
7. Ali, J.M.; Hoang, N.H.; Hussain, M.A.; Dochain, D. Review and classification of recent observers applied in chemical process systems. *Comput. Chem. Eng.* **2015**, *76*, 27–41.
8. Jaulin, L.; Kieffer, M.; Didrit, O.; Walter, E. Interval analysis. In *Applied Interval Analysis*; Springer: London, UK, 2001; pp. 11–43.
9. Makino, K.; Berz, M. Taylor models and other validated functional inclusion methods. *Int. J. Pure Appl. Math.* **2003**, *6*, 239–316.
10. Rumschinski, P.; Borchers, S.; Bosio, S.; Weismantel, R.; Findeisen, R. Set-base dynamical parameter estimation and model invalidation for biochemical reaction networks. *BMC Syst. Biol.* **2010**, *4*, 69. [CrossRef] [PubMed]

11. Sahlodin, A.M.; Chachuat, B. Convex/concave relaxations of parametric ODEs using Taylor models. *Comput. Chem. Eng.* **2011**, *35*, 844–857. [CrossRef]
12. Blanchini, F.; Miani, S. *Set-Theoretic Methods in Control*; Springer: Cham, Switzerland, 2008.
13. Schweppe, F. Recursive state estimation: Unknown but bounded errors and system inputs. *IEEE Trans. Autom. Control* **1968**, *13*, 22–28. [CrossRef]
14. Gouzé, J.L.; Rapaport, A.; Hadj-Sadok, M.Z. Interval observers for uncertain biological systems. *Ecol. Model.* **2000**, *133*, 45–56. [CrossRef]
15. Mazenc, F.; Dinh, T.N.; Niculescu, S.I. Robust interval observers and stabilization design for discrete-time systems with input and output. *Automatica* **2013**, *49*, 3490–3497. [CrossRef]
16. Efimov, D.; Perruquetti, W.; Raïssi, T.; Zolghadri, A. On interval observer design for time-invariant discrete-time systems. In Proceedings of the 2013 IEEE European Control Conference (ECC), Zurich, Switzerland, 17–19 July 2013; pp. 2651–2656.
17. Chisci, L.; Garulli, A.; Zappa, G. Recursive state bounding by parallelotopes. *Automatica* **1996**, *32*, 1049–1055. [CrossRef]
18. Alamo, T.; Bravo, J.M.; Camacho, E.F. Guaranteed state estimation by zonotopes. *Automatica* **2005**, *41*, 1035–1043. [CrossRef]
19. Maksarov, D.; Norton, J. Computationally efficient algorithms for state estimation with ellipsoidal approximations. *Int. J. Adapt. Control Signal Process.* **2002**, *16*, 411–434. [CrossRef]
20. Mahadevan, R.; Edwards, J.S.; Doyle, F.J., III. Dynamic flux balance analysis of diauxic growth in *Escherichia coli*. *Biophys. J.* **2002**, *83*, 1331–1340. [CrossRef]
21. Hjersted, J.L.; Henson, M.A.; Mahadevan, R. Genome-scale analysis of Saccharomyces cerevisiae metabolism and ethanol production in fed-batch culture. *Biotechnol. Bioeng.* **2007**, *97*, 1190–1204. [CrossRef] [PubMed]
22. Ghorbaniaghdam, A.; Chen, J.; Henry, O.; Jolicoeur, M. Analyzing clonal variation of monoclonal antibody-producing CHO cell lines using an in silico metabolomic platform. *PLoS ONE* **2014**, *9*, e90832. [CrossRef]
23. Budman, H.; Patel, N.; Tamer, M.; Al-Gherwi, W. A dynamic metabolic flux balance based model of fed-batch fermentation of bordetella pertussis. *Biotechnol. Prog.* **2013**, *29*, 520–531. [CrossRef] [PubMed]
24. Wilhelm, S.; Manjunath, B. tmvtnorm: A package for the truncated multivariate normal distribution. *Sigma* **2010**, *2*, 1–25. [CrossRef]
25. Botev, Z.I. The normal law under linear restrictions: Simulation and estimation via minimax tilting. *arXiv* **2016**, arXiv:1603.04166.
26. Akbari, A.; Barton, P.I. An improved multi-parametric programming algorithm for flux balance analysis of metabolic networks. *J. Optim. Theory Appl.* **2018**, *178*, 502–537. [CrossRef]
27. Borrelli, F.; Bemporad, A.; Morari, M. Geometric algorithm for multiparametric linear programming. *J. Optim. Theory Appl.* **2003**, *118*, 515–540. [CrossRef]
28. Oberdieck, R.; Diangelakis, N.A.; Nascu, I.; Papathanasiou, M.M.; Sun, M.; Avraamidou, S.; Pistikopoulos, E.N. On multi-parametric programming and its applications in process systems engineering. *Chem. Eng. Res. Des.* **2016**, *116*, 61–82. [CrossRef]
29. Murabito, E.; Simeonidis, E.; Smallbone, K.; Swinton, J. Capturing the essence of a metabolic network: A flux balance analysis approach. *J. Theor. Biol.* **2009**, *260*, 445–452. [CrossRef] [PubMed]
30. Shen, X.; Budman, H. A method for tackling primal multiplicity of solutions of dynamic flux balance models. *Comput. Chem. Eng.* **2020**, *143*, 107070. [CrossRef]
31. Song, Y.; Grizzle, J.W. The extended Kalman filter as a local asymptotic observer for nonlinear discrete-time systems. In Proceedings of the 1992 IEEE American Control Conference, Chicago, IL, USA, 24–26 June 1992; pp. 3365–3369.
32. Herceg, M.; Kvasnica, M.; Jones, C.N.; Morari, M. Multi-parametric toolbox 3.0. In Proceedings of the 2013 IEEE European Control Conference (ECC), Zurich, Switzerland, 17–19 July 2013; pp. 502–510.
33. Delos, V.; Teissandier, D. Minkowski sum of HV-polytopes in Rn. *arXiv* **2014**, arXiv:1412.2562.
34. Meadows, A.L.; Karnik, R.; Lam, H.; Forestell, S.; Snedecor, B. Application of dynamic flux balance analysis to an industrial Escherichia coli fermentation. *Metab. Eng.* **2010**, *12*, 150–160. [CrossRef]

Review

How to Tackle Underdeterminacy in Metabolic Flux Analysis? A Tutorial and Critical Review

Philippe Bogaerts [1,*,†] **and Alain Vande Wouwer** [2,*,†]

1   3BIO-BioControl, Université Libre de Bruxelles, Avenue F.D. Roosevelt, 50-CP 165/61,
    1050 Brussels, Belgium
2   Systems, Estimation, Control and Optimization (SECO), University of Mons, 7000 Mons, Belgium
*   Correspondence: philippe.bogaerts@ulb.be (P.B.); alain.vandewouwer@umons.ac.be (A.V.W.)
†   These authors contributed equally to this work.

Abstract: Metabolic flux analysis is often (not to say almost always) faced with system underdeterminacy. Indeed, the linear algebraic system formed by the steady-state mass balance equations around the intracellular metabolites and the equality constraints related to the measurements of extracellular fluxes do not define a unique solution for the distribution of intracellular fluxes, but instead a set of solutions belonging to a convex polytope. Various methods have been proposed to tackle this underdeterminacy, including flux pathway analysis, flux balance analysis, flux variability analysis and sampling. These approaches are reviewed in this article and a toy example supports the discussion with illustrative numerical results.

Keywords: flux variability analysis; flux balance analysis; sampling; metabolic network; elementary flux modes

Citation: Bogaerts, P.;
Vande Wouwer, A. How to Tackle
Underdeterminacy in Metabolic Flux
Analysis? A Tutorial and Critical
Review. *Processes* **2021**, *9*, 1577.
https://doi.org/10.3390/pr9091577

Academic Editor: Andrzej Pawłowski

Received: 31 July 2021
Accepted: 27 August 2021
Published: 2 September 2021

Publisher's Note: MDPI stays neutral
with regard to jurisdictional claims in
published maps and institutional affiliations.

1. Introduction

Computational approaches for studying the flux distribution inside metabolic networks of microbial strains or mammalian cell lines have gained a tremendous importance in biotechnology. Indeed, the production of high-added value biochemicals is based on large-scale cultures of genetically engineered strains, and the determination of the flux distribution provides insight into the biosynthesis pathways, the impact of metabolic engineering and the influence of the culture conditions. Different approaches have been developed to compute this flux distribution, which are based on a common assumption that the intracellular metabolites do not accumulate, or in other words, that the cell is in a metabolic pseudo-steady state [1]. This assumption leads to a system of mass-balance equations of the form:

$$N\underline{v} = 0 \qquad v_i \geq 0 \qquad \forall i \tag{1}$$

where $N \in \mathbb{R}^{n_s \times n_v}$ is the stoichiometric matrix (and the incidence matrix of the graph representing the metabolic network), $\underline{v} \in \mathbb{R}^{n_v}$ is the vector of intracellular fluxes (in mmol/gDW/h), which are assumed positive (i.e., to have a net direction), and n_s is the number of intracellular metabolites. N is assumed full-row rank, thus defining n_s independent mass balance equations. This system of equations expresses the zero balance in each internal node of the metabolic network, and imposes a set of linear equality constraints, which are not sufficient to determine a unique solution for the flux vector $\underline{v}$. This system of equations is often supplemented by additional mass balance equations expressing the link between the intracellular fluxes and the measurements of external fluxes (uptake or production of extracellular metabolites):

$$N_m \underline{v} = \underline{v}_m \tag{2}$$

where $\underline{v}_m \in \mathbb{R}^{n_m}$. Even though this additional information allows restricting the solution space, it is usually not sufficient to define a unique solution. More precisely, a subset of the fluxes might be exactly calculable [2] while only intervals of values for the remaining fluxes can be computed. In general, the system of equations under consideration can be formulated as:

$$A_e \underline{v} = \underline{b}_e \tag{3}$$

$$A_i \underline{v} \leq \underline{b}_i \tag{4}$$

where the equality constraints (1) and (2) are put together in (3), and Equation (4) contains the positivity constraints as well as other bound constraints, e.g., upper bounds on some of the fluxes, corresponding to prior knowledge or biological assumptions. The matrices $A_e \in \mathbb{R}^{n_e \times n_v}$ and $A_i \in \mathbb{R}^{n_i \times n_v}$, correspond to n_e equality constraints and n_i inequality constraints, respectively. To tackle the underdeterminacy, several approaches have been proposed in recent years, which can be grouped into two distinct strategies:

1. Dealing with the underdeterminacy—this strategy is adopted in several methods where minimal and maximal bounds on the admissible fluxes are determined. This category of methods includes Flux Pathway Analysis (FPA), where convex analysis is used to decompose the admissible flux distributions into Elementary Flux Modes (EFMs) or Extreme Pathways [3,4], Flux Variability Analysis (FVA), which is a Linear-Programming (LP)-based method determining the range of admissible fluxes [5], Flux Spectrum Approach (FSA), which is another LP-based method taking insufficient and uncertain measurements into account [6]. Random sampling of the admissible solution set allows determining the marginal probability density functions of the fluxes [7–10], and statistical methods based on the maximum entropy principle can be used to infer intracellular flux distributions [11,12].

2. Reducing or eliminating the underdeterminacy—this strategy consists in adding constraints in various ways, e.g., including more measurements of the extracellular fluxes or, possibly, measurements of the intracellular fluxes using specific techniques such as ^{13}C tracing [13,14] and parallel labeling [15], leading to the sophisticated procedures of ^{13}C MFA. Alternatively, additional constraints can be introduced by formulating biological assumptions either based on prior knowledge and/or experimental observations [16,17] or systematic procedures to determine active constraints [18]. The use of thermodynamic constraints can be important in relation with reaction reversibility and the limitation of the solution space [19]. Moreover, thermodynamical constraints can prevent infeasible loops in a metabolic network as demonstrated in [20]. Underdeterminacy can also be reduced (or even eliminated) through the formulation of an optimization problem originating from the assumption of an optimal metabolic behavior of the cells. This approach corresponds to Flux Balance Analysis (FBA) [21,22], which uses an objective function expressed as a linear combination of selected fluxes. Recently, the increasing availability of metabolite profiling data obtained through gas and liquid chromatography combined with mass spectroscopy has also allowed the integration of time-course absolute quantitative metabolomics in unsteady-state (or dynamic) FBA [23,24]. In the usual situation where FBA still leads to an underdetermined system with an infinite number of flux distributions that optimize the cost function, variants of FBA have been proposed in order to define a unique solution, e.g., the geometric approach developed in [25] that searches for the minimal flux distribution satisfying the given objective. Assuming that fluxes correlate with enzyme levels, this specific flux distribution would correspond to the minimization of the amount of enzymes required to satisfy the objective defined in FBA. Ultimately, the concept of Most Accurate Fluxes [26] allows computing a unique flux distribution, hence eliminating the system underdeterminacy, with a very low computational load and without any assumption regarding an optimal biological behavior.

In the following, we review some of these methods and their implementation with a toy example, which provides a numerical illustration of the main concepts. The Matlab code of this example is provided in the Supplementary Materials associated to this article.

2. A Toy Example

Despite its small size, the metabolic network (see Figure 1) that is considered to illustrate the several methods introduced in the previous section presents many representative features, e.g., several intracellular metabolites, extracellular substrates, and intra- and extra-cellular products. This network is described by the following reactions:

$$S_1 \xrightarrow{v_1} M_1$$

$$S_2 \xrightarrow{v_2} M_2$$

$$S_2 \xrightarrow{v_3} M_3$$

$$M_1 \xrightarrow{v_4} M_4 + M_5$$

$$M_4 \xrightarrow{v_5} M_5$$

$$M_2 \xrightarrow{v_6} M_5$$

$$M_5 \xrightarrow{v_7} P_1$$

$$M_1 + M_3 \xrightarrow{v_8} P_2$$

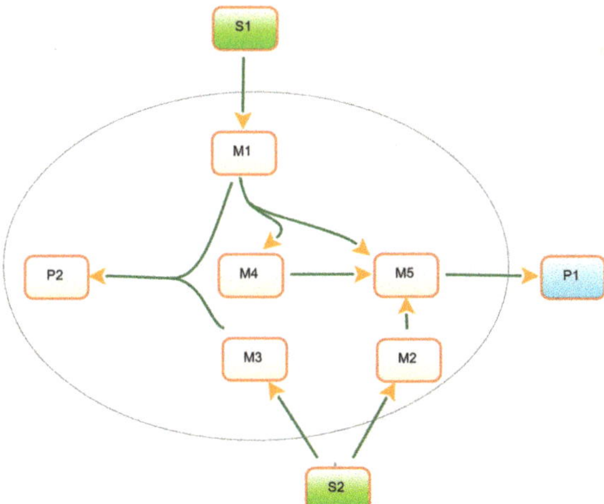

Figure 1. Simple metabolic network. M_i, $i = 1, \ldots, 5$ are the internal metabolites, S_i, $i = 1, 2$ are the extracellular substrates, and P_i, $i = 1, 2$ are the extracellular and intracellular product, respectively.

The quasi-steady state assumption for the internal metabolites M_i, $i = 1, \ldots, 5$, yields a system of algebraic mass-balance equations in the form of Equation (1)

$$\begin{bmatrix} 1 & 0 & 0 & -1 & 0 & 0 & 0 & -1 \\ 0 & 1 & 0 & 0 & 0 & -1 & 0 & 0 \\ 0 & 0 & 1 & 0 & 0 & 0 & 0 & -1 \\ 0 & 0 & 0 & 1 & -1 & 0 & 0 & 0 \\ 0 & 0 & 0 & 1 & 1 & 1 & -1 & 0 \end{bmatrix} \begin{bmatrix} v_1 \\ v_2 \\ v_3 \\ v_4 \\ v_5 \\ v_6 \\ v_7 \\ v_8 \end{bmatrix} = N\underline{v} = 0 \qquad v_i \geq 0 \qquad \forall i \qquad (5)$$

The stoichiometric matrix N is full row rank (i.e., 5) and the system of equations has 3 degrees of freedom (5 independent equations for 8 unknown fluxes). In an ideal measurement configuration, we consider that the 3 extracellular fluxes can be measured, for instance:

$$\begin{bmatrix} 1 & 0 & 0 & 0 & 0 & 0 & 0 & 0 \\ 0 & 1 & 1 & 0 & 0 & 0 & 0 & 0 \\ 0 & 0 & 0 & 0 & 0 & 0 & 1 & 0 \end{bmatrix} \underline{v} = N_m \underline{v} = \begin{bmatrix} v_{m1} \\ v_{m2} \\ v_{m3} \end{bmatrix} = \begin{bmatrix} 3.5 \\ 2.7 \\ 1.8 \end{bmatrix} = \underline{v}_m \tag{6}$$

so that a unique solution, i.e., flux distribution, can be found:

$$\underline{v} = \begin{bmatrix} 3.5000 \\ 0.0667 \\ 2.6333 \\ 0.8667 \\ 0.8667 \\ 0.0667 \\ 1.8000 \\ 2.6333 \end{bmatrix} \tag{7}$$

In the sequel, we will consider various situations with less measurements, and alternative methods to deal with, reduce or eliminate the underdeterminacy.

3. Dealing with the Underdeterminacy

The solution space of Equation (5) can be described using the concept of EFMs [3,27] which represent minimal, non-decomposable pathways connecting substrates to products. Every flux in the metabolic network can be described as a convex combination of EFMs:

$$\underline{v} = \sum_{i=1}^{n_{EFM}} \mu_i \underline{e}_i = E\underline{\mu}, \quad \mu_i \geq 0 \tag{8}$$

where n_{EFM} is the number of EFMs $\underline{e}_i$.

The EFMs can be computed using readily available software such as Metatool [28], EFMtool [29] or FluxModeCalculator [30]. The main issue associated to this computation is the combinatorial explosion of the number of EFMs with the network size (a network of less than 100 reactions can have tens of thousands of EFMs), and the fact that the computation involves some form of enumeration, which requires large computer memory space and computation time. Alternative procedures have therefore been proposed to compute minimal sets of EFMs without enumerating all of them [31]. For the small network under consideration in this review, this computation is trivial and leads to

$$E = \begin{bmatrix} \underline{e}_1 & \underline{e}_2 & \underline{e}_3 \end{bmatrix} = \begin{bmatrix} 0 & 1 & 0.5 \\ 1 & 0 & 0 \\ 0 & 1 & 0 \\ 0 & 0 & 0.5 \\ 0 & 0 & 0.5 \\ 1 & 0 & 0 \\ 1 & 0 & 1 \\ 0 & 1 & 0 \end{bmatrix} \tag{9}$$

The three EFMs define a polyhedral cone in the positive orthant, which contains all possible flux distributions. They correspond to a minimal bioreaction system, which provides an input–output representation of the cell metabolism (this kind of representation can be very useful to derive reduced-order macroscopic representations of the culture system [32–34], but this subject will be addressed later on in this paper; see Section 5.4).

$$S_2 \rightarrow P_1$$
$$S_1 + S_2 \rightarrow P_2 \tag{10}$$
$$0.5S_1 \rightarrow P_1$$

We already know that if three measurements are available, such as the ones defined in Equation (2), the solution is unique and the coefficient vector $\begin{bmatrix} \mu_1 & \mu_2 & \mu_3 \end{bmatrix}^T = \begin{bmatrix} 0.0667 & 2.6333 & 1.7333 \end{bmatrix}^T$. If less measurements are available, for instance only v_{m1} and v_{m2}, then the EFM basis of the linear system

$$\begin{bmatrix} N & \underline{0} \\ N_m & -\underline{v}_m \end{bmatrix} \begin{bmatrix} \underline{v} \\ 1 \end{bmatrix} = \underline{0} \tag{11}$$

leads to the so-called extreme rays $\underline{f}_i$ [35]

$$\begin{bmatrix} \underline{f}_1 & \underline{f}_2 \end{bmatrix} = \begin{bmatrix} 3.5 & 3.5 \\ 0 & 2.7 \\ 2.7 & 0 \\ 0.8 & 3.5 \\ 0.8 & 3.5 \\ 0 & 2.7 \\ 1.6 & 9.7 \\ 2.7 & 0 \end{bmatrix} \tag{12}$$

which defines a pointed polyhedral cone that is a subspace of the previous solution cone. Moreover, the flux spectrum $F_0 = \{\underline{v} : v_i^{\min} \leq v_i \leq v_i^{\max}\}$ with $v_i^{\min} = \min\{f_{ki}, k = 1, \cdots, p\}$ and $v_i^{\max} = \max\{f_{ki}, k = 1, \cdots, p\}$ is easily deduced, giving

$$\begin{bmatrix} v_1 = 3.5 \\ 0 \leq v_2 \leq 2.7 \\ 0 \leq v_3 \leq 2.7 \\ 0.8 \leq v_4 \leq 3.5 \\ 0.8 \leq v_5 \leq 3.5 \\ 0 \leq v_6 \leq 2.7 \\ 1.6 \leq v_7 \leq 9.7 \\ 0 \leq v_8 \leq 2.7 \end{bmatrix} \tag{13}$$

which indeed encloses the unique solution found with an additional measurement.

An alternative and straightforward way to compute the flux spectrum is provided by Flux Variability Analysis (FVA) [5], which consists in formulating a double optimization problem (minimization/maximization) of the flux distribution under the constraints provided by the metabolic network stoichiometry, the measurements, and any other additional biological constraints.

$$\left. \begin{aligned} v_i^{\min} &= \min_{\underline{v}} v_i \\ v_i^{\max} &= \max_{\underline{v}} v_i \end{aligned} \right\} \forall i \in [1, n_v] \tag{14}$$

$$s.t. \quad \begin{cases} A_e \underline{v} = \underline{b}_e \\ A_i \underline{v} \leq \underline{b}_i \end{cases}$$

The unique solution to this problem, if bounded and feasible, can be computed using linear programming, as available in many software libraries (e.g., *linprog* in *Matlab* or other LP solvers such as CPLEX interfaced in the language of your choice such as Python or Julia) or dedicated environments (e.g., COBRA [36] and CellNetAnalyzer [37]). In our application example, we use *linprog* to compute the flux spectrum in the situation where only the first measurement is available (v_{m1}) and a constraint is imposed in the form $v_2 + v_3 \leq 5$.

$$A_e = \begin{bmatrix} & & & & N & & & \\ 1 & 0 & 0 & 0 & 0 & 0 & 0 & 0 \end{bmatrix} \quad \underline{b_e} = \begin{bmatrix} 0 & 0 & 0 & 0 & 0 & 3.5 \end{bmatrix}^T$$

$$A_i = \begin{bmatrix} & & & & -I_8 & & & \\ 0 & 1 & 1 & 0 & 0 & 0 & 0 & 0 \end{bmatrix} \quad \underline{b_i} = \begin{bmatrix} 0 & 0 & 0 & 0 & 0 & 0 & 0 & 0 & 5 \end{bmatrix}^T \tag{15}$$

Thus, the spectrum F_0 is given by

$$\begin{bmatrix} v_1 = 3.5 \\ 0 \leq v_2 \leq 5 \\ 0 \leq v_3 \leq 3.5 \\ 0 \leq v_4 \leq 3.5 \\ 0 \leq v_5 \leq 3.5 \\ 0 \leq v_6 \leq 5 \\ 0 \leq v_7 \leq 12 \\ 0 \leq v_8 \leq 3.5 \end{bmatrix} \tag{16}$$

The application of FVA can be delicate when the measurements are corrupted by noise, as the constraints imposed by the metabolic network and/or the bounds on the fluxes could become incompatible with the measurement information. Several approaches take account of the measurement uncertainty, such as Flux Spectrum Analysis [6] or Adaptive Flux Variability Analysis [38], which relax the constraints to allow for a feasible solution. Here, we consider a variation of our simple example where only the first measurement $v_{m1} = 10.5$ would be available and 3 constraints would be imposed, i.e., $v2 \leq 5$, $v5 \leq 5$, $v8 \leq 5$. In this case, the matrices become

$$A_e = \begin{bmatrix} & & & & N & & & \\ 1 & 0 & 0 & 0 & 0 & 0 & 0 & 0 \end{bmatrix} \quad \underline{b_e} = \begin{bmatrix} 0 & 0 & 0 & 0 & 0 & 10.5 \end{bmatrix}^T$$

$$A_i = \begin{bmatrix} & & & & -I_8 & & & \\ 0 & 1 & 0 & 0 & 0 & 0 & 0 & 0 \\ 0 & 0 & 0 & 0 & 1 & 0 & 0 & 0 \\ 0 & 0 & 0 & 0 & 0 & 0 & 0 & 1 \end{bmatrix} \quad \underline{b_i} = \begin{bmatrix} 0 & 0 & 0 & 0 & 0 & 0 & 0 & 0 & 5 & 5 & 5 \end{bmatrix}^T \tag{17}$$

but unfortunately the LP solver returns the message that no feasible solution can be found. What is happening? In fact, the solution of the FVA problem without the knowledge of the first measurement v_{m1} returns the following spectrum:

$$\begin{bmatrix} 0 \leq v_1 \leq 10 \\ 0 \leq v_2 \leq 5 \\ 0 \leq v_3 \leq 5 \\ 0 \leq v_4 \leq 5 \\ 0 \leq v_5 \leq 5 \\ 0 \leq v_6 \leq 5 \\ 0 \leq v_7 \leq 15 \\ 0 \leq v_8 \leq 5 \end{bmatrix} \tag{18}$$

which shows that the maximum admissible value of v_1 is 10. Hence, the noisy measurement $v_{m1} = 10.5$ is incompatible with this upper bound, and it is not possible to include it as such in the equality constraints. A way round this issue is to introduce inequality constraints in the form

$$(1 - e)v_{m1} \leq v_1 \leq (1 + e)v_{m1} \tag{19}$$

where e represents a relative uncertainty. In our example, we could choose $e = 5\%$, which is the smallest uncertainty leading to the matrices modification

$$A_e = \begin{bmatrix} N \end{bmatrix} \qquad \underline{b}_e = \begin{bmatrix} 0 & 0 & 0 & 0 & 0 \end{bmatrix}^T$$

$$A_i = \begin{bmatrix} -I_8 \\ 0 & 1 & 0 & 0 & 0 & 0 & 0 & 0 \\ 0 & 0 & 0 & 0 & 1 & 0 & 0 & 0 \\ 0 & 0 & 0 & 0 & 0 & 0 & 0 & 1 \\ -1 & 0 & 0 & 0 & 0 & 0 & 0 & 0 \\ 1 & 0 & 0 & 0 & 0 & 0 & 0 & 0 \end{bmatrix} \quad \underline{b}_i = \begin{bmatrix} 0 & 0 & 0 & 0 & 0 & 0 & 0 & 0 & 0 & 5 & 5 & 5 & -9.975 & 11.025 \end{bmatrix}^T \tag{20}$$

and a feasible solution

$$\begin{bmatrix} 9.975 \leq v_1 \leq 10 \\ 0 \leq v_2 \leq 5 \\ 4.975 \leq v_3 \leq 5 \\ 4.975 \leq v_4 \leq 5 \\ 4.975 \leq v_5 \leq 5 \\ 0 \leq v_6 \leq 5 \\ 9.95 \leq v_7 \leq 15 \\ 4.975 < v_8 < 5 \end{bmatrix} \tag{21}$$

On another matter, it can be convenient to eliminate the equality constraints and to formulate the problem in terms of inequality constraints only in a space of reduced dimension [39]. To this end, we can use the kernel (or null space) of the matrix A_e. If $A_0 \in \mathbb{R}^{n_v \times (n_v - n_e)}$ is a matrix whose columns form a basis of this kernel ($n_q = n_v - n_e$ is the nullity of the kernel for a full row rank matrix A_e), then any solution of $A_e \underline{v} = \underline{b}_e$ (Equation (3)) can be expressed as

$$\underline{v} = \underline{v}_0 + A_0 \underline{q} \tag{22}$$

where $\underline{v}_0$ is a particular solution to Equation (3) and the vector $\underline{q} \in \mathbb{R}^{n_q}$ allows the reformulation of the inequality constraints as

$$A_i^0 \underline{q} \leq \underline{b}_i^0 \tag{23}$$

with $A_i^0 = A_i A_0$ and $\underline{b}_i^0 = \underline{b}_i - A_i \underline{v}_0$.

A particular solution $\underline{v}_0$ can for instance be obtained by solving the following problem (in this case $\underline{v}_0$ is the flux vector with minimum Euclidean norm in the set of solutions)

$$\underline{v}_0 = \min_{\underline{v}} \underline{v}^T \underline{v}$$
$$s.t. \quad \begin{cases} A_e \underline{v} = \underline{b}_e \\ A_i \underline{v} \leq \underline{b}_i \end{cases} \tag{24}$$

To illustrate this, we return to our original example and consider the situation where only the first measurement $v_{m1} = 3.5$ is available and a constraint is imposed in the form $v_2 + v_3 \leq 5$. $\underline{v}_0$ can be computed using quadratic programming, e.g., *quadprog* in Matlab

$$\underline{v}_0 = \begin{bmatrix} 3.500 & 0 & 2.625 & 0.875 & 0.875 & 0 & 1.750 & 2.625 \end{bmatrix}^T \tag{25}$$

The null space of the equality constraint matrix $A_e \in \mathbb{R}^{6 \times 8}$ can be computed using *null* in Matlab and is given by

$$A_0 = \begin{bmatrix} 0 & 0.2019 & -0.2846 & 0.2846 & 0.2846 & 0.2019 & 0.7711 & -0.2846 \\ 0 & 0.5994 & 0.2627 & -0.2627 & -0.2627 & 0.5994 & 0.0740 & 0.2627 \end{bmatrix}^T \tag{26}$$

and, in turn

$$A_i^0 = \begin{bmatrix} 0 & -0.2019 & 0.2846 & -0.2846 & -0.2846 & -0.2019 & -0.7711 & 0.2846 & -0.0827 \\ 0 & -0.5994 & -0.2627 & 0.2627 & 0.2627 & -0.5994 & -0.0740 & -0.2627 & 0.8621 \end{bmatrix}^T \tag{27}$$

and

$$\underline{b}_i^0 = \begin{bmatrix} 3.500 & 0 & 2.625 & 0.875 & 0.875 & 0 & 1.750 & 2.625 & 2.375 \end{bmatrix}^T \tag{28}$$

The application of FVA in the q-space (a reduced space of dimension $n_q = 2$), i.e., with inequality constraints only

$$\left. \begin{array}{l} q_i^{\min} = \min\limits_{\underline{q}} q_i \\[2mm] q_i^{\max} = \max\limits_{\underline{q}} q_i \end{array} \right\} \forall i \in [1, n_q] \tag{29}$$

$$s.t. \quad A_i^0 \underline{q} \leq \underline{b}_i^0$$

gives a spectrum Q_0

$$\begin{bmatrix} -2.3454 \leq q_1 \leq 12.9102 \\ -2.3698 \leq q_2 \leq 3.9938 \end{bmatrix} \tag{30}$$

In the reduced q-space, the system of inequality constraints defines half hyperplanes whose intersection consists of a convex polytope that contains all the admissible flux distributions $\underline{q}$. Uniformly sampling this convex polytope allows subsequently computing the marginal probability density functions (or marginal distributions) of each flux.

In our toy example, the rejection algorithm [40] can be applied, which boils down to uniformly sample each coordinate q_i ($\forall i \in [1, n_q]$) on $[q_i^{\min}, q_i^{\max}]$. The obtained sample $\underline{q}$ is kept if it satisfies the inequality constraints $A_i^0 \underline{q} \leq b_i^0$, otherwise it is rejected. The procedure is repeated until the desired number of samples is reached. Figure 2 shows 10^4 samples obtained with the rejection algorithm. Despite its simplicity and the genuine uniform distribution that it provides, this algorithm cannot be used with high dimensional spaces and irregular shaped polytopes given that the fraction of rejected samples increases dramatically with the number of metabolic fluxes considered in the network.

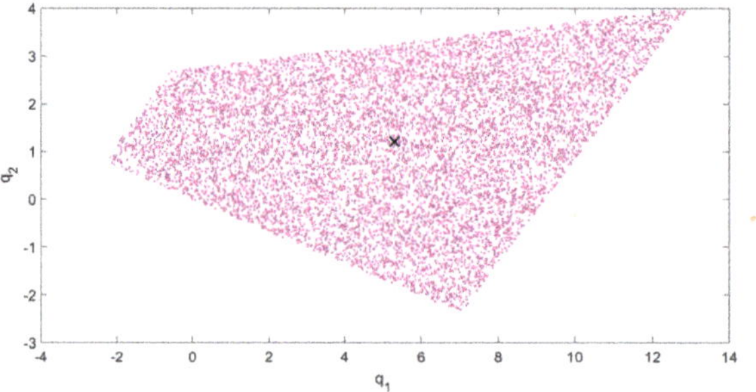

Figure 2. Rejection algorithm in the q-space (mean = X).

Other algorithms have been (and are still) developed to circumvent this problem [41,42]. Among the oldest and simplest methods, hit-and-run algorithms [43] consist of Markov Chain Monte Carlo methods that sample the convex polytope via some specific random walk. While they can be used with high dimensional spaces, their main drawback is that the samples often get stuck in some part of the polytope when this latter exhibits an irregular shape, which is generally the case. Figure 3 represents the marginal distributions of each flux in the v-space (transforming the $\underline{q}$ samples into $\underline{v}$ samples through Equation (22)), inferred from 10^4 samples obtained with the rejection algorithm and with a specific hit-and-run algorithm (namely, the random direction algorithm). Both results are almost identical.

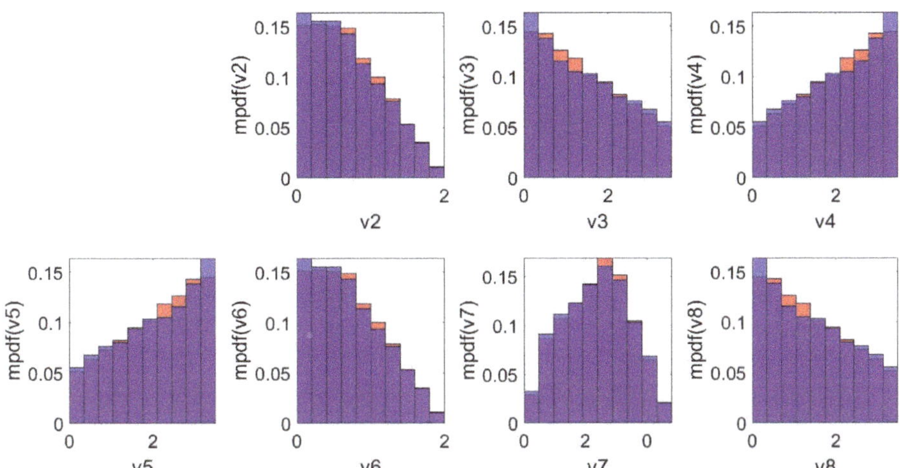

Figure 3. Marginal probability distribution in the original v-space, inferred from 10^4 samples with the rejection algorithm (red) and a hit-and-run algorithm (blue).

4. Reducing or Eliminating the Underdeterminacy

A straightforward way to reduce or eliminate underdeterminacy is to include additional measurements, either extracellular as shown in Equation (6) for our simple example, or intracellular using ^{13}C tracing [13,14] or the integration of time-course absolute quantitative metabolomics [23,24], which would amount to directly measure some of the internal fluxes v_i, $i = 1, \ldots, 8$ in the toy example. However, this implies additional time-consuming, delicate, and costly experiments and equipment. If sufficient additional measurements are not available, a candidate flux distribution can be provided by Flux Balance Analysis (FBA) [21,22], which assumes some optimal behavior of the cell, such as maximum cell growth rate or maximum ATP production rate, and formulates a linear programming problem

$$\hat{\underline{v}} = \arg\min_{v} \underline{\lambda}^T \underline{v}$$
$$s.t. \begin{cases} A_e \underline{v} = \underline{b}_e \\ A_i \underline{v} \leq \underline{b}_i \end{cases}$$

(31)

If the LP is feasible and bounded, then the cost function $J = \underline{\lambda}^T \underline{v}$ has a unique minimum value J^*, but the corresponding flux distribution $\hat{\underline{v}}$ is not necessarily unique, still implying underdeterminacy. In this latter case, it is possible to combine FBA (31) with FVA by subsequently solving a set of $2n_v$ LPs

$$\left. \begin{aligned} v_i^{\min} = \min_{\underline{v}} v_i \\ v_i^{\max} = \max_{\underline{v}} v_i \end{aligned} \right\} \forall i \in [1, n_v]$$
$$s.t. \begin{cases} A_e \underline{v} = \underline{b}_e \\ \underline{\lambda}^T \underline{v} = J^* \\ A_i \underline{v} \leq \underline{b}_i \end{cases}$$

(32)

This FVA problem includes an additional equality constraint enforcing the value $J^* = \underline{\lambda}^T \underline{v}$ determined in the first FBA step.

This can be illustrated with our toy example, first by applying FBA with the assumption that v_7 is maximum. In this case, $\hat{v}_7 = 12$ and the corresponding flux distribution is given by

$$\underline{\hat{v}} = \begin{bmatrix} 3.50 & 5.00 & 0 & 3.50 & 3.50 & 5.00 & 12.0 & 0 \end{bmatrix}^T \tag{33}$$

which is confirmed to be the unique solution by applying FVA (which produces a flux spectrum which reduces to the single value $\hat{v}$).

If we repeat this exercise with the maximization of v_8, we find $\hat{v}_8 = 3.5$, but the corresponding flux distribution is not unique and belongs to the spectrum

$$\begin{bmatrix} v_1 = 3.5 \\ 0 \le v_2 \le 1.5 \\ v_3 = 3.5 \\ v_4 = 0 \\ v_5 = 0 \\ 0 \le v_6 \le 1.5 \\ 0 \le v_7 \le 1.5 \\ v_8 = 3.5 \end{bmatrix} \tag{34}$$

5. An Overview of Important Topics

In this section, we address a few important questions when dealing with the analysis of metabolic networks. Some of them have a direct impact on the underdeterminacy of the metabolic flux analysis, such as the topology and size of the metabolic network or the selection of the extra- or intra-cellular measurements. On the contrary, other issues, such as the dynamic evolution of the extracellular fluxes, might have no particular influence on this underdeterminacy, but will impact the computational procedure and the visualization of the results, and this is why we include them in the global picture.

5.1. How to Select the Size/Detail of the Metabolic Network?

The structure and size of the metabolic network can be represented by an incidence graph, whose topological properties are important for the solution of Equations (1)–(4). The analysis of these properties, such as determinacy, redundancy, balanceability, and calculability, can be assessed using software tools such as CellNetAnalyser [37] or COPASI [44], which were the first to propose such functionalities. The size of the network will also have a tremendous influence on the number of EFMs. A large network will probably imply that the EFMs are no longer enumerable due to memory and computation limitations. Hence, the importance of procedures to compute only subsets of EFMs such as [31], and of concepts such as the giant strong component (GSC), which represents the largest fully connected part of a metabolic network as introduced originally in [45]. Indeed, the GSC usually contains less than one-third of the nodes of the network, but key metabolites, and is more feasible for analysis of flux distribution and computation of EFMs. Another concept of interest is the introduction of minimal cut sets (MCSs), which represent sets of reactions whose removal will disable certain network functions [46], and which have been shown to be the EFMs of a dual network [47]. MCSs can be used to study the observability of reaction rates in metabolic flux analyses. The selection of the network will therefore be a compromise between describing the metabolic features of interest and the tractability of the computation procedure underlying the flux determination. An open research question is the selection of the right metabolic network for the derivation of low-order dynamic models (as introduced in the following Section 5.4). Should a reduction be operated a priori based on biological assumptions and simplifications or only a posteriori, in the course of the derivation of the dynamic macroscopic model?

5.2. Dynamic Metabolic Flux Interval Analysis

To date, we have considered that the measured specific extracellular fluxes are constant. This is typically the case in the early exponential growth phase of batch cultures or in the steady state of continuous cultures. However, there are other situations where the fluxes are time varying following changes in the environmental conditions (substrate excess or depletion, accumulation of byproducts). This can occur in fed-batch cultures, the transient phase between batch and continuous modes, or the transient phase between different setpoints in continuous operation. The previous analysis can be extended to these situations by considering a time-scale separation where the bioreactor environment is the slow subsystem and the cells or micro-organisms the fast subsystem, which can therefore be considered in a pseudo steady-state. The dynamics of the extracellular substrates S and products P can be described by a set of mass balance equations

$$\frac{dS}{dt} = -v_S X - D(S - S_{in})$$
$$\frac{dP}{dt} = v_P X - D(P - P_{in})$$
(35)

where S and P represent the extracellular substrate and product concentrations, respectively, v_S is the vector of specific uptake rates, v_P the vector of specific production rates, $D = F_{in}/V$ is the dilution rate (ratio of the inlet flow rate $F_{in}(t)$ to the culture volume $V(t)$).

A straightforward approach consists in smoothing the extracellular concentration evolutions, computing the derivatives of the smoothed signals and evaluating the uptake and production rates using Equation (35) and the knowledge about the transportation terms (functions of the dilution rate and inlet concentrations S_{in} and P_{in}). This approach was originally developed for MFA with no underdeterminacy or even overdeterminacy. One of the earlier reports can be found in [48] where the lysine fermentation process by Corynebacterium glutamicum is studied and the cell metabolic state is estimated online based on a small metabolic network with 11 fluxes. This work is extended in [49], where HEK cells are cultivated in perfusion and the metabolic fluxes are estimated online using a medium-size metabolic network of 40 reactions. Other notable accounts include [50], where Escherichia coli cultivations shifting from carbon limitation to nitrogen limitation and vice versa are studied, and [51], where a human cell line is analyzed and the authors compare Dynamic MFA (DMFA) to a flux average approach where the culture is divided in phases over which constant (average) fluxes are considered.

When underdeterminacy prevails, the approaches previously reviewed can be extended to take account of the dynamic evolution of the extracellular fluxes as well.

In [52,53], Dynamic Flux Balance Analysis (DFBA) is introduced and applied to the analysis of diauxic growth of Escherichia coli on glucose and acetate. Two optimization approaches can be considered: (a) a sequence of static LP problems corresponding to the subdivision of the batch time into time intervals over which the fluxes are assumed constant, or (b) a global dynamic optimization formulated over the total batch duration that can be solved using multiple shooting and orthogonal collocation to be converted into a NLP. The latter approach allows the consideration of an integrated (over the system trajectory) optimality objective or an end-of-batch objective, as well as nonlinear constraints on the fluxes resulting from a priori knowledge about kinetic expressions, but results in a significantly more complex problem. DFBA is nowadays a popular approach, which has led to many interesting applications (see for instance [54–58]) and the emergence of software tools such as DFBLAB [59].

In [60], Flux Spectrum Approach (FSA), which is a method based on the formulation of a set of min/max LP problems taking account of a range of measurement errors, is applied to the analysis of the evolution over time of the fluxes in small metabolic network of the CHO metabolism and data borrowed from [35].

In [61], a linear objective function subject to elementary modes as constraints is optimized to determine the fluxes in the metabolic network of Corynebacterium glutamicum at different temporal phases of fermentation. The use of convex analysis and EFMs is investigated in [35] to determine intervals for the metabolic fluxes of CHO cells in batch cultures, which are decomposed into several time periods corresponding to different phases of the cell life cycle. This work is extended to more detailed metabolic networks in [62] and to the dynamic evolution of the flux spectrum, without assuming a decomposition in phases, in [34] where convex analysis is applied to hybridoma cultures switching from batch to perfusion mode.

5.3. How to Represent the Accumulation of Internal Metabolites?

Besides the possible time evolution of the extracellular fluxes discussed in the previous subsection, another dynamic phenomenon might need special attention: the accumulation of an intracellular metabolite over time, implying that the traditional assumption of quasi steady-state is no longer valid. This situation is, for instance, observed in cultures of photosynthetic micro-organisms that can accumulate various components, such as carbohydrates and lipids. It is also well-known that yeasts, such as Saccharomyces cerevisiae, are able to accumulate some carbohydrates, e.g., trehalose, that play a role of carbon and energy reserve, as well as of stress protectant against harmful environmental conditions [63]. Abandoning the quasi steady state assumption for some of the intracellular metabolites boils down to removing their corresponding rows in the stoichiometric matrix N involved in Equation (1), hence increasing the number of degrees of freedom that characterizes the system underdeterminacy. To compensate for the mass balance equations withdrawn from (1), the kinetics of accumulation and/or reuse of the intracellular metabolites should be included in mass balance ODEs. However, the lack of intracellular measurements along time usually hampers the identification of these intracellular reaction rates. In [64], the authors propose the DRUM (Dynamic Reduction of Unbalanced Metabolism) modeling framework that consists in defining subsets of balanced intracellular metabolites that are interconnected via linking metabolites. These latter may accumulate or be reused. The subsets of balanced metabolites are reduced to macroscopic reactions, based on Elementary Flux Mode analysis, for which simple kinetic models are derived that allow building mass balance ODEs for the linking metabolites, biomass production, substrate consumption and product excretion. This methodology is applied to the lipid and carbohydrate accumulation in the microalgae Tisochrysis lutea. The modeling approach proposed in [65] can be applied to any metabolic network whose kinetics can be locally linearized. As in [64], the authors also consider subnetworks of fast reactions (involving metabolites that are assumed at quasi steady state) connected via metabolites consumed at low rates. Based on this time scale separation, the methodology allows reducing high dimensional linearized models, while accounting for the accumulation of some metabolites, as well as for the dilution effect due to biomass growth. In [66], a FBA-based simulator of Saccharomyces cerevisiae fed-batch cultures is proposed, using the assumption that the intracellular alpha-ketoglutarate is unbalanced. Its dynamics are described with a linear combination of the glucose and nitrogen specific uptake rates whose models involve the alpha-ketoglutarate concentration.

5.4. Model Reduction to Macroscopic Scale

Macroscopic models mainly predict biomass growth, the consumption of external substrates and the secretion of external products. Their structural simplicity allows their use for bioprocess optimization, control and online monitoring. To these purposes, it can be worthy to reduce metabolic models to a macroscopic scale. In [67,68], the authors propose a systematic methodology to build macroscopic reaction rate models via the definition of additional constraints whose number corresponds exactly to the number of degrees of freedom, i.e., the difference between the number of metabolic fluxes and the number of available equality constraints, hence removing the system underdeterminacy. These constraints express linear combinations of the metabolic fluxes as nonlinear functions of the extracellular concentrations, which correspond to the macroscopic reaction rate models. Another method that has often been used to define macroscopic reactions is the Elementary Flux Mode analysis [32–34,69]. As discussed above, EFMs are the shortest pathways from substrates to products. However, the main drawback with EFMs is that their total number dramatically increases with the network size. To overcome this problem, different solutions have been proposed. To avoid the exhaustive enumeration of all EFMs, refs [31,70] propose fast algorithms that randomly compute minimal sets of EFMs. In [71], the number of EFMs is reduced through a projection from the flux space to the yield space. Refs. [72,73] select EFMs, via ranking or controlled random search algorithms, using a multi-criteria objective function that combines prediction error, model size and efficiency of the EFMs (investment required to produce the enzymes). Another approach is made of Lumped Hybrid Cybernetic Models in which EFMs are grouped into clusters, each of them being associated to an average EFM [74]. In [75], the column generation method is used to determine a (non-necessarily unique) minimal subset of EFMs, which consists in solving iteratively two levels of optimization problems: the master problem is a quadratic optimization problem that identifies the macroscopic flux values using a subset of EFMs, and the subproblem is a linear problem for identifying EFMs that improve the model prediction in the master problem. Based on extensions of Dynamic Metabolic Flux Analysis introduced in [76], that only uses concentration measurements and avoids any numerical differentiation, refs. [77,78] select reduced sets of EFMs via a geometrical reduction (excluding EFMs with a cosine-similarity algorithm) followed by a multi-objective genetic algorithm that minimizes the prediction error and the size of the EFMs subset. A linear optimization problem has been formulated in [79] for selecting the best subset of EFMs based on a relaxation criterion. The methodology is extended in [80] and includes a more efficient selection procedure for the minimal subset of EFMs. Note that once the macroscopic reactions have been deduced from the metabolic network, it remains to identify their kinetic models. To that purpose, general kinetic models and systematic identification procedures can be very useful [81,82]. Model reduction methodologies based on subsets of balanced metabolites interconnected via linking metabolites that may accumulate within cells [64,65], have also been introduced in the previous section. Finally, independently of any EFM computation, macroscopic models may also consist of ODEs describing the mass balances for the biomass and the extracellular species, in which the reaction rates are computed at each time point by solving a FBA problem [16,17,66]. In [66], the underdeterminacy of the FBA problem is taken into account within the set of mass balance ODEs by computing the minimum and maximum admissible values of the ethanol concentration associated to the respective minimum and maximum admissible values of the specific ethanol production rate that are computed through FVA. Hence, the underdeterminacy at the level of FBA leads to corridors of admissible values along time for the concentrations of some macroscopic components, typically the extracellular products that are secreted.

5.5. How to Handle the Measurement Errors?

In the toy example, the adverse effect of an error on the measured extracellular flux was alleviated by relaxing an equality constraint and reformulating it as two inequality constraints (see Equation (19)). Such errors are frequent in practice, as it is necessary to compute the specific extracellular fluxes from the measurements of the evolution of the biomass and the extracellular concentrations. If we consider, for simplicity, the exponential growth phase of a batch culture, this can be formulated in the following way:

$$\begin{aligned} \frac{dS}{dt} &= -\underline{v}_S X \\ \frac{dP}{dt} &= \underline{v}_P X \\ \frac{dX}{dt} &= \mu X \end{aligned} \tag{36}$$

The solution of these equations are in the form of a linear regression

$$\underline{S}(t) = -\frac{v_S}{\mu}X(t) + \left(\underline{S}(0) + \frac{v_S}{\mu}X(0)\right) = \underline{a}_1 X(t) + \underline{b}_1$$

$$\underline{P}(t) = \frac{v_P}{\mu}X(t) + \left(\underline{P}(0) - \frac{v_P}{\mu}X(0)\right) = \underline{a}_2 X(t) + \underline{b}_2 \tag{37}$$

$$\ln(X(t)) = \mu t + \ln(X(0)) = a_3 t + b_3$$

where the regressors $\underline{a}_i$ give the estimation of the specific fluxes. If the environmental conditions evolve over time, it will be necessary to consider the inflows and outflows of the bioreactor as in dynamic model (35), and the variation of the growth rate of the biomass. This will imply the evaluation of the time derivatives using smoothing and numerical differentiation as explained in Section 5.2, or the formulation of the fluxes as piecewise linear functions without the need for numerical differentiation as proposed by [76]. Whatever the numerical procedure, experimental and numerical errors will always corrupt the information about the specific fluxes, which could in turn become incompatible with the constraints imposed by the metabolic network and prior knowledge about the system biology. This has to be taken into account by some form of constraints relaxation such as developed in Flux Spectrum Analysis [6] or Adaptive Flux Variability Analysis [38], where the uncertainty on the extracellular fluxes is represented as an interval around the measured values. In [38], minimum values for these uncertainties are determined by solving a sequence of optimization problems. The impact of these uncertainties on the solution, i.e., on the extent of the intervals for the intracellular fluxes, will largely depend on the structure and size of the metabolic network. This point is, however, still an open research question.

5.6. Some Further Perspectives on Sampling Algorithms

The ongoing research on sampling algorithms aims at improving their computational efficiency, convergence properties and ability to extensively explore the convex polytope of admissible flux distributions. To avoid the above mentioned problem of the samples that often become stuck in some part of the polytope when using hit-and-run algorithms [43], other methods have been proposed such as the artificial centering hit-and-run method (ACHR) [83] and the optimized general parallel sampler (OPTGP) [9]. Ref. [84] showed that the ACHR algorithms have convergence problems with high-dimensional polytopes, and introduced rounding methods with better performances. These methods were further improved in [10], leading to the coordinated hit-and-run with rounding (CHRR) method that computes the largest ellipsoid inscribed in the polytope and the rounding transformation of this ellipsoid into a unit ball. This latter transformation is then applied to the convex polytope whose sampling therefore becomes much more efficient in terms of computational time and convergence. Regarding these criteria, refs. [41,42] have shown that CHRR outperforms ACHR and OPTGP. Recently, refs. [39,85] have proposed the DISCOPOLIS (DIscrete Sampling of COnvex POlytopes via Linear program Iterative Sequences) algorithm that, instead of being a Markov Chain Monte Carlo method, provides (subsets of) samples that are independent of each other. This allows obtaining larger ranges of admissible flux values. Given the increasing complexity of the available metabolic networks, involving thousands of fluxes, there is still a need for the development and improvement of sampling algorithms with a reasonably low computational time and extensive exploration abilities for irregular shaped polytopes. Another difficult challenge consists in sampling non-convex solution spaces that are observed when using thermodynamical constraints for preventing infeasible loops in the metabolic network [20,84].

6. Conclusions

This paper reviews and applies to a toy example (the Matlab code of this example is provided in the Supplementary Materials associated to this article) some methods that can be used with underdetermined problems in metabolic flux analysis. These methods are grouped in two different strategies, the former consisting in simply dealing with the underdeterminacy without trying to reduce it, and the latter consisting in reducing or even eliminating the underdeterminacy. The choice between these two strategies depends on the specific problem to be solved and on the goals of the analysis. On the one hand, one could be interested by a unique solution that would be representative of a specific metabolic behavior in the considered cell line, e.g., by adding additional constraints describing these specific conditions and/or by defining an appropriate optimal behavior through FBA. On the other hand, one could be interested by the diversity of the possible metabolic behaviors resulting from the underdeterminacy, e.g., by analyzing the marginal distributions of each flux obtained from a sampling method. The point of utmost importance is to be aware of the system underdeterminacy. For example, even if one solution is provided by an algorithm used to solve the linear program of a FBA problem, that solution is not necessarily unique and a quick check via FVA could highlight that the system remains underdetermined. Many of the methods that have been presented are actually complementary, as illustrated in the abovementioned coupling of FBA with FVA. System underdeterminacy is a direct consequence of biological complexity and of the limited access to intracellular measurements. Developing methods to analyze underdetermined systems and/or to reduce their underdeterminacy in diverse and complementary ways will therefore remain an important research topic in the future. The interested reader might also consider the recent review [86], which includes an in-depth discussion of kinetic approaches for modeling cell metabolism.

Supplementary Materials: The following material is available online at https://www.mdpi.com/article/10.3390/pr9091577/s1, code S1: toy_example.zip.

Author Contributions: Conceptualization, P.B. and A.V.W.; methodology, P.B. and A.V.W.; software, P.B. and A.V.W.; investigation, P.B. and A.V.W.; writing—original draft preparation, P.B. and A.V.W.; writing—review and editing, P.B. and A.V.W.; visualization, P.B. and A.V.W. All authors have read and agreed to the published version of the manuscript.

Funding: This research received no external funding.

Conflicts of Interest: The authors declare no conflict of interest.

References

1. Stephanopoulos, G.; Aristidou, A.A.; Nielsen, J. *Metabolic Engineering: Principles and Methodologies*; Elsevier Science: San Diego, CA, USA, 1998.
2. Stelling, J.; Klamt, S.; Bettenbrock, K.; Schuster, S.; Gilles, E.D. Metabolic network structure determines key aspects of functionality and regulation. *Nature* **2002**, *420*, 190–193. [CrossRef]
3. Schuster, S.; Hilgetag, C. On elementary flux modes in biochemical reaction systems at steady state. *J. Biol. Syst.* **1994**, 2, 165–182. [CrossRef]
4. Klamt, S.; Stelling, J. Two approaches for metabolic pathway analysis? *Trends Biotechnol.* **2003**, *21*, 64–69. [CrossRef]
5. Mahadevan, R.; Schilling, C. The effects of alternate optimal solutions in constraint-based genome-scale metabolic models. *Metab. Eng.* **2003**, *5*, 264–276. [CrossRef] [PubMed]
6. Llaneras, F.; Picó, J. An interval approach for dealing with flux distributions and elementary modes activity patterns. *J. Theor. Biol.* **2007**, *246*, 290–308. [CrossRef] [PubMed]
7. Thiele, I.; Price, N.D.; Vo, T.D.; Palsson, B.Ø. Candidate metabolic network states in human mitochondria impact of diabetes, ischemia, and diet. *J. Biol. Chem.* **2005**, *280*, 11683–11695. [CrossRef]
8. Saa, P.A.; Nielsen, L.K. ll-ACHRB: A scalable algorithm for sampling the feasible solution space of metabolic networks. *Bioinformatics* **2016**, *32*, 2330–2337. [CrossRef] [PubMed]
9. Megchelenbrink, W.; Huynen, M.; Marchiori, E. optGpSampler: An improved tool for uniformly sampling the solution-space of genome-scale metabolic networks. *PLoS ONE* **2014**, *9*, e86587. [CrossRef] [PubMed]
10. Haraldsdóttir, H.S.; Cousins, B.; Thiele, I.; Fleming, R.M.; Vempala, S. CHRR: Coordinate hit-and-run with rounding for uniform sampling of constraint-based models. *Bioinformatics* **2017**, *33*, 1741–1743. [CrossRef] [PubMed]
11. De Martino, A.; De Martino, D. An introduction to the maximum entropy approach and its application to inference problems in biology. *Heliyon* **2018**, *4*, e00596. [CrossRef]
12. De Martino, D.; Andersson, A.M.; Bergmiller, T.; Guet, C.C.; Tkačik, G. Statistical mechanics for metabolic networks during steady state growth. *Nat. Commun.* **2018**, *9*, 2988. [CrossRef] [PubMed]

13. Ahn, W.S.; Antoniewicz, M.R. Metabolic flux analysis of CHO cells at growth and non-growth phases using isotopic tracers and mass spectrometry. *Metab. Eng.* **2011**, *13*, 598–609. [CrossRef] [PubMed]
14. Long, C.P.; Antoniewicz, M.R. High-resolution 13 C metabolic flux analysis. *Nat. Protoc.* **2019**, *14*, 2856–2877. [CrossRef]
15. Crown, S.B.; Antoniewicz, M.R. Parallel labeling experiments and metabolic flux analysis: Past, present and future methodologies. *Metab. Eng.* **2013**, *16*, 21–32. [CrossRef]
16. Richelle, A.; Gziri, K.M.; Bogaerts, P. A methodology for building a macroscopic FBA-based dynamical simulator of cell cultures through flux variability analysis. *Biochem. Eng. J.* **2016**, *114*, 50–64. [CrossRef]
17. Bogaerts, P.; Gziri, K.M.; Richelle, A. From MFA to FBA: Defining linear constraints accounting for overflow metabolism in a macroscopic FBA-based dynamical model of cell cultures in bioreactor. *J. Process Control* **2017**, *60*, 34–47. [CrossRef]
18. Nikdel, A.; Braatz, R.D.; Budman, H.M. A systematic approach for finding the objective function and active constraints for dynamic flux balance analysis. *Bioprocess Biosyst. Eng.* **2018**, *41*, 641–655. [CrossRef] [PubMed]
19. Soh, K.C.; Hatzimanikatis, V. Constraining the flux space using thermodynamics and integration of metabolomics data. In *Metabolic Flux Analysis*; Springer Science+Business Media: New York, NY, USA, 2014; pp. 49–63.
20. De Martino, D. Thermodynamics of biochemical networks and duality theorems. *Phys. Rev. E* **2013**, *87*, 052108. [CrossRef]
21. Raman, K.; Chandra, N. Flux balance analysis of biological systems: Applications and challenges. *Br. Bioinform.* **2009**, *10*, 435–449. [CrossRef]
22. Orth, J.D.; Thiele, I.; Palsson, B.Ø. What is flux balance analysis? *Nat. Biotechnol.* **2010**, *28*, 245–248. [CrossRef]
23. Willemsen, A.M.; Hendrickx, D.M.; Hoefsloot, H.C.; Hendriks, M.M.; Wahl, S.A.; Teusink, B.; Smilde, A.K.; van Kampen, A.H. MetDFBA: Incorporating time-resolved metabolomics measurements into dynamic flux balance analysis. *Mol. BioSyst.* **2015**, *11*, 137–145. [CrossRef] [PubMed]
24. Bordbar, A.; Yurkovich, J.T.; Paglia, G.; Rolfsson, O.; Sigurjónsson, Ó.E.; Palsson, B.O. Elucidating dynamic metabolic physiology through network integration of quantitative time-course metabolomics. *Sci. Rep.* **2017**, *7*, 46249. [CrossRef] [PubMed]
25. Smallbone, K.; Simeonidis, E. Flux balance analysis: A geometric perspective. *J. Theor. Biol.* **2009**, *258*, 311–315. [CrossRef] [PubMed]
26. Gziri, K.M.; Bogaerts, P. Determining a unique solution to underdetermined metabolic networks via a systematic path through the Most Accurate Fluxes. *IFAC-PapersOnLine* **2019**, *52*, 352–357. [CrossRef]
27. Zanghellini, J.; Ruckerbauer, D.E.; Hanscho, M.; Jungreuthmayer, C. Elementary flux modes in a nutshell: Properties, calculation and applications. *Biotechnol. J.* **2013**, *8*, 1009–1016. [CrossRef]
28. Pfeiffer, T.; Sanchez-Valdenebro, I.; Nuno, J.; Montero, F.; Schuster, S. METATOOL: For studying metabolic networks. *Bioinformatics* **1999**, *15*, 251–257. [CrossRef]
29. Terzer, M.; Stelling, J. Large-scale computation of elementary flux modes with bit pattern trees. *Bioinformatics* **2008**, *24*, 2229–2235. [CrossRef] [PubMed]
30. Van Klinken, J.B.; Willems van Dijk, K. FluxModeCalculator: An efficient tool for large-scale flux mode computation. *Bioinformatics* **2016**, *32*, 1265–1266. [CrossRef]
31. Jungers, R.M.; Zamorano, F.; Blondel, V.D.; Vande Wouwer, A.; Bastin, G. Fast computation of minimal elementary decompositions of metabolic flux vectors. *Automatica* **2011**, *47*, 1255–1259. [CrossRef]
32. Provost, A.; Bastin, G. Dynamic metabolic modelling under the balanced growth condition. *J. Process Control* **2004**, *14*, 717–728. [CrossRef]
33. Zamorano, F.; Vande Wouwer, A.; Jungers, R.M.; Bastin, G. Dynamic metabolic models of CHO cell cultures through minimal sets of elementary flux modes. *J. Biotechnol.* **2013**, *164*, 409–422. [CrossRef] [PubMed]
34. Fernandes de Sousa, S.; Bastin, G.; Jolicoeur, M.; Vande Wouwer, A. Dynamic metabolic flux analysis using a convex analysis approach: Application to hybridoma cell cultures in perfusion. *Biotechnol. Bioeng.* **2016**, *113*, 1102–1112. [CrossRef] [PubMed]
35. Provost, A. *Metabolic Design of Dynamic Bioreaction Models*; Faculté des Sciences Appliquées, Université catholique de Louvain: Louvain-la-Neuve, Belgium, 2006.
36. Becker, S.A.; Feist, A.M.; Mo, M.L.; Hannum, G.; Palsson, B.Ø.; Herrgard, M.J. Quantitative prediction of cellular metabolism with constraint-based models: The COBRA Toolbox. *Nat. Protoc.* **2007**, *2*, 727–738. [CrossRef] [PubMed]
37. Klamt, S.; Saez-Rodriguez, J.; Gilles, E.D. Structural and functional analysis of cellular networks with CellNetAnalyzer. *BMC Syst. Biol.* **2007**, *1*, 1–13. [CrossRef]
38. Abbate, T.; Dewasme, L.; Vande Wouwer, A.; Bogaerts, P. Adaptive flux variability analysis of HEK cell cultures. *Comput. Chem. Eng.* **2020**, *133*, 106633. [CrossRef]
39. Bogaerts, P.; Rooman, M. DISCOPOLIS: An algorithm for uniform sampling of metabolic flux distributions via iterative sequences of linear programs. *IFAC-PapersOnLine* **2019**, *52*, 269–274. [CrossRef]
40. Rubinstein, R. Generating random vectors uniformly distributed inside and on the surface of different regions. *Eur. J. Oper. Res.* **1982**, *10*, 205–209. [CrossRef]
41. Herrmann, H.A.; Dyson, B.C.; Vass, L.; Johnson, G.N.; Schwartz, J.M. Flux sampling is a powerful tool to study metabolism under changing environmental conditions. *NPJ Syst. Biol. Appl.* **2019**, *5*, 1–8. [CrossRef]
42. Fallahi, S.; Skaug, H.J.; Alendal, G. A comparison of Monte Carlo sampling methods for metabolic network models. *PLoS ONE* **2020**, *15*, e0235393. [CrossRef]

43. Smith, R.L. Efficient Monte Carlo procedures for generating points uniformly distributed over bounded regions. *Oper. Res.* **1984**, *32*, 1296–1308. [CrossRef]

44. Hoops, S.; Sahle, S.; Gauges, R.; Lee, C.; Pahle, J.; Simus, N.; Singhal, M.; Xu, L.; Mendes, P.; Kummer, U. COPASI—A complex pathway simulator. *Bioinformatics* **2006**, *22*, 3067–3074. [CrossRef]

45. Ma, H.W.; Zeng, A.P. The connectivity structure, giant strong component and centrality of metabolic networks. *Bioinformatics* **2003**, *19*, 1423–1430. [CrossRef] [PubMed]

46. Klamt, S.; Gilles, E.D. Minimal cut sets in biochemical reaction networks. *Bioinformatics* **2004**, *20*, 226–234. [CrossRef] [PubMed]

47. Ballerstein, K.; von Kamp, A.; Klamt, S.; Haus, U.U. Minimal cut sets in a metabolic network are elementary modes in a dual network. *Bioinformatics* **2012**, *28*, 381–387. [CrossRef]

48. Takiguchi, N.; Shimizu, H.; Shioya, S. An on-line physiological state recognition system for the lysine fermentation process based on a metabolic reaction model. *Biotechnol. Bioeng.* **1997**, *55*, 170–181. [CrossRef]

49. Henry, O.; Kamen, A.; Perrier, M. Monitoring the physiological state of mammalian cell perfusion processes by on-line estimation of intracellular fluxes. *J. Process Control* **2007**, *17*, 241–251. [CrossRef]

50. Lequeux, G.; Beauprez, J.; Maertens, J.; Van Horen, E.; Soetaert, W.; Vandamme, E.; Vanrolleghem, P.A. Dynamic metabolic flux analysis demonstrated on cultures where the limiting substrate is changed from carbon to nitrogen and vice versa. *J. Biomed. Biotechnol.* **2010**, *2010*, 621645. [CrossRef] [PubMed]

51. Niklas, J.; Schräder, E.; Sandig, V.; Noll, T.; Heinzle, E. Quantitative characterization of metabolism and metabolic shifts during growth of the new human cell line AGE1. HN using time resolved metabolic flux analysis. *Bioprocess Biosyst. Eng.* **2011**, *34*, 533–545. [CrossRef]

52. Varma, A.; Palsson, B.O. Stoichiometric flux balance models quantitatively predict growth and metabolic by-product secretion in wild-type Escherichia coli W3110. *Appl. Environ. Microbiol.* **1994**, *60*, 3724–3731. [CrossRef]

53. Mahadevan, R.; Edwards, J.S.; Doyle III, F.J. Dynamic flux balance analysis of diauxic growth in Escherichia coli. *Biophys. J.* **2002**, *83*, 1331–1340. [CrossRef]

54. Hjersted, J.L.; Henson, M.A. Optimization of fed-batch Saccharomyces cerevisiae fermentation using dynamic flux balance models. *Biotechnol. Prog.* **2006**, *22*, 1239–1248. [CrossRef] [PubMed]

55. Meadows, A.L.; Karnik, R.; Lam, H.; Forestell, S.; Snedecor, B. Application of dynamic flux balance analysis to an industrial Escherichia coli fermentation. *Metab. Eng.* **2010**, *12*, 150–160. [CrossRef]

56. Hanly, T.J.; Henson, M.A. Dynamic flux balance modeling of microbial co-cultures for efficient batch fermentation of glucose and xylose mixtures. *Biotechnol. Bioeng.* **2011**, *108*, 376–385. [CrossRef]

57. Grafahrend-Belau, E.; Junker, A.; Eschenröder, A.; Müller, J.; Schreiber, F.; Junker, B.H. Multiscale metabolic modeling: Dynamic flux balance analysis on a whole-plant scale. *Plant Physiol.* **2013**, *163*, 637–647. [CrossRef]

58. Emenike, V.N.; Schenkendorf, R.; Krewer, U. Model-based optimization of biopharmaceutical manufacturing in Pichia pastoris based on dynamic flux balance analysis. *Comput. Chem. Eng.* **2018**, *118*, 1–13. [CrossRef]

59. Gomez, J.A.; Höffner, K.; Barton, P.I. DFBAlab: A fast and reliable MATLAB code for dynamic flux balance analysis. *BMC Bioinform.* **2014**, *15*, 1–10. [CrossRef]

60. Llaneras, F.; Picó, J. A procedure for the estimation over time of metabolic fluxes in scenarios where measurements are uncertain and/or insufficient. *BMC Bioinform.* **2007**, *8*, 1–25. [CrossRef]

61. Gayen, K.; Venkatesh, K.V. Analysis of optimal phenotypic space using elementary modes as applied to Corynebacterium glutamicum. *BMC Bioinform.* **2006**, *7*, 1–13. [CrossRef]

62. Zamorano, F.; Vande Wouwer, A.; Bastin, G. A detailed metabolic flux analysis of an underdetermined network of CHO cells. *J. Biotechnol.* **2010**, *150*, 497–508. [CrossRef]

63. Richelle, A.; Fickers, P.; Bogaerts, P. Macroscopic modelling of baker's yeast production in fed-batch cultures and its link with trehalose production. *Comput. Chem. Eng.* **2014**, *61*, 220–233. [CrossRef]

64. Baroukh, C.; Muñoz-Tamayo, R.; Steyer, J.P.; Bernard, O. DRUM: A new framework for metabolic modeling under non-balanced growth. Application to the carbon metabolism of unicellular microalgae. *PLoS ONE* **2014**, *9*, e104499.

65. López Zazueta, C.; Bernard, O.; Gouzé, J.L. Dynamical reduction of linearized metabolic networks through quasi steady state approximation. *AIChE J.* **2019**, *65*, 18–31. [CrossRef]

66. Plaza, J.; Bogaerts, P. FBA-based simulator of Saccharomyces cerevisiae fed-batch cultures involving an internal unbalanced metabolite. *IFAC-PapersOnLine* **2019**, *52*, 169–174. [CrossRef]

67. Haag, J.E.; Vande Wouwer, A.; Bogaerts, P. Systematic procedure for the reduction of complex biological reaction pathways and the generation of macroscopic equivalents. *Chem. Eng. Sci.* **2005**, *60*, 459–465. [CrossRef]

68. Haag, J.E.; Vande Wouwer, A.; Bogaerts, P. Dynamic modeling of complex biological systems: A link between metabolic and macroscopic description. *Math. Biosci.* **2005**, *193*, 25–49. [CrossRef]

69. Niu, H.; Amribt, Z.; Fickers, P.; Tan, W.; Bogaerts, P. Metabolic pathway analysis and reduction for mammalian cell cultures—Towards macroscopic modeling. *Chem. Eng. Sci.* **2013**, *102*, 461–473. [CrossRef]

70. Machado, D.; Soons, Z.; Patil, K.R.; Ferreira, E.C.; Rocha, I. Random sampling of elementary flux modes in large-scale metabolic networks. *Bioinformatics* **2012**, *28*, i515–i521. [CrossRef]

71. Song, H.S.; Ramkrishna, D. Reduction of a set of elementary modes using yield analysis. *Biotechnol. Bioeng.* **2009**, *102*, 554–568. [CrossRef] [PubMed]

72. Soons, Z.I.; Ferreira, E.C.; Rocha, I. Selection of elementary modes for bioprocess control. *IFAC Proc. Vol.* **2010**, *43*, 156–161. [CrossRef]

73. Soons, Z.I.; Ferreira, E.C.; Rocha, I. Identification of minimal metabolic pathway models consistent with phenotypic data. *J. Process Control* **2011**, *21*, 1483–1492. [CrossRef]

74. Song, H.S.; Ramkrishna, D.; Pinchuk, G.E.; Beliaev, A.S.; Konopka, A.E.; Fredrickson, J.K. Dynamic modeling of aerobic growth of Shewanella oneidensis. Predicting triauxic growth, flux distributions, and energy requirement for growth. *Metab. Eng.* **2013**, *15*, 25–33. [CrossRef] [PubMed]

75. Oddsdóttir, H.Æ.; Hagrot, E.; Chotteau, V.; Forsgren, A. On dynamically generating relevant elementary flux modes in a metabolic network using optimization. *J. Math. Biol.* **2015**, *71*, 903–920. [CrossRef]

76. Leighty, R.W.; Antoniewicz, M.R. Dynamic metabolic flux analysis (DMFA): A framework for determining fluxes at metabolic non-steady state. *Metab. Eng.* **2011**, *13*, 745–755. [CrossRef]

77. Hebing, L.; Neymann, T.; Thüte, T.; Jockwer, A.; Engell, S. Efficient generation of models of fed-batch fermentations for process design and control. *IFAC-PapersOnLine* **2016**, *49*, 621–626. [CrossRef]

78. Hebing, L.; Neymann, T.; Engell, S. Application of dynamic metabolic flux analysis for process modeling: Robust flux estimation with regularization, confidence bounds, and selection of elementary modes. *Biotechnol. Bioeng.* **2020**, *117*, 2058–2073. [CrossRef]

79. Abbate, T.; de Sousa, S.F.; Dewasme, L.; Bastin, G.; Vande Wouwer, A. Inference of dynamic macroscopic models of cell metabolism based on elementary flux modes analysis. *Biochem. Eng. J.* **2019**, *151*, 107325. [CrossRef]

80. Maton, M.; Bogaerts, P.; Vande Wouwer, A. Selection of a Minimal Suboptimal Set of EFMs for Dynamic Metabolic Modelling. In Proceedings of the IFAC PapersOnLine 11th IFAC Symposium on Advanced Control of Chemical Processes, Venice, Italy, 13–16 June 2021.

81. Haag, J.; Vande Wouwer, A.; Remy, M. A general model of reaction kinetics in biological systems. *Bioprocess Biosyst. Eng.* **2005**, *27*, 303–309. [CrossRef]

82. Richelle, A.; Bogaerts, P. Systematic methodology for bioprocess model identification based on generalized kinetic functions. *Biochem. Eng. J.* **2015**, *100*, 41–49. [CrossRef]

83. Kaufman, D.E.; Smith, R.L. Direction choice for accelerated convergence in hit-and-run sampling. *Oper. Res.* **1998**, *46*, 84–95. [CrossRef]

84. De Martino, D.; Mori, M.; Parisi, V. Uniform sampling of steady states in metabolic networks: heterogeneous scales and rounding. *PLoS ONE* **2015**, *10*, e0122670. [CrossRef] [PubMed]

85. Bogaerts, P.; Rooman, M. DISCOPOLIS 2.0: A new recursive version of the algorithm for uniform sampling of metabolic flux distributions with linear programming. In Proceedings of the IFAC PapersOnLine 11th IFAC Symposium on Advanced Control of Chemical Processes, Venice, Italy, 13–16 June 2021.

86. Yasemi, M.; Jolicoeur, M. Modelling Cell Metabolism: A Review on Constraint-Based Steady-State and Kinetic Approaches. *Processes* **2021**, *9*, 322. [CrossRef]

 *processes*

MDPI

Article

Advanced Kinetic Modeling of Bio-co-polymer Poly(3-hydroxybutyrate-*co*-3-hydroxyvalerate) Production Using Fructose and Propionate as Carbon Sources

Stefanie Duvigneau [1,*], Robert Dürr [2], Jessica Behrens [1,2] and Achim Kienle [1,2]

[1] Institute for Automation Engineering, Otto von Guericke University Magdeburg, 39106 Magdeburg, Germany; jessica.behrens@ovgu.de (J.B.); kienle@mpi-magdeburg.mpg.de (A.K.)
[2] Process Synthesis and Process Dynamics, Max Planck Institute for Dynamics of Complex Technical Systems, 39106 Magdeburg, Germany; duerr@mpi-magdeburg.mpg.de
* Correspondence: stefanie.duvigneau@ovgu.de; Tel.: +49-391-67-50-222

Abstract: Biopolymers are a promising alternative to petroleum-based plastic raw materials. They are bio-based, non-toxic and degradable under environmental conditions. In addition to the homopolymer poly(3-hydroxybutyrate) (PHB), there are a number of co-polymers that have a broad range of applications and are easier to process in comparison to PHB. The most prominent representative from this group of bio-copolymers is poly(3-hydroxybutyrate-*co*-3-hydroxyvalerate) (PHBV). In this article, we show a new kinetic model that describes the PHBV production from fructose and propionic acid in *Cupriavidus necator* (*C. necator*). The developed model is used to analyze the effects of process parameter variations such as the CO_2 amount in the exhaust gas and the feed rate. The presented model is a valuable tool to improve the microbial PHBV production process. Due to the coupling of CO_2 online measurements in the exhaust gas to the biomass production, the model has the potential to predict the composition and the current yield of PHBV in the ongoing process.

Keywords: bioplastic; copolymerization; polyhydroxyalkanoate; kinetic modeling

Citation: Duvigneau, S.; Dürr, R.; Behrens, J.; Kienle, A. Advanced Kinetic Modeling of Bio-co-polymer Poly(3-hydroxybutyrate-*co*-3-hydroxyvalerate) Production Using Fructose and Propionate as Carbon Sources. *Processes* **2021**, *9*, 1260. https://doi.org/10.3390/pr9081260

Academic Editors: Philippe Bogaerts and Alain Vande Wouwer

Received: 29 June 2021
Accepted: 15 July 2021
Published: 21 July 2021

Publisher's Note: MDPI stays neutral with regard to jurisdictional claims in published maps and institutional affiliations.

1. Introduction

One suitable alternative to conventional petroleum-based plastics is that of the group of polyhydroxyalkanoates (PHAs) [1,2]. These polyesters stand out because of their favorable processing properties, e.g., their melting behavior or different blending options [3]. Furthermore, these are produced microbially by many bacteria and some archaea using a wide variety of non-fossil carbon sources [1]. There is a repertoire of possible and cheap substrates, such as those of inexpensive sugars in waste streams from the manufacturing industry (juice production, sugar cane processing), volatile fatty acids (VFAs) from biogas plants and wastewater in sewage treatment plants and even CO_2 [4–11]. In addition to the diverse possibilities of microbially producing bio-based PHAs, this plastic raw material has another important property: PHAs are degradable under environmental conditions [12]. However, from an economic point of view, the industrial production of PHAs is about five times more expensive than the production of petroleum-based polymers [13]. In addition to an improved extraction and processing of the polymers, a large part of the costs can be saved through optimized bioprocesses with increased PHA yield. This can be achieved by the incorporation of the sophisticated experimental investigation of different process modes or the optimization of substrates and feeding strategies with mathematical modeling [2,14,15]. Furthermore, model approaches represent the basic component in the development of advanced process control and intensification strategies [16,17].

In the research area of PHA production, a large number of models can be found that greatly differ in terms of modeling approach and complexity. Due to the complexity and variability of the bioprocess, there is no universal tool for predicting product yields regardless of the producing organism, bioreactor or process conditions such as temperature or pH [18]. Some approaches appear promising due to their simplicity and are able to reproduce the concentration curves in a qualitative manner, while they contain only little metabolic information [19–21]. Other approaches take the metabolism into account, however, due to their complexity, they can only be used to a limited extent for model-based process control intensification and are difficult to transfer to other PHA producers [22–27].

Many of the modeling approaches focus on the microbial production of the best-known representative from the group of PHAs: poly(3-hydroxybutyrate) (PHB). From an industrial point of view, however, the experimentally well-investigated bio-co-polymer poly(3-hydroxybutyrate-*co*-3-hydroxyvalerate) (PHBV) is more interesting, because of its lower melting temperature, higher elongation-to-break values and higher biocompatibility in comparison to PHB [28]. However, only a few model approaches already exist to investigate microbial PHBV production [21,29]. PHBV is also the target product of the present work. In order to develop a universal simulation tool, the mathematical model must contain a balanced amount of metabolic information. Such a modeling approach is rarely found in the literature: in [21], the description of the metabolism was reduced to central points with respect to PHA production (e.g., acetyl-CoA production), which can be found in many organisms in mixed cultures.

In the mathematical model presented here, we applied a time-dependent, kinetic parameter for the formation of residual biomass from fructose and propionic acid in *C. necator*, in order to map the dynamics of the present metabolic activity without detailed metabolic information. As the researchers at the University of Antioquia (Colombia) already have shown [30,31], the online measurement for the CO_2 content in exhaust gas serves as an excellent measure of the dynamic growth rate. Here, we also apply this correlation and hence, the model is a suitable candidate for the online estimation of PHBV product yields. In addition, it can be used to predict the polymer composition, since 3-hydroxybutyrate (3HB) and 3-hydroxyvalerate (3HV) amounts in the chains are considered in the model. For the parameter adaption, two data sets from aerobic PHA production in *C. necator* were used, and one experimental setup with only fructose as a carbon source and the other with fructose and propionic acid as carbon sources. In the last part of this manuscript, the model is used in a simulation study to investigate the influence of the feed rate for the propionic acid and of constant CO_2 in the exhaust gas on the bio-co-polymer yield and the 3HV proportion in the polymer.

The presented model approach is a suitable candidate for the development of a soft sensor for the online prediction of the polymer yield and composition. This opens new options to increase the flexibility, productivity and quality of the PHBV production process. With further model extensions, e.g., by coupling to polymerization kinetics [32], it will be possible to additionally estimate the chain length distribution during the process to obtain more information about polymer properties.

2. Experimental Methods

2.1. Mircoorganism and Cultivation Conditions

C. necator (H16, DSM 428) obtained from DSMZ GmbH Braunschweig was used for the fermentations. Bacteria were precultured in a shake flask with 10 vol% LB medium (Carl Roth, Karlsruhe, Germany) at 30 °C and 150 rpm. After reaching an optical density of 4 at 600 nm, the bacteria were transferred to an shake flask filled with 10 vol% of M81 medium supplemented with 20 g/L fructose and 1.5 g/L ammonium chloride. The recipe for the Medium 81 can be found in [23] or on the web page of the DSMZ. The M81 preculture was grown until an OD of 4.8 and used as an inoculum for the bioreactors. The fermentations were performed in a DASGIP parallel bioreactor system (Eppendorf AG, Juelich, Germany) with an inoculation OD of 0.4. During the experiments, the pH was kept at 6.8 and the dissolved oxygen (DO) was 70%. The DO measurements were performed with sensors from Mettler Toledo (Gießen, Germany). In the case of fructose as a single carbon source, the pH-control was performed with 2 M H_2SO_4. During the reactor experiment with fructose and propionic acid, the pH was stabilized with 20 g/L solution propionic acid as shown in [33]. The detection of the exhaust gas composition was done with the GA4 module of the DASGIP parallel bioreactor system (Eppendorf AG, Juelich, Germany). The initial conditions for the reactor experiments are shown in Table A1. All bioreactor experiments were performed with M81 media at 30 °C.

2.2. Determination of Total Biomass

For the determination of total biomass (TBM), 1 ml culture broth was centrifuged for 10 min at $9600 \times g$ and 4 °C (VWR MicroStar 17R, Pennsylvania, PA, USA). In a second step, the cell pellet was dried over night at 80 °C and weighted.

2.3. Enzyme Assay

By using enzymatic test kits (Kit No. 5390 and No. 10139106035, R-Biopharm AG, Darmstadt, Germany) and following the manufacturer's instructions, ammonium and fructose concentrations were determined from the supernatant of the sample.

High-Pressure Liquid Chromatography

Concentrations of 3HB and 3HV were determined by applying the procedure published in [34] using an Agilent 1100 high performance liquid chromatography (HPLC). For this, 1 mL of the culture broth was alkaline digested as reported in [35]. The samples were filtered through a 0.25 µm nylon membrane and 10 µL were loaded on the reverse phase column (Inertsil 100A ODS-3, 5 µm poresize, 250 × 4.6 mm, MZ-Analysentechnik GmbH, Mainz, Germany) and isocratically eluted with 1 mL·min^{-1} at 60 °C with 92% low concentrated H_2SO_4 (0.025% solution, Carl Roth, Karlsruhe, Germany) and 8% acetonitrile (Carl Roth, Karlsruhe, Germany). The 3HB and 3HV concentrations in the polymer chains of the samples were determined by using crotonic (Carl Roth, Karlsruhe, Germany) and 2-pentenoic acid standard samples (Sigma Aldrich, St. Louis, MO, USA), respectively. In parallel, a PHBV sample (12% 3HV, Sigma-Aldrich /Merck, Darmstadt, Germany) with known concentration must be measured to calculate the conversion yields Y_{HB} and Y_{HB} [34]:

$$Y_{HB} = 2 \cdot \frac{c_{CA}}{c_{HB}},\tag{1}$$

$$Y_{HV} = 2 \cdot \frac{c_{PA}}{c_{HV}}.\tag{2}$$

Here, the dilution ratio (D) is 2, c_{HB} is the known HB and c_{HV} the known 3HV concentration of the PHBV test sample. Due to the standard measurement of crotonic acid c_{CA} and 2-pentenoic acid c_{PA}, the conversion yields Y_{HB} and Y_{HV} can be determined, respectively. Detection takes place with a photodiode-array detector (G7115A, Agilent, Waldbronn, Germany) at 210 nm.

3. Kinetic Modeling Approach

The description of the formation and degradation in the microbial PHA production is an important building block for the complete production process. There are already a number of model candidates for the formation of PHB [18,20,23,36] that can accurately reflect the development of the homopolymer concentration over time. Compared to the homopolymer PHB, the copolymer PHBV has significantly improved processing properties. However, so far, only simple kinetic approaches for the formation of PHBV were developed [21]. Furthermore, the model from Špoljarić and colleagues was developed for the conversion of fatty acid methyl esters (FAMES) from biofuel to PHBV using lumped metabolic pathways [19]. The model presented here describes the formation and degradation of 3HB and 3HV in the polymer chains using fructose and propionate, two carbon sources that frequently occur in inexpensive residues or can be produced from them, e.g., by using waste streams from juice, cheese and paper production. In our model approach, detailed metabolic reaction pathways were not taken into account to keep the model structure as simple as possible.

The following assumptions were made for the model:

- A simple mass-action kinetic is assumed for the dynamics of the substrates;
- Propionate has an inhibiting effect and decelerates the growth of bacteria [33];
- The conversion of PHA into enzymatically active biomass (residual) is not affected by external propionic acid concentrations, as this is an internal process;
- *C. necator* begins to produce PHA already before nitrogen is depleted [18]. This behavior is considered in the model via an inhibitory term with nitrogen by a Michaelis–Menten kinetics approach (see Equation (3), term inh_2);
- Steric effects in the granules prevent the further production of PHA after reaching a total amount of 89 % of the total biomass (TBM, $P_{t,max}$) [36,37].

In the following, a set of ordinary differential equations for the dynamics of the system with fructose and propionic acid as substrates and residual biomass, 3HB and 3HV in the polymer chains as products is described. The dynamic state equation for the fructose concentration is given as

$$\frac{dc_{fru}}{dt} = -k_1 \cdot b_{CO_2}(t) \cdot c_{res} \cdot c_{fru} \cdot c_n \cdot inh_1$$
$$- k_4 \cdot c_{res} \cdot c_{fru} \cdot inh_2 \cdot inh_3$$
$$- k_7 \cdot b_{CO_2}(t) \cdot c_{fru} \cdot c_{res}$$
$$- D \cdot c_{fru}$$

$$(3)$$

with:

$$inh_1 = max\left(0, 1 - \frac{c_p}{c_{p,inh}}\right), inh_2 = max\left(0, 1 - \frac{c_n}{c_n + c_{n,sw}}\right), inh_3 = max\left(0, 1 - \frac{P_t}{P_{t,max}}\right).$$

Fructose can be metabolized for biomass production with the rate parameter k_1, the accumulation of 3HB in the polymer with k_4 or the conversion to CO_2 with k_7. The growth of biomass through fructose is controlled by the activity coefficient $b_{CO_2}(t)$ based on the CO_2 ratio in the exhaust gas and inhibited by the concentration of propionate with the term inh_1. At a concentration of 1.5 g/L propionic acid ($c_{p,inh}$), the substrate uptake for biomass is completely inhibited [33]. Since CO_2 in the exhaust gas is often defined as a proportion of the gas composition, we chose the relative CO_2 proportion to describe the metabolic activity $b_{CO_2}(t)$ as follows:

$$b_{CO_2}(t) = \frac{CO_{2,out}(t)}{CO_{2,in}}.$$

$$(4)$$

The metabolic activity is described by the quotient of $CO_{2,out}$ in the exhaust gas and $CO_{2,in}$ in the fresh inlet air. Since *C. necator* is a PHA producer of group 2 according to Novak et al. [18], the build-up of 3HB from fructose begins when there is still a small amount of ammonium in the medium. This effect is modeled by the term inh_2. As described in [36], steric effects at high polymer concentrations inhibit the conversion of substrates to PHA (term inh_3). According to literature values [1], the maximum achievable amount $P_{t,max}$ is 0.89 (89 % of the total biomass).

The inhibitory steric effect is given as the ratio between overall HA concentration and total biomass concentration:

$$P_t = \frac{(c_{hb} + c_{hv})}{(c_{hb} + c_{hv} + c_{res})}.$$

$$(5)$$

Finally, the dilution factor in the fed-batch process:

$$D = \frac{F_{in}}{V},$$

$$(6)$$

is the ratio of the feed flow rate F_{in} and reactor volume V.

For the computational study, a volume balance is necessary:

$$\frac{dV}{dt} = F_{in}.$$

(7)

As for fructose, a state equation can be set up for propionate dynamics:

$$\begin{aligned}
\frac{dc_p}{dt} = & -k_2 \cdot b_{CO_2}(t) \cdot c_{res} \cdot c_p \cdot c_n \cdot \mathrm{inh}_1 \\
& - (k_5 + k_6) \cdot c_{res} \cdot c_p \cdot \mathrm{inh}_2 \cdot \mathrm{inh}_3 \\
& - k_8 \cdot b_{CO_2}(t) \cdot c_p \cdot c_{res} \\
& + D \cdot \left(c_{p,in} - c_p\right).
\end{aligned}$$

(8)

It describes the consumption of propionate for biomass with a rate coefficient k_2, CO_2 with k_8 and 3HB production with k_5. In addition to the generation of 3HB, propionate can also be converted into 3HV (k_8). In fed-batch mode, a propionate solution according to Table A1 is fed to the system with the feed flow rate F_{in}.

For growth, organisms need ammonium. The state equation for the ammonium dynamics is:

$$\begin{aligned}
\frac{dc_n}{dt} = & -c_{res} \cdot c_n \cdot b_{CO_2}(t) \cdot \left(k_1 \cdot c_{fru} + k_2 \cdot c_p\right) \cdot \mathrm{inh}_1 \\
& - k_3 \cdot c_{res} \cdot c_n \cdot (c_{hb} + c_{hv}) \\
& - D \cdot c_n.
\end{aligned}$$

(9)

In addition to the ammonium uptake for biomass growth by consuming external carbon sources (first term in Equation (9)), ammonium is needed to convert the biopolymer to catalytically active biomass with the degradation rate parameter k_3.

The dynamical behavior of residual (non-PHA, catalytically active) biomass is described as follows:

$$\begin{aligned}
\frac{dc_{res}}{dt} = & c_{res} \cdot c_n \cdot \left[\mathrm{inh}_1 \cdot \left(k_1 \cdot c_{fru} + k_2 \cdot c_p\right) \cdot b_{CO_2} + k_3 \cdot (c_{hb} + c_{hv})\right] \\
& - D \cdot c_{res}.
\end{aligned}$$

(10)

Residual biomass is produced through the consumption of external carbon sources such as fructose and propionate and the conversion of 3HB and 3HV from the polymer chains in the presence of ammonium.

The following ordinary differential equations (ODEs) account for the dynamics of the monomers 3HB and 3HV in the polymer chains:

$$\begin{aligned}
\frac{dc_{hb}}{dt} = & c_{res} \cdot \mathrm{inh}_2 \cdot \mathrm{inh}_3 \cdot \left(k_4 \cdot c_{fru} + k_5 \cdot c_p\right) \\
& - k_3 \cdot c_{res} \cdot c_n \cdot c_{hb} - D \cdot c_{hb}.
\end{aligned}$$

(11)

$$\begin{aligned}
\frac{dc_{hv}}{dt} = & k_6 \cdot c_{res} \cdot c_p \cdot \mathrm{inh}_2 \cdot \mathrm{inh}_3 \\
& - k_3 \cdot c_{res} \cdot c_n \cdot c_{hv} - D \cdot c_{hv}.
\end{aligned}$$

(12)

For the accumulation and breakdown of the biopolymer (3HB and 3HV), CO_2 formation is negligible, since the metabolic reaction pathways produce only little CO_2 compared to the breakdown of sugars and organic acids into catalytically active components of the total biomass.

3.1. Numerical Solution

3.1.1. Interpolation for Volume, $CO_{2,out}$ and Feed Rate

For the simulation of the model, the temporal evolution of the CO_2 amount in the exhaust gas from reactor experiments is required. Since the online measurement often fluctuates and does not provide a smooth curve, the data were interpolated to integrate them into the ODE model. For this, a smoothing spline interpolation was carried out with the *MATLAB* command *csaps*. For both experiments, different smoothing factors were evaluated. A smoothing factor of 0.2 was selected for the experiment with fructose as the only carbon source (data set 1), while the data from the reactor experiment with fructose and propionic acid as carbon sources (data set 2) achieved a smooth and well-fitted curve with a smoothing factor of 0.02. For the interpolation, the splines are evaluated at the sampling points. The evaluation was carried out with the *MATLAB* command *ppval*. The curves and online data are shown in Figure 1 for both data sets.

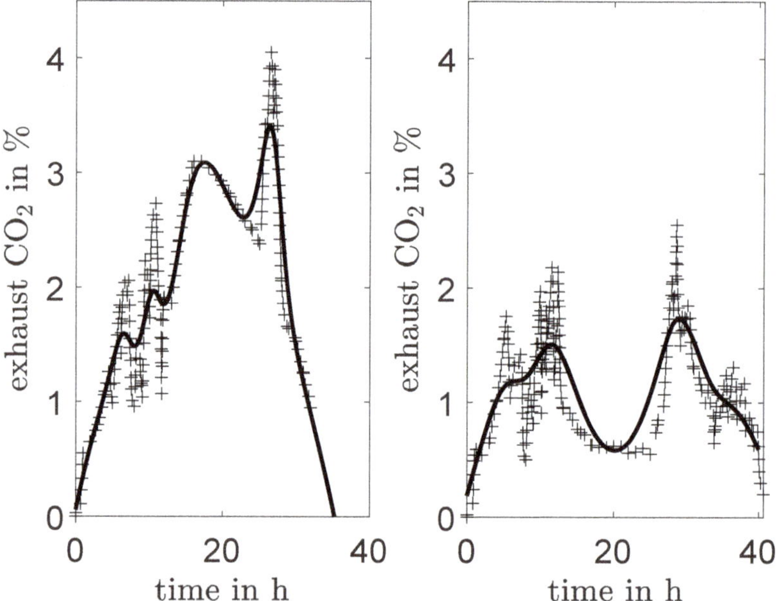

Figure 1. Exhaust CO_2 for data set 1 (fructose as carbon source, **left** panel, +) and data set 2 (fructose and propionic acid as carbon sources, **right** panel, +) and the interpolations (solid lines).

The feeding of the odd carbon source propionic acid was achieved by pH control. If the pH increases, a certain amount of propionic acid with 20 g/L in the feed is added to the bioreactor to stabilize the pH at 6.8. The pre-implemented PI controller of the DASGIP parallel bioreactor system (Eppendorf AG, Jülich, Germany) was used for this purpose. As for the activity factor, frequent fluctuations are observed because of the special pH-dependent feeding strategy and thus the feed rate for propionic acid was interpolated and evaluated in the same way as the CO_2 amount in the exhaust gas (Figure 2). Here, a factor of 0.02 delivered a smooth curve.

Furthermore, an interpolation of the volume was necessary for the fed-batch experiment with fructose and propionic acid as carbon sources (data set 2). For this purpose, a polynomial of order 10 was determined with the *MATLAB* command *polyfit* and evaluated with *polyval* at the sampling times. As seen in Figure 3, volume reduction by sampling was also taken into account. Hence, a decrease in volume was recorded despite an average feed rate of approximately 20 mL/h between 15 and 25 h (see Figure 2). In the experiment with fructose as the single carbon source, the reactor was operated in batch mode. Since it was assumed that the system was ideally mixed, the changes in concentration were only caused by internal sinks and sources (no substrate was pumped in), and the volume reduction due to sampling can be neglected in data set 1. All simulations, interpolations and evaluations were carried out with *MATLAB 2019b*.

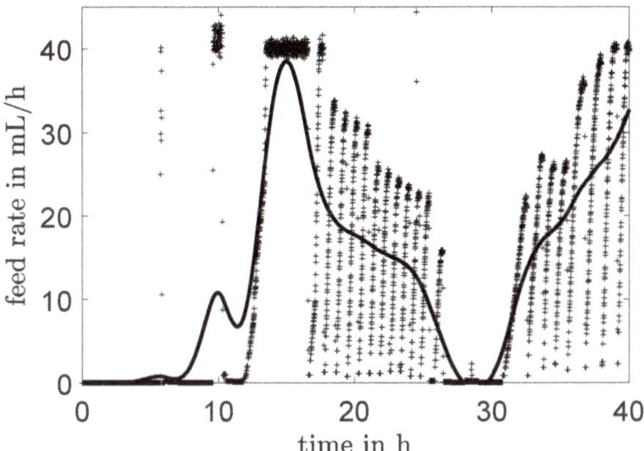

Figure 2. Experimental feed rate and polynomial for the propionic acid inlet of data set 2 (fructose and propionic acid as carbon sources).

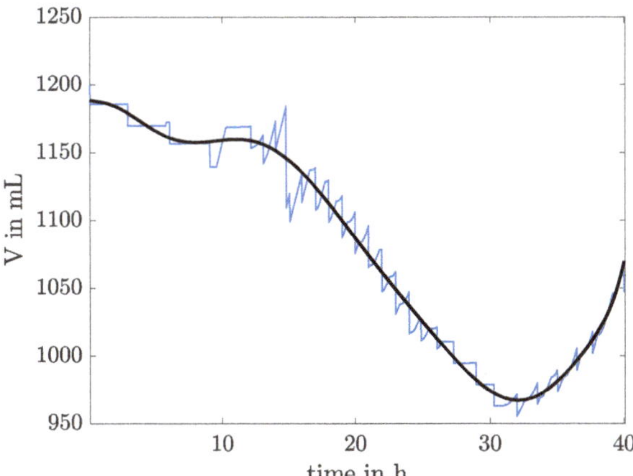

Figure 3. Experimental volume (small blue line) and polynomial (black curve) of data set 2 (fructose and propionic acid as carbon sources).

3.1.2. Parameter Identification

For parameter identification, the following objective function was minimized:

$$ESS = \sum_{i=1}^{n} \left(\frac{x_{exp}(t_i) - x_{sim}(t_i)}{max(x_{exp})} \right)^2. \tag{13}$$

Here, the error between the simulated x_{sim} and experimental data x_{exp} at time point t_i is determined and weighted with the maximum value in the experimental data set.

The kinetic parameters were determined using the algorithm *fmincon* in *MATLAB 2019b*. The ODEs were numerically solved with the algorithm *ode15s* with a relative tolerance of 10^{-9}. To prevent a sub-optimal initial parameter set a multi-start approach with N = 10,000 was applied. To further validate the resulting parameter set obtained by the local optimization strategy, the global optimization algorithm differential evolution (DE) was selected [38]. The resulting set of parameters can be found in Table A1.

To prove the parameter identifiability, profile likelihoods are determined for the parameter set [39]. Appendix B shows all profile likelihoods and the corresponding likelihood-based confidence intervals. All profile likelihoods show a distinct minimum at the estimated parameter set and thus all parameters and the model itself are (locally) identifiable. The confidence intervals are obtained by calculating the value of a χ^2-distribution with a confidence level $\alpha = 0.95$ and one degree of freedom as proposed in [39].

4. Results

4.1. Identification Using Different Data Sets

The kinetic model was adapted to two data sets from bioreactor experiments with a working volume of 1.2 L. In the first data set, fructose was the only carbon source that was metabolized under aerobic conditions. The second data set was also obtained under aerobic conditions with fructose and propionic acid as carbon sources. Here, propionic acid was added via a pH-regulated feed as proposed by Kim and coworkers [33] aiming for a constant pH value of 6.8. From the available data of these sets, the online data for the CO_2 content in the inflow and in the exhaust gas, the feed rate for the propionic acid and the exact volume considering the sampling volume were used in the case of data set 2. In the case of the data set 1, online data for the CO_2 content in the inflow and in the exhaust gas were also used, but a constant volume and a batch mode ($F_{in} = 0$) were considered. The CO_2 content in the exhaust gas, the feed stream for the propionic acid and the reactor volume were approximated as described in Section 3.1.1. The smoothed measurement data were used for the model simulation. Furthermore, the concentrations for total biomass, biopolymer, fructose and propionic acid were determined offline in both data sets (see Section 2).

The dynamic behavior for the conversion of the substrates from data set 1 (only fructose) can be reproduced well with the present model (solid lines, Figure 4a). The model shows larger deviations for the substrates from data set 2 (fructose and propionic acid), especially in the last time segment from 25 h (dashed lines, Figure 4b). On the one hand, this is due to the approximation of the inflow rate for propionic acid (see Figure 2), and on the other hand, the measurement of the propionic acid in the medium becomes more difficult. It seems to be, that there are more and more apoptosis fragments, e.g., matrix, RNA and proteins in the culture supernatant that disrupt the signal obtained by HPLC (own experimental findings). These fragments could be avoided by elaborate sample preparation before HPLC measurement, e.g., additional filtration, boiling procedures or the supplementation of organic acids. Furthermore, the chromatographic peak consisting of propionic acid can be be fractionated and separated from impurities by a second HPLC run with an adjusted mobile phase.

The model for the case of fructose as a single substrate can reproduce the production and depletion of total biomass and 3HB with sufficient accuracy (Figure 5a). The same applies to the case with fructose and propionic acid as carbon sources (Figure 5b). In particular, the degradation of 3HB and 3HV in the polymer chains after an NH_4Cl shot at 24 h can be mapped very well by the model. This property is important when working with waste streams which also contain nitrogen sources and which are added during the ongoing process in order to keep the carbon in excess. Since the model can reproduce the product concentrations very well, it can further be used for a simulation study.

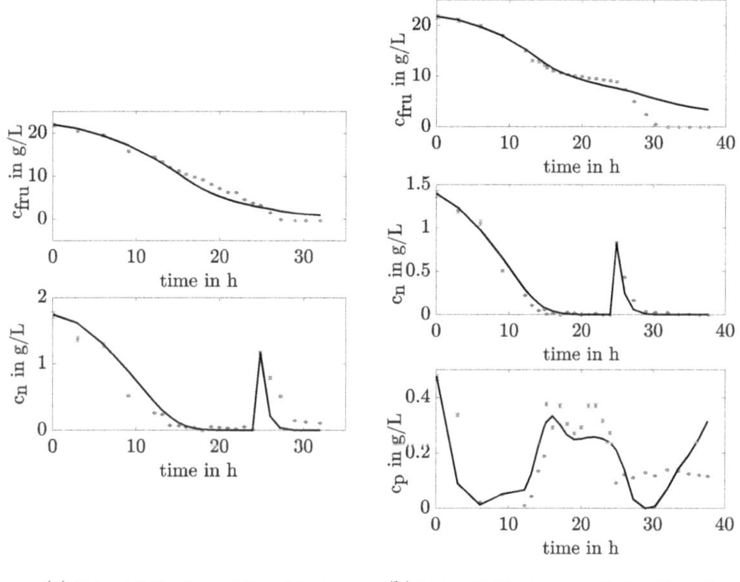

(**a**) *Data set 1 (fructose only)—substrates*   (**b**) *Data set 2 (fructose + propionic acid)—substrates*

Figure 4. Consumption of substrates for two feeding scenarios: (**a**) fructose as a single carbon source without additional feeding (batch); (**b**) fructose and propionic acid as carbon sources with pH-dependent propionic acid feeding.

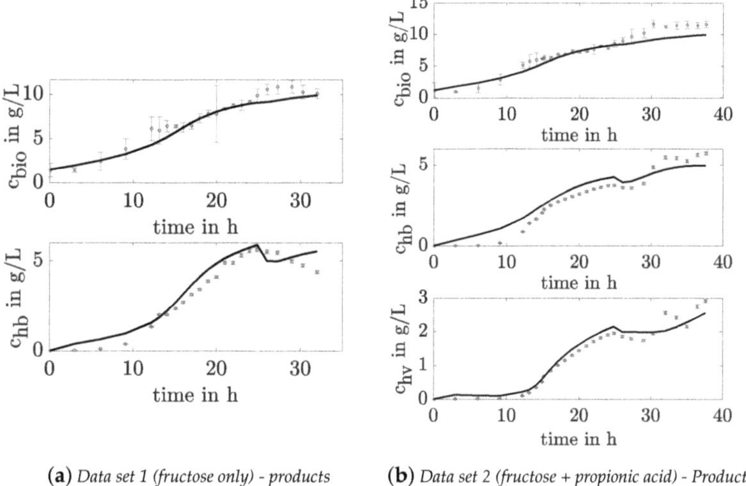

(**a**) *Data set 1 (fructose only) - products*   (**b**) *Data set 2 (fructose + propionic acid) - Products*

Figure 5. Production and degradation of polyhydroxyalkanoate (PHA) monomers (c_{hb}, c_{hv}) and total biomass ($c_{bio} = c_{res} + c_{hb} + c_{hv}$).

4.2. Computational Study

In the following, the identified model was used to investigate the maximum product concentrations by applying different constant CO_2 and propionic acid feed profiles. A constant feed rate can easily be implemented on every bioreactor system with continuous pumps. The constant CO_2 in the exhaust gas represents a complex control task, as CO_2 is produced by the bacteria themselves (autogenous CO_2). However, the research work in [40] shows the feasibility of this control task which can also be used in future validation experiments.

In the simulation study (Figure 6), a constant CO_2 proportion in the exhaust gas and a constant feed rate for propionate were assumed for the process time. The propionic acid concentration in the feed was set to 20 g/L as in data set 2. Figure 6a shows the maximum total 3HB and 3HV in the polymer chains P_t* as a function of CO_2 in the exhaust gas and the feed rate for propionic acid. The inhibitory area is clearly visible on the left side. In this range, the propionic acid concentration in the medium becomes too high so that growth is inhibited. In the case of increased CO_2 values in the exhaust gas, the feed rate can also be increased without triggering growth inhibition. This behavior can be justified as follows: a higher exhaust gas value for CO_2 produced by the microorganisms (autogenous CO_2) is triggered by an increased uptake of substrates from the medium. In *C. necator*, the degradation of pentoses and hexoses takes place via the Entner–Doudoroff (ED) pathway. Furthermore, the tricarboxylic acid cycle (TCA) is mainly responsible for the generation of energy and precursor molecules for biomass synthesis from the precursors of metabolic sugar degradation and organic acids. Both the ED pathway and the TCA generate CO_2 as a by-product, which can be found in the exhaust gas values of the simulation study. With other words, increased autogenous CO_2 in the exhaust gas can be translated into stronger residual biomass growth with increased substrate uptake. As a result, the growth inhibition with increased CO_2 in the exhaust gas only occurs at higher feeding rates for propionic acid. In addition to the beneficial effect of higher exhaust CO_2, the total biopolymer concentration decreases in the non-inhibitory area. This effect occurs because the CO_2 output is related to the higher residual biomass growth and hence, the substrates were less translated into biopolymers.

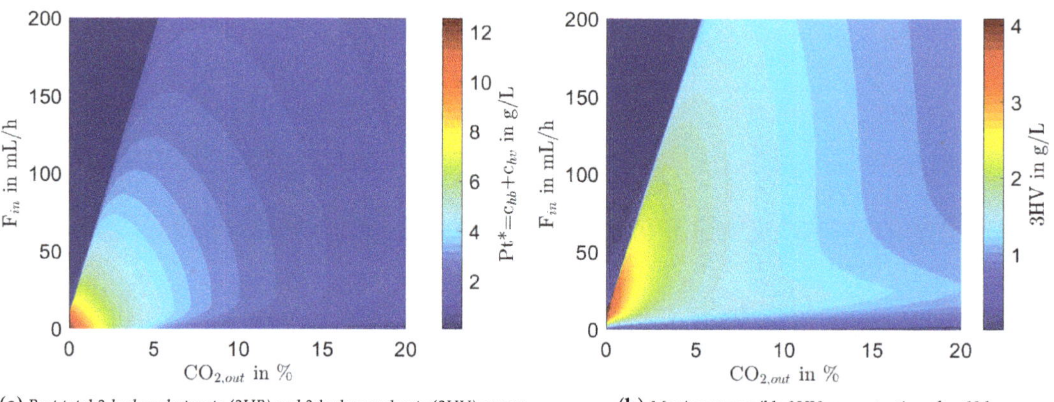

(**a**) *Best total 3-hydroxybutyrate (3HB) and 3-hydroxyvalerate (3HV) concentration*

(**b**) *Maximum possible 3HV concentration after 60 h*

Figure 6. Simulation study with constant values for CO_2 and F_{in}.

The highest total biopolymer concentration of 12.5 g/L within 60 h simulation time can be achieved without feeding propionic acid to the system and with a very low CO_2 content in the exhaust gas. However, without an active feed rate, no 3HV will be produced and the 3HV content is decisive for improved processing of copolymers compared to their homopolymers. The maximum HV concentration in the polymer is shown in Figure 6b for different constant autogenous CO_2 and propionic acid feeding rates. Here, the inhibiting region can be seen again due to a high propionate concentration in the medium. For a high 3HV concentration, our model predicts feed rates between 12 and 40 mL/h depending on the CO_2 in the exhaust. In general, less than 2.5 % CO_2 in the exhaust gas leads to higher 3HV concentrations. As our simulation results show, the feed rate for propionic acid must be chosen very carefully because of their strong correlation with the CO_2 value: overly high feed rates at higher CO_2 in the exhaust gas lead to less residual biomass and a decreased 3HV concentration (see Figure A1). Further characteristic values of the simulation study are shown in Figure A1 (e.g., residual biomass at the maximum total 3HB and 3HV concentration, total monomer/total biomass ratio).

Three exemplary time courses for different production goals were illustrated in Figure 7. In case A, the dynamic behavior to achieve a high total polymer concentration (3HB + 3HV) with the given initial conditions is shown. For this purpose, the feed rate was set to 0 mL/h and the CO_2 amount to 1 %. For case B, the same CO_2 value as in case A was applied together with a feed rate of 25 mL/h to show an example time course for a high 3HV concentration. In addition to the two preferred fermentation results (cases A and B, Figure 7), the inhibitory case was also shown (case C, Figure 7) by increasing the feed rate to 105 mL/h.

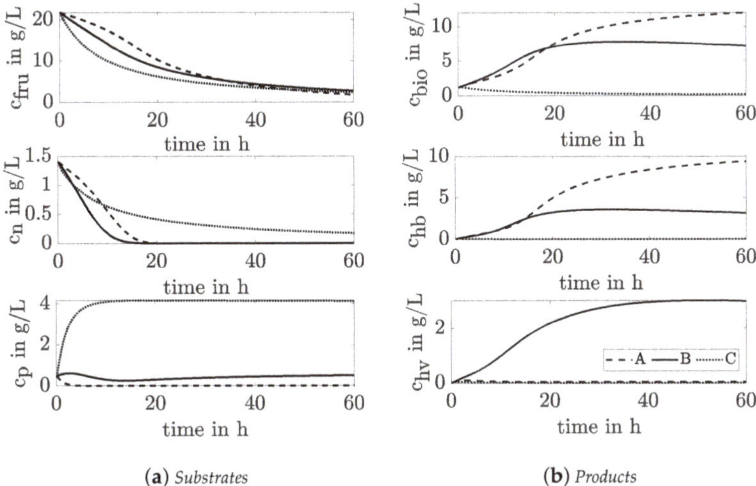

(**a**) *Substrates* (**b**) *Products*

Figure 7. Three example cases of dynamical behavior applying constant exhaust CO_2 and feed rate for propionic acid (profile study 1). Legend: A, maximum total biopolymer concentration ($c_{hb} + c_{hv}$); B, high 3HV concentration (c_{hv}); and C, inhibition caused by propionic acid (c_p).

5. Concluding Remarks

In this manuscript, a model approach is presented which enables the integration of online data for the estimation of the yield and the composition of the copolymer PHBV in *C. necator*. Compared to other approaches, no genome-scale metabolic networks or reduced variants are necessary [23,41], since it is a pure kinetic approach that describes changes in the metabolism due to a CO_2-dependent biomass production rate that changes over time. Despite the lack of detailed metabolic information, our kinetic model can display the data sets with fructose as a single substrate and fructose and propionate as substrate with high accuracy. The model approach has similar complexity as the models presented in Koller et al. [36] and Dias et al. [21], whereby the metabolic activity that controls the biomass growth can be described by the CO_2 profile in the exhaust gas. Because of the compact model structure and smart coupling to available online measurements the approach can be applied as a soft sensor for the prediction of biomass, substrates, product yield and the composition of the biopolymer (e.g., 3HV/3HB ratio). First steps in this direction where already successfully applied for a vinasse-molasse PHA process [30,31]. In comparison to the model approach used in [30,31], our more complex model represents an excellent basis to predict 3HV and 3HB amounts in the polymer chains. In addition to the application options during the process, CO_2 profiles can be estimated in optimization studies with predefined concentrations profiles for the substrates and products. As a first step, we investigated in our simulation study the effect of constant CO_2 fractions in the exhaust gas and constant feed rates for propionic acid on biopolymer yield and composition. The lower the feed rate was set and the less CO_2 was in the exhaust gas, the more PHA was produced but with less 3HV content in the polymer. A suitable feed rate for propionic acid input was predicted to be between 12 and 40 mL/h in order to achieve high 3HV concentration in the final co-biopolymer. Furthermore, the model approach can be used for the design of observers and state estimators for the reconstruction of non-measurable states. A first example in this direction was recently presented by Carius and coworkers [16]. Here, an unscented Kalman filter and a moving horizon estimation based on a hybrid cybernetic PHB model [23] was designed and evaluated.

Future work should focus on the design of model-based soft sensor approaches, since this will enable the reliable online estimation of the PHBV content using measurements of the exhaust gas online without the need for additional expensive hardware sensors. Furthermore, the transferability of the kinetic model to other PHA producers must be researched. Here, the focus should be on PHA producers, which are already producing PHA under growth conditions, e.g., *C. necator* DSM 515, as the model was designed for this group of bacteria. Furthermore, it should be checked whether a characteristic CO_2 profile occurs during the accumulation of other copolymer building blocks, such as 4-hydroxybutyrate. Finally, a control concept should be developed which is able to keep CO_2 in the exhaust gas on a desired level over a longer period of time. The work of Shang and colleagues [40] shows that it is in principle possible to control a process parameter that is strongly influenced or caused by the bacteria. The CO_2 amount in the exhaust gas was adjusted in the work of Shang and coworkers in order to investigate the inhibitory effects of CO_2. Such an effect has not yet been taken into account in the model presented in this manuscript, but should be considered in future model extensions.

Overall, our model approach provides the basis for a broad range of possible future applications and will help make the production process of biopolymers more reliable and less expensive.

Author Contributions: S.D. and R.D. conceived and designed research. S.D. and J.B. conducted the experiments. S.D. analyzed and visualized the data and wrote the original draft of the manuscript. R.D. and A.K. supervised the work and edited the original draft. All authors have read and agreed to the published version of the manuscript.

Funding: We would like to acknowledge the EU-programme ERDF (European Regional Development Fund) for the funding of the project DIGIPOL.

Institutional Review Board Statement: Not applicable

Informed Consent Statement: Not applicable

Data Availability Statement: The data presented in this study are available on request from the corresponding author.

Conflicts of Interest: The authors declare no conflict of interest.

Appendix A. Additional Figures—Profile Study

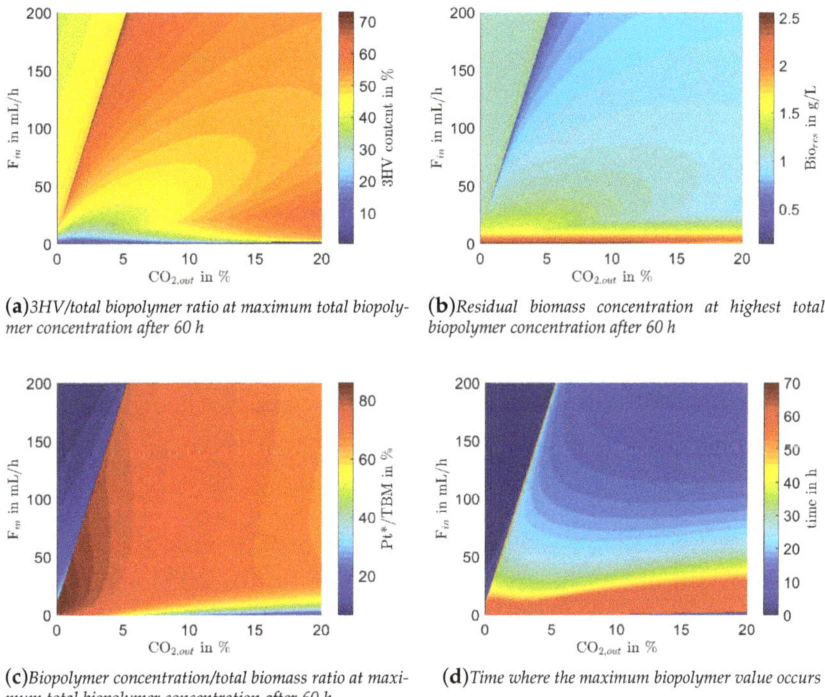

(**a**)*3HV/total biopolymer ratio at maximum total biopolymer concentration after 60 h*

(**b**)*Residual biomass concentration at highest total biopolymer concentration after 60 h*

(**c**)*Biopolymer concentration/total biomass ratio at maximum total biopolymer concentration after 60 h*

(**d**)*Time where the maximum biopolymer value occurs*

Figure A1. Simulation study with constant values for CO_2 and F_{in}—3HV content (**a**), residual biomass (**b**), PHA/BTM ratio (**c**), time (**d**).

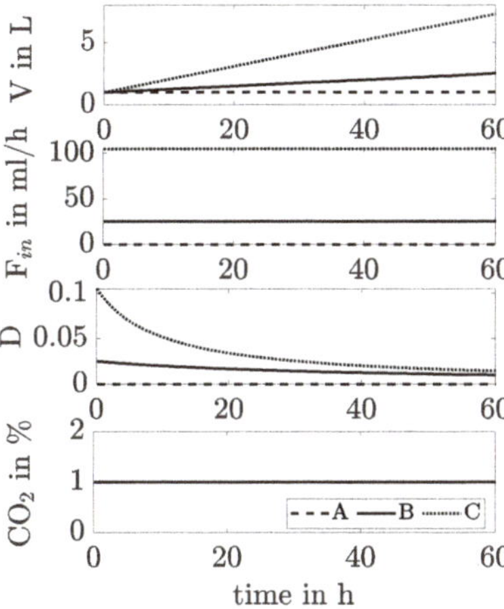

Figure A2. Reactor volume V, feed rate F_{in}, dilution rate D and CO_2 in the exhaust for the three example cases. Legend: A, maximum total biopolymer concentration; B, high 3HV concentration; and C, inhibition caused by propionic acid.

Appendix B. Parameter Identifiability

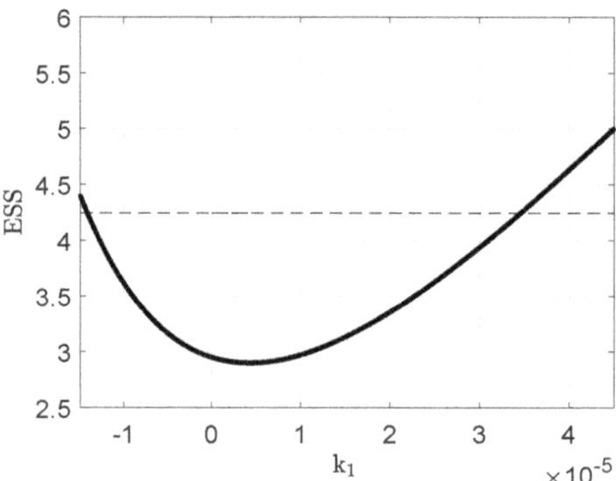

Figure A3. Profile likelihood of k_1. Horizontal dashed line represents the explained sum of squares (ESS) value of the 95% confidence interval.

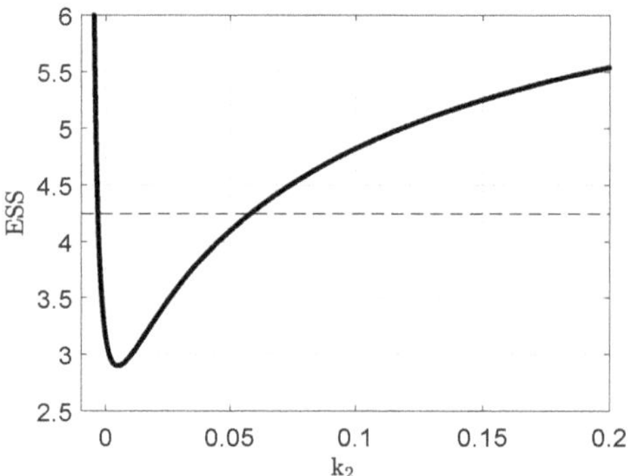

Figure A4. Profile likelihood of k_2. Horizontal dashed line represents the ESS value of the 95% confidence interval.

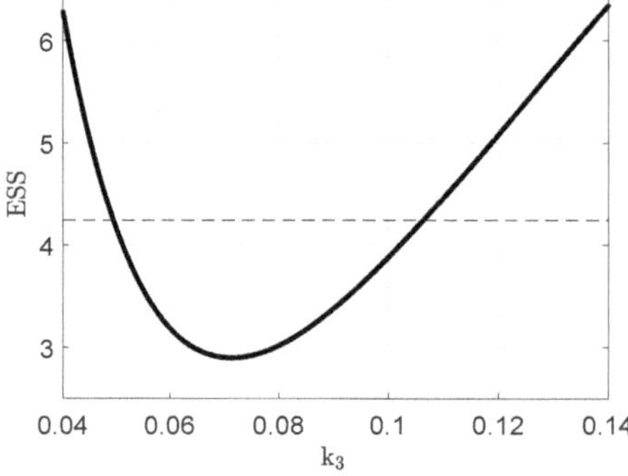

Figure A5. Profile likelihood of k_3. Horizontal dashed line represents the ESS value of the 95% confidence interval.

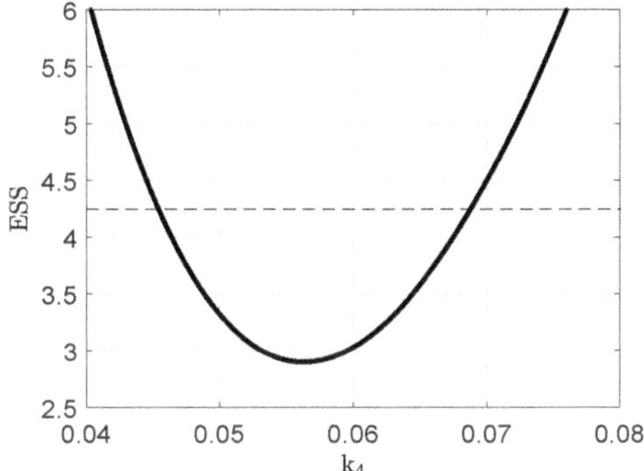

Figure A6. Profile likelihood of k_4. Horizontal dashed line represents the ESS value of the 95% confidence interval.

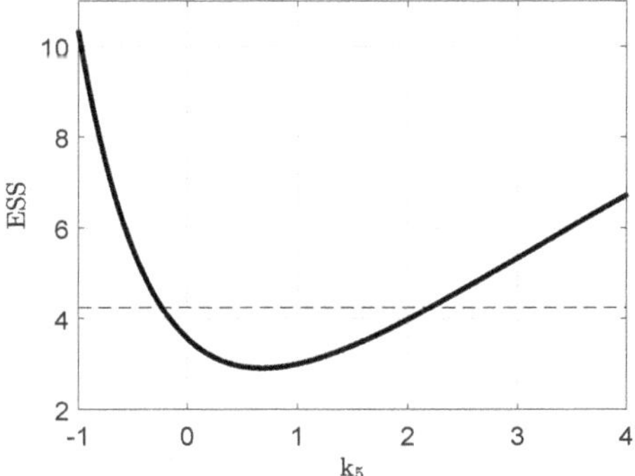

Figure A7. Profile likelihood of k_5. Horizontal dashed line represents the ESS value of the 95% confidence interval.

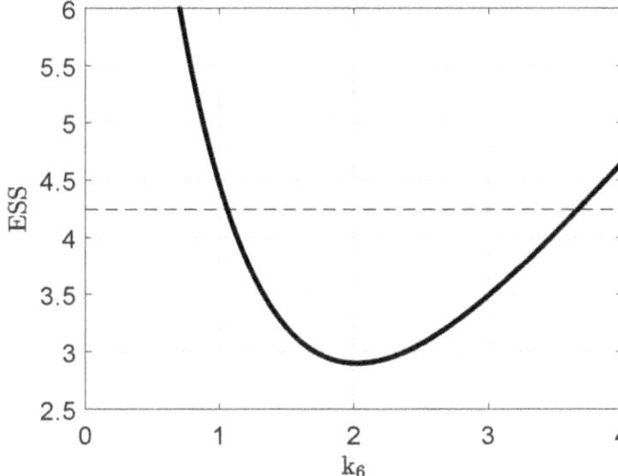

Figure A8. Profile likelihood of k_6. Horizontal dashed line represents the ESS value of the 95% confidence interval.

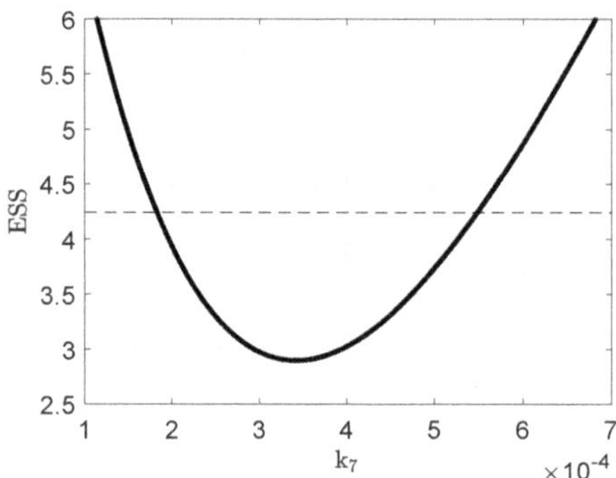

Figure A9. Profile likelihood of k_7. Horizontal dashed line represents the ESS value of the 95% confidence interval.

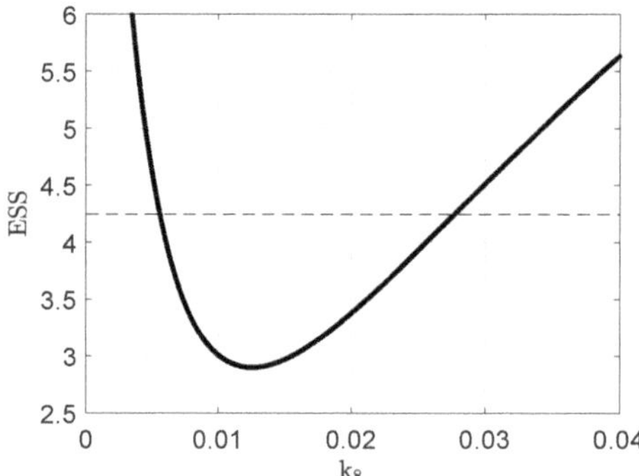

Figure A10. Profile likelihood of k_8. Horizontal dashed line represents the ESS value of the 95% confidence interval.

Appendix C. Parameter Values and Initial Conditions

Table A1. Kinetic parameters, states and variables.

Parameter	Unit	Description	Value
fitted			
k_1	$(L^2/(g^2 h))$	consumption of fructose and ammonium for growth	4.34×10^{-6}
k_2	$(L^2/(g^2 h))$	consumption of propionic acid and ammonium for growth	0.0048
k_3	$(L^2/(g^2 h))$	HA consumption	0.0713
k_4	$(L/(g\ h))$	consumption of fructose for 3HB accumulation	0.0563
k_5	$(L/(g\ h))$	consumption of propionic acid for 3HB accumulation	0.6803
k_6	$(L/(g\ h))$	consumption of propionic acid for 3HV accumulation	2.0208
k_7	$(L^2/(g^2 h))$	fructose consumption for maintenance	3.4184×10^{-4}
k_8	$(L^2/(g^2 h))$	propionic acid consumption for maintenance	0.0125

Table A1. *Cont.*

Parameter	Unit	Description	Value
fixed			
$c_{p,inh}$	(g/L)	inhibitory propionic acid concentration	1.5 [33]
$c_{n,sw}$	(g/L)	Michalis–Menten rate for ammonium	0.2
$c_{p,in}$	(g/L)	propionic acid concentration in the feed	20
$c_{fru}(0)$	(g/L)	initial fructose concentration	
		data set 1/data set 2	21.96/21.75
$c_p(0)$	(g/L)	initial propionic acid concentration	
		data set 1/data set 2	0/0.48
$c_n(0)$	(g/L)	initial ammonium concentration	
		data set 1/data set 2	1.74/1.40
$c_{res}(0)$	(g/L)	initial residual biomass concentration	
		data set 1/ data set 2	1.47 / 1.16
$c_{hb}(0)$	(g/L)	initial 3HB concentration	
		data set 1/ data set 2	0.03 / 0.03
$c_{hv}(0)$	(g/L)	initial 3HV concentration	
		data set 1/ data set 2	0 / 0.01
P_t	(g/L)	HA (3HB+3HV) concentration	time dependent
$CO_{2,out}$	(%)	exhaust CO_2	measurement
$CO_{2,in}$	(%)	inlet CO_2	measurement
b_{CO_2}	(-)	CO_2 dependent metabolic activity	time dependent
D	(-)	dilution rate	time dependent
F_{in}	(%)	feed rate for propionic acid	pH controlled [33]
$P_{t,max}$	(g/L)	maximum concentration of PHA	0.89 [37]

References

1. Tan, G.Y.A.; Chen, C.L.; Li, L.; Ge, L.; Wang, L.; Razaad, I.M.N.; Li, Y.; Zhao, L.; Mo, Y.; Wang, J.Y. Start a research on biopolymer polyhydroxyalkanoate (PHA): A review. *Polymers* **2014**, *6*, 706–754. [CrossRef]
2. Koller, M. A review on established and emerging fermentation schemes for microbial production of polyhydroxyalkanoate (PHA) biopolyesters. *Fermentation* **2018**, *4*, 30. [CrossRef]
3. Bugnicourt, E.; Cinelli, P.; Lazzeri, A.; Alvarez, V. Polyhydroxyalkanoate (PHA): Review of synthesis, characteristics, processing and potential applications in packaging. *Express Polym. Lett.* **2014**, *8*, 791–808. [CrossRef]
4. Nielsen, C.; Rahman, A.; Rehman, A.U.; Walsh, M.K.; Miller, C.D. Minireview Food waste conversion to microbial polyhydroxyalkanoates. *Microb. Biotechnol.* **2017**, *10*, 1338–1352. [CrossRef] [PubMed]
5. Koller, M.; Atli, A.; Dias, M.; Reiterer, A.; Braunegg, G. *Plastics from Bacteria—Microbial PHA Production from Waste Raw Materials*; Springer: Berlin/Heidelberg, Germany, 2010; Volume 14, pp. 85–119. [CrossRef]
6. Riedel, S.L.; Jahns, S.; Koenig, S.; Bock, M.C.E.; Brigham, C.J.; Bader, J.; Stahl, U. Polyhydroxyalkanoates production with *Ralstonia eutropha* from low quality waste animal fats. *J. Biotechnol.* **2015**, *214*, 119–127. [CrossRef] [PubMed]
7. Brigham, C.J.; Riedel, S.L. The potential of polyhydroxyalkanoate production from food wastes. *Appl. Food Biotechnol.* **2019**, *6*, 7–18. [CrossRef]
8. Garcia-Gonzalez, L.; De Wever, H. Acetic acid as an indirect sink of CO_2 for the synthesis of polyhydroxyalkanoates (PHA): Comparison with PHA production processes directly using CO_2 as feedstock. *Appl. Sci.* **2018**, *8*, 1416. [CrossRef]
9. Jia, Q.; Xiong, H.; Wang, H.; Shi, H.; Sheng, X.; Sun, R.; Chen, G. Production of polyhydroxyalkanoates (PHA) by bacterial consortium from excess sludge fermentation liquid at laboratory and pilot scales. *Bioresour. Technol.* **2014**, *171*, 159–167. [CrossRef]
10. Bengtsson, S.; Karlsson, A.; Alexandersson, T.; Quadri, L.; Hjort, M.; Johansson, P.; Morgan-Sagastume, F.; Anterrieu, S.; Arcos-Hernandez, M.; Karabegovic, L.; et al. A process for polyhydroxyalkanoate (PHA) production from municipal wastewater treatment with biological carbon and nitrogen removal demonstrated at pilot-scale. *New Biotechnol.* **2017**, *35*, 42–53. [CrossRef]
11. Sabapathy, P.C.; Devaraj, S.; Meixner, K.; Anburajan, P.; Kathirvel, P.; Ravikumar, Y.; Zabed, H.M.; Qi, X. Recent developments in Polyhydroxyalkanoates (PHAs) production—A review. *Bioresour. Technol.* **2020**, *306*, 123–132. [CrossRef]
12. Meereboer, K.W.; Misra, M.; Mohanty, A.K. Review of recent advances in the biodegradability of polyhydroxyalkanoate (PHA) bioplastics and their composites. *Green Chem.* **2020**, *22*, 5519–5558. [CrossRef]
13. Gholami, A.; Mohkam, M.; Rasoul-Amini, S.; Younes, G. Industrial production of polyhydroxyalkanoates by bacteria: Opportunities and challenges. *Minerva Biotecnol.* **2016**, *28*, 59–74.
14. Atli, A.; Koller, M.; Scherzer, D. Continuous production of poly ([R]-3-hydroxybutyrate) by *Cupriavidus necator* in a multistage bioreactor cascade. *Biotechnol. Prod. Process. Eng.* **2011**, *91*, 295–304. [CrossRef]

15. Penloglou, G.; Vasileiadou, A.; Chatzidoukas, C.; Kiparissides, C. Model-based intensification of a fed-batch microbial process for the maximization of polyhydroxybutyrate (PHB) production rate. *Bioprocess Biosyst. Eng.* **2017**, *40*, 1247–1260. [CrossRef] [PubMed]
16. Carius, L.; Pohlodek, J.; Morabito, B.; Franz, A.; Mangold, M.; Findeisen, R.; Kienle, A. Model-based State Estimation Based on Hybrid Cybernetic Models. *IFAC-PapersOnLine* **2018**, *51*, 197–202. [CrossRef]
17. Morabito, B.; Kienle, A.; Findeisen, R.; Carius, L. Multi-mode Model Predictive Control and Estimation for Uncertain Biotechnological Processes. *IFAC-PapersOnLine* **2019**, *52*, 709–714. [CrossRef]
18. Novak, M.; Koller, M.; Braunegg, G.; Horvat, P. Mathematical Modelling as a Tool for Optimized PHA Production. *Chem. Biochem. Eng.* **2015**, *29*, 183–220. [CrossRef]
19. Špoljaric´, I.V.; Lopar, M.; Koller, M.; Muhr, A.; Salerno, A.; Reiterer, A.; Malli, K.; Angerer, H.; Strohmeier, K.; Schober, S.; et al. Mathematical modeling of poly[(R)-3-hydroxyalkanoate] synthesis by *Cupriavidus necator* DSM 545 on substrates stemming from biodiesel production. *Bioresour. Technol.* **2013**, *133*, 482–494. [CrossRef] [PubMed]
20. Dias, J.M.; Serafim, L.S.; Lemos, P.C.; Reis, M.A.; Oliveira, R. Mathematical modelling of a mixed culture cultivation process for the production of polyhydroxybutyrate. *Biotechnol. Bioeng.* **2005**, *92*, 209–222. [CrossRef]
21. Dias, J.M.; Oehmen, A.; Serafim, L.S.; Lemos, P.C.; Reis, M.A.; Oliveira, R. Metabolic modelling of polyhydroxyalkanoate copolymers production by mixed microbial cultures. *BMC Syst. Biol.* **2008**, *2*, 1–21. [CrossRef]
22. Duvigneau, S.; Dürr, R.; Carius, L.; Kienle, A. Hybrid Cybernetic Modeling of Polyhydroxyalkanoate Production in *Cupriavidus necator* using Fructose and Acetate as Substrates. *IFAC-PapersOnLine* **2020**, *53*, 16872–16877. [CrossRef]
23. Franz, A.; Song, H.S.; Ramkrishna, D.; Kienle, A. Experimental and theoretical analysis of poly(beta-hydroxybutyrate) formation and consumption in *Ralstonia eutropha*. *Biochem. Eng. J.* **2011**, *55*, 49–58. [CrossRef]
24. Katoh, T.; Yuguchi, D.; Yoshii, H.; Shi, H.D.; Shimizu, K. Dynamics and modeling on fermentative production of poly(β-hydroxybutyric acid) from sugars via lactate by a mixed culture of *Lactobacillus delbrueckii* and *Alcaligenes eutrophus*. *J. Biotechnol.* **1999**, *67*, 113–134. [CrossRef]
25. Duvigneau, S.; Dürr, R.; Kranert, L.; Wilisch-Neumann, A.; Carius, L.; Findeisen, R.; Kienle, A. Hybrid Cybernetic Modeling of the Microbial Production of Polyhydroxyalkanoates Using Two Carbon Sources. *Comput. Aided Chem. Eng.* **2021**, *50*, 1969–1974. [CrossRef]
26. Franz, A.; Dürr, R.; Kienle, A. Population balance modeling of biopolymer production in cellular systems. *IFAC Proc. Vol.* **2014**, *47*, 1705–1710. [CrossRef]
27. Dürr, R.; Franz, A.; Kienle, A. Combination of limited measurement information and multidimensional population balance models. *IFAC-PapersOnLine* **2015**, *48*, 261–266. [CrossRef]
28. Strong, P.; Laycock, B.; Mahamud, S.; Jensen, P.; Lant, P.; Tyson, G.; Pratt, S. The Opportunity for High-Performance Biomaterials from Methane. *Microorganisms* **2016**, *4*, 11. [CrossRef]
29. Jiang, Y.; Hebly, M.; Kleerebezem, R.; Muyzer, G.; van Loosdrecht, M.C. Metabolic modeling of mixed substrate uptake for polyhydroxyalkanoate (PHA) production. *Water Res.* **2011**, *45*, 1309–1321. [CrossRef]
30. García, C.; Alcaraz, W.; Acosta-Cárdenas, A.; Ochoa, S. Application of process system engineering tools to the fed-batch production of poly(3-hydroxybutyrate-co-3-hydroxyvalerate) from a vinasses–molasses Mixture. *Bioprocess Biosyst. Eng.* **2019**, *42*, 1023–1037. [CrossRef]
31. Ochoa, S.; García, C.; Alcaraz, W. Real-time optimization and control for polyhydroxybutyrate fed-batch production at pilot plant scale. *J. Chem. Technol. Biotechnol.* **2020**, *95*, 3221–3231. [CrossRef]
32. Dürr, R.; Duvigneau, S.; Kienle, A. Microbial Production of Polyhydroxyalkanoates—Modeling of Chain Length Distribution. *Comput. Aided Chem. Eng.* **2021**, *50*, 1975–1981. [CrossRef]
33. Kim, J.H.; Kim, B.G.; Choi, C.Y. Effect of propionic acid on poly (β-hydroxybutyric-co-β-hydroxyvaleric) acid production by *Alcaligenes eutrophus*. *Biotechnol. Lett.* **1992**, *14*, 903–906. [CrossRef]
34. Duvigneau, S.; Kettner, A.; Carius, L.; Griehl, C.; Findeisen, R.; Kienle, A. Fast , inexpensive , and reliable HPLC method to determine monomer fractions in poly (3-hydroxybutyrate-co-3-hydroxyvalerate). *Appl. Microbiol. Biotechnol.* **2021**, *105*, 4743–4749. [CrossRef] [PubMed]
35. Satoh, H.; Sakamoto, T.; Kuroki, Y.; Kudo, Y.; Mino, T. Application of the Alkaline-Digestion-HPLC Method to the Rapid Determination of Polyhydroxyalkanoate in Activated Sludge. *J. Water Environ. Technol.* **2016**, *14*, 411–421. [CrossRef]
36. Koller, M.; Horvat, P.; Hesse, P.; Bona, R.; Kutschera, C.; Atlić, A.; Braunegg, G. Assessment of formal and low structured kinetic modeling of polyhydroxyalkanoate synthesis from complex substrates. *Bioprocess Biosyst. Eng.* **2006**, *29*, 367–377. [CrossRef] [PubMed]
37. Mohidin Batcha, A.F.; Prasad, D.M.; Khan, M.R.; Abdullah, H. Biosynthesis of poly(3-hydroxybutyrate) (PHB) by *Cupriavidus necator* H16 from jatropha oil as carbon source. *Bioprocess Biosyst. Eng.* **2014**, *37*, 943–951. [CrossRef] [PubMed]
38. Storn, R.; Price, K. Differential Evolution—A Simple and Efficient Heuristic for global Optimization over Continuous Spaces. *J. Glob. Optim.* **1997**, *11*, 341–359. [CrossRef]
39. Raue, A.; Kreutz, C.; Maiwald, T.; Bachmann, J.; Schilling, M.; Klingmüller, U.; Timmer, J. Structural and practical identifiability analysis of partially observed dynamical models by exploiting the profile likelihood. *Bioinformatics* **2009**, *25*, 1923–1929. [CrossRef] [PubMed]

40. Shang, L.; Jiang, M.; Ryu, C.H.; Chang, H.N.; Cho, S.H.; Lee, J.W. Inhibitory effect of carbon dioxide on the fed-batch culture of *Ralstonia eutropha*: Evaluation by CO_2 pulse injection and autogenous CO_2 methods. *Biotechnol. Bioeng.* **2003**, *83*, 312–320. [CrossRef]
41. Park, J.M.; Kim, T.Y.; Lee, S.Y. Genome-scale reconstruction and in silico analysis of the *Ralstonia eutropha* H16 for polyhydroxyalkanoate synthesis, lithoautotrophic growth, and 2-methyl citric acid production. *BMC Syst. Biol.* **2011**, *5*, 101. [CrossRef]

Article

Dynamic Model for Biomass and Proteins Production by Three *Bacillus Thuringiensis* ssp *Kurstaki* Strains

Tatiana Segura Monroy [1], Nouha Abdelmalek [2], Souad Rouis [3], Mireille Kallassy [4], Jihane Saad [4,5], Joanna Abboud [4,5], Julien Cescut [5], Nadia Bensaid [2], Luc Fillaudeau [1] and César Arturo Aceves-Lara [1,*]

[1] Toulouse Biotechnology Institute, Bio & Chemical Engineering, Université de Toulouse, F-31077 Toulouse, France; seguramo@insa-toulouse.fr (T.S.M.); luc.fillaudeau@insa-toulouse.fr (L.F.),
[2] Laboratoires Pharmaceutiques MédiS, B.P 206 Nabeul 8000, Tunisia; nouha_abdelmalek@yahoo.com (N.A.); nadia.bensaid@labomedis.com (N.B.)
[3] Centre of Biotechnology of Sfax, B.P 1177 Sfax 3018, Tunisia; souad.rouis@cbs.rnrt.tn
[4] Faculty of Sciences, Saint-Joseph University, Riad El Solh, Beirut 1004 2020, Lebanon; mireille.kallassy@usj.edu.lb (M.K.); jihane.saad@inrae.fr (J.S.); joanna.abboud-1@inrae.fr (J.A.)
[5] Toulouse White Biotechnology (UMS INRAE/INSA/CNRS), 135 Avenue de Rangueil, F-31077 Toulouse, France; Julien.Cescut@inrae.fr
* Correspondence: seguramo@etud.insa-toulouse.fr

Abstract: *Bacillus thuringiensis* is a microorganism used for the production of biopesticides worldwide. In the present paper, different kinetic models were analyzed to study and compare three different strains of *Bt* ssp *kurstaki* (LIP, BLB1, and HD1). Bioperformances (vegetative cell, spore, substrate, and protein) and successive culture phases (oxidative growth, limitation and sporulation, and protein release) were depicted with an overarching aim to estimate total protein productivity, yield, and titer. In the end, two models were calibrated using experimental dataset (11 batches culture in 3 L bioreactor with semisynthetic medium), subsequently validated, and statistically compared. Both models satisfactorily followed the dynamics of the experimental data. Finally, a dynamic model was selected following the Akaike information criterion (AIC).

Keywords: *B. thuringiensis kurstaki*; biopesticides; kinetic parameters; dynamic model

Citation: Monroy, T.S.; Abdelmalek, N.; Rouis, S.; Kallassy, M.; Saad, J.; Abboud, J.; Cescut, J.; Bensaid, N.; Fillaudeau, L.; Aceves-Lara, C.A. Dynamic Model for Biomass and Proteins Production by Three *Bacillus Thuringiensis* ssp *Kurstaki* Strains. *Processes* **2021**, *9*, 2147. https://doi.org/10.3390/pr9122147

Academic Editor: Philippe Bogaerts

Received: 27 October 2021
Accepted: 24 November 2021
Published: 28 November 2021

Publisher's Note: MDPI stays neutral with regard to jurisdictional claims in published maps and institutional affiliations.

1. Introduction

B. thuringiensis is a facultative anaerobic Gram-positive sporulating bacterium, frequently used in the production of some biopesticides and as a source of genes for transgenic expression in plants [1]. It usually inhabits different environments, among which soil, settled dust, insects, water, and others have been identified [2]. *B. thuringiensis* has been shown to be toxic to various organisms, such as lepidopterans, coleopterans, dipterans, or nematodes, but is considered safe for mammals. Thus, the products based on *B. thuringiensis* (*Bt*) provide effective and environmentally benign control of several insects in agricultural, forestry, and disease-vector applications [3]. This insecticidal activity is mainly due to the production of some intracellular inclusions (called σ-endotoxins) during the sporulation phase of *B. thuringiensis* cells.

Most of the biopesticides distributed in the world are mainly based on *Btk.* HD1 strain. However, two recent strains, identified as *Btk.* LIP (from Lebanese soil), and BLB1 (from Tunisian soil), have been isolated and described to be more efficient than HD1 [4], and, therefore, will be studied in this work.

Due to the several changes of cell physiology during the σ-endotoxins production bioprocess (exponential growth, formation of inclusion, formation of spore, lysis), *B. thuringiensis* culture is considered a laborious process. Although one possibility to optimize *B. thuringiensis* culture is through mathematical models, there are not many mathematical models that describe the dynamics of the growth phases of *B. thuringiensis* culture [5]. Holmberg and Sievanen [6] proposed a model based on Monod kinetics to describe the relationship between cell growth and toxin production. Later, Rivera et al. (1999) [7] showed the Monod model limitations when they tried to describe the biomass diversity of *B. thuringiensis*. In their model, they assumed the presence of two kind of cells in the *B. thuringiensis* culture; those available to multiply, and those that have become spores. Thus, they divided them between the biomass of the vegetative cells and the biomass of the spore-forming cells. In addition, they used the Monod model to describe the relationship between vegetative cell growth and substrate concentration, as did Holmberg and Sievanen [6].

Furthermore, Popovic et al. [8] proposed a model that considered a minimum level of poly-β-hydroxy butyric acid (PHB) required in cells at the beginning of sporulation for efficient sporulation and endotoxins productions. Additionally, they used Contois kinetics to describe the growth of the cells, considering that this model fits better to the experimental data than the expression of Monod.

Therefore, this work proposes a dynamic model for *B. thuringiensis*. Section 2 describes experimental data and the dynamic model for *B. thuringiensis*. The simulation results and the performance evaluation are shown in Section 3. Finally, conclusions and perspectives close this paper.

2. Materials and Methods

The materials and methods used to generate the set of experimental data are described in this section. Then, experimental data are introduced, and assumption and formulation of models are explained.

2.1. Microorganism and Culture Media

Three *B. thuringiensis* ssp. *kurstaki* strains were used in the present work: a Lebanese strain LIP [4], a Tunisian strain BLB1 [9], and HD1 strain used as a reference (industrial gold standard) [10]. Luria broth (LB) medium was used for inoculum production, whereas a semisynthetic medium (SSM), defined by Sarrafzadeh et al. [11], was used for fermentation assays. Their compositions ($g \cdot L^{-1}$) are described in Table 1. For the SSM, concentrated glucose (Sol 2) and all salts solutions (Sol 3–5) were prepared and sterilized separately and added before inoculation to the rest of the medium (Sol 1) previously sterilized.

Table 1. Semisynthetic and Luria broth media composition ($g \cdot L^{-1}$).

Sol	Components	Semisynthetic (SSM)	LB
	Peptone	-	10
	NaCl	-	5
	Yeast Extract	0.5	5
	Casein acid hydrolysate	4.5	-
1	$(NH_4)_2SO_4$	6	-
	K_2HPO_4	1.4	-
	KH_2PO_4	1.4	-
2	Glucose	5	-
3	$MgSO_4, 7H_2O$	0.61	-
4	$CaCl_2, 2H_2O$	0.332	-
5	$MnSO_4, H_2O$	0.006	-

2.2. Inoculum Preparation

Inocula were prepared by transferring cells from nutrient agar slants into 10 mL of LB medium and incubated overnight at 30 °C in a rotary shaker set at 200–230 rpm. Aliquots corresponding to an initial OD600 = 0.15 were used to inoculate 1 L Erlenmeyer flasks containing 100 mL LB medium. After 10–12 h of incubation at 30 °C, in a rotary shaker set at 200–230 rpm, the OD600 was determined. The culture broth was used to inoculate the bioreactor containing the studied media to start with an initial OD600 of 0.15.

2.3. Culture Conditions

Several fermentations were conducted over 48 h in batch mode at 30 °C in a 3 L Biostat B plus fermenter (Sartorius, Göttingen, Germany) containing 1.8 L of the SSM medium and with continuous regulation of pH at 6.8 using 1 M H_2SO_4 and 3 M NaOH. Dissolved oxygen was continuously monitored by an optical oxygen sensor and maintained at 25% pO2-sat with a constant aeration rate (VVM = 10 with Qair = 0.18 min·L^{-1}) and variable stirring (from 250 to 1200 rpm). Foaming was controlled using an antifoam (Emultrol DFM DV-14 FG), through the fermentation process.

2.4. Analysis and Sampling Strategy

Several samples were collected from the *Bt* broth during experiments, and substrate, biomass, and product analyses (glucose, cell and spore counting, protein) were conducted to determine biokinetics.

2.4.1. Detection of Sporulation

The diverse cell states were distinguished, during the fermentation process, based on their morphological differences and the refractile nature of the endospores, using a phase contrast microscope (ZeissPrima Pro, Paris, France, ×100 oil).

2.4.2. Biomass Analyses

Cells and Spores Counts

The follow-up of the biopesticides production was checked by estimating viable cell counts (VC) and spore counts (SC) by plate counts. To determine VC and SC, the withdrawn samples were serially diluted, spread on LB plates and incubated at 30 °C for 16–18 h. For determining SC, the appropriately diluted samples were heated at 85 °C for 15 min and cooled for 5 min before spreading onto LB plates. Number of colonies should be between 20 and 300 to be acceptable. All analyses were realized in triplicate.

Cell Dry Weight

A known amount of sample (1 to 20 mL) was filtered via nitrocellulose membrane (0.2 µm) and the membrane was then dried at 70 °C (24 h). Biomass dry weight is determined by differential weighing of the filter before the filtration and after filtration and drying.

Quantification of Proteins Production

In order to estimate the concentration of total proteins (mainly composed of δ-endotoxin) produced during the fermentation, 1 mL sample was centrifuged at 13,000 rpm for 5 min at 4 °C. The supernatant was collected for other analysis and the pellet was washed twice with cold NaCl 1M and four times with cold water. The protein crystal was then dissolved in 0.05 N NaOH for 2 to 3 h at 30 °C in a rotary shaker (200 rpm). The suspension was then centrifuged at 13,000 rpm for 5 min, and the supernatant containing the solubilized proteins was recuperated.

The concentration of the proteins in this supernatant was determined by Bradford assay [12] using bovine serum albumin (BSA) as a protein standard. Absorbances were measured after 10 min at 595 nm (2300 EnSpire Multilabel Plate Reader). The obtained value was the average of three measures of the same sample (microwell plate). Considering our protocol, protein concentration estimates the total protein production after separation but not specifically the δ- endotoxin, even if it is the dominant fraction.

Sugar Analysis

The sugar concentrations were determined using HPLC-UV. The HPLC assays were performed using an Ultimate 3000 RSLC/MWD/RI/CAD. A mobile phase of 5 mM H_2SO_4 with a flow rate of 0.6 mL·min^{-1} was used. The mobile phase was filtered and degassed through a 0.2 µm cellulose nitrate membrane. The samples and standards were also filtered before injection into the HPLC.

2.5. Experimental Data

Between three to four batch cultures per strain were carried out (Table 2). Two batches per strain were used to perform parameter calibration, and between one and two batches were used to validate the models. The experimental datasets for each strain are presented in Figures 1–3. It is relevant to indicate that in batch 07, dry matter measurement was estimated by OD600nm for exponential growth phase.

Table 2. Batches culture carried out per strain.

Strain	Batch
Btk. HD1	B03, B04, B07
Btk. BLB1	B01, B02, B05, B06
Btk. LIP	B08, B09, B10

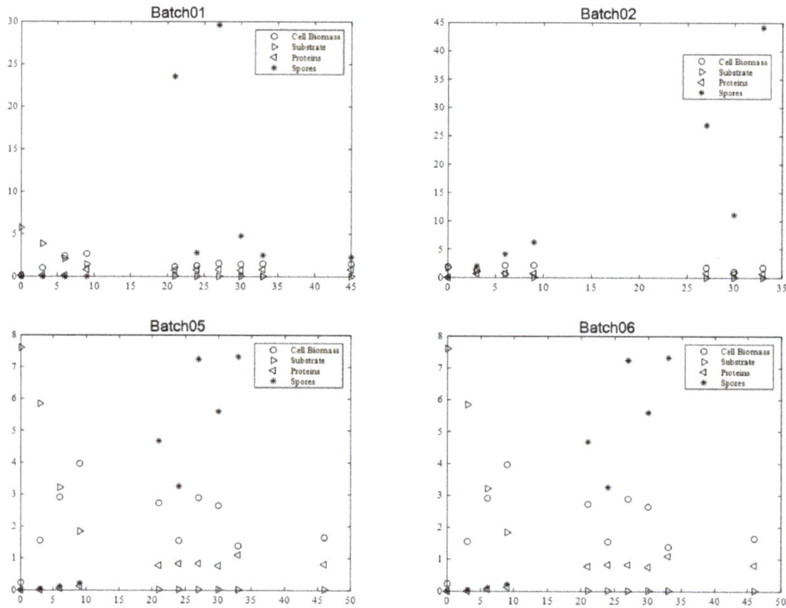

Figure 1. Evolution of cell biomass, spores, substrate, and protein content as a function of time with BLB1 strain (batches: 01, 02, 05, and 06).

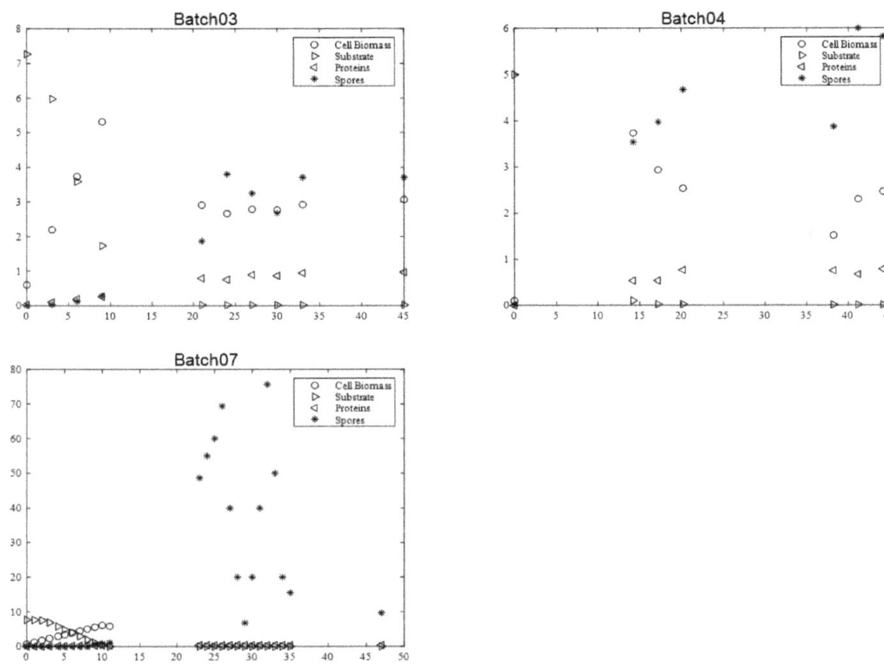

Figure 2. Evolution of cell biomass, spores, substrate, and protein content as a function of time with HD1 strain (batches: 03, 04, and 07).

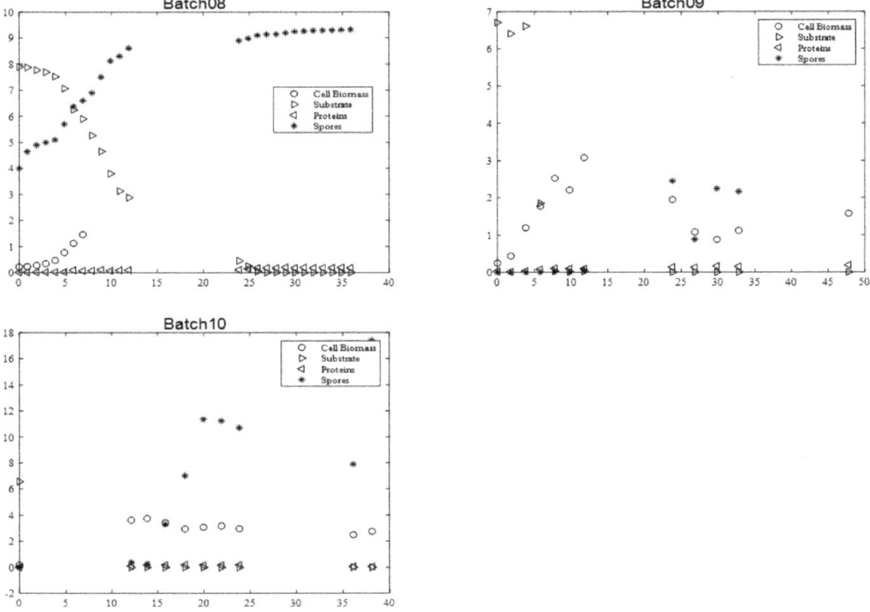

Figure 3. Evolution of cell biomass, spores, substrate, and protein content as a function of time with LIP strain (batches: 08, 09, and 10).

During all cultures, several milestones should be identified: (i) the maximum cell biomass production, (ii) substrate depletion, (iii) full sporulation informing about proteins production, and (iv) its release in supernatant due to full cell lysis.

2.5.1. *Btk.* BLB1 Strain

For the BLB1 strain, data were taken from four batches (01, 02, 05, and 06). Figure 1 presents the graphical experimental data. Batch 11 was used to validate the model.

The maximum biomass concentration was reached after approximately 10 h culture for the four batches, and then began to decrease. As expected, this time approximately coincided with the substrate depletion, corresponding to a glucose concentration close to $0 \, g \cdot L^{-1}$. In addition, the concentration of spores began to increase at this moment, since the limitation of the substrate induced their formation. After 20 h, sporulation reached a plateau value. Cell lysis was fully achieved after 30 h, as indicated by protein release into supernatant. Finally, protein content reached around $0.8–1 \, g \cdot L^{-1}$ with BLB1 stain.

2.5.2. *Btk.* HD1 Strain

Similar data and milestones were obtained for the HD1 strain (Figure 2). Around 10 h, maximum biomass concentration and substrate depletion were reached. However, in batch 07, no values were recorded for the biomass concentration after exponential growth phase due to technical misplaced measurements. After 20 h, sporulation rate was achieved, and protein content plateaued after 30 h. Final protein content was around 0.8–1 g/L, except for batch 07 (0.4 g/L).

2.5.3. *Btk.* LIP Strain

Figure 3 presents identical variables and leads to identify the same milestones and critical time as described above with *Btk.* BLB1 and HD1 strains. LIP strain exhibited the lowest protein concentration, close to 0.2 g/L. This result could be explained by a lower cell lysis rate; therefore, protein crystals were not released in the supernatant.

2.5.4. Model Assumptions

The main features of dynamic models include the key parameters describing bioperformances (vegetative cell, spores, substrate, proteins) and associated kinetics, considering successive phases (oxidative growth, limitation, sporulation, and protein release), during bioproduction. The mass balance equations on each compound are shown in Equations (1), (2), (4)–(7). Equation (1) represents the evolution of biomass concentration with respect to time, while the relationship between bacterial growth and substrate consumption is shown in Equations (1) and (2).

$$\frac{dX}{dt} = \mu * X - k_d * X \tag{1}$$

$$\frac{dS}{dt} = -\frac{\mu * X}{Y1} \tag{2}$$

where X is the biomass concentration $(g \cdot L^{-1})$, S is the concentration of substrate $(g \cdot L^{-1})$, Y1 is the yield coefficient between the biomass and the substrate (gBiomass/gGlucose), and k_d is the death rate (h^{-1}).

The cell growth process is represented by the Contois expression, as follows:

$$\mu = \mu_{max} \frac{S}{(Kc * X) + S} \tag{3}$$

where μ is the specific growth rate (h^{-1}) and μ_{max} is the maximum specific growth rate (h^{-1}), a constant defined for a substrate concentration; X1 is the concentration of biomass $(g \cdot L^{-1})$; S1 is the concentration of glucose $(g \cdot L^{-1})$; and Kc is a saturation constant.

Equations (4)–(7) show the mass balance for proteins and spores. In the first model, proteins and spores are correlated with biomass (Model 1). In the second model, α and β parameters were used to unassociated proteins and spores from biomass production (Model 2). The corresponding models are shown in Equations (4) and (5) (Model 1) and Equations (6) and (7) (Model 2).

Model 1

$$\frac{dPro}{dt} = \frac{X1*Y2}{Y1} \tag{4}$$

$$\frac{dSpo}{dt} = \frac{X1*Y3}{Y1} \tag{5}$$

Model 2

$$\frac{dPro}{dt} = \frac{X1*Y2}{Y1} + \alpha \tag{6}$$

$$\frac{dSpo}{dt} = \frac{X1*Y3}{Y1} + \beta \tag{7}$$

where Pro is the protein concentration ($g \cdot L^{-1}$), Spo is the spores concentration ($CFU \cdot 10^{-8}/mL$), Y2 ($gPro \cdot gGlucose^{-1} \cdot h^{-1}$) and Y3 ($CFU \cdot 10^{-5} \cdot g\ Glucose^{-1} \cdot h^{-1}$) are yield coefficients, and $\alpha(g \cdot L^{-1} \cdot h^{-1})$ is a constant.

The set of equations were simulated using MATLAB®(R2019a).

Three statistical criteria were used to analyze the fit of the models using experimental datasets. These parameters were the coefficient of determination (R2), the root mean square errors (RSME), and the correction of Akaike information criterion (AICc). The expressions of these parameters are presented in Equations (8)–(11), respectively.

$$R^2 = \frac{SSR}{SST} \tag{8}$$

SSR is sum of squared regression, and SST is sum of squared total.

$$RMSE = \frac{(X - \overline{X})^T W (X - \overline{X})}{n - p} \tag{9}$$

where n represents the number of data, p the number of parameters, W the weighting matrix, and X and $\overline{X}$ are the data and estimated data, respectively [13].

$$AICc = AIC + \frac{2p(p+1)}{n - p - 1} \tag{10}$$

$$AIC = 2p + n(\ln(2\pi) + \ln(SSE) - \ln(n) + 1) \tag{11}$$

The main parameter used to determine the model that best fits the data is the AICc parameter [13]. The AIC criterion is one of the most popular for the comparison of models because it considers the number of parameters, the number of data, and the residuals, making it a parameter that balances the complexity of the model and the fit of the data [14]. Additionally, the parameter correction (AICc) gives accurate results for a larger number of datasets. Thus, the model with the lowest value for AICc is selected to represent the experimental data more adequately.

Parameter calibration was carried out in MATLAB using a particle swarm optimization (PSO) algorithm. This method, as its name implies, is inspired by the behavior of swarms of insects in nature. Thus, for a set of variables to be optimized, the method begins by placing random particles in the search space, but then a series of rules are established considering each parameter and the set of parameters ("swarm") globally. Thus, the variables are optimized quite well, and few computational resources are spent, becoming a fast method in convergence, and simple in application [15].

3. Results

This section presents the results obtained through various simulations carried out in MATLAB. It is divided into two main sections: model calibration and model validation.

Initially, the parameters of each model were estimated (six for Model 1 and eight for Model 2) with the experimental data of two batches for each strain. Subsequently, the parameters found were used to observe the behavior of the four state variables (biomass, glucose, proteins, and spores concentrations) in two batches for the BLB1 strain and one batch for the HD1 and LIP strains. Likewise, the results obtained were compared with the experimental data. In this way, the validation of the parameters found in the calibration phase was carried out. Moreover, to compare models 1 and 2, a series of statistical parameters were calculated from which the selection of the model that best fits the experimental data is facilitated.

3.1. Model Calibration BLB1 Strain

Kinetics parameters of the *B. thuringiensis* culture were calibrated for three strains: BLB1 (Table 3), HD1 (Table 5), and LIP (Table 7). The maximum specific growth rate (μmax) was between 1.15 h$-$1 for the BLB1 strain (Model 1) and 0.39 h^{-1} for the LIP strain (Model 2). These results are within ranges similar to those reported by Holmberg and Sievanen (1980), who reported values between 1.90 and 0.17 h^{-1} [6], and Atehortúa et al. (2007), who reported values between 0.80 and 0.58 h^{-1} [16]. The death rate (k_d) was between 0.0458 (BLB1 strain) and 0.0184 h^{-1} (LIP strain), which coincided with previous results in the literature (between 0 and 0.13 h^{-1}) [6]. The yield coefficient between the biomass and the substrate (Y1) was between 0.49 (gBiomass·gGlucose^{-1}) and 0.96 (gBiomass·gGlucose^{-1}) for all strains.

Table 3. Optimized parameter values from BLB1 strain.

Parameter	BLB1	
	Model 1	**Model 2**
μ max (h^{-1})	1.1490	1.0720
Kc	4.7450	4.1250
Kd (h^{-1})	0.0437	0.0439
Y1 gBiomass·gGlucose^{-1}	0.7136	0.7141
Y2 (gPro/gGlucose*h)	0.0067	0.0067
Y3 (CFU*10^{-5}/gGlucose*h)	0.0537	0.0524
Alpha (g/L*h)	-	0.0001
Beta (CFU*10^{-5}/L*h)	-	0.0001

The comparison between Models 1 and 2 and the experimental data for the BLB1 strain are shown in Figure 4. According to the figure, Models 1 and 2 did not have very noticeable differences; therefore, the alpha and beta parameters of Model 2 did not have a great impact on the modeling. Both models showed a satisfactory fit to the experimental data; however, the statistical study will give precise information on the best model. It is important to note that the quantification of the spore concentration and protein concentration are more subject to systematic error than biomass and glucose, which may explain the discrepancies between models and measurements.

Model 1 had higher values for μmax and Kc, although other parameters remained with similar values. Furthermore, for the constants alpha and beta, very low values of 0.0001 were obtained, which confirmed that both models are similar.

Table 4 presents the statistical coefficients that allowed comparing Model 1 and Model 2 for BLB1 strain. The statistical coefficients showed a good fit to the experimental data. Spores concentration was the variable with the worst fit. These results are supported by Figure 4, since the furthest experimental values of the two models can be observed in it.

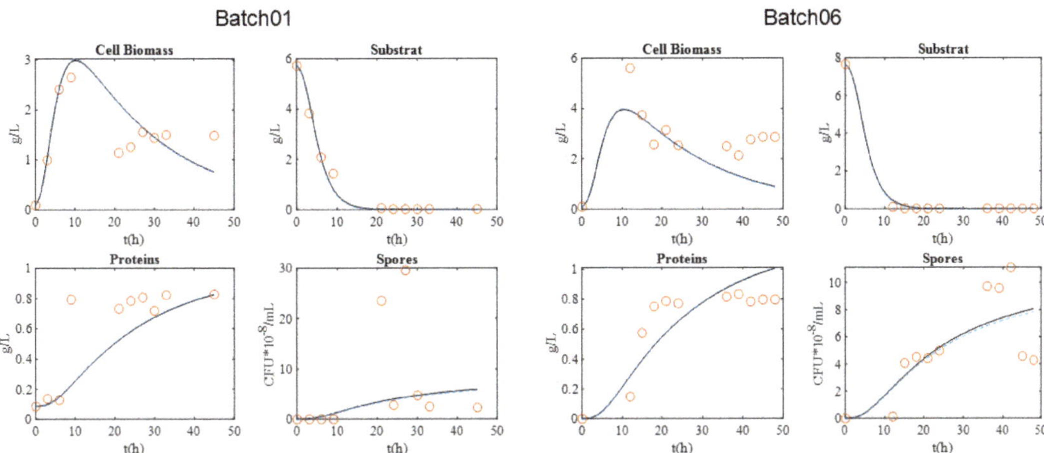

Figure 4. Calibration of the two models with the dataset from BLB1 strain. (o) Experimental data; (-) Model 1, (- -) Model 2.

Table 4. Statistical evaluation of the two models with BLB1 strain.

	Model	Criterion	Biomass	Glucose	Proteins	Spores
Batch01	1	R2	0.6902	0.9863	0.7088	0.1350
		RMSE	0.4539	0.2445	0.2138	10.2994
		AICc	29.5540	17.1797	14.4991	91.9945
	2	R2	0.6935	0.9835	0.7088	0.1357
		RMSE	0.4552	0.2711	0.2125	10.3321
		AICc	149.6104	139.2478	134.3794	212.0580
Batch06	1	R2	0.5663	0.9976	0.7329	0.5756
		RMSE	1.1693	0.1154	0.1633	2.2508
		AICc	41.2816	−9.6689	−2.0279	55.6878
	2	R2	0.5686	0.9981	0.7337	0.5762
		RMSE	1.1699	0.0996	0.1638	2.2597
		AICc	96.2913	42.1035	53.0457	110.7744

As far as the determination coefficient (R2), values greater than 0.98 were observed for glucose measurements, which indicates that this variable has the best fit. However, as mentioned above, the parameter of greatest interest is the AICc since it takes into account several important aspects. The model that presented the lowest AICc values for the four variables was Model 1. This model does not include any extra constants, which makes it a less complex model than Model 2, but also predicts the behavior of the variables studied.

3.2. Model Calibration HD1 Strain

The results for the HD1 strain are shown in Figure 5 and Table 5. It was shown that there are big differences between Model 1 and Model 2. Protein concentration and spore concentration were the variables in which these differences were most visible according to Figure 5, which makes sense since the alpha and beta constants present in Model 2 have a direct influence on these two variables. Additionally, both models presented a very good fit for the biomass and substrate concentration.

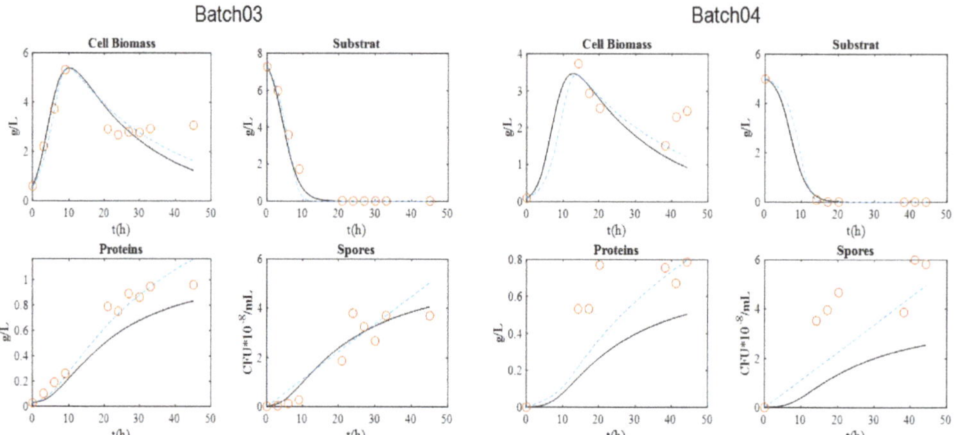

Figure 5. Calibration of the two models with the dataset from HD1 strain. (o) Experimental data; (-) Model 1; (- -) Model 2.

Table 5. Optimized parameter values from HD1 strain.

Parameter	HD1	
	Model 1	**Model 2**
μ max (h^{-1})	0.5985	0.4024
Kc	1.4700	0.5140
Kd (h^{-1})	0.0458	0.0352
Y1 gBiomass·gGlucose^{-1}	0.9612	0.8333
Y2 (gPro/gGlucose*h)	0.0056	0.0050
Y3 (CFU*10^{-5}/gGlucose*h)	0.0281	0.0001
Alpha (g/L*h)	-	0.0062
Beta (CFU*10^{-5}/L*h)	-	0.1116

The optimized parameters showed higher values in Model 1 than in Model 2 for almost all parameters. Although the values of µmax were lower than those found for the BLB1 strain, parameters such as Y1 and alpha and beta constants were higher than those obtained with the BLB1 strain.

The results of the statistical parameters for the calibration of the HD1 strain are presented in Table 6. In a similar way to the BLB1 strain, the variable that best fits the models according to the coefficient of determination was glucose, even reaching a value of 1 for batch 4 and Model 2. In general, the R2 and RMSE coefficients showed a better fit of the models with the experimental data for the HD1 strain than the previously analyzed BLB1 strain. In fact, for HD1 strain, the experimental data of the spore concentration have a better fit than those obtained for the BLB1 strain.

Since for the HD1 strain, Model 1 showed the lowest AICc values, this model was considered as the most appropriate to predict the behavior of the HD1 strain according to the results of the calibration. This means that Model 1 showed the best parsimony.

3.3. Model Calibration LIP Strain

Figure 6 shows the results obtained for the LIP strain. Graphically, Model 1 and Model 2 showed great similarities except for protein concentration. The values obtained for the parameters and the statistical coefficients reflect these differences.

Table 6. Statistical evaluation of the two models with HD1 strain.

	Model	Criterion	Biomass	Glucose	Proteins	Spores
Batch03	1	R2	0.7086	0.9876	0.9763	0.9122
		RMSE	0.7287	0.3705	0.1677	0.4863
		AICc	39.0235	25.4968	9.6408	30.9351
	2	R2	0.7575	0.9803	0.9546	0.8486
		RMSE	0.6305	0.4203	0.0870	0.6881
		AICc	156.1279	148.0185	116.5131	157.8768
Batch04	1	R2	0.7063	0.9992	0.7036	0.73007
		RMSE	0.77276	0.0544	0.3128	2.7302
		AIC	14.6311	−22.5237	1.9751	32.3051
	2	R2	0.7666	1,0000	0.6469	0.7191
		RMSE	0.6203	0.0112	0.2140	1.5365
		AICc	−56.4420	−112.6332	−71.3404	−43.7434

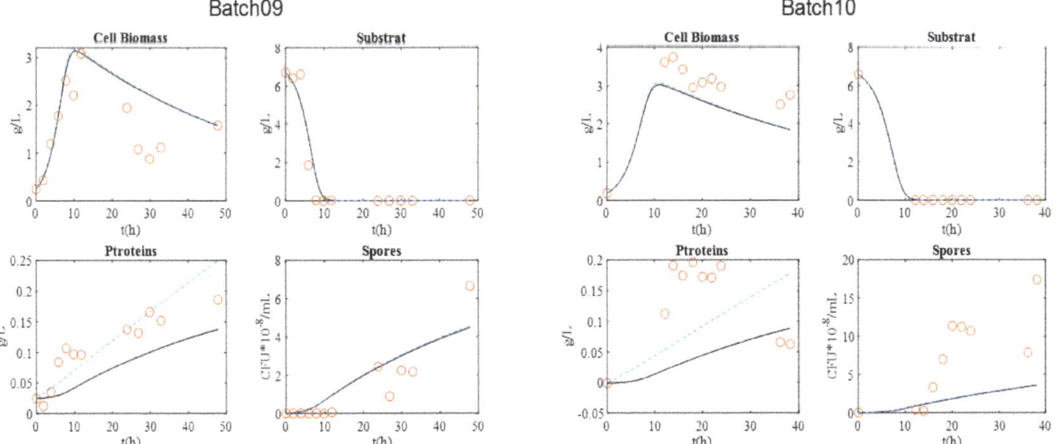

Figure 6. Calibration of the two models with the dataset from LIP strain. (o) Experimental data; (-) Model 1; (- -) Model 2.

Table 7 summarizes the parameter values of both models. It is noteworthy that the proteins/substrate yield coefficient (Y2) showed very low values.

Table 7. Optimized parameter values from LIP strain.

Parameter	LIP	
	Model 1	Model 2
μ max (h^{-1})	0.3966	0.3916
Kc	0.6899	0.5794
Kd (h^{-1})	0.0189	0.0193
Y1 gBiomass·gGlucose^{-1}	0.4866	0.4956
Y2 (gPro/gGlucose*h)	0.0005	0.0001
Y3 (CFU*10^{-5}/gGlucose*h)	0.0213	0.0218
Alpha (g/L*h)	-	0.0042
Beta (CFU*10^{-5}/L*h)	-	0.0002

Moreover, the calibrated parameters showed a higher value for the alpha parameter than the beta one. Therefore, the beta parameter, has a very small value and little influence on Model 2.

Table 8 indicates the values of the statistical parameters for the LIP strain. The statistical parameters showed a good fit of the models, presenting very low values of the determination coefficient only for the protein concentration in batch 10. Glucose continues to be the variable that has the best fits between two models. Additionally, as in the BLB1 and HD1 cases, Model 1 obtained the lowest values for AICc, making it the model that could best predict the behavior of the LIP strain.

Table 8. Statistical evaluation of the two models with LIP strain.

	Model	Criterion	Biomass	Glucose	Proteins	Spores
Batch09	1	R2	0.6610	0.9387	0.8130	0.7747
		RMSE	0.6795	0.7169	0.0472	0.9691
		AICc	23.7638	25.0484	−40.2561	32.2820
	2	R2	0.6653	0.9397	0.8377	0.7731
		RMSE	0.6992	0.7088	0.0289	0.9773
		AICc	59.6479	59.9742	−16.7948	67.6853
Batch10	1	R2	0.9528	1.0000	0.0002	0.6794
		RMSE	0.6082	0.0083	0.1146	7.0190
		AICc	35.4078	−50.4395	2.0181	84.3253
	2	R2	0.9507	1.0000	0.0002	0.6796
		RMSE	0.5819	0.0039	0.0920	6.9800
		AICc	154.5226	54.2584	117.6327	204.2139

3.4. Model Validation

Several datasets of each strain, different from those used in the calibration of the models, were used to validate the results obtained previously. Batch 2 and 5 were used to validate the parameters obtained for the BLB1 strain, the data from batch 7 were used for the HD1 strain, and, finally, batch 8 helped validate the parameters of the LIP strain.

3.4.1. BLB1 Strain

Figure 7 shows the results of the validation for the BLB1 strain, and Table 9 shows the respective statistical coefficients. For both experiments, the calibrated parameters fit the experimental data very well. According to Figure 7, Models 1 and 2 behaved similarly and there were no noticeable differences.

Table 9. Statistical evaluation of the two models with BLB1 strain.

	Model	Criterion	Biomass	Glucose	Proteins	Spores
Batch02	1	R2	0.1518	0.9195	0.3212	0.6722
		RMSE	0.6181	0.2109	0.4249	17.9155
		AIC	11.5082	−3.5438	6.2621	58.6431
	2	R2	0.1484	0.9077	0.3225	0.6716
		RMSE	0.6275	0.2275	0.4229	17,9704
		AICc	−56.2800	−70.4822	−61.8033	−9.3141
Batch05	1	R2	0.7510	0.9828	0.9047	0.9029
		RMSE	0.5389	0.4323	0.1302	1.0764
		AICc	32.9878	28.5794	4.5721	46.8259
	2	R2	0.7524	0.9814	0.9053	0.9029
		RMSE	0.5423	0.4433	0.1289	1.0345
		AICc	153.1159	149.0829	124.3779	166.0305

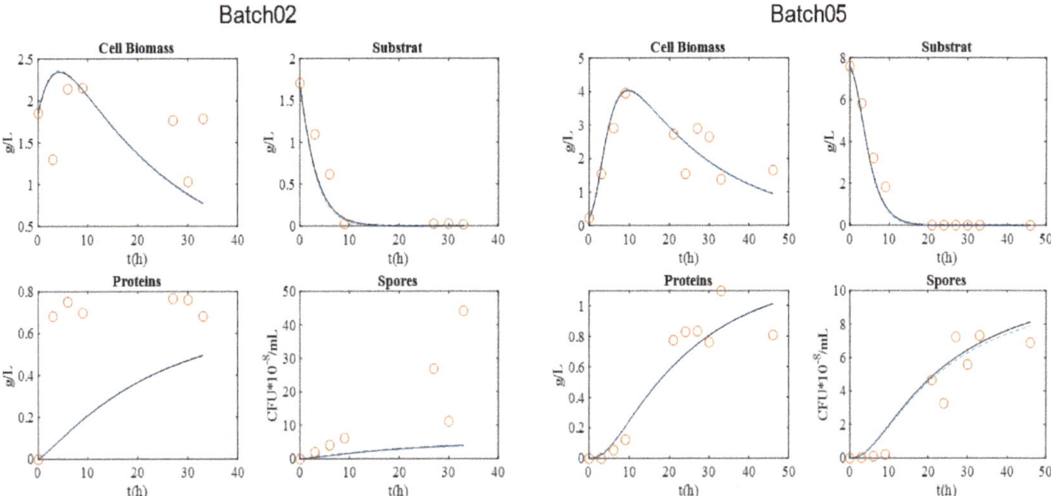

Figure 7. Validation of the two models with the dataset from BLB1 strain. (o) Experimental data; (-) for Model 1; (- -) for Model 2.

Statistical analysis showed that Model 1 fit better than Model 2 because it has the lowest values of the AICc coefficient. As seen graphically, the statistical parameters showed less adjustment for some variables of batch 02 than for batch 05.

3.4.2. HD1 Strain

Data from batch 7 were used for validation of the parameters obtained in HD1 strain calibration. The results are shown in Figure 8 and Table 10. As said before, no values were recorded for the biomass concentration after exponential growth phase due to technical misplaced measurements, which is reflected in Figure 8. However, simulations showed that biomass during exponential growth phase and the other state variables fit adequately. The set of statistical coefficients that express the effectiveness of the models is expressed in Table 10.

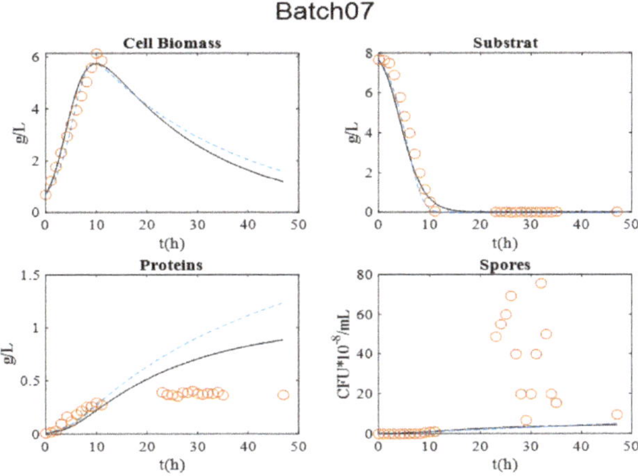

Figure 8. Validation of the two models with the dataset from HD1 strain. (o) Experimental data; (-) for Model 1; (- -) for Model 2.

Table 10. Statistical evaluation of the two models with HD1 strain.

	Model	Criterion	Biomass	Glucose	Proteins	Spores
Batch07	1	R2	NC *	0.9940	0.7991	0.4551
		RMSE	NC *	0.3078	0.2263	30.1202
		AICc	NC *	−55.7705	−71.7791	182.5655
	2	R2	NC *	0.9981	0.7720	0.3628
		RMSE	NC *	0.1337	0.3701	30.1186
		AICc	NC *	−91.0931	−38.1445	190.6123

NC *: Not calculated.

Both models fit quite well for glucose concentration and followed the dynamics of spores concentration. The statistical coefficients showed the worst results for spores. As demonstrated previously, according to the AICc criterion, Model 1 should be the one used to represent the data of the HD1 strain.

3.4.3. LIP Strain

Figure 9 show the results of the validation for the LIP strain. Similar to batch 7 (HD1 strain), batch 8, which corresponds to the LIP strain, showed a biomass measurement problem. However, glucose concentration was well represented by the two models. Although the models followed the dynamics for the concentration in proteins and spores, a slight lag was evident for the spores.

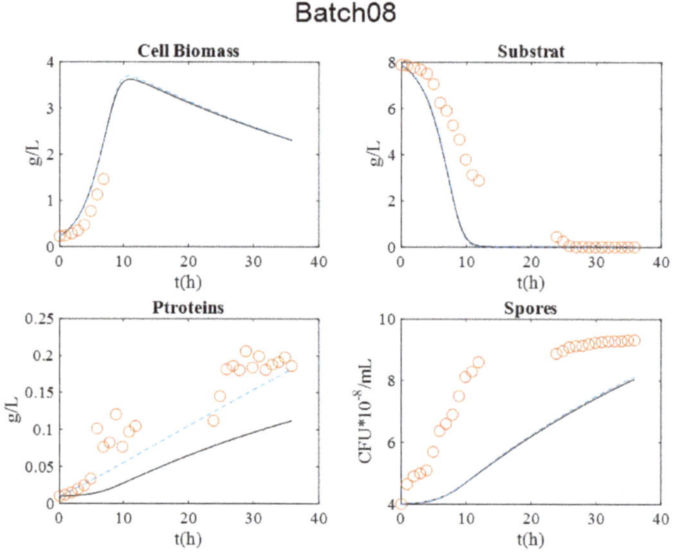

Figure 9. Validation of the two models with the dataset from LIP strain. (o) Experimental data; (-) Model 1; (- -) Model 2.

Table 11 presents results for statistics parameters in LIP validation. As for the other two strains, the glucose data showed the best fit. However, the coefficient of determination for proteins and spores showed a good fit of the models. As in all the cases studied in this report, the most suitable model to predict and represent the experimental data is Model 1, according to the AICc criterion.

Table 11. Statistical evaluation of the two models with LIP strain.

	Model	Criterion	Biomass	Glucose	Proteins	Spores
Batch08	1	R2	NC *	0.9883	0.8321	0.7054
		RMSE	NC *	0.3716	0.0817	2.4370
		AICc	NC *	−45.9879	−124.7646	51.8161
	2	R2	NC *	0.9878	0.8937	0.7042
		RMSE	NC *	0.3813	0.0351	2.4099
		AICc	NC *	−36.5917	−160.6215	59.2837

NC *: Not calculated.

4. Conclusions

The objectives of approach were to model bioperformances (vegetative cell, spore, substrate, and protein) considering different *B. thuringiensis* ssp. *kurstaki* strains and successive culture phases (oxidative growth, limitation and sporulation, protein release). Our overarching aim to estimate total proteins production (mainly composed of δ-endotoxin) was successfully achieved. Initially, the bibliographic research allowed understanding of the context and the different phenomena involved in the study of the *B. thuringiensis* culture, such as the particular life cycle of these microorganisms and the importance of the endotoxins produced.

B. thuringiensis is an important microorganism for the biopesticide market worldwide. The experimental simulations developed in the present study and based on *B. thuringiensis* cultures made it possible to analyze the behavior of the concentration in biomass, substrate, proteins, and spores and adjust two models to the experimental datasets. The calibration of both models allowed to calculate the kinetic parameters of the culture, and the experimental data presented a good fit. Likewise, the models were validated in a satisfactory way.

For the selection of the best model, the AICc criterion was used, which, for all batches, showed better results for Model 1 due to its parsimony. Additionally, although the BLB1 strain showed the highest maximum specific growth rate (μmax), the HD1 strain presented the highest biomass/substrate yield coefficient values (Y1), in opposition to the LIP strain which presented the lowest values for this yield. As for the production of proteins, mainly used for insecticidal toxicity, the BLB1 strain presented the highest concentration and proteins/substrate yield coefficient (Y2), while the LIP strain showed the lowest values for this yield.

Author Contributions: Conceptualization, C.A.A.-L. and L.F.; methodology, T.S.M. and C.A.A.-L.; software, T.S.M. and C.A.A.-L.; validation, T.S.M. and C.A.A.-L.; formal analysis, T.S.M., N.A., S.R., M.K., L.F. and C.A.A.-L.; investigation, T.S.M. and C.A.A.-L.; resources, N.A., S.R., M.K., J.S., J.A., J.C., N.B. and L.F.; data curation, N.A., S.R., M.K., J.S., J.A., J.C., N.B. and L.F.; writing—original draft preparation, T.S.M. and C.A.A.-L.; writing—review and editing, T.S.M., N.A., S.R., M.K., L.F. and C.A.A.-L.; visualization, T.S.M. and C.A.A.-L.; supervision, N.A., S.R., M.K., L.F. and C.A.A.-L.; project administration, S.R. and L.F.; funding acquisition, S.R. and L.F. All authors have read and agreed to the published version of the manuscript.

Funding: This work was conducted under the IPM-4-Citrus project, funded by the European Commission (ID:734921; Period: From 1 April 2017 to 31 March 2021; https://cordis.europa.eu/project/rcn/207633/factsheet/en, accessed on 19 November 2021).

Institutional Review Board Statement: Not applicable.

Informed Consent Statement: Not applicable.

Conflicts of Interest: The authors declare no conflict of interest.

References

1. Schnepf, E.; Crickmore, N.; Van Rie, J.; Lereclus, D.; Baum, J.; Feitelson, J.; Zeigler, D.R.; Dean, D.H. Bacillus thuringiensis and Its Pesticidal Crystal Proteins. *Microbiol. Mol. Biol. Rev.* **1998**, *62*, 775–806. [CrossRef] [PubMed]
2. Iriarte, J.; Porcar, M.; Lecadet, M.-M.; Caballero, P. Isolation and Characterization of Bacillus thuringiensis Strains from Aquatic Environments in Spain. *Curr. Microbiol.* **2000**, *40*, 402–408. [CrossRef] [PubMed]
3. Rowe, G.E.; Margaritis, A. Bioprocess design and economic analysis for the commercial production of environmentally friendly bioinsecticides from *Bacillus thuringiensis* HD-1*kurstaki*. *Biotechnol. Bioeng.* **2004**, *86*, 377–388. [CrossRef] [PubMed]
4. El Khoury, M.; Azzouz, H.; Chavanieu, A.; Abdelmalak, N.; Chopineau, J.; Awad, M.K. Isolation and characterization of a new *Bacillus thuringiensis* strain Lip harboring a new cry1Aa gene highly toxic to Ephestia kuehniella (Lepidoptera: Pyralidae) larvae. *Arch. Microbiol.* **2014**, *196*, 435–444. [CrossRef] [PubMed]
5. Navarro-Mtz, A.K.; Pérez-Guevara, F. Construction of a biodynamic model for Cry protein production studies. *AMB Express* **2014**, *4*, 79. [CrossRef] [PubMed]
6. Holmberg, A.; Sievanen, R. Fermentation of Bacillus thuringiensis for Exotoxin Production: Process Analysis Study. *Biotechnol. Bioeng.* **1980**, *22*, 1707–1724. [CrossRef]
7. Rivera, D.; Margaritis, A.; de Lasa, H. A sporulation kinetic model for batch growth of *B. thuringiensis*. *Can. J. Chem. Eng.* **1999**, *77*, 903–910. [CrossRef]
8. Popovic, M.K.; Liu, W.-M.; Iannotti, E.L.; Bajpai, R.K. A Mathematical Model for Vegetative Growth of *Bacillus Thuringiensis*. *Eng. Life Sci.* **2001**, *1*, 85–90. [CrossRef]
9. Saadaoui, I.; Rouis, S.; Jaoua, S. A new Tunisian strain of Bacillus thuringiensis kurstaki having high insecticidal activity and δ-endotoxin yield. *Arch. Microbiol.* **2009**, *191*, 341–348. [CrossRef] [PubMed]
10. Carozzi, N.B.; Kramer, V.C.; Warren, G.W.; Evola, S.; Koziel, M.G. Prediction of insecticidal activity of *Bacillus thuringiensis* strains by polymerase chain reaction product profiles. *Appl. Environ. Microbiol.* **1991**, *57*, 3057–3061. [CrossRef] [PubMed]
11. Sarrafzadeh, M.H.; Guiraud, J.P.; Lagneau, C.; Gaven, B.; Carron, A.; Navarro, J.-M. Growth, Sporulation, δ-Endotoxins Synthesis, and Toxicity During Culture of *Bacillus thuringiensis* H14. *Curr. Microbiol.* **2005**, *51*, 75–81. [CrossRef] [PubMed]
12. Bradford, O.F. Adaptation of the Bradford protein assay to membrane-bound proteins by solubilizing in glucopyranoside detergents. *Anal. Biochem.* **1987**, *162*, 11–17.
13. Robles-Rodriguez, C.; Bideaux, C.; Gaucel, S.; Laroche, B.; Gorret, N.; Aceves-Lara, C. Reduction of metabolic models by polygons optimization method applied to Bioethanol production with co-substrates. *IFAC Proc. Vol.* **2014**, *47*, 6198–6203. [CrossRef]
14. Burnham, K.P.; Anderson, D.R. Multimodel Inference: Understanding AIC and BIC in Model Selection. *Sociol. Methods Res.* **2004**, *33*, 261–304. [CrossRef]
15. Robles-Rodriguez, C.E.; Bideaux, C.; Roux, G.; Molina-Jouve, C.; Aceves-Lara, C.A. Soft-Sensors for Lipid Fermentation Variables Based on PSO Support Vector Machine (PSO-SVM). In Proceedings of the 13th International Conference on Distributed Computing and Artificial Intelligence, Sevilla, Spain, 1–3 June 2016; pp. 175–183.
16. Atehortúa, P.; Álvarez, H.; Orduz, S. Modeling of growth and sporulation of Bacillus thuringiensis in an intermittent fed batch culture with total cell retention. *Bioprocess Biosyst. Eng.* **2007**, *30*, 447–456. [CrossRef] [PubMed]

*processes*

MDPI

Article

Global Stability Analysis of a Bioreactor Model for Phenol and Cresol Mixture Degradation

Neli Dimitrova [1,*,†] **and Plamena Zlateva** [1,2,†]

1   Institute of Mathematics and Informatics, Bulgarian Academy of Sciences, Acad. G. Bonchev Str. Block 8, 1113 Sofia, Bulgaria
2   Institute of Robotics, Bulgarian Academy of Sciences, Acad. G. Bonchev Str. Block 2, 1113 Sofia, Bulgaria; plamzlateva@abv.bg
*   Correspondence: nelid@math.bas.bg
†   These authors contributed equally to this work.

Abstract: We propose a mathematical model for phenol and p-cresol mixture degradation in a continuously stirred bioreactor. The model is described by three nonlinear ordinary differential equations. The novel idea in the model design is the biomass specific growth rate, known as sum kinetics with interaction parameters (SKIP) and involving inhibition effects. We determine the equilibrium points of the model and study their local asymptotic stability and bifurcations with respect to a practically important parameter. Existence and uniqueness of positive solutions are proved. Global stabilizability of the model dynamics towards equilibrium points is established. The dynamic behavior of the solutions is demonstrated on some numerical examples.

Keywords: mathematical model; continuous bioreactor; biodegradation; phenol and p-cresol mixture; SKIP model; equilibrium points; stability analysis; global stabilizability; numerical simulation

Citation: Dimitrova, N.; Zlateva, P. Global Stability Analysis of a Bioreactor Model for Phenol and Cresol Mixture Degradation. *Processes* **2021**, *9*, 124. https://doi.org/10.3390/pr9010124

Received: 18 November 2020
Accepted: 5 January 2021
Published: 8 January 2021

1. Introduction

Organic chemical mixtures are among the most persistent environmental pollutants. Different aromatic compounds such as phenol, cresols, nitrophenols, benzene, etc. coexist as complex mixtures in wastewaters from petroleum refineries, coal mining and other industrial chemical sources [1]. Biological degradation has recently become a viable technology for remediation of organic pollutants as an alternative to the traditional physical and chemical methods that can be costly and produce hazardous products. Most of the current research has been directed to the isolation and study of microbial species with high-degradation activity and capabilities of degrading chemical compounds. The review paper [2] reports on hundreds of isolated bacteria capable of degrading aromatic compounds, among them different strains of *Aspergillus awamori, Arthrobacter, Burkholderia, Mycobacterium, Pseudomonas, Rhodococcus, Staphylococcus, Trametes hirsute* etc. The biodegradation of one or all chemical components depends on the composition of the particular mixture and the used microorganisms [3–5]. The adequate analysis of interactions between the compounds and their influence on microbial growth is very important for understanding the simultaneous metabolism of phenolic mixtures [6].

Most research on microbial potentials to degrade chemical pollutants has been performed on a laboratory scale. Based on batch processes various mathematical biodegradation kinetic models have been recently developed and widely used. Among them are Monod's, Haldane's (known also as Andrews), sum kinetic models, sum kinetics with interaction parameter (SKIP) models, etc. [7,8]. It is known that Monod's and Haldane's models are appropriate for single substrate utilization. The Monod model describes the biodegradation rate in dependence of the biomass concentration. When a substrate inhibits its own degradation then Haldane's model is more appropriate. In [9] the Haldane equation modified with a Monod-like switching function is proposed and applied to the biological removal of mixtures of phenolic compounds in sequential batch bioreactors. In [10] the aerobic biodegradability of phenol, resorcinol and 5-methylresorcinol and their different two-component mixtures is investigated and various kinetic models are tested to obtain the best curve fit.

In the case when a mixture of two or more substrates occurs, the sum kinetic and SKIP models predict better the outcome of biodegradation experiments. The latter have been proposed for the first time in [11] and widely used by many researchers. The (no-interaction) sum kinetics model for cell growth is usually represented as a sum of the specific growth rates on each substrate, e.g., as a sum of Monod- and/or Haldane-type specific growth rates. These models were evaluated in [12,13] for biodegradation of benzene, toluene and phenol mixtures using *Pseudomonas putida* F1 and *Burkholderia* sp. strain JS150 and found that the interactions between these substrates could not be described by sum kinetics models. On the contrary, the SKIP model predicts better the outcome of the mixed-culture experiments. This is due the fact that the SKIP models extend the sum kinetics models by incorporating interaction parameters to describe more accurately the biodegradation of the chemical mixture.

The biodegradation of benzene, toluene and phenol is studied in [14] by adaptation of *Pseudomonas putida* F1 ATCC 700007. For the substrate mixtures, a SKIP model is used. The latter provides an excellent prediction of the growth kinetics and the interactions between these substrates.

In [15] biodegradation kinetics of different multiple substrate mixtures of mono-aromatic volatile organic carbon (VOCs) such as toluene, ethyl benzene and o-xylene are studied. A general mixed-substrate biodegradation model is developed which can describe the biodegradation kinetics of common industrial VOCs when present as a mixture, incorporating parameters for interaction effects.

The paper [16] examines biodegradation kinetics of styrene and ethylbenzene, independently and as binary mixtures, using a series of aerobic batch degradation. The SKIP model and the purely competitive enzyme kinetics model are employed to evaluate any interactions. The SKIP model is found to more accurately describe the interactions.

Here, we propose a mathematical model for biodegradation of phenol and 4-methylphenol (*p*-cresol) in a *continuously stirred tank bioreactor*, in which the biodegradation kinetics is described by a SKIP model. The bioreactor model presents an extension of the growth kinetic model proposed in [17]. There, the growth behavior and degradation capacity of *Aspergillus awamori* NRRL 3112 microbial strain on the binary mixture phenol/*p*-cresol are investigated. Based on laboratory experiments, the growth kinetic model is first evaluated by a sum kinetic model involving Haldane's specific growth rate. An alternative model is then formulated by adding interaction parameters into the sum kinetics model to produce the SKIP model. It is shown that the SKIP model describes better the degradation patterns in the biological system.

The paper is organized as follows. The next Section 2 presents a short description of the proposed mathematical model. Section 3 includes steady states computations. Local stability analysis and bifurcations of the equilibrium points are presented in Section 4. Section 5 reports on general and important properties of the model solutions and provides results on the global stabilizability of the system towards an interior equilibrium point. The last Section 6 presents numerical examples as illustration of the theoretical studies on the model dynamics.

2. Model Description

We consider the following mathematical model for phenol and *p*-cresol mixture degradation in a continuously stirred bioreactor

$$\frac{dX(t)}{dt} = \left(\mu(S_{ph}, S_{cr}) - D \right) X(t) \tag{1}$$

$$\frac{dS_{ph}(t)}{dt} = -k_{ph}\, \mu(S_{ph}, S_{cr}) X(t) + D(S_{ph}^0 - S_{ph}(t)) \tag{2}$$

$$\frac{dS_{cr}(t)}{dt} = -k_{cr}\, \mu(S_{ph}, S_{cr}) X(t) + D(S_{cr}^0 - S_{cr}(t)), \tag{3}$$

where $\mu(S_{ph}, S_{cr})$ is the specific growth rate, presented by

$$\mu(S_{ph}, S_{cr}) = \frac{\mu_{max(ph)} S_{ph}}{k_{s(ph)} + S_{ph} + \frac{S_{ph}^2}{k_{i(ph)}} + I_{cr/ph} S_{cr}} + \frac{\mu_{max(cr)} S_{cr}}{k_{s(cr)} + S_{cr} + \frac{S_{cr}^2}{k_{i(cr)}} + I_{ph/cr} S_{ph}}. \tag{4}$$

The definition of the state variables X, S_{ph} and S_{cr} as well as of the model parameters is given in Table 1. The numerical values in the last column are validated by laboratory experiments and given in [17].

The specific growth rate $\mu(S_{ph}, S_{cr})$ represents a SKIP (sum kinetics with interaction parameters) model for biological degradation of the chemical compounds. The interaction parameters $I_{cr/ph}$ and $I_{ph/cr}$ indicate the degree to which substrate *p*-cresol affects the biodegradation of substrate phenol, and substrate phenol affects the biodegradation of substrate *p*-cresol, respectively. The larger value of $I_{cr/ph}$ (see Table 1) indicates that *p*-cresol inhibits the utilization of phenol much more than phenol inhibits the utilization of *p*-cresol. The kinetic function $\mu(S_{ph}, S_{cr})$ also involves inhibition terms $\frac{S_{ph}^2}{k_{i(ph)}}$ and $\frac{S_{cr}^2}{k_{i(cr)}}$ for cell growth on phenol and *p*-cresol, respectively. Obviously, $\mu(S_{ph}, 0)$ and $\mu(0, S_{cr})$ are the well-known Andrews (or Haldane) model functions, which are unimodal and achieve their maximum at $S_{ph} = \sqrt{k_{s(ph)} k_{i(ph)}}$ and $S_{cr} = \sqrt{k_{s(cr)} k_{i(cr)}}$ respectively.

The influent concentrations S_{ph}^0, S_{cr}^0 and the dilution rate D are the parameters that can be manipulated by the operator of the bioreactor. In our analysis we assume that S_{ph}^0 and S_{cr}^0 are constant and consider the dilution rate D as a varying control parameter. Clearly, $D > 0$ should be fulfilled.

The same model (1)–(3) has been considered in [18] using a more simple specific growth rate function $\mu(S_{ph}, S_{cr})$ which does not involve the inhibition terms $\frac{S_{ph}^2}{k_{i(ph)}}$ and $\frac{S_{cr}^2}{k_{i(cr)}}$ for cell growth on phenol and *p*-cresol. Adding these terms makes the dynamics (1)–(3) more complicated, but as shown in [17], see also [5], the SKIP model (4) describes the trend of experimental data much better than other kinetic models.

Table 1. Model variables and parameters.

	Definitions	Values
X	biomass concentration $[g/dm^3]$	–
S_{ph}	phenol concentration $[g/dm^3]$	–
S_{cr}	p-cresol concentration $[g/dm^3]$	–
D	dilution rate $[h^{-1}]$	–
S_{ph}^0	influent phenol concentration $[g/dm^3]$	0.7
S_{cr}^0	influent p-cresol concentration $[g/dm^3]$	0.3
k_{ph}	metabolic coefficient $[S_{ph}/X]$	11.7
k_{cr}	metabolic coefficient $[S_{cr}/X]$	5.8
$k_{i(ph)}$	inhibition constant for cell growth on phenol $[g/dm^3]$	0.61
$k_{i(cr)}$	inhibition constant for cell growth on cresol $[g/dm^3]$	0.45
$I_{ph/cr}$	interaction coefficient indicating the degree to which phenol affects the p-cresol biodegradation	0.3
$I_{cr/ph}$	interaction coefficient indicating the degree to which p-cresol affects the phenol biodegradation	8.6
$\mu_{max(ph)}$	maximum specific growth rate on phenol as a single substrate $[h^{-1}]$	0.23
$\mu_{max(cr)}$	maximum specific growth rate on p-cresol as a single substrate $[h^{-1}]$	0.17
$k_{s(ph)}$	saturation constant for cell growth on phenol $[g/dm^3]$	0.11
$k_{s(cr)}$	saturation constant for cell growth on p-cresol $[g/dm^3]$	0.35

3. Existence of Equilibrium Points

We shall investigate existence of the model equilibrium points in dependence of the control parameter D.

The equilibrium points of (1)–(3) are solutions of the following system of algebraic equations

$$\left(\mu(S_{ph}, S_{cr}) - D \right) X = 0 \tag{5}$$

$$-k_{ph}\,\mu(S_{ph}, S_{cr})X + D(S_{ph}^0 - S_{ph}) = 0 \tag{6}$$

$$-k_{cr}\,\mu(S_{ph}, S_{cr})X + D(S_{cr}^0 - S_{cr}) = 0. \tag{7}$$

Obviously, the point $E_0 = (0, S_{ph}^0, S_{cr}^0)$ (with $X = 0$) is an equilibrium point of the model for all $D > 0$.

We are looking now for solutions of (5)–(7) assuming that $X \neq 0$.

After multiplying Equation (6) by $-k_{cr}$, Equation (7) by k_{ph} and summing the latter, we obtain

$$- k_{cr}(S_{ph}^0 - S_{ph}) + k_{ph}(S_{cr}^0 - S_{cr}) = 0. \tag{8}$$

Let us express S_{ph} from (8) as a function of S_{cr}. Denoting

$$K = \frac{k_{ph}}{k_{cr}}, \quad S^0 = S_{ph}^0 - KS_{cr}^0, \tag{9}$$

We obtain

$$S_{ph} = S_{ph}^0 - \frac{k_{ph}}{k_{cr}}\left(S_{cr}^0 - S_{cr} \right) = S^0 + KS_{cr}. \tag{10}$$

After replacing the latter presentation of S_{ph} into the equation $\mu(S_{ph}, S_{cr}) = D$ from (5) we obtain an equation with respect to S_{cr} of the form

$$\mu(S^0 + KS_{cr}, S_{cr}) = D,$$

or equivalently

$$\frac{\mu_{max(ph)}\left(S^0 + KS_{cr}\right)}{k_{s(ph)} + S^0 + KS_{cr} + \frac{1}{k_{i(ph)}}(S^0 + KS_{cr})^2 + I_{cr/ph}S_{cr}}$$

$$+ \frac{\mu_{max(cr)}S_{cr}}{k_{s(cr)} + S_{cr} + \frac{1}{k_{i(cr)}}S_{cr}^2 + I_{ph/cr}(S^0 + KS_{cr})} = D.$$

Straightforward calculations lead to a polynomial equation of the form

$$A_1 S_{cr}^4 + A_2 S_{cr}^3 + A_3 S_{cr}^2 + A_4 S_{cr} + A_5 = 0, \tag{11}$$

where

$$A_1 = -D \cdot \frac{1}{k_{i(cr)}} \cdot \frac{1}{k_{i(ph)}};$$

$$A_2 = \mu_{max(ph)}\frac{K}{k_{i(cr)}} + \mu_{max(cr)}\frac{1}{k_{i(ph)}}$$
$$- D\left[(1 + I_{ph/cr}K)\frac{1}{k_{i(ph)}} + \frac{1}{k_{i(cr)}}\left(I_{cr/ph} + K + 2KS^0\frac{1}{k_{i(ph)}}\right)\right];$$

$$A_3 = \mu_{max(ph)}\left[S^0\frac{1}{k_{i(cr)}} + K(1 + I_{ph/cr}K)\right] + \mu_{max(cr)}\left[K + I_{cr/ph} + 2KS^0\frac{1}{k_{i(ph)}}\right]$$
$$- D\left[\frac{1}{k_{i(ph)}}(k_{s(cr)} + I_{ph/cr}S^0) + (1 + I_{ph/cr}K)\left(K + I_{cr/ph} + 2KS^0\frac{1}{k_{i(ph)}}\right)\right.$$
$$\left. + \frac{1}{k_{i(cr)}}\left(k_{s(ph)} + S^0 + \frac{1}{k_{i(ph)}}S^{02}\right)\right];$$

$$A_4 = \mu_{max(ph)}\left[S^0(1 + I_{ph/cr}K) + K(k_{s(cr)} + I_{ph/cr}S^0)\right]$$
$$+ \mu_{max(cr)}\left(k_{s(ph)} + S^0 + \frac{1}{k_{i(ph)}}S^{02}\right)$$
$$- D\left[(k_{s(cr)} + I_{ph/cr}S^0)\left(K + I_{cr/ph} + 2KS^0\frac{1}{k_{i(ph)}}\right)\right.$$
$$\left. + (1 + I_{ph/cr}K)\left(k_{s(ph)} + S^0 + \frac{1}{k_{i(ph)}}S^{02}\right)\right];$$

$$A_5 = \left[\mu_{max(ph)}S^0 - D\left(k_{s(ph)} + S^0 + \frac{1}{k_{i(ph)}}S^{02}\right)\right](k_{s(cr)} + I_{ph/cr}S^0).$$

All coefficients A_i, $i = 1, 2, \ldots, 5$, depend on the parameter D.

Obviously, if $A_5 = 0$, then Equation (11) possesses a solution $S_{cr} = 0$. We have

$$A_5 = 0 \iff D = D_{cr} := \frac{S^0\mu_{max(ph)}k_{i(ph)}}{S^{02} + k_{i(ph)}S^0 + k_{i(ph)}k_{s(ph)}} = \mu(S^0, 0). \tag{12}$$

The latter value of D_{cr} is biologically reasonable only if $S^0 > 0$. Using the numerical values of the model coefficients in the last column of Table 1, we obtain

$$S^0 = S_{ph}^0 - KS_{cr}^0 \approx 0.09483 > 0,$$

and so,

$$D_{cr} = \mu(S^0, 0) \approx 0.09933.$$

This means that for $D = D_{cr}$ there exists an equilibrium point with $S_{cr} = 0$. Further from (10) we compute the component of $S_{ph} = S^0$, and from (7) we get the corresponding component of $X = \dfrac{S^0_{cr}}{k_{cr}}$. Thus, at $D = D_{cr}$ there exists a steady state

$$E_1 = E_1(D_{cr}) = \left(\frac{S^0_{cr}}{k_{cr}}, S^0, 0 \right) = (0.05172,\ 0.09483,\ 0). \tag{13}$$

Considering the cubic equation $A_1 S^3_{cr} + A_2 S^2_{cr} + A_3 S_{cr} + A_4 = 0$ at $D = D_{cr}$ (i.e., with $A_5 = 0$), numerical computations produce the following roots of the latter equation

$$-4.484933737, \quad 0.2614282531 \pm i\, 0.2468184467,$$

so, the real root is negative and cannot serve as a component of the model equilibrium point.

If $D \neq D_{cr}$ then Equation (11) may possess up to 4 real positive solutions with respect to S_{cr}. If there exists at least one positive solution of (11), say S^*_{cr}, such that $S^*_{cr} < S^0_{cr}$ for some values of D, we shall have an interior (with positive components) equilibrium of the form

$$E^* = (X^*, S^*_{ph}, S^*_{cr}), \quad S^*_{ph} = S^0 + K S^*_{cr} < S^0_{ph}, \quad X^* = \frac{S^0_{cr} - S^*_{cr}}{k_{cr}} = \frac{S^0_{ph} - S^*_{ph}}{k_{ph}}. \tag{14}$$

Remark 1. *If we express S_{cr} from (8) as a function of S_{ph} and denote $\hat{K} = \dfrac{k_{cr}}{k_{ph}} = \dfrac{1}{K}$, $\hat{S}^0 = S^0_{cr} - \hat{K} S^0_{ph}$, then we shall have $S_{cr} = \hat{S}^0 + \hat{K} S_{ph}$. Similar calculations as above will produce a polynomial equation of the form $\hat{A}_1 S^4_{ph} + \hat{A}_2 S^3_{ph} + \hat{A}_3 S^2_{ph} + \hat{A}_4 S_{ph} + \hat{A}_5 = 0$, where the coefficients $\hat{A}_i$ are similar to A_i, $i = 1, 2, \ldots, 5$, within $\hat{S}^0$ and $\hat{K}$ instead of S^0 and K, respectively. In this case we have*

$$\hat{A}_5 = \left(\mu_{max,cr} - D \left(k_{s(cr)} + \hat{S}^0 + \frac{\hat{S}^{0^2}}{k_{i(cr)}} \right) \right) (k_{s(ph)} + I_{cr/ph} \hat{S}^0).$$

Obviously, $\hat{A}_5 = 0$ at $\hat{D} = \mu(0, \hat{S}^0)$. But in this case $\hat{S}^0 = -\frac{1}{K} S^0 \approx -0.047 < 0$, thus there is no value of D at which $S_{ph} = 0$ is a root of the polynomial $\sum_{i=1}^{5} \hat{A}_i S^{5-i}_{ph} = 0$. As we shall see in the following, this is the case with the equilibrium component S_{ph}.

Numerical computations show that if $D > D_{cr}$ then there are no positive real roots of Equation (11). Therefore, we can expect interior (coexistence) equilibria of the form E^* if $D \in (0, D_{cr})$, in case that the equilibrium components with respect to S_{cr} satisfy the inequality $S_{cr} \leq S^0_{cr}$. Further we obtain numerically the following results:

- There exists a value $D = D^{(1)}_{cr} \approx 0.0745599$, so that Equation (11) possesses a double root $S_{cr} \approx 0.04327$ for $D = D^{(1)}_{cr}$.

- If $D < D^{(1)}_{cr}$ then there are no positive roots of (11) which are less than or equal to S^0_{cr}.

- Denote $D^{(2)}_{cr} := \mu(S^0_{ph}, S^0_{cr}) \approx 0.08651 < D_{cr}$. If $D \in \left(D^{(1)}_{cr}, D^{(2)}_{cr} \right)$ then there are two positive roots of (11) which are less than S^0_{cr}.

- If $D \in \left(D^{(2)}_{cr}, D_{cr} \right)$, $D_{cr} = \mu(S^0, 0) \approx 0.09933$, then there is only one positive root of (11) which is less than S^0_{cr}.

The left plot in Figure 1 shows the graph of the function $\mu(S^0 + KS_{cr}, S_{cr})$ for $S_{cr} \in [0, S^0_{cr}] = [0, 0.3]$; the horizontal dash lines correspond to the values of $D_{cr}^{(1)}$, $D_{cr}^{(2)}$ and D_{cr}.

Therefore, the model (1)–(3) possesses two interior equilibrium points depending on the values of D. Denote them by

$$E_2 = E_2(D) = \left(X^{(2)}, S_{ph}^{(2)}, S_{cr}^{(2)}\right), \quad D \in \left(D_{cr}^{(1)}, D_{cr}\right);$$
$$E_3 = E_3(D) = \left(X^{(3)}, S_{ph}^{(3)}, S_{cr}^{(3)}\right), \quad D \in \left(D_{cr}^{(1)}, D_{cr}^{(2)}\right), \quad \text{with } S_{cr}^{(3)} > S_{cr}^{(2)}.$$

Numerical computations also produce the following results:

$$E_2(D_{cr}) = E_1 = (0.05172,\ 0.09483,\ 0), \qquad D_{cr} = \mu(S^0, 0) = 0.09933;$$

$$E_2(D_{cr}^{(1)}) = E_3(D_{cr}^{(1)}) = (0.04426, 0.18211, 0.04327), \quad D_{cr}^{(1)} = 0.0745599;$$

$$E_3(D_{cr}^{(2)}) = E_0 = (0, S_{ph}^0, S_{cr}^0) = (0,\ 0.7,\ 0.3), \qquad D_{cr}^{(2)} = \mu(S_{ph}^0, S_{cr}^0) = 0.08651.$$

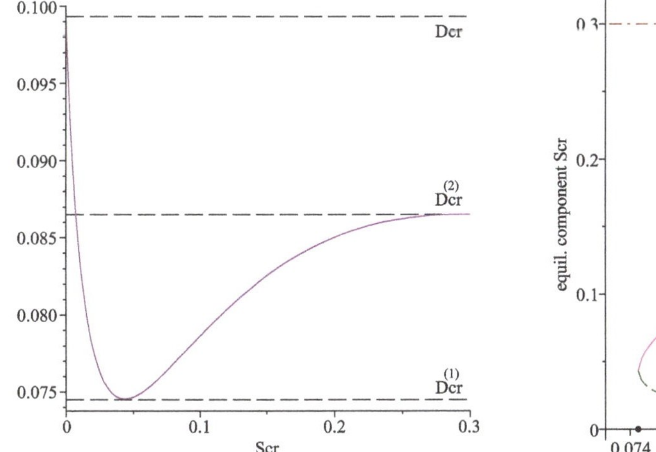
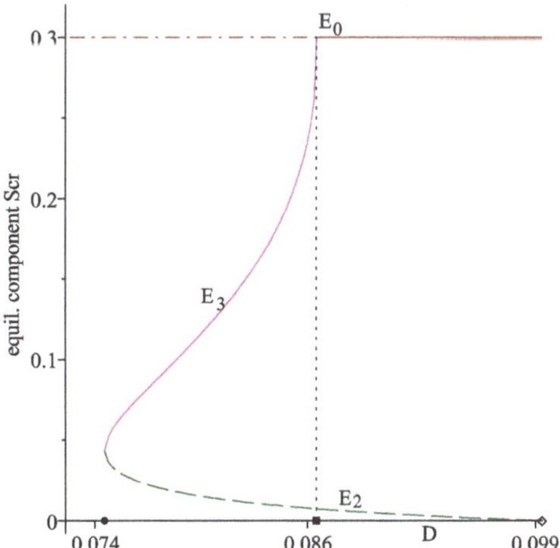

Figure 1. (**Left**): graph of the function $\mu(S^0 + KS_{cr}, S_{cr})$ for $S_{cr} \in [0, S^0_{cr}]$. (**Right**): the equilibrium components $S_{cr}^{(2)}$ (dash line) and $S_{cr}^{(3)}$ (solid line), parameterized on D. The horizontal dash-dot&solid line passes trough S_{cr}^0. On the horizontal axis, the solid circle denotes $D_{cr}^{(1)}$, the solid box denotes $D_{cr}^{(2)}$, the diamond denotes D_{cr}. The vertical dot line passes through $D_{cr}^{(2)}$.

Figure 1 (right plot) and Figure 2 visualize the components S_{cr}, S_{ph} and X of the equilibria E_0, E_2 and E_3. In the three plots, the components of the equilibrium point E_0 are marked by horizontal dash-dot&solid lines, the components of E_2 are marked by dash lines and the ones of E_3 are shown by solid lines.

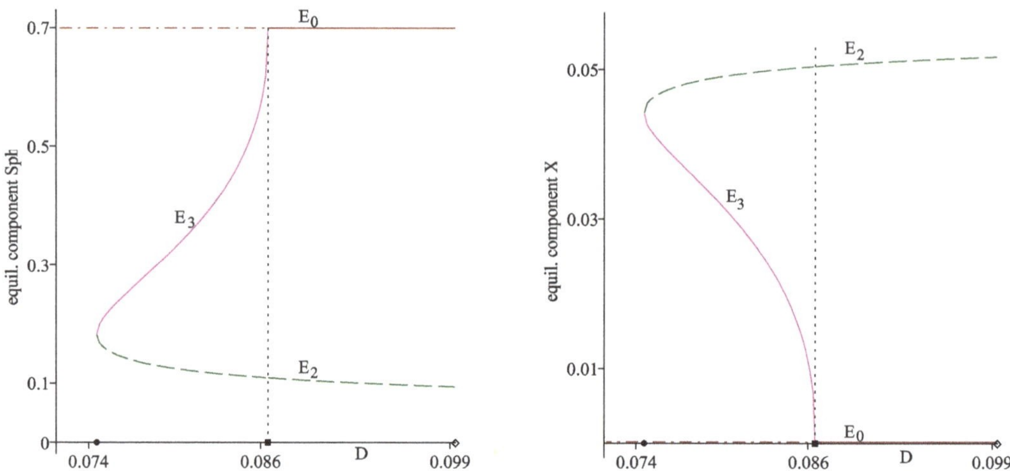

Figure 2. (**Left**): the equilibrium components $S_{ph}^{(2)}$ (dash line) and $S_{ph}^{(3)}$ (solid line), parameterized on D. The horizontal dash-dot&solid line passes trough S_{ph}^{0}. (**Right**): the equilibrium components $X^{(2)}$ (dash line) and $X^{(3)}$ (solid line), parameterized on D. The horizontal dash-dot&solid line passes trough 0. On the horizontal axis (left and right plot), the solid circle denotes $D_{cr}^{(1)}$, the solid box denotes $D_{cr}^{(2)}$, the diamond denotes D_{cr}. The vertical dot line passes through $D_{cr}^{(2)}$.

4. Local Stability of the Equilibrium Points

In this section we shall study the conditions for local asymptotic stability of the model equilibrium points.

It is well known that an equilibrium point is locally asymptotically stable, if all eigenvalues of the Jacobi matrix evaluated at this equilibrium have negative real parts, cf. e.g., [19]. The eigenvalues of the Jacobi matrix coincide with the roots of the corresponding characteristic polynomial.

To simplify notations, in the following we shall sometimes write μ instead of $\mu(S_{ph}, S_{cr})$. The Jacobi matrix J related to the model Equations (1)–(3) has the form

$$
J = \begin{pmatrix}
\mu(S_{ph}, S_{cr}) - D & \frac{\partial \mu}{\partial S_{ph}} X & \frac{\partial \mu}{\partial S_{cr}} X \\
-k_{ph}\mu(S_{ph}, S_{cr}) & -k_{ph}\frac{\partial \mu}{\partial S_{ph}} X - D & -k_{ph}\frac{\partial \mu}{\partial S_{cr}} X \\
-k_{cr}\mu(S_{ph}, S_{cr}) & -k_{cr}\frac{\partial \mu}{\partial S_{ph}} X & -k_{cr}\frac{\partial \mu}{\partial S_{cr}} X - D
\end{pmatrix}.
$$

The characteristic polynomial corresponding to J is defined by $det(J - \lambda I_3)$, where λ is any complex number and I_3 is the (3×3)–identity matrix

$$
det(J - \lambda I_3) = \begin{vmatrix}
\mu(S_{ph}, S_{cr}) - D - \lambda & \frac{\partial \mu}{\partial S_{ph}} X & \frac{\partial \mu}{\partial S_{cr}} X \\
-k_{ph}\mu(S_{ph}, S_{cr}) & -k_{ph}\frac{\partial \mu}{\partial S_{ph}} X - D - \lambda & -k_{ph}\frac{\partial \mu}{\partial S_{cr}} X \\
-k_{cr}\mu(S_{ph}, S_{cr}) & -k_{cr}\frac{\partial \mu}{\partial S_{ph}} X & -k_{cr}\frac{\partial \mu}{\partial S_{cr}} X - D - \lambda
\end{vmatrix}.
$$

Multiplying the second row of the above determinant by $-\dfrac{k_{cr}}{k_{ph}}$ and adding the latter to the third row, we obtain

$$
det(J - \lambda I_3) = \begin{vmatrix} \mu(S_{ph}, S_{cr}) - D - \lambda & \frac{\partial \mu}{\partial S_{ph}} X & \frac{\partial \mu}{\partial S_{cr}} X \\[2mm] -k_{ph}\mu(S_{ph}, S_{cr}) & -k_{ph}\frac{\partial \mu}{\partial S_{ph}} X - D - \lambda & -k_{ph}\frac{\partial \mu}{\partial S_{cr}} X \\[2mm] 0 & \frac{k_{cr}}{k_{ph}}(D + \lambda) & -D - \lambda \end{vmatrix}.
$$

Straightforward calculations deliver the following characteristic polynomial

$$
det(J - \lambda I_3) = (D + \lambda)^2 \left[\mu(S_{ph}, S_{cr}) - D - \lambda - X\left(k_{ph}\frac{\partial \mu}{\partial S_{ph}} + k_{cr}\frac{\partial \mu}{\partial S_{cr}} \right) \right]. \tag{15}
$$

Denote by $J(E_i)$ the Jacobian matrix evaluated at the equilibrium point E_i, $i = 0, 1, 2, 3$. It follows from (15) that $\lambda_{1,2} = -D < 0$ are always eigenvalues of $J(E_i)$, $i = 0, 1, 2, 3$. The third eigenvalue λ_3 is determined from the second multiplier of (15).

Proposition 1.

(i)   If $D < D_{cr}^{(2)} = \mu(S_{ph}^0, S_{cr}^0)$ then the equilibrium point $E_0 = \left(0, S_{ph}^0, S_{cr}^0\right)$ (with $X = 0$) is locally asymptotically unstable (a saddle).

(ii)   If $D > D_{cr}^{(2)}$ then E_0 is locally asymptotically stable (a stable node).

(iii)   At $D = D_{cr}^{(2)}$ the equilibrium E_0 is neither stable, nor unstable: $J(E_0)$ possesses a zero eigenvalue, $\lambda_3 = 0$, thus $D_{cr}^{(2)}$ is a bifurcation parameter value.

(iv)   The equilibrium point $E_1 = E_1(D_{cr}) = \left(\dfrac{S_{cr}^0}{k_{cr}}, S^0, 0\right)$, (see (13)), is locally asymptotically unstable.

Proof. (i)–(iii) We obtain from (15)

$$
det(J(E_0) - \lambda I_3) = (D + \lambda)^2 (\mu(S_{ph}^0, S_{cr}^0) - D - \lambda),
$$

thus the third root λ_3 satisfies

$$
\lambda_3 = \mu(S_{ph}^0, S_{cr}^0) - D \begin{cases} > 0, & \text{if } D < D_{cr}^{(2)} = \mu(S_{ph}^0, S_{cr}^0), \\[1mm] < 0, & \text{if } D > D_{cr}^{(2)}, \\[1mm] = 0, & \text{if } D = D_{cr}^{(2)}. \end{cases}
$$

(iv) The characteristic polynomial corresponding to the equilibrium E_1 is presented by

$$
det(J(E_1) - \lambda I_3) = -(D_{cr} + \lambda)^2 \left(S_{cr}^0 \left(K\frac{\partial \mu}{\partial S_{ph}}(S^0, 0) + \frac{\partial \mu}{\partial S_{cr}}(S^0, 0) \right) + \lambda \right).
$$

The third root λ_3 of the latter polynomial is computed numerically and is equal to

$$
\lambda_3 = -S_{cr}^0 \left(K\frac{\partial \mu}{\partial S_{ph}}(S^0, 0) + \frac{\partial \mu}{\partial S_{cr}}(S^0, 0) \right) \approx -(-0.7574) > 0,
$$

which means that $E_1(D_{cr})$ is a saddle equilibrium point. This proves the proposition. $\square$

The equilibrium components $S_{ph}^{(i)}$ and $S_{cr}^{(i)}$ of the equilibria E_i, $i = 2, 3$, satisfy the equation $\mu(S_{ph}, S_{cr}) = D$, so that from (15) we obtain

$$det(J(E_i) - \lambda I_3) = -(D + \lambda)^2 \left[\lambda + X^{(i)} \left(k_{ph} \frac{\partial \mu}{\partial S_{ph}} (S_{ph}^{(i)}, S_{cr}^{(i)}) + k_{cr} \frac{\partial \mu}{\partial S_{cr}} (S_{ph}^{(i)}, S_{cr}^{(i)}) \right) \right],$$
$$i = 2, 3.$$

The third root $\lambda_3^{(i)} = -X^{(i)} \left(k_{ph} \frac{\partial \mu}{\partial S_{ph}} (S_{ph}^{(i)}, S_{cr}^{(i)}) + k_{cr} \frac{\partial \mu}{\partial S_{cr}} (S_{ph}^{(i)}, S_{cr}^{(i)}) \right)$ is found numerically by computing the right-hand side expression on a discrete mesh of values for D, where $D \in (D_{cr}^{(1)}, D_{cr})$ for E_2, and $D \in (D_{cr}^{(1)}, D_{cr}^{(2)})$ for E_3. Figure 3 visualizes the three eigenvalues of $J(E_2)$ and $J(E_3)$. One can see that the eigenvalues of $J(E_3)$ are negative (right plot), and $J(E_2)$ possesses one real positive eigenvalue (left plot). Moreover, one eigenvalue of $J(E_2)$ approaches zero at $D = D_{cr}^{(1)}$, and one eigenvalue of $J(E_3)$ approaches zero at $D = D_{cr}^{(1)}$ and $D = D_{cr}^{(2)}$, thus $D_{cr}^{(1)}$ and $D_{cr}^{(2)}$ are bifurcation parameter values.

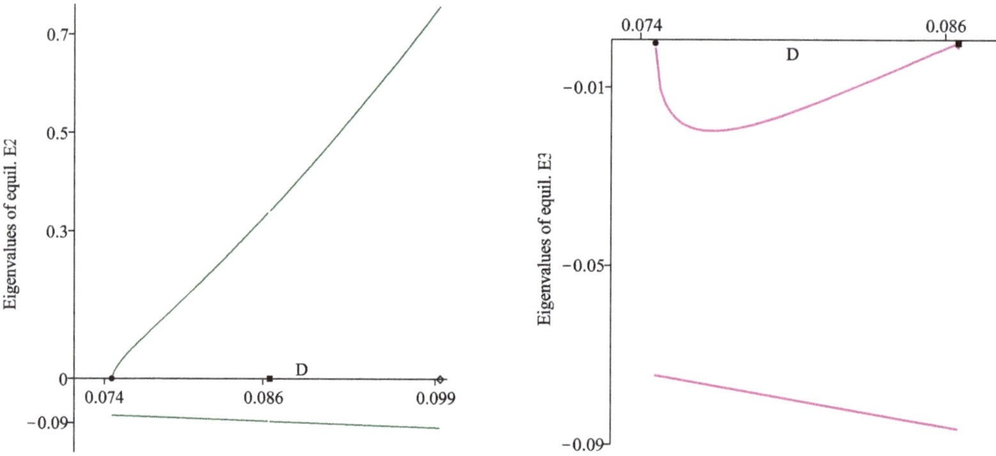

Figure 3. Eigenvalues corresponding to the equilibrium points E_2 (**left**) and E_3 (**right**), parameterized on D. On the horizontal axis, the solid circle denotes $D_{cr}^{(1)}$, the solid box denotes $D_{cr}^{(2)}$, the diamond denotes D_{cr}.

We summarize the above results in the next proposition.

Proposition 2.

(i)   *The equilibrium E_2, defined for $D \in (D_{cr}^{(1)}, D_{cr})$, is locally asymptotically unstable (a saddle).*

(ii)   *The equilibrium E_3, defined for $D \in (D_{cr}^{(1)}, D_{cr}^{(2)})$, is locally asymptotically stable (a stable node).*

(iii)   *At $D = D_{cr}^{(1)}$, the two interior equilibrium points, E_2 and E_3, are 'born', thus $D_{cr}^{(1)}$ is a bifurcation value of the parameter D. At $D = D_{cr}^{(1)}$ the steady states E_2 and E_3 undergo a saddle-node bifurcation.*

(iv)   *At $D = D_{cr}^{(2)}$ the equilibrium points E_3 and E_0 coalesce and exchange stability for $D > D_{cr}^{(2)}$. Thus, at $D = D_{cr}^{(2)}$ the steady states E_3 and E_0 undergo a transcritical bifurcation.*

Figure 1 (right plot) and Figure 2 also visualize the stability of E_0, E_2 and E_3: the solid lines correspond to the components of the stable equilibria, the dash and the dash-dot lines mark the components of the unstable equilibria. Therefore, these three plots can also be considered as bifurcation diagrams: a saddle node bifurcation occurs at the parameter value $D = D_{cr}^{(1)}$, and $D = D_{cr}^{(2)}$ serves as a transcritical bifurcation point.

5. Global Stabilizability of the Model Dynamics

First we prove that the model (1)–(3) exhibits the standard properties that we would expect from a bioreactor model, namely uniqueness and positiveness of solutions for non-negative initial conditions.

Theorem 1. *Consider the model (1)–(3) and assume that $X(0) \geq 0$, $S_{ph}(0) \geq 0$, $S_{cr}(0)) \geq 0$.*

(i)   If $X(0) = 0$ then all model solutions tend to the equilibrium point $E_0 = (0, S_{ph}^0, S_{cr}^0)$.

(ii)   If $X(0) > 0$ then $X(t) > 0$, $S_{ph}(t) > 0$, $S_{cr}(t) > 0$ for all $t > 0$.

(iii)   All solutions are uniformly bounded for all $t \geq 0$.

Proof. *(i)* Let $X(0) = 0$ and $S_{ph}(0) \geq 0$, $S_{cr}(0) \geq 0$ be satisfied. It follows that $X(t) = 0$ for all $t \geq 0$ due to uniqueness of solutions of the Cauchy problem. Then the model (1)–(3) reduces to

$$
\begin{aligned}
\frac{dS_{ph}(t)}{dt} &= D(S_{ph}^0 - S_{ph}(t)) \\
\frac{dS_{cr}(t)}{dt} &= D(S_{cr}^0 - S_{cr}(t)).
\end{aligned}
$$

The latter equations imply that $S_{ph}(t)$ and $S_{cr}(t)$ converge exponentially to S_{ph}^0 and S_{cr}^0 respectively. The plane $X = 0$ is invariant for the model.

(ii)–(iii) Assume that $X(0) > 0$, $S_{ph}(0) \geq 0$, $S_{cr}(0)) \geq 0$. It follows from Equation (1) that

$$
\frac{dX}{X} = \int_0^t (\mu(S_{ph}(\tau), S_{cr}(\tau)) - D)d\tau,
$$

$$
X(t) = X(0)e^{\int_0^t (\mu(S_{ph}(\tau), S_{cr}(\tau)) - D)d\tau} > 0 \text{ for each } t \geq 0.
$$

Denote $\Sigma_1(t) = S_{ph}(t) + k_{ph}X(t) - S_{ph}^0$. Then Equations (1) and (2) imply

$$
\frac{d}{dt}\Sigma_1(t) = \frac{dS_{ph}}{dt} + k_{ph}\frac{dX}{dt} = D\left(S_{ph}^0 - (S_{ph} + k_{ph}X)\right) = -D\Sigma_1(t),
$$

which means that $\Sigma_1(t) = e^{-Dt}\Sigma_1(0)$, thus $\lim_{t \to \infty} \Sigma_1(t) = 0$, or equivalently

$$
\lim_{t \to \infty} \left(S_{ph}(t) + k_{ph}X(t)\right) = S_{ph}^0.
$$

Since $X(t) > 0$ for all $t > 0$ this means $S_{ph}(t) > 0$ for all $t > 0$ as well. Moreover, $X(t)$ and $S_{ph}(t)$ are uniformly bounded.

Similarly, using Equations (1) and (3) and denoting $\Sigma_2(t) = S_{cr}(t) + k_{cr}X(t) - S_{cr}^0$ we obtain $\Sigma_2(t) = e^{-Dt}\Sigma_2(0)$, which means that

$$
\lim_{t \to \infty} (S_{cr}(t) + k_{cr}X(t)) = S_{cr}^0. \tag{16}
$$

Therefore, $S_{cr}(t) > 0$ for all $t > 0$ and $S_{cr}(t)$ is uniformly bounded for $t \geq 0$. Hence, the model solutions $X(t)$, $S_{ph}(t)$, $S_{cr}(t)$ exist for all time $t \geq 0$. This completes the proof of Theorem 1. $\square$

In the following we shall prove the global asymptotic stabilizability of system (1)–(3) when the control parameter D belongs to the interval $\left(D_{cr}^{(1)}, D_{cr}^{(2)}\right)$, with $D_{cr}^{(2)} = \mu(S_{ph}^0, S_{cr}^0)$.

Similarly to the proof of Theorem 1, denote $\Sigma_3(t) = S_{ph}(t) - KS_{cr}(t) - S^0$, where K and S^0 are defined in (9). After multiplying Equation (3) by $-\dfrac{k_{ph}}{k_{cr}}$ and adding to Equation (2) we obtain

$$
\begin{aligned}
\frac{d}{dt}\Sigma_3(t) &= \frac{d}{dt}\Big(S_{ph}(t) - KS_{cr}(t)\Big) = D\Big(S_{ph}^0 - S_{ph}(t) - KS_{cr}^0 + KS_{cr}(t)\Big) \\
&= D\Big((S_{ph}^0 - KS_{cr}^0) - (S_{ph}(t) - KS_{cr}(t))\Big) \\
&= -D\Big(S^0 - (S_{ph}(t) - KS_{cr}(t))\Big) = -D\Sigma_3(t).
\end{aligned}
$$

This means that $\Sigma_3(t) = e^{-Dt}\Sigma_3(0)$, $\Sigma_3(0) \geq 0$, so $\lim_{t\to\infty}\Sigma_3(t) = 0$. Then system (1)–(3) may be written in the form

$$
\begin{aligned}
\frac{d}{dt}\Sigma_3(t) &= -D\Sigma_3(t) \\
\frac{d}{dt}X(t) &= \Big(\mu(S^0 + KS_{cr}(t), S_{cr}(t)) - D\Big)X(t) \\
\frac{d}{dt}S_{cr}(t) &= -k_{cr}\mu(S^0 + KS_{cr}(t), S_{cr}(t))X(t) + D(S_{cr}^0 - S_{cr}(t)).
\end{aligned}
$$

Since $\lim_{t\to\infty}\Sigma_3(t) = 0$, the positive ω-limit set of any solution of system (1)–(3) is contained in the set

$$
\Omega_3 = \Big\{(X, S_{ph}, S_{cr}) : X > 0, S_{ph} \geq 0, S_{cr} \geq 0, \Sigma_3 = 0\Big\}.
$$

Using the theory of the asymptotically autonomous systems (cf. [20,21]) it follows that all trajectories forming the ω-limit set of any solution of (1)–(3) with initial conditions in Ω_3 are solutions of the following limiting system

$$
\begin{aligned}
\frac{dX(t)}{dt} &= (\mu(S^0 + KS_{cr}(t), S_{cr}(t)) - D)X(t) \\
\frac{dS_{cr}(t)}{dt} &= -k_{cr}\,\mu(S^0 + KS_{cr}(t), S_{cr}(t))X(t) + D(S_{cr}^0 - S_{cr}(t)).
\end{aligned}
\tag{17}
$$

We consider Equation (17) on the set

$$
\Omega_2 = \Big\{(X, S_{cr}) : X > 0, S_{cr} \geq 0, S^0 + KS_{cr} \geq 0\Big\}.
$$

Denote for simplicity $\mu_{cr}(S_{cr}) = \mu(S^0 + KS_{cr}, S_{cr})$. Then obviously $\mu_{cr}(S_{cr}^0) = \mu(S^0 + KS_{cr}^0, S_{cr}^0) = \mu(S_{ph}^0, S_{cr}^0) = D_{cr}^{(2)}$ holds true.

Let us choose an arbitrary value $\bar{D} \in \left(D_{cr}^{(1)}, D_{cr}^{(2)}\right)$, and consider the following system obtained from (17) after substituting $D = \bar{D}$ in the latter:

$$
\frac{dX(t)}{dt} = (\mu_{cr}(S_{cr}(t)) - \bar{D})X(t)
\tag{18}
$$

$$
\frac{dS_{cr}(t)}{dt} = -k_{cr}\,\mu_{cr}(S_{cr}(t))X(t) + \bar{D}(S_{cr}^0 - S_{cr}(t)).
\tag{19}
$$

Let us recall, that at $D = \bar{D}$ there are two interior equilibria of the model (1)–(3),

$$E_2(\bar{D}) = (X^{(2)}(\bar{D}), S_{ph}^{(2)}(\bar{D}), S_{cr}^{(2)}(\bar{D})) \;\; \text{and}$$

$$E_3(\bar{D}) = (X^{(3)}(\bar{D}), S_{ph}^{(3)}(\bar{D}), S_{cr}^{(3)}(\bar{D})) \;\; \text{with} \;\; S_{cr}^{(2)}(\bar{D}) < S_{cr}^{(3)}(\bar{D}).$$

Denote

$$\bar{E} = (\bar{X}, \bar{S}_{cr}) = \left(X^{(3)}(\bar{D}), S_{cr}^{(3)}(\bar{D}) \right).$$

Obviously, $\bar{E}$ is an equilibrium point of (18) and (19).
We make the following assumption.

Assumption 1. *There exist points S_{cr}^- and S_{cr}^+ such that $0 < S_{cr}^- < S_{cr}^+ < S_{cr}^0$ and $\mu_{cr}(S_{cr})$ is monotone increasing for all $S_{cr} \in (S_{cr}^-, S_{cr}^+)$.*

Assumption 1 identifies the equilibrium $\bar{E}$ with the projection of $E_3(\bar{D})$ in the plane $S_{ph} - KS_{cr} = S^0$. If we choose for S_{cr}^- the S_{cr}-component of the double root of Equation (11), i.e., $S_{cr}^- = S_{cr}^{(2)}(D_{cr}^{(1)}) - S_{cr}^{(3)}(D_{cr}^{(1)})$, and $S_{cr}^+ - S_{cr}^0$, then $\mu_{cr}(S_{cr})$ is monotone increasing in (S_{cr}^-, S_{cr}^+), see the left plot in Figure 1.

Based on the above considerations, the problem for global stabilizability of the model (1)–(3) is reduced to proving the global stabilizability of the well known basic bioreactor (chemostat) model (17), which is well studied in the literature, see e.g., [20,22–24] and the references therein. The next Theorem 2 is also a corollary from Theorem 2.1 in [25]. We present the proof here for reader's convenience.

Theorem 2. *Let Assumption 1 be fulfilled. Assume that $\bar{D} \in \left(D_{cr}^{(1)}, D_{cr}^{(2)} \right)$. Then for any initial point $(X(0), S_{cr}(0)) \in \Omega_2$ the corresponding solution of (18) and (19) converges asymptotically towards the equilibrium point $\bar{E}$.*

Proof. Let us fix an arbitrary initial point $(X(0), S_{cr}(0)) \in \Omega_2$.

First we shall show that there exists time $T > 0$, such that $S_{cr}(t) < S_{cr}^0$ for all $t > T$. Assume that $S_{cr}(t) \geq S_{cr}^0$ holds true for each $t > 0$. Then we have from (19) that

$$\frac{dS_{cr}(t)}{dt} = -k_{cr}\,\mu_{cr}(S_{cr})X(t) + \bar{D}(S_{cr}^0 - S_{cr}(t)) < 0.$$

Barbălat's Lemma [26] implies

$$0 = \lim_{t \to \infty} \frac{dS_{cr}(t)}{dt} = \lim_{t \to \infty} \left(-k_{cr}\,\mu_{cr}(S_{cr})X(t) + \bar{D}(S_{cr}^0 - S_{cr}(t)) \right),$$

which leads to $S_{cr}(t) \to S_{cr}^0$ and $X(t) \to 0$ as $t \to \infty$. Further we have that $\mu_{cr}(\bar{S}_{cr}) = \bar{D} < D_{cr}^{(2)} = \mu_{cr}(S_{cr}^0)$. The continuity of $\mu_{cr}(\cdot)$ and the relation $S_{cr}(t) \to S_{cr}^0$ as $t \to \infty$ imply that there exists a number $\delta > 0$ such that

$$\mu_{cr}(S_{cr}(t)) - \bar{D} = \mu_{cr}(S_{cr}(t)) - \mu_{cr}(\bar{S}_{cr}) \geq \delta$$

for all sufficiently large t. Then it follows $\dfrac{dX(t)}{dt} = (\mu_{cr}(S_{cr}(t)) - \bar{D})X(t) \geq \delta X(t)$ for all sufficiently large t, which contradicts the boundedness of $X(t)$. Hence, there exists a sufficiently large $T > 0$ with $S_{cr}(T) \leq S_{cr}^0$. If the equality $S_{cr}(T) = S_{cr}^0$ holds true, then we have

$$\frac{dS_{cr}}{dt}(T) = -k_{cr}\,\mu_{cr}(S_{cr}(T))X(T) + \bar{D}(S_{cr}^0 - S_{cr}(T))$$
$$= -k_{cr}\,\mu_{cr}(S_{cr}(T))X(T) < 0.$$

The last inequality shows that $S_{cr}(t) < S_{cr}^0$ for each $t > T$.

Let us fix an arbitrary $\gamma \in \left(0, (\mu_{cr}(S_{cr}^0) - \mu_{cr}(\bar{S}_{cr}))/2\right)$. (Note that $\mu_{cr}(S_{cr})$ is monotone increasing.) The continuity of μ_{cr} implies that there exists $\varepsilon > 0$ such that $\mu_{cr}(\bar{S}_{cr}) + \gamma < \mu_{cr}(S_{cr})$ for each $S_{cr} \in \left[S_{cr}^0 - (1+k_{cr})\varepsilon, S_{cr}^0\right)$. It follows from (16) that there exists time $T_\varepsilon > 0$ so that $X(t)$ and $S_{cr}(t)$ satisfy

$$S_{cr}^0 - \varepsilon < S_{cr}(t) + k_{cr}X(t) < S_{cr}^0 + \varepsilon \quad \text{for each } t \geq T_\varepsilon. \tag{20}$$

Assume now that $X(\bar{t}) \leq \varepsilon$ for some $\bar{t} \geq T_\varepsilon$; then we obtain from (20)

$$S_{cr}^0 > S_{cr}(\bar{t}) \geq S_{cr}^0 - k_{cr}X(\bar{t}) - \varepsilon \geq S_{cr}^0 - (1+k_{cr})\varepsilon,$$

i.e., $S_{cr}(\bar{t}) \in \left[S_{cr}^0 - (1+k_{cr})\varepsilon, S_{cr}^0\right)$. Hence,

$$\frac{d}{dt}X(\bar{t}) = (\mu_{cr}(S_{cr}(\bar{t})) - \bar{D})X(\bar{t}) = (\mu_{cr}(S_{cr}(\bar{t})) - \mu_{cr}(\bar{S}_{cr}))X(\bar{t}) \geq \gamma X(\bar{t}) > 0.$$

It follows then that $X(t) \geq e^{(t-\bar{t})\gamma}X(\bar{t})$. If there exists $t_1 \geq \bar{t}$ such that $X(t_1) = \varepsilon$, then at every time $t_2 \geq t_1$ with $X(t_2) = \varepsilon$ we have $\dfrac{d}{dt}X(t_2) = (\mu_{cr}(S_{cr}(t_2)) - \bar{D})X(t_2) \geq \gamma\varepsilon > 0$. Hence there exists time $T_1 > T$ such that $X(t) \geq \varepsilon$ for each $t \geq T_1$.

The above considerations mean that the ω-limit set of the corresponding trajectory of (18) and (19) lies in the set

$$\{(X, S_{cr}) : \ X \geq \varepsilon, \ 0 \leq S_{cr} \leq S_{cr}^0\}.$$

For $X > 0$ and $S_{cr} \in (0, S_{cr}^0)$ we define the following Lyapunov function

$$V = V(X, S_{cr}) = \int_{\bar{X}}^{X} \frac{\eta - \bar{X}}{\eta}\,d\eta + \int_{\bar{S}_{cr}}^{S_{cr}} \frac{\bar{X}(\mu_{cr}(\xi) - \bar{D})}{\bar{D}(S_{cr}^0 - \xi)}\,d\xi.$$

The derivative $\dfrac{d}{dt}V$ of V along the solutions of (18) and (19) is presented by

$$\begin{aligned}
\frac{d}{dt}V &= \frac{X - \bar{X}}{X}(\mu_{cr}(S_{cr}) - \bar{D})X + \frac{\bar{X}(\mu_{cr}(S_{cr}) - \bar{D})}{\bar{D}(S_{cr}^0 - S_{cr})}\left(-k_{cr}\mu_{cr}(S_{cr})X + \bar{D}(S_{cr}^0 - S_{cr})\right) \\
&= X(\mu_{cr}(S_{cr}) - \bar{D})\left(1 - \frac{\bar{X}k_{cr}\mu_{cr}(S_{cr})}{\bar{D}(S_{cr}^0 - S_{cr})}\right) \\
&= X(\mu_{cr}(S_{cr}) - \bar{D})\left(1 - \frac{S_{cr}^0 - \bar{S}_{cr}}{S_{cr}^0 - S_{cr}} \cdot \frac{\mu_{cr}(S_{cr})}{\bar{D}}\right) \leq 0
\end{aligned}$$

for each $S_{cr} \in (0, S_{cr}^0)$ and $X > 0$. Applying LaSalle's invariance principle it follows that each trajectory of (18) and (19) approaches the equilibrium point $\bar{E}$, i.e., $\bar{E}$ is globally asymptotically stable. This proves the theorem. $\square$

It follows from Propositions 1 and 2 that when the control input D takes values $D > D_{cr}^{(2)} = \mu(S_{ph}^0, S_{cr}^0)$ then the model (1)–(3) possesses two equilibrium points—the wash-out equilibrium $E_0 = (0, S_{ph}^0, S_{cr}^0)$ and the interior equilibrium E_2, such that E_0 is locally asymptotically stable and E_2 is locally asymptotically unstable. Using the reduced model (17) it can be shown that the restriction $\bar{E}_0 = (0, S_{cr}^0)$ of the wash-out equilibrium E_0 is globally asymptotically stable if $D > D_{cr}^{(2)} = \mu_{cr}(S_{cr}^0)$. Although the proof can be extracted from the more general Lemma 2.2 in [24], we present it below for completeness.

Theorem 3. *Assume that $D > D_{cr}^{(2)}$ holds true. Then for any initial point $(X(0), S_{cr}(0)) > 0$ the corresponding solution of (17) converges asymptotically towards the equilibrium $\bar{E}_0 = (0, S_{cr}^0)$.*

Proof. Choose some $\bar{D}_0 > D_{cr}^{(2)}$ and consider system (18) and (19), where $\bar{D}$ is replaced by $\bar{D}_0$. Assume that $\lim_{t \to \infty} X(t) = X^* > 0$. Then Barbălat's Lemma [26] applied to Equation (18) implies $0 = \lim_{t \to \infty} \frac{d}{dt} X(t) = \lim_{t \to \infty} (\mu_{cr}(S_{cr}(t)) - \bar{D}_0) X^*$, which means that $\lim_{t \to \infty} \mu_{cr}(S_{cr}(t)) = \bar{D}_0 > \mu_{cr}(S_{cr}^0)$. From the continuity of $\mu_{cr}(\cdot)$ it follows that there exists time $T > 0$ and a positive number δ such that $\mu_{cr}(S_{cr}(t)) - \mu_{cr}(S_{cr}^0) \geq \delta$ for all $t \geq T$. The latter inequality leads to $\frac{d}{dt} X(t) \geq \delta X(t)$, a contradiction with the boundedness of $X(t)$. Therefore, $X(t) \to 0$ as $t \to \infty$. From the theory of the asymptotically autonomous systems (cf. [20,21]) it follows that the dynamics (18) and (19) can be reduced to the limiting equation $\frac{d}{dt} S_{cr}(t) = \bar{D}_0 (S_{cr}^0 - S_{cr}(t))$, which implies $\lim_{t \to \infty} S_{cr}(t) = S_{cr}^0$, and this completes the proof. $\square$

6. Dynamic Behavior of the Model Solutions: Numerical Simulation

In this section we present two numerical examples that illustrate the dynamic behavior of the model solutions.

Example 1. $D = 0.08 \in (D_{cr}^{(1)}, D_{cr}^{(2)})$

In this case there exist two positive (coexistence) equilibrium points

$$E_2 = (0.0491, 0.126, 0.0153) \quad \text{and} \quad E_3 = (0.0318, 0.328, 0.116),$$

such that E_2 is locally asymptotically unstable, E_3 is the globally asymptotically stable equilibrium point according to Theorem 2. The wash-out equilibrium $E_0 = (0, 0.7, 0.3)$ is locally asymptotically unstable.

The left plot in Figure 4 visualizes the convergence of the solutions towards the corresponding equilibrium components of E_3 using two different starting points. The right plot of Figure 4 as well as Figure 5 visualize projections of the trajectories in the phase planes (X, S_{cr}), (X, S_{ph}) and (S_{ph}, S_{cr}) respectively with three different initial points, denoted by circles. The corresponding projections of the invariant planes are marked by dash lines in the three plots.

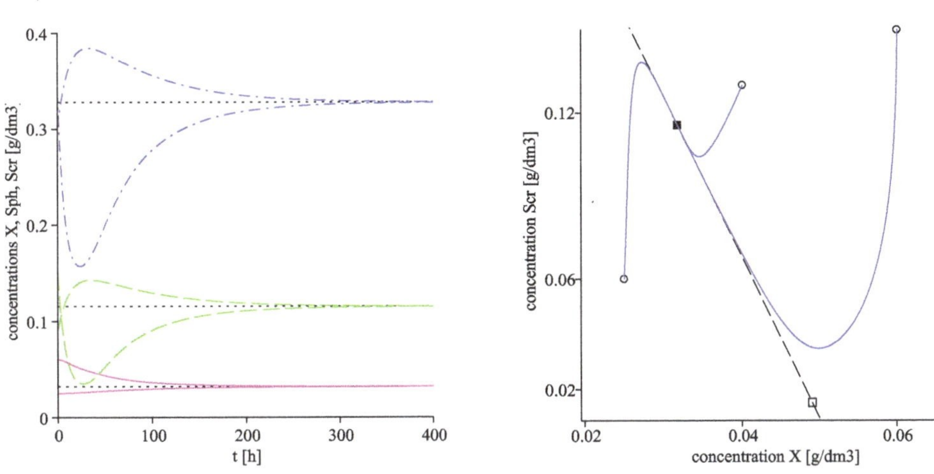

Figure 4. $D = 0.08$. (**Left**): time evolution of solutions X (solid line), S_{ph} (dash-dot line) and S_{cr} (dash line); the horizontal dot lines pass through the corresponding components of the equilibrium point E_3. (**Right**): projections of the trajectories in the (X, S_{cr})-phase plane with three different initial points, denoted by circles. The corresponding equilibrium components of E_3 are marked by a solid box, of E_2 are denoted by a box. The dash line presents the a projection of invariant plane $S_{cr} + k_{cr} X = S_{cr}^0$.

Example 2. $D = 0.095 > D_{cr}^{(2)} \approx 0.0865$

In this case there exists only one interior equilibrium point $E_2 = (0.0514, 0.0987, 0.00191)$ which is locally asymptotically unstable. The wash-out equilibrium $E_0 = (0, 0.7, 0.3)$ is the globally asymptotically stable steady state according to Theorem 3.

The global stability of E_0 for large values of the control parameter D means total wash-out of the biomass X and thus no detoxification of the bioreactor medium.

The left plot in Figure 6 visualizes the convergence of the solutions towards the corresponding components of E_0 using two different starting points. The right plot of Figure 6 as well as Figure 7 visualize projections of the trajectories in the phase planes (X, S_{cr}), (X, S_{ph}) and (S_{ph}, S_{cr}) respectively with three different initial points, marked by circles. The latter three plots also visualize the projections of the invariant planes in the corresponding phase planes.

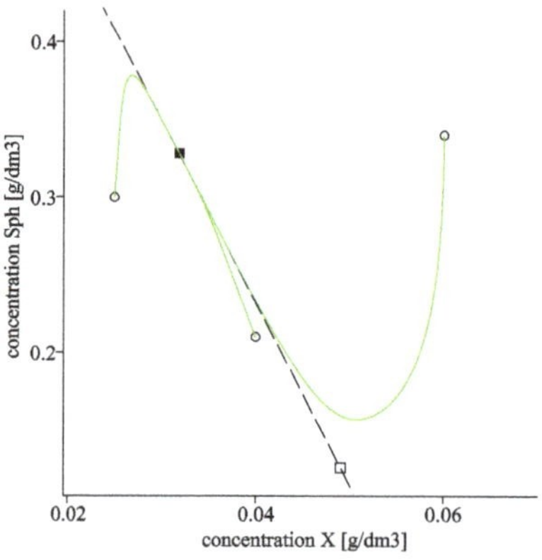
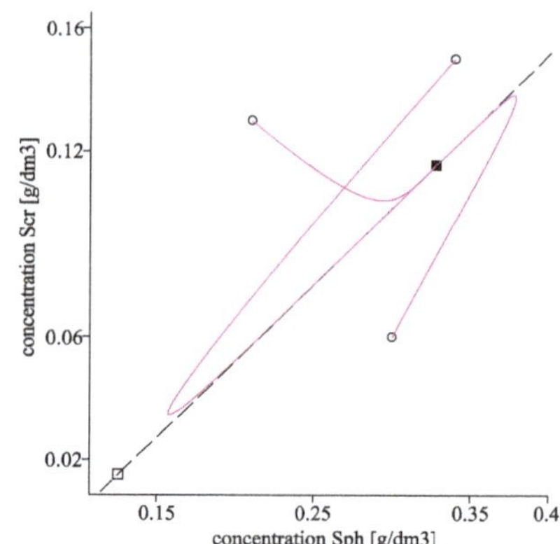

Figure 5. $D = 0.08$. Projections of the trajectories in the (X, S_{ph})-phase plane (**left**) and in the (S_{ph}, S_{cr})-phase plane (**right**) with three different initial points, denoted by circles. The corresponding equilibrium components of E_3 are marked by solid boxes, of E_2 are denoted by boxes. The dash lines present projections of the invariant planes $S_{ph} + k_{ph}X = S_{ph}^0$ (left) and $S_{ph} - KS_{cr} = S^0$ (right).

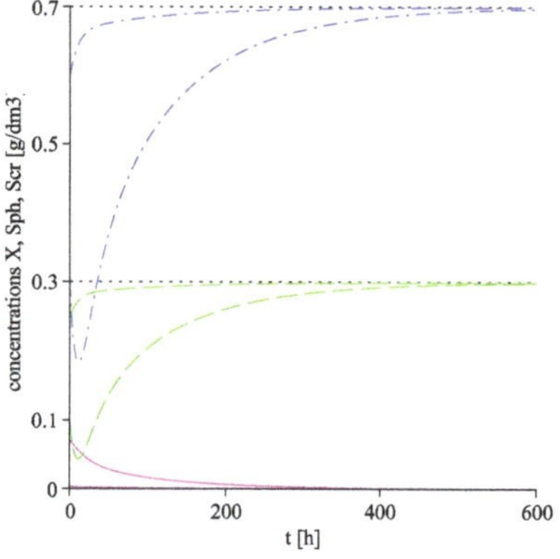
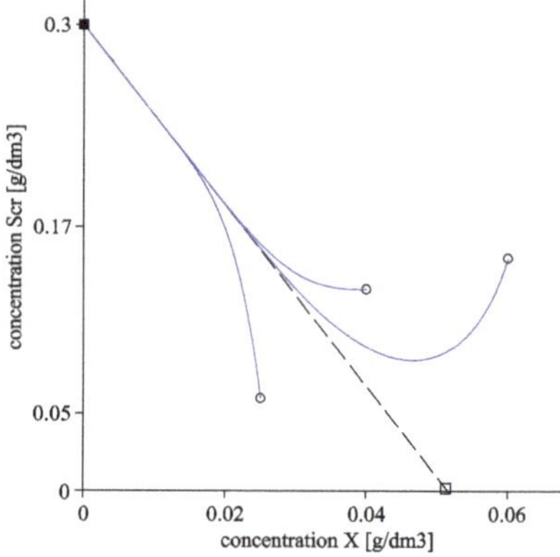

Figure 6. $D = 0.095$. (**Left**): time evolution of solutions X (solid line), S_{ph} (dash-dot line) and S_{cr} (dash line); the horizontal dot lines pass through the corresponding components of the equilibrium point E_0. (**Right**): projections of the trajectories in the (X, S_{cr})-phase plane with three different initial points, denoted by circles. The corresponding equilibrium components of E_0 are marked by a solid box, of E_2 are denoted by a box. The dash line presents a projection of the invariant plane $S_{cr} + k_{cr}X = S_{cr}^0$.

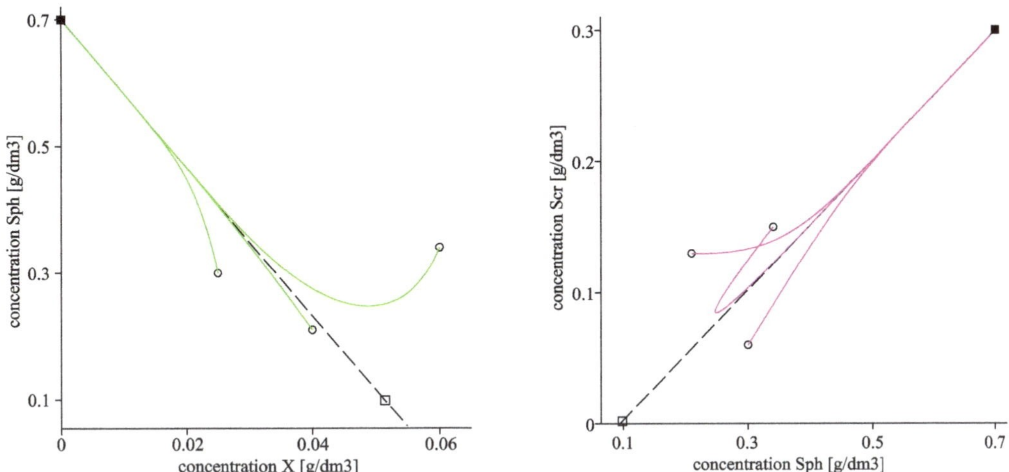

Figure 7. $D = 0.095$. Projections of the trajectories in the (X, S_{ph})-phase plane (**left**) and in the (S_{ph}, S_{cr})-phase plane (**right**) with three different initial points, denoted by circles. The corresponding components of E_0 are marked by solid boxes, of E_2 are denoted by boxes. The dash lines present projections of the invariant planes $S_{ph} + k_{ph}X = S_{ph}^0$ (left) and $S_{ph} - KS_{cr} = S^0$ (right).

7. Conclusions

We perform a mathematical analysis of a dynamic model, describing phenol and 4-methylphenol (p-cresol) biodegradation in a continuously stirred tank bioreactor. The model is described by three nonlinear ordinary differential equations and presents an extension of the batch growth model given in [17] to perform the ability of *Aspergillus awamori* strain to degrade the mixture of phenol and p-cresol. The novel idea is the usage of *sum kinetic interaction parameters* in the analytic expression of the microorganisms specific growth rate $\mu(S_{ph}, S_{cr})$ in the medium, as well as inhibition terms with respect to both phenol and p-cresol concentrations. The advantages of using such kind of specific growth rates is validated by practical laboratory experiments [17], see also [5]. To our knowledge, such kind of dynamic models, describing biodegradation in continuous biorectors (chemostats), are not studied in the literature until now.

We compute the equilibrium points of the model and investigate their local asymptotic stability as well as existence of bifurcations in dependence of the input control parameter D, the dilution rate. It is shown that an equilibrium $E_0 = \left(0, S_{ph}^0, S_{cr}^0\right)$, corresponding to total wash-out of the biomass in the bioreactor, exists for all $D > 0$. We find values of D such that two interior (coexistence) equilibria E_2 and E_3 do exist: E_3 is defined for $D \in \left(D_{cr}^{(1)}, D_{cr}^{(2)}\right)$, and E_2 exists if $D \in \left(D_{cr}^{(1)}, D_{cr}\right)$. Local stability analysis shows that E_2 is locally asymptotically unstable and E_3 is locally asymptotically stable where they exist, E_0 is locally asymptotically stable for $D > D_{cr}^{(2)}$. Two types of bifurcations of the equilibria occur, a saddle node bifurcation at $D = D_{cr}^{(1)}$ where E_2 and E_3 coalesce, and a transcritical bifurcation at $D = D_{cr}^{(2)}$, where E_0 coincides with E_3 and E_3 disappears for $D > D_{cr}^{(2)}$. Practically, the bifurcation values $D_{cr}^{(1)}$ and $D_{cr}^{(2)}$ of D should be carefully avoided, because small nearby perturbations may cause destabilization of the process, leading to total wash-out of the biomass. Most of the computations are carried out numerically due to the complicated expression of the model function $\mu(\cdot)$ and the large number of model parameters. The computations are performed in the computer algebra system *Maple*. The most important property of the model solutions—existence, uniqueness and uniform boundedness—is established theoretically in Theorem 1. We also prove (Theorem 2) the global asymptotic stability of the interior equilibrium point E_3 when D takes values within certain bounds, $D \in \left(D_{cr}^{(1)}, D_{cr}^{(2)} \right)$, $D_{cr}^{(2)} = \mu(S_{ph}^0, S_{cr}^0)$. The existence of these bounds for D is not restrictive in practical applications, since the dilution rate D is proportional to the speed of the pumping mechanism which feeds the bioreactor, thus there always exist a lower and an upper bound for D [27]. Choosing D in the interval $\left(D_{cr}^{(1)}, D_{cr}^{(2)} \right)$ ensures practically long-term sustainability of the bioremediation process in the bioreactor. On the other hand, large values of the dilution rate D, $D > D_{cr}^{(2)} = \mu(S_{ph}^0, S_{cr}^0)$, may cause total wash-out of the biomass in the reactor and may lead to process breakdown. This is due to the fact that the wash-out equilibrium $E_0 = (0, S_{ph}^0, S_{cr0})$ is the global attractor of the dynamics (Theorem 3). The dynamic behavior of the model solutions is illustrated by some numerical examples for different values of the dilution rate.

Author Contributions: Conceptualization, N.D. and P.Z.; methodology, N.D.; software, N.D.; theoretical investigation, N.D.; validation, N.D. and P.Z.; data curation, P.Z.; writing—original draft preparation, N.D.; writing—review and editing, N.D. and P.Z. Both authors have read and agreed to the published version of the manuscript.

Funding: This research received no external funding.

Acknowledgments: This work has been partially supported by the National Scientific Program "Information and Communication Technologies for a Single Digital Market in Science, Education and Security (ICTinSES)", contract No DO1–205/23.11.2018, financed by the Ministry of Education and Science in Bulgaria. The work of the first author has been partially supported by grant No BG05M2OP001-1.001-0003, financed by the Science and Education for Smart Growth Operational Program (2014–2020) in Bulgaria and co-financed by the European Union through the European Structural and Investment Funds.

Conflicts of Interest: The authors declare no conflict of interest.

References

1. Singh, P.; Jain, R.; Srivastava, N.; Borthakur, A.; Pal, D.B.; Singh, R.; Madhav, S.; Srivastava, P.; Tiwary, D.; Mishra, P.K. Current and emerging trends in bioremediation of petrochemical waste: A review. *Crit. Rev. Environ. Sci. Technol.* **2017**, *47*, 155–201. [CrossRef]
2. Seo, J.-S.; Keum, Y.-S.; Li, Q.X. Bacterial degradation of aromatic compounds. *Int. J. Environ. Res. Public Health* **2009**, *6*, 278–309. [CrossRef] [PubMed]
3. Sharma, N.K.; Philip, L.; Bhallamudi, S.M. Aerobic degradation of phenolics and aromatic hydrocarbons in presence of cyanide. *Bioresour. Technol.* **2012**, *121*, 263–273. [CrossRef] [PubMed]
4. Tomei, M.C.; Annesini, M.C. Biodegradation of phenolic mixtures in a sequencing batch reactor: A kinetic study. *Environ. Sci. Pollut. Res.* **2008**, *15*, 188–195. [CrossRef] [PubMed]
5. Yemendzhiev, H.; Zlateva, P.; Alexieva, Z. Comparison of the biodegradation capacity of two fungal strains toward a mixture of phenol and cresol by mathematical modeling. *Biotechnol. Biotechnol. Equip.* **2012**, *26*, 3278–3281. [CrossRef]
6. Kietkwanboot, A.; Chaiprapat, S.; Müller, R.; Suttinun, O. Biodegradation of phenolic compounds present in palm oil mill effluent as single and mixed substrates by Trameteshirsuta AK04. *J. Environ. Sci. Heal. Part A Toxic/Hazard. Subst. Environ. Eng.* **2020**, *55*, 989–1002. [CrossRef]
7. Liu, J.; Jia, X.; Wen, J.; Zhou, Z. Substrate interactions and kinetics study of phenolic compounds biodegradation by Pseudomonas sp. cbp1-3. *Biochem. Eng. J.* **2012**, *67*, 156–166. [CrossRef]
8. Kumar, S.; Arya, D.; Malhotra, A.; Kumar, S.; Kumar, B. Biodegradation of dual phenolic substrates in simulated wastewater by Gliomastixindicus MTCC 3869. *J. Environ. Chem. Eng.* **2013**, *1*, 865–874. [CrossRef]
9. Angelucci, D.M.; Annesini, M.C.; Tomei, M.C. Modelling of biodegradation kinetics for binary mixtures of substituted phenols in sequential bioreactors. *Chem. Eng. Trans.* **2013**, *32*, 1081–1086. [CrossRef]
10. Lepik, R.; Tenno, T. Biodegradability of phenol, resorcinol and 5-methylresorcinol as single and mixed substrates by activated sludge. *Oil Shale* **2011**, *28*, 425–446. [CrossRef]
11. Yoon, H.; Klinzing, G.; Blanch, H.W. Competition for mixed substrate by microbial populations. *Biotechnol. Bioeng.* **1977**, *19*, 1193–1210. [CrossRef] [PubMed]
12. Reardon, K.F.; Mosteller, D.C.; Rogers, J.D.B. Biodegradation kinetics of benzene, toluene, and phenol as single and mixed substrates for *Pseudomonas Putida* F1. *Biotechnol. Bioeng.* **2000**, *69*, 385–400. [CrossRef]
13. Reardon, K.F.; Mosteller, D.C.; Rogers, J.B.; DuTeau, N.M.; Kim, K.-H. Biodegradation kinetics of aromatic hydrocarbon mixtures by pure and mixed bacterial cultures. *Environ. Health* **2002**, *110*, 1005–1011. [CrossRef] [PubMed]

14. Abuhamed, T.; Bayraktar, E.; Mehmetoğlu, T.; Mehmetoğlu, Ü. Kinetics model for growth of *Pseudomonas Putida* F1 Benzene, Toluene Phenol Biodegrad. *Process Biochem.* **2004**, *39*, 983–988. [CrossRef]
15. Datta, A.; Philip, L.; Bhallamudi, S.M. Modeling the biodegradation kinetics of aromatic and aliphatic volatile pollutant mixture in liquid phase. *Chem. Eng. J.* **2014**, *241*, 288–300. [CrossRef]
16. Hazrati, H.; Shayegan, J.; Seyedi, S.M. Biodegradation kinetics and interactions of styrene and ethylbenzene as single and dual substrates for a mixed bacterial culture. *J. Environ. Health Sci. Eng.* **2015**, *13*, 72, [CrossRef] [PubMed]
17. Yemendzhiev, H.; Gerginova, M.; Zlateva, P.; Stoilova, I.; Krastanov, A.; Alexieva, Z. Phenol and cresol mixture degradation by *Aspergillus awamori* strain: Biochemical and kinetic substrate interactions. *Proc. ECOpole* **2008**, *2*, 153–159. Available online: https://ecesociety.com/proceedings-of-ecopole-peco (accessed on 7 January 2021).
18. Dimitrova, N.; Zlateva, P. Stability Analysis of a Model for Phenol and Cresol Mixture Degradation. In Proceedings of the IOP Conference Series: Earth and Environmental Science (EES), Xiamen, China, 7–9 June 2019; Volume 356, p. 012209, [CrossRef]
19. Wiggins, S. *Introduction to Applied Nonlinear Dynamical Systems and Chaos*; Springer: New York, NY, USA, 1990; [CrossRef]
20. Smith, H.L.; Waltman, P. *The Theory of the Chemostat: Dynamics of Microbial Competition*; Cambridge University Press: Cambridge, UK, 1995; [CrossRef]
21. Thieme, H.R. Convergence results and a Poincaré–Bendixon trichotomy for asymptotically autonomous differential equations. *J. Math. Biol.* **1992**, *30*, 755–763. [CrossRef]
22. Harmand, J.; Lobry, C.; Rapaport, A.; Sari, T. *The Chemostat: Mathematical Theory of Microorganism Cultures*; Volume 1 of Chemical Engineering Series; Wiley: Hoboken, NJ, USA, 2017; [CrossRef]
23. Hsu, S.B. Limiting behavior of competing species. *SIAM J. Appl. Math.* **1978**, *34*, 760–763. [CrossRef]
24. Wolkowicz, G.S.K.; Lu, Z. Global dynamics of a mathematical model of competition in the chemostat: General response function and differential death rates. *SIAM J. Appl. Math.* **1992**, *52*, 222–233. [CrossRef]
25. Dimitrova, N.S.; Krastanov, M.I. Model-based optimization of biogas production in an anaerobic biodegradation process. *Comput. Math. Appl.* **2014**, *68*, 986–993. [CrossRef]
26. Gopalsamy, K. *Stability and Oscillations in Delay Differential Equations of Population Dynamics*; Kluwer Academic Publishers: Dordrecht, The Netherlands, 1992; [CrossRef]
27. Grognard, F.; Bernard, O. Stability analysis of a wastewater treatment plant with saturated control. *Water Sci. Technol.* **2006**, *53*, 149–157. [CrossRef] [PubMed]

Article

Design, Implementation and Simulation of a Small-Scale Biorefinery Model

Mihaela Sbarciog †, Viviane De Buck, Simen Akkermans, Satyajeet Bhonsale, Monika Polanska and Jan F. M. Van Impe *

BioTeC+, Chemical and Biochemical Process Technology and Control, Department of Chemical Engineering, KU Leuven, 9000 Gent, Belgium; mihaelaiuliana.sbarciog@kuleuven.be (M.S.); viviane.debuck@kuleuven.be (V.D.B.); simen.akkermans@kuleuven.be (S.A.); satyajeetsheetal.bhonsale@kuleuven.be (S.B.); monika.polanska@kuleuven.be (M.P.)
* Correspondence: jan.vanimpe@kuleuven.be
† Current address: 3BIO-BioControl, Université Libre de Bruxelles, 1050 Brussels, Belgium.

Abstract: Second-generation biomass is an underexploited resource, which can lead to valuable products in a circular economy. Available locally as food waste, gardening and pruning waste or agricultural waste, second-generation biomass can be processed into high-valued products through a flexi-feed small-scale biorefinery. The flexi-feed and the use of local biomass ensure the continuous availability of feedstock at low logistic costs. However, the viability and sustainability of the biorefinery must be ensured by the design and optimal operation. While the design depends on the available feedstock and the desired products, the optimisation requires the availability of a mathematical model of the biorefinery. This paper details the design and modelling of a small-scale biorefinery in view of its optimisation at a later stage. The proposed biorefinery comprises the following processes: steam refining, anaerobic digestion, ammonia stripping and composting. The models' integration and the overall biorefinery operation are emphasised. The simulation results assess the potential of the real biowaste collected in a commune in Flanders (Belgium) to produce oligosaccharides, lignin, fibres, biogas, fertiliser and compost. This represents a baseline scenario, which can be subsequently employed in the evaluation of optimised solutions. The outlined approach leads to better feedstocks utilisation and product diversification, raising awareness on the impact and importance of small-scale biorefineries at a commune level.

Keywords: biomass; biorefinery design; process integration; scheduling; simulation

Citation: Sbarciog, M.; De Buck, V.; Akkermans, S.; Bhonsale, S.; Polanska, M.; Van Impe, J.F.M. Design, Implementation and Simulation of a Small-Scale Biorefinery Model. *Processes* **2022**, *10*, 829. https://doi.org/10.3390/pr10050829

Academic Editors: Philippe Bogaerts and Alain Vande Wouwer

Received: 31 January 2022
Accepted: 15 April 2022
Published: 22 April 2022

Publisher's Note: MDPI stays neutral with regard to jurisdictional claims in published maps and institutional affiliations.

1. Introduction

Lignocellulosic biomass represents an abundant and sustainable carbon-rich feedstock that can be employed to produce, e.g., (bio-)chemicals, biofuels, fibres and nutrients. As the urge to employ sustainable processing methods and feedstocks is becoming ever greater with regard to climate change, interest in how to optimally process this feedstock has steadily increased over the past decade. To obtain valuable and useful products, the dense crystalline structure of lignocellulosic biomass needs to be broken down. The three main building-blocks of lignocellulosic biomass are: *(i)* lignin, *(ii)* cellulose and *(iii)* hemicellulose (see Figure 1). While cellulose is an extremely dense structure, mainly consisting of $\beta(1 \rightarrow 4)$-D-Glucose units, structured in a helical strand, hemicellulose and lignin are more diverse, containing a multitude of different compounds, the most important of which are represented in Figure 1.

The breakdown and conversion of lignocellulosic biomass in useful and value-added products occurs in a biorefinery. In [1], the authors presented a classification system to categorise biorefineries based on: *(i)* which platforms or intermediate products they used, *(ii)* the products they produced, *(iii)* the used feedstocks and *(iv)* the used processes. So-called first-generation biorefineries used food products, such as corn and wheat, as feedstock and converted these starch-rich and uniform feedstocks into bulk products such as biofuels and other platform chemicals. In essence, these biorefineries were the biomass-based counterpart of regular petroleum refineries (where raw fossil oil is refined into value-added products). However, as they used food as feedstock and they competed with the food industry for arable land, these enterprises were faced with heavy criticism [1,2].

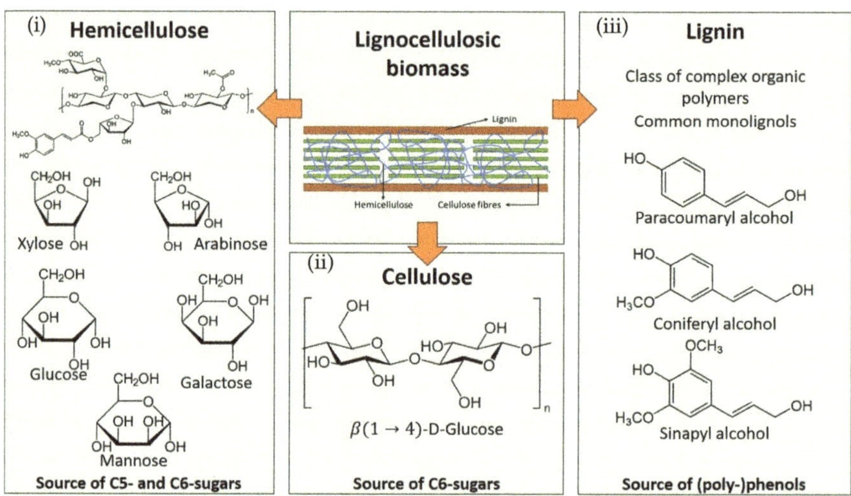

Figure 1. Schematic representation of lignocellulosic biomass and its three main building-blocks: *(i)* hemicellulose, *(ii)* cellulose and *(iii)* lignin (taken from [3]).

Second-generation biorefineries were developed as a response to this major flaw of first-generation biorefineries. In a second-generation biorefinery, the feedstock that is processed is a waste, or a non-food stream [1,2,4]. In particular, the usage of biowaste has become of great interest lately, as biowaste biorefineries additionally act as waste processing facilities [5], thus playing a central role in the circular economy. When regarding the application potential of biorefineries on a local level, their ability to revalorise biowaste in locally desired products is a compelling advantage [6,7]. (Lignocellulosic) Waste streams, however, have the major disadvantage that their composition, due to their waste nature, is no longer uniform [8,9]. Moreover, when only considering local biowaste streams, their periodic yields are also limited. Combined, these two disadvantages make biowaste streams unsuitable for the production of bulk products such as biofuels. Unlike first-generation biorefineries, the processes employed in a local and small-scale biorefinery need to be more robust and flexible, as the feedstock's composition may be variable. Additionally, the selected processes should render high-value products, as the overall feedstock throughput of the biorefinery will be limited due to its local nature [10]. In addition, Refs. [4,9] indicated that the production of fuels by biorefineries is inherently unsustainable. The combination of both considerations has led to a steady increase in interest regarding the production of high-value-added products from biowaste streams using biorefinery systems.

Even though the advantages of small-scale biorefineries have been indicated by multiple research studies, their implementation is lagging behind [6,7]. In particular, with regards to planning and designing a suitable small-scale and locally adapted biorefinery, decision-makers and investors are faced with a multitude of challenges and uncertainties. These uncertainties can often lead to the absence of the required investments to build biorefinery facilities. This, on its turn, leads again to an increased lack of trust in the considered systems [11,12]. (Online) Decision support tools (DST) are a convenient alternative for aiding decision-makers in designing the most suitable biorefinery systems for their particular settings as well as to increase their confidence in the proposed process layout [11,12].

An important part of developing a small-scale and locally embedded biorefinery consists of selecting the most suitable processes for converting locally available biowaste streams into value-added and desirable products [3]. Mathematical models of the considered processes allow for an easy assessment of the performance of the investigated processing chain. Moreover, mathematical models can be used to optimise the process flow-sheet itself as well as the processing conditions. Mathematical models of biorefinery conversion processes can range from static to dynamic models. Whereas the first type of models consider a steady-state process, i.e., time has no influence on the states of the process, a dynamic model allows for a more accurate prediction of the process's states and outputs, which are inherently influenced by the time variation of the inputs and disturbances acting on the process. Especially when considering batch processes, such as steam refining and composting, the duration of the process has a major influence on the process outcome. An additional advantage of dynamic models is that they allow for optimising the way a biorefinery is operated, i.e., imposing an (on-line) optimal control system on the biorefinery [13,14]. Static process models would only allow for a yield-based process optimisation. With regard to optimising the process flow-sheet itself, often, the so-called superstructure modelling system is employed [15–17]. To increase the extendability and user-friendliness of these superstructure models (which often consider a vast amount of different processes and/or process combinations), a simplified and generalised modelling framework is employed for all the considered processes. In this contribution, a similar superstructure model is being developed; however, as the scope of the foreseen biorefinery is limited to local and small-scale processing of biowaste streams, the number of processes selected using expert knowledge is limited. Moreover, as the eventual goal is to submit the proposed design to a process and a control optimisation, dynamic models have been selected for the considered processes. The developed model can be used in the first stage to assess the potential of locally available biowaste for producing value-added products. Subsequently, it may be employed to optimise the entire operation of the biorefinery in view of maximising the production of desired products. The model can be integrated together with a bioinventory tool, which provides a survey of the available biowaste, and an optimisation tool, which may take into account also sustainable indicators, into a decision support tool that allows for the design of small-scale, flexi-feed, sustainable biorefineries in a local setting.

The remainder of this contribution consists of a thorough discussion on the design of a local small-scale biorefinery in Section 2, followed by an in-depth discussion in Section 3 on the employed models for each process that was incorporated in the design, the models' integration and the biorefinery operation. Section 4 focuses on the obtained results from simulating the biorefinery operation for the biowaste available in a commune in the Flanders region. Finally, conclusions, as well as some remarks with regard to future research, are drawn in Section 5.

2. Biorefinery Design

A biorefinery can be simply considered to consist of three parts that are linked together: *(i)* the input or feedstock part, *(ii)* the process part and *(iii)* the output or product part. All three parts come with their own set of distinct (strategical) design choices that need to be considered when designing a biorefinery system (see Figure 2) [18]. Moreover, the choices made in one part of the biorefinery system will inevitably influence the choices that have to be made with regard to the other parts. To illustrate this, when the biorefinery is designed with the ethos of producing a certain product, e.g., biofuel, only a small set of suitable conversion processes and feedstocks will remain eligible for selection.

A local and small-scale biorefinery should tailor for local needs: local feedstocks should be converted into locally desired products. As the biorefinery system proposed in this contribution should additionally function as a waste-processing facility, the overall biorefinery system is designed starting from the feedstock part.

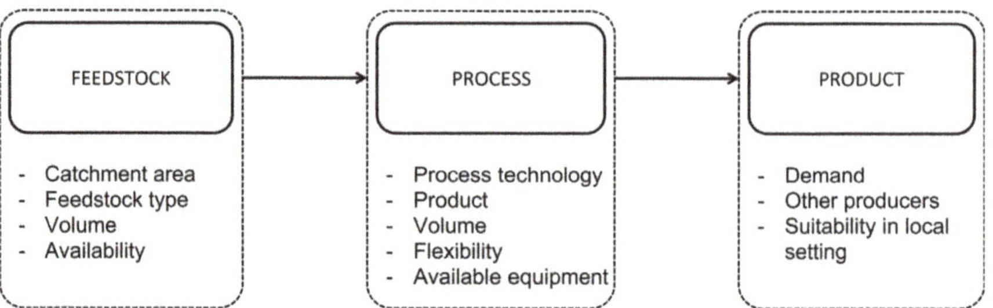

Figure 2. Simplified outline of a biorefinery with its three parts (feedstock, process and product), and their respective main strategic decisions (taken from [18]).

When designing a biorefinery from the feedstock side, one of the main aspects to consider is which feedstock is supposed to be processed by the facility. The most optimal conversion processes for a dense and highly crystalline feedstock will differ considerably from those employed when a soft feedstock is being processed. Additionally, the selection of certain conversion processes over others will have an influence on which volumes the biorefinery can handle and/or which products the facility will render.

2.1. Designing a Local Small-Scale Biorefinery

The two main lignocellulosic feedstocks that should be considered in a Flemish setting are kitchen waste and wood waste obtained from garden and landscape maintenance. Based on expert knowledge, a set of flexible and robust processes which are able to process either one or both of the considered feedstock streams were selected. The concatenation of the proposed conversion processes was obtained by employing a reverse-engineering design approach. More specifically, for each of the proposed feedstocks and for all of the proposed conversion processes, which products could be obtained was assessed. Based on the selected (set of) products, the required processes were coupled in such a way that the net flow of waste streams, i.e., undervalorised outflows of organic material, was minimised.

2.2. Proposed Integrated Small-Scale Biorefinery for a Flemish Setting

Figure 3 displays the eventually obtained flow-sheet of the small-scale biorefinery that will be presented and modelled in this contribution. Note that this flow-sheet represents the overarching, or superstructure, of the proposed small-scale and flexi-feed biorefinery in a Flemish setting. Depending on which products the biorefinery operators are interested in and/or their local circumstances with regard to equipment, the flow-sheet can be adapted to fit these local needs.

The main processes have been selected based on the foreseen feedstocks that could be processed by the biorefinery, i.e., two major lignocellulosic feedstocks that can be found in the case study region of Flanders: kitchen waste and (postconsumer) wood waste. Four main processes have been selected based on expert knowledge: *(i)* steam refining, *(ii)* anaerobic digestion, *(iii)* ammonia stripping and *(iv)* composting.

The four main processes are concatenated in such a way that the waste/output streams of each process are, on their turn, maximally utilised. This overarching integrated biorefinery design has been defined using expert knowledge.

Wood waste obtained from garden and landscape maintenance (i.e., pruning waste) is initially submitted to a pretreatment step consisting of chipping and sieving. Small enough wood chips are processed using steam refining, whereas bigger pieces (on average, this residue stream accounts for 5% of the total wood waste) are processed directly via a composting process.

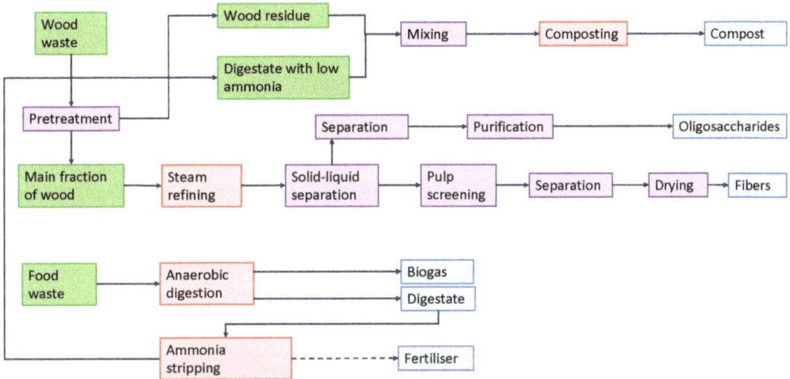

Figure 3. Schematic representation of the small-scale biorefinery modelled in this contribution. (Intermediate) Feedstocks are represented in green, main processes are represented in red, secondary processes are represented in purple, and (intermediate) products are represented in blue.

A steam refiner is, in essence, a mill consisting of two grinding wheels spinning in opposite directions between which the wood chips are ground down. To facilitate this process, steam is injected in the mill to weaken the crystalline structure of the lignocellulosic chips and to extract small molecules from the wood chips. The two main product categories of a steam refining process are solid fibres and a liquid extract containing a broad range of components, e.g., oligosaccharides. The obtained fibres can be further processed into paper, while the extract stream can be further refined to nutrients, tensides, etc.

Kitchen waste is a fairly soft feedstock and therefore does not require severe conversion processes like wood waste does. In the region of Flanders, kitchen waste that is collected via home-to-home collection rounds is of compostable quality. However, when aiming to maximise the potential of this particular feedstock, other conversion processes can take place prior to the final composting step. As kitchen waste is a nitrogen-rich feedstock, it lends itself perfectly to being processed using an anaerobic digestion step, followed by an ammonia stripping step.

An anaerobic digester is a continuous process during which the input stream is broken down, in the absence of oxygen, by bacteria into biogas and digestate. Biogas is a mix of methane (CH_4), carbon dioxide (CO_2) and hydrogen (H_2), whereas digestate is a nitrogen-rich waste stream. Anaerobic digestion is one of the most commonly applied biorefinery processes. In particular, countries such as Germany and France are at the forefront with regard to biomethane facilities, with 232 and 131 biomethane plants, respectively [19]. The quality of the produced biogas is defined by the amount of CH_4 in the output stream, i.e., the more methane, the higher the quality of the biogas. The methane percentage in biogas can be optimised by either adjusting the process parameters (see also Section 3.2) or adjusting the composition of the feedstock streams that are entering the biorefinery. The latter can, for instance, be artificially obtained by mixing two feedstocks together [20].

Even though anaerobic digestion is a flexible process, which has been employed for the conversion of manure, wastewater sludge and other recalcitrant organic waste streams [21], it has the disadvantage that the conversion process is coupled with the production of a steady and nitrogen-rich waste stream: the digestate. As nitrogen is not dissimilated during the digestion process, it accumulates in the digester's nongaseous output stream. In the case study region of Belgium, processing such high volumes of a nitrogen-rich waste stream cannot be accomplished without removing the bulk of the nitrogen content from the waste stream. Moreover, the nitrogen still present in the digestate stream can be used to produce a nitrogen fertiliser. Unbound nitrogen (i.e., in its NH_3/NH_4^+ state) can be easily removed from a liquid stream at increased pH using air stripping.

The remaining digestate stream, now with low nitrogen content, still represents a constant and relatively high flow of organic material that can still be further processed. The final step in the proposed biorefinery design consists of a composting step of the digestate combined with the residual wood waste that could not be processed using the steam refining process.Whereas the anaerobic digestion and ammonia stripping are operated in a continuous mode, the composting process is operated as a batch process. Switching from a continuous production line to a batch-operated production line requires a decoupling, often obtained by the usage of holding tanks. Holding tanks can store the continuously produced input of the batch process until the batch reactor is available again. This topic is detailed in the next section.

3. Materials and Methods

3.1. Steam Refining

The steam refining model used in this biorefinery setup is a kinetic model developed in [22,23] for the treatment of birch wood at temperatures in the range $[180, 240]\ °C$, with the concept of a wood biorefinery in view. This implies the selective separation of the three main wood components (lignin, cellulose/xylose and hemicellulose/glucan), which may be further used for the production of high-value components. Consequently, the model includes the three main processes, i.e., delignification, degradation and conversion of xylan and degradation and conversion of glucan, which are briefly described below. The efficiency of the treatment is determined by the experiment temperature and duration. A schematic representation of the steam refining process is illustrated in Figure 4.

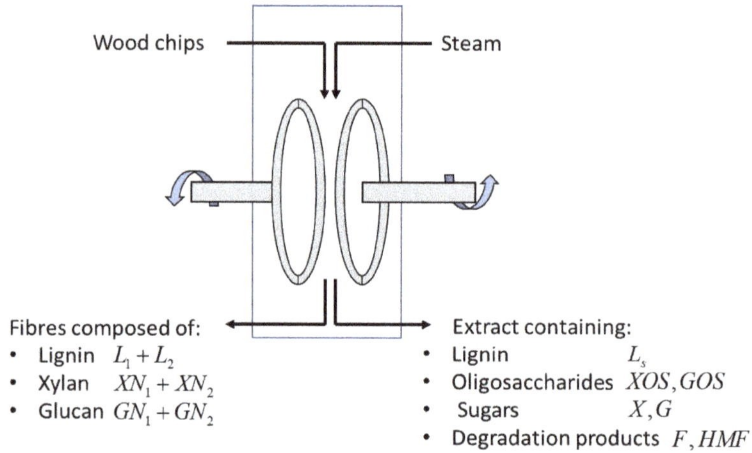

Figure 4. Schematic representation of a steam refiner.

The delignification process consists of a solubilisation reaction and a condensation reaction. In the solubilisation reaction, the lignin in the solid, which is divided into a hard-to-remove fraction L_1 and an easy-to-remove fraction L_2, is converted into solubilised lignin L_s. In the condensation reaction, a part of the solubilised lignin returns to the solid phase as condensed lignin L_c. Hence, the amount of lignin in the solid phase is determined by the sum $L_1 + L_2 + L_c$, while the amount of lignin in the extract is determined by L_s. The delignification process is mathematically described by:

$$\frac{dL_1}{dt} = -k_{s1}L_1$$

$$\frac{dL_2}{dt} = -k_{s2}L_2 \tag{1}$$

$$\frac{dL_s}{dt} = k_{s1}L_1 + k_{s2}L_2 - k_cL_s$$

$$\frac{dL_c}{dt} = k_cL_s$$

where k_{s1}, k_{s2} and k_c respectively represent the temperature-dependent kinetic rates of the hard degradable lignin fraction, easily degradable lignin fraction and condensed lignin. The numerical values of the kinetic rates and the initial hard and easily degradable fractions for the considered temperature range are given in [22] (Table 3).

The degradation and conversion of xylan and glucan follow the same pathway: polymers (xylan and glucan) are converted into their corresponding monomers (xylose and glucose) via intermediate oligosaccharides. Subsequently, the monomers are degraded to furfural, 5-hydroxymethylfurfural and other degradation products. Similar to the delignification process, it is assumed in [23] that both xylan and glucan in the solid consist of a fast-degrading fraction and a slow-degrading fraction. Considering first-order reactions, the conversion of xylan is described by

$$\frac{dXN_1}{dt} = -k_{x1} \cdot XN_1$$

$$\frac{dXN_2}{dt} = -k_{x2} \cdot XN_2$$

$$\frac{dXOS}{dt} = k_{x1} \cdot XN_1 + k_{x2} \cdot XN_2 - k_{x3} \cdot XOS \tag{2}$$

$$\frac{dX}{dt} = k_{x3} \cdot XOS - (k_{x4} + k_{x5}) \cdot X$$

$$\frac{dF}{dt} = k_{x4} \cdot X - k_{x6} \cdot F$$

while the conversion of glucan is given by

$$\frac{dGN_1}{dt} = -(k_{g1} + k_{g6}) \cdot GN_1$$

$$\frac{dGN_2}{dt} = -(k_{g2} + k_{g7}) \cdot GN_2$$

$$\frac{dGOS}{dt} = k_{g1} \cdot GN_1 + k_{g2} \cdot GN_2 - (k_{g3} + k_{g8}) \cdot GOS \tag{3}$$

$$\frac{dG}{dt} = k_{g3} \cdot GOS - (k_{g4} + k_{g9}) \cdot G$$

$$\frac{dHMF}{dt} = k_{g4} \cdot G - k_{g5} \cdot HMF$$

In (2), XN_1 and XN_2 denote respectively the fast- and slow-degrading xylan fractions, XOS denotes the xylo-oligosaccharides, X represents xylose and F represents furfural, while k_{xi} with $i = 1 \ldots 6$ are the temperature-dependent kinetic rates. Initial fractions of xylan in the solid as well as the parameters of the kinetic rates may be found in [23] (Tables 3 and 4). In (3), GN_1 and GN_2 denote respectively the fast- and slow-degrading glucan fractions, GOS denotes the gluco-oligosaccharides, G represents glucose and HMF represents 5-hydroxymethylfurfural, while k_{gj} with $j = 1 \ldots 9$ are the temperature-dependent kinetic rates. Initial fractions of glucan in the solid as well as the parameters of the kinetic rates may be found in [23] (Tables 5 and 6). Note that the states in (1) are expressed in percentages with respect to the original amount of lignin in the wood, which represents 22.36% of the dry wood mass (Table 1 in [22]), while the states in (2) and (3) are expressed in g per kg of dry wood.

3.2. Anaerobic Digestion

The anaerobic digestion model employed here is developed in [24]. It is an extension of the well-known ADM1 model [21], with the amendments made in [25], to accommodate the food waste digestion experimentally observed.

ADM1 is the most complex model describing the anaerobic digestion process. It includes the following conversion steps [21,26]: (i) disintegration of the composite material; (ii) hydrolysis of carbohydrates, proteins and lipids into their corresponding building blocks; (iii) acidogenesis, in which monosaccharides, amino-acids and long-chain fatty acids are fermented and short-chain organic acids, hydrogen, carbon dioxide and ammonia are produced; (iv) acetogenesis, in which various metabolic products of the previous degradation stages are mainly broken down into acetic acid, hydrogen and carbon dioxide, and (v) methanogenesis, in which mainly acetic acid and hydrogen are converted into methane. These conversion steps occur simultaneously and involve a variety of microorganisms. The process takes place in a continuous stirred tank reactor as schematically illustrated in Figure 5: waste is continuously supplied in the influent with the flow rate q (m^3/day), and an equal flow is withdrawn from the reactor such that the liquid volume V_{liq} remains constant; the produced biogas (composed of methane, carbon dioxide and hydrogen) leaves the reactor with the flow rate q_{gas}.

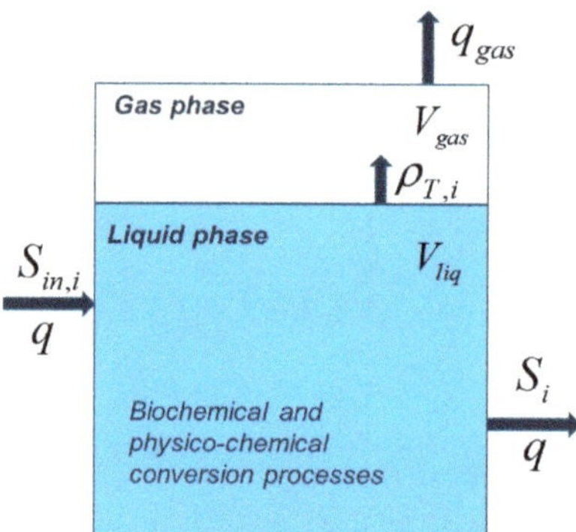

Figure 5. Anaerobic digestion system.

The ADM1 model consists of the mass balance of components in the solid-liquid and gas phases as summarised by

$$\frac{dS_i}{dt} = \frac{q}{V_{liq}}(S_{in,i} - S_i) + \sum_j \rho_j \nu_{ij} - \rho_{T,j} \tag{4}$$

$$\frac{dX_i}{dt} = \frac{q}{V_{liq}}(X_{in,i} - X_i) + \sum_j \rho_j \nu_{ij} \tag{5}$$

$$\frac{dS_{gas,i}}{dt} = -\frac{q_{gas}}{V_{gas}}S_{gas,i} + \frac{V_{liq}}{V_{gas}}\rho_{T,j} \tag{6}$$

where: S_i denotes the concentration of the soluble component i, X_i denotes the concentration of the particulate component i and $S_{gas,i}$ denotes the concentration in gas of either methane, carbon dioxide or hydrogen; $S_{in,i}/X_{in,i}$ is the concentration in the influent of soluble/particulate component i; ρ_j represents the reaction rate of process j, while ν_{ij} represents the stoichiometric coefficient of component i on the process j; $\rho_{T,j}$ represents the transfer rate to the gas j and V_{gas} denotes the headspace volume. Note that the only soluble components which are transferred to the gas phase are methane, hydrogen and inorganic carbon. Except for nitrogen and carbon concentrations expressed in kmol/m^3, all the other concentrations are expressed as kgCOD/m^3.

In addition to the soluble matter, particulate matter and gas components, ADM1 includes balances for anions, cations and ion states, allowing for an accurate calculation of the process pH. The implementation reported by [25] consists of 35 differential equations and 8 algebraic equations, while the implementation [24] used in this work considers the acetate oxidation pathway, which is proved to make significant contributions to methane production and, in some cases, become more important than the acetoclastic methanogenesis. Compared to the original ADM1 implementation,

- two new processes are included, namely the acetate degradation by a new biomass group of acetate oxidisers and the decay of the new biomass group;
- hydrogen ions' concentration used to compute pH is a state of the model [27];
- some parameters are re-estimated to account for the digestion of waste with high nitrogen content such as food waste.

The main food waste characteristics and their translation into inlet concentrations of ADM1 model are also found in [24].

3.3. Ammonia Stripping

The ammonia stripping process [28–30] takes place in a closed vessel, where air is continuously sparged at its bottom and its content is continuously agitated. Ammonia is transferred from the liquid to the air bubbles, which leave the vessel. To facilitate the phase transition, the pH of the system must be increased, which requires the addition of a base. After the stripping, acid is added to re-establish the pH. The air outflow rich in ammonia is supplied to ammonia scrubbing process (not considered here), which allows the retrieval of ammonia as salt that can be used as fertiliser. A schematic representation of the ammonia stripping process is illustrated in Figure 6.

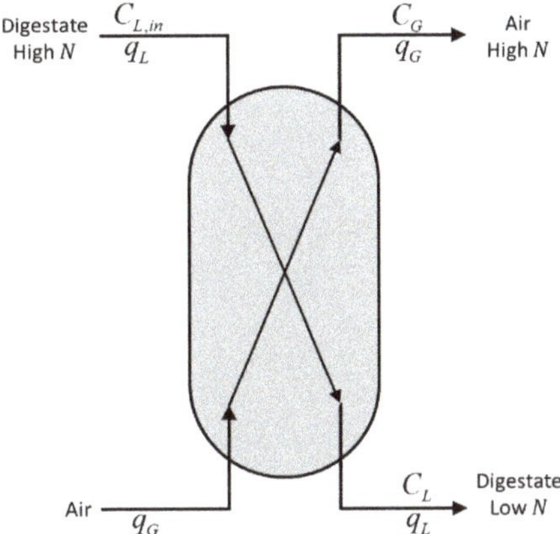

Figure 6. Schematic representation of the ammonia stripping process.

The mathematical model [28–30] describing the process relies on the mass balance for ammonia:

$$V_L \frac{dC_L}{dt} + \varepsilon_G V_L \frac{dC_G}{dt} = q_L(C_{L,in} - C_L) + q_G(C_{G,in} - C_G) \tag{7}$$

where q_G is the volumetric air flow rate (m^3/s), $C_{G,in}$ and C_G are the ammonia concentrations in the air entering and leaving the vessel (kmol/m^3), V_L is the liquid volume (m^3), $C_{L,in}$ and C_L are the ammonia concentrations in the liquid phase entering and leaving the vessel (kmol/m^3) and ε_G is the air holdup or the volume fraction of the air bubbles entrained in the liquid (dimensionless). Since [29]

- no ammonia is present in the influent air ($C_{G,in} = 0$);

- ammonia accumulation in the air bubbles is insignificant;
- the equilibrium equation between air and liquid for any gas is given by Henry's law;
- and the output stripping gas is probably not close to saturation,

the mass balance (7) reduces to

$$V_L \frac{dC_L}{dt} = q_L(C_{L,in} - C_L) - q_G K_H C_L \tag{8}$$

where K_H denotes Henry's constant (dimensionless).

The efficiency of the ammonia stripping process is calculated as

$$\text{Removed ammonia}(\%) = \left(1 - \frac{C_L}{C_{L,in}}\right) \times 100 \tag{9}$$

3.4. Composting

Composting refers to the aerobic degradation of organic matter into valuable inert material (compost), which can be used as fertiliser for soil and plants.

The model of composting in biocells proposed in [31] relies on three main biochemical phenomena: (i) hydrolysis of the insoluble substrate, (ii) aerobic degradation of soluble substrate and (iii) biomass decay. The model describes the evolution of the main process variables (concentrations expressed in mol C/m^3): soluble substrate S, representing the material that can be directly degraded; insoluble substrate I, representing the waste which needs to be hydrolysed before conversion; biomass concentration X; liquid part L; inert material (compost) M, and oxygen concentration O. As the biocell is a closed system, oxygen has to be supplied to maintain the biomass growth and the overall conversion.

The mass balance equations read:

$$\frac{dS}{dt} = -\frac{1}{Y_S}\mu(S,O)X + K_h I \tag{10}$$

$$\frac{dI}{dt} = -K_h I + \frac{1}{Y_I} b X \tag{11}$$

$$\frac{dX}{dt} = \mu(S,O)X - bX \tag{12}$$

$$\frac{dL}{dt} = \frac{1}{Y_L}\mu(S,O)X \tag{13}$$

$$\frac{dM}{dt} = -\left(1 - \frac{1}{Y_S} + \frac{1}{Y_L}\right)\mu(S,O)X + \left(1 - \frac{1}{Y_I}\right)bX \tag{14}$$

$$\frac{dO}{dt} = -\frac{1}{Y_O}\mu(S,O)X + FO_{in} \tag{15}$$

where

$$\mu(S,O) = \mu_{max}\frac{S}{K_S + S}\frac{O}{K_O + O} \tag{16}$$

describes the biomass growth, which is limited by the availability of soluble substrate and oxygen. The numerical values of the parameters [31] are given in Table 1.

Table 1. Parameters' values for the composting model [31].

Parameter	Value	Parameter	Value
μ_{max}	$0.72\,\text{h}^{-1}$	Y_S	0.53
K_S	$0.2573\,\text{mol/m}^3$	Y_I	1.02
K_O	$2.822\,\text{mol/m}^3$	Y_L	1.34
K_h	$18 \times 10^{-4}\,\text{h}^{-1}$	Y_O	1.12
b	$0.1368\,\text{h}^{-1}$	O_{in}	$8\,\text{mol/m}^3$

3.5. Models' Integration and Scheduling

In a biorefinery, several processes are connected in a cascade, where the output of one process is fed as input to another process. Additionally, some processes may be operated in continuous mode, while others may be operated in batch or fed-batch mode. Hence, simulating the biorefinery operation and evaluating its efficiency requires models' integration and scheduling.

The biorefinery designed in this work consists of two branches, and its operation is indicated in Figure 7. The first branch comprises only one process, the steam refining process. Steam refining is a process which is operated in batch mode. Depending on the availability of wood waste and storage capacity, reactor size and considerations on energy use, the treatment of wood waste could be accomplished in real life in either one batch or repetitive batches. On the second branch, anaerobic digestion is a process operated in continuous mode, composting is a process operated in batch mode while ammonia stripping could be operated either continuously or in batch mode. For simplicity, we consider that the ammonia stripping process is operated in continuous mode, at the same rate as the anaerobic digestion process ($q = q_L$). To buffer the transition between the continuous and batch operations, the low ammonia digestate exiting the ammonia stripping process is collected for a period of time $\triangle t$ (days) in a tank, whose content is then transferred to a composting cell. Composting is a slow process which might not end in $\triangle t$ days, the next period for filling in the buffer tank. Hence, several composting cells can be used in parallel, as shown in Figure 7.

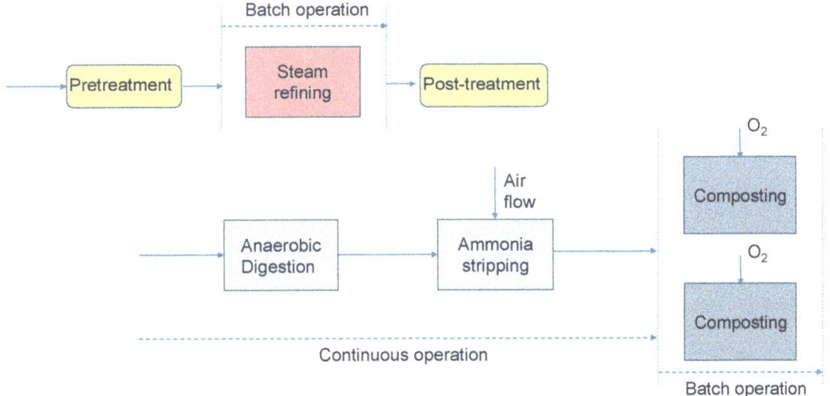

Figure 7. Biorefinery operation.

Assuming that no biochemical reaction takes place in the tank, its dynamics in terms of the states of the composting model are given by

$$\frac{dV}{dt} = q$$
$$\frac{dS}{dt} = \frac{q}{V} \cdot (S_{in} - S)$$
$$\frac{dI}{dt} = \frac{q}{V} \cdot (I_{in} - I) \tag{17}$$
$$\frac{dM}{dt} = \frac{q}{V} \cdot (M_{in} - M)$$

where V represents the volume of digestate accumulated in the tank, S, I and M are respectively the concentrations of soluble, insoluble and inert matter in the tank, while S_{in}, I_{in} and M_{in} are respectively the concentrations of soluble, insoluble and inert matter in the influent digestate.

Generally, the models employed to simulate a biorefinery are built for each specific process with the main goal of characterising its dynamics with a certain degree of detail. Thus, connecting two such models requires a good mapping of the first model's outputs to the second model's inputs. While anaerobic digestion is a complex process with a detailed characterisation of organic matter, its connection with the ammonia stripping model is straightforward, as the concentration of ammonia appears both in the effluent of the anaerobic digestion model and the influent of the ammonia stripping model ($S_{NH_3} = C_{L,in}$). Assuming that no reaction takes place during ammonia stripping, the next step is to match the mix of digestate with wood to the states of the composting process.

In the first stage, the characteristics of digestate and the characteristics of wood are individually mapped to the states of the composting model. The conversion factors are respectively shown in Tables 2 and 3. Then, the initial states of the composting model are calculated as follows:

- Compute the volume of digestate collected from the anaerobic digestion process as $V_{AD} = q \cdot \triangle t$, where $\triangle t$(days) is the period of digestate collection and q (m^3/day) is the volumetric flow rate the anaerobic digester was operated with in the interval $\triangle t$. Note that $V_{AD} = V$, the volume in the tank at the time instant $\tau = \triangle t$;
- Compute the volume of wood to be mixed with the digestate as $V_w = w/\rho$, where w (kg) is the mass of the wood and ρ (kg/m^3) is its density;
- The concentrations of the soluble substrate, insoluble substrate and inert material entering the composting process are respectively given by

$$S_0 = \frac{x \cdot V_{AD} + 22.15 \cdot w}{V_{AD} + V_w} \tag{18}$$

$$I_0 = \frac{y \cdot V_{AD} + 12.75 \cdot w}{V_{AD} + V_w} \tag{19}$$

$$M_0 = \frac{z \cdot V_{AD}}{V_{AD} + V_w} \tag{20}$$

where x, y and z (mol C/L) are respectively the concentrations of the soluble substrate, insoluble substrate and inert material in the tank at time instant $\tau = \triangle t$, i.e., $S(\tau)$, $I(\tau)$ and $M(\tau)$ given by (17). In (17), $S_{in}(t)$, $I_{in}(t)$ and $M_{in}(t)$ are respectively the concentrations of the soluble substrate, insoluble substrate and inert material in the low-ammonia digestate entering the tank, which are calculated at each time instant using the conversion coefficients shown in Table 3. The coefficients in (18)–(20) (mol C/kg wood) are calculated based on data in Table 2.

Table 2. Conversion factors of wood characteristics into units of the composting model.

Component	Chemical Formula	Composting	g/(kg Wood) [22]	mol C/(kg Wood)
Rhamnose	$C_6H_{12}O_6$	S	1.00	0.0333
Galactose	$C_6H_{12}O_6$	S	6.70	0.223
Mannose	$C_6H_{12}O_6$	S	17.60	0.586
Xylose	$C_6H_{12}O_6$	S	209.30	6.97
Glucose	$C_6H_{12}O_6$	S	430.50	14.3
Klason lignin	$C_{81}H_{92}O_{28}$	I	177.00	9.47
Acid-soluble lignin	$C_{81}H_{92}O_{28}$	I	46.60	2.49
Acetyl group	COOH	I	35.40	0.786
Extractives	/	/	17.20	/
Others	/	/	58.70	/

Table 3. Conversion factors from anaerobic digestion to composting.

| State | | Conversion Factor | Reference |
ADM1	Composting	mol C/kgCOD	
S_{su}	S	31.3	[25]
S_{aa}	S	27.2	own calculation, [32]
S_{fa}	S	21.7	[25]
S_{va}	S	24.0	[25]
S_{bu}	S	25.0	[25]
S_{pro}	S	26.8	[25]
S_{ac}	S	30.0	[25]
X_{ch}	S	31.3	[25]
X_{pr}	S	27.2	own calculation, [32]
X_{li}	S	22.3	own calculation, [32]
X_c	I	25.2	own calculation, [32]
X_{su}	I	27.2	own calculation ($C_5H_7O_2N$)
X_{aa}	I	27.2	own calculation ($C_5H_7O_2N$)
X_{fa}	I	27.2	own calculation ($C_5H_7O_2N$)
X_{c4}	I	27.2	own calculation ($C_5H_7O_2N$)
X_{pro}	I	27.2	own calculation ($C_5H_7O_2N$)
X_{ac}	I	27.2	own calculation ($C_5H_7O_2N$)
X_{h2}	I	27.2	own calculation ($C_5H_7O_2N$)
X_{ac2}	I	27.2	own calculation ($C_5H_7O_2N$)
X_I	M	30	[25]

4. Results and Discussion

The simulation results are based on the amounts and types of biowaste collected in the commune De Pinte in Flanders (Belgium). Via home-to-home collection, food waste is gathered every two weeks [18], while the wood waste is collected only once a year. Based on data given in Table 4, which shows the seasonality of the waste, a total amount of 395.508 tonnes of food waste is collected yearly. The yearly wood waste amount is 98.060 tonnes.

Table 4. Average amount of food waste collected per month in De Pinte [18].

| Period | Collection Day | |
	Day 1 [kg/month]	Day 2 [kg/month]
October–March	9557	18,340
April–September	15,260	22,773

As illustrated in Figures 3 and 7, the wood waste needs pretreatment before entering the steam refining process, while the reactor content needs posttreatment at the end of the process to retrieve the products of interest. No dynamic models are employed for these treatments but static blocks, which correct the amounts based on the experimental evidence. The pretreatment steps include washing, chipping, sieving and drying. It is assumed that after sieving, 95% of the wood chips has the proper size for steam refining, while the remaining 5% represents the wood residue which is processed via composting. Before entering the steam refining process, the small-sized wood chips need to be dried, the treatment in which the wood mass reduces by 10%. Consequently, the amount of wood processed via steam refining (denoted as the main fraction of wood in Figure 3) is 83,841.3 kg, while the wood residue amount equals 4903 kg.

The steam refining conversion is influenced by the temperature and the length of the experiment. Figure 8 shows the evolution of the model states for a temperature $T = 200\,^\circ$C and various experiment lengths. For the evaluation included below, an experiment length of 10 min is selected. This choice is motivated by the fact that no priority is given to any of the products of interest: lignin (solubilised lignin L_s), fibres ($L + XN + GN$, where each component is the sum of the fast- and slow-degradable fractions) and oligosaccharides ($XOS + GOS$). However, for longer experiments, the degradation of the products of interest occurs: the solubilised lignin degrades to condensed lignin, which is of no practical interest, while the oligosaccharides are converted into degradation products such as furfural (F) and 5-hydroxymethylfurfural (HMF).

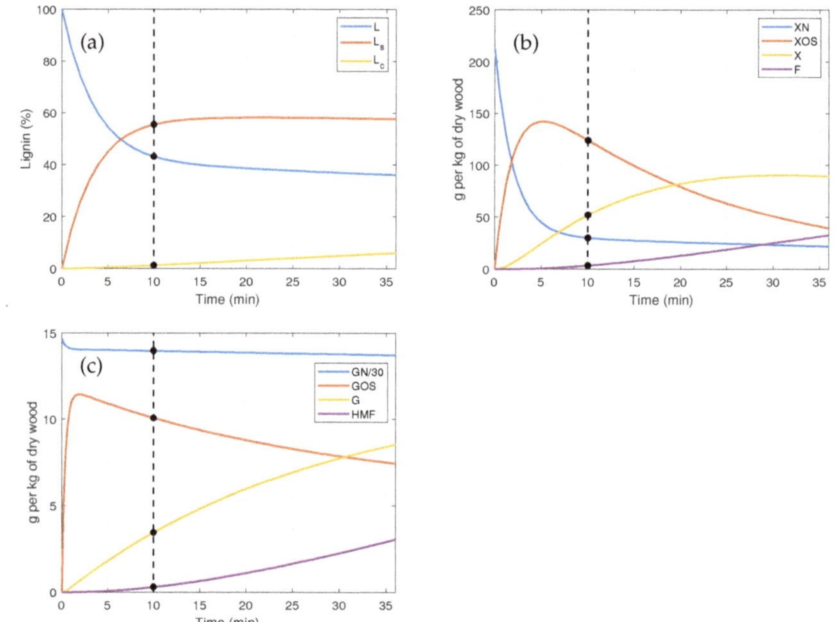

Figure 8. Steam refining: (**a**) Solubilised and solid-state lignin. (**b**) Xylan and its derivatives. (**c**) Glucan and its derivatives.

Figure 9 shows the conversion of the wood waste and the amounts of the products obtained, respectively, after steam refining (indicated in blue boxes) and after post-treatment (indicated in red boxes). These amounts are calculated based on the products' yields corresponding to a treatment duration of ten minutes and the amount of dry matter entering the process. The yields (see Figure 8) are as follows: 43.2% and 55.5% of original lignin content of wood, respectively, for the lignin remaining in the solid phase (L) and the solubilised lignin (L_s), 30 g/kg dry matter and 418.7 g/kg dry matter, respectively, for the xylan (XN) and glucan (GN), 124 g/kg dry matter and 10 g/kg dry matter, respectively, for xylo-oligosaccharides (XOS) and gluco-oligosaccharides (GOS). It is assumed that during postprocessing, 3% of fibres and 5% of oligosaccharides are lost. Note that the extract contains also monosaccharides (xylose and glucose) and can be used as a waste stream to feed another process or can be processed by anaerobic digestion.

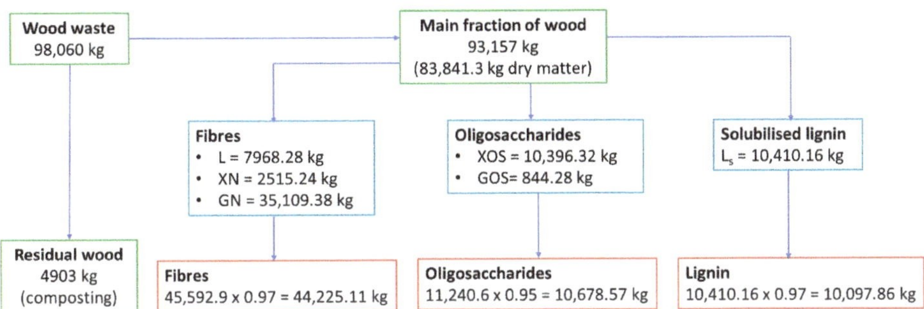

Figure 9. Products and their amounts obtained from wood steam refining: green boxes indicate feedstock, blue boxes show the products' amounts after steam refining treatment, and red boxes show the products' amounts after post-treatment.

The entire food waste collected throughout the year is processed by anaerobic digestion. Since this waste is available continuously, the simulation of the anaerobic digestion process is also performed for one year, with a constant supply of waste equal to 1083.6 kg/day. Similar to [24], it is assumed that the waste has the water mass, which implies that the digestor is operated with a constant flow $q = 1.0836$ m^3/day. Since low hydraulic retention times (defined as the ratio between the liquid volume and the feed flow rate) may lead to the reactor wash-out, in this simulation, the reactor liquid volume is selected as $V_{liq} = 12$ m^3 and the gas volume is chosen as $V_{gas} = 3$ m^3. Figure 10 shows the obtained outflow rate of biogas, the outflow rate of methane, the volumetric production of methane and the composition of biogas for the entire operation span. The effluent of the digestor is sent to the ammonia stripping process. The same liquid volume and the same feed flow rate as for the anaerobic digestion is assumed for this process. Air is continuously supplied such that efficiency of the removal in the range [80, 90]% is achieved. Figure 11 illustrates the influent and effluent ammonia concentrations and the corresponding removed amounts.

The digestate with low ammonia content is collected for a period of 100 days in a storage tank. At the end of the collection period, the content of the storage tank is loaded in a biocell for composting, which receives a continuous air supply during the operation such that the oxygen is not limiting the growth of the aerobic microorganisms. Two biocells are used; the first one is loaded on days 100 and 300, the second biocell is loaded on days 200 and 400. The volume of digestate loaded in the composting cells on days 100, 200 and 300 amounts to 108.36 m^3, while the volume loaded on day 400 is 71.46 m^3, as it was collected only for 65 days. Figure 12 shows the evolution of the states of interest of the composting process taking place in each of the two employed biocells (one column corresponds to one cell). Note that the model predicts the content of compost expressed in mol C/m^3, while in the evaluation, one is interested in the amount of produced compost. For this, the conversion factor 1 mol C compost = 25.7 g compost is used, which is determined based on the chemical formula assumed for compost ($C_{204}H_{325}O_{85}N_{77}S$). The evaluation of the second branch of the proposed biorefinery is illustrated in Figure 13.

Overall, it may be concluded that the waste produced during one year in the commune De Pinte can be bioprocessed into 44,225.11 kg of fibres, 10,678.57 kg of oligosaccharides, 10,097.86 kg of lignin, 113,730 m^3 of biogas, among which there is 56,728 m^3 of methane, and 30,887.3 kg of compost. Additionally, nitrogen fertilizer could be produced from the removed ammonia. Although the obtained amounts are not obtained from an optimised operation, the proposed biorefinery design allows for the production of several high-value-added products.

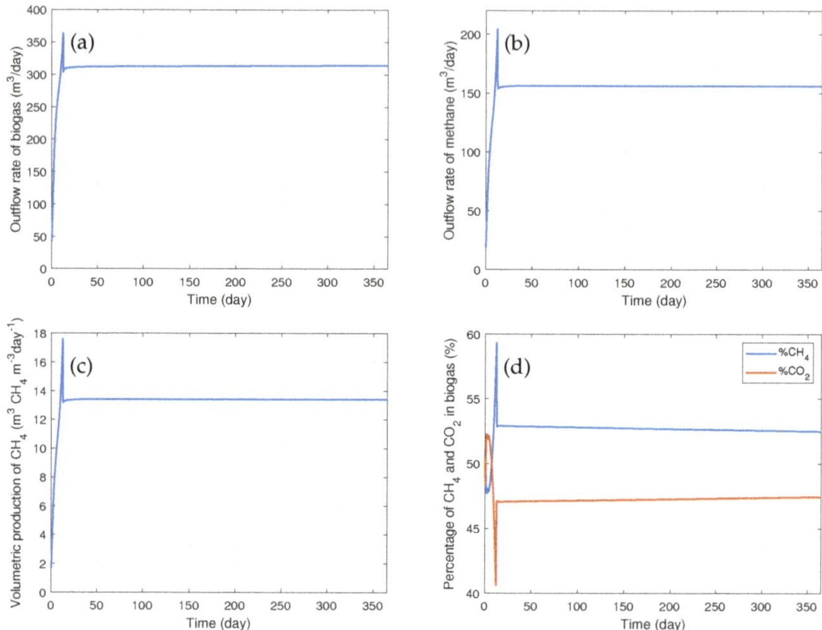

Figure 10. Anaerobic digestion: (**a**) Outflow rate of biogas. (**b**) Outflow rate of methane. (**c**) Volumetric production of methane. (**d**) Methane percentage in biogas.

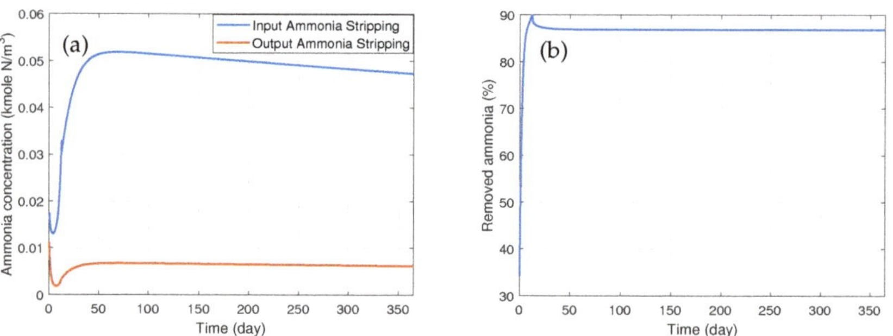

Figure 11. Ammonia stripping: (**a**) Influent and effluent ammonia concentrations. (**b**) Efficiency of the stripping process.

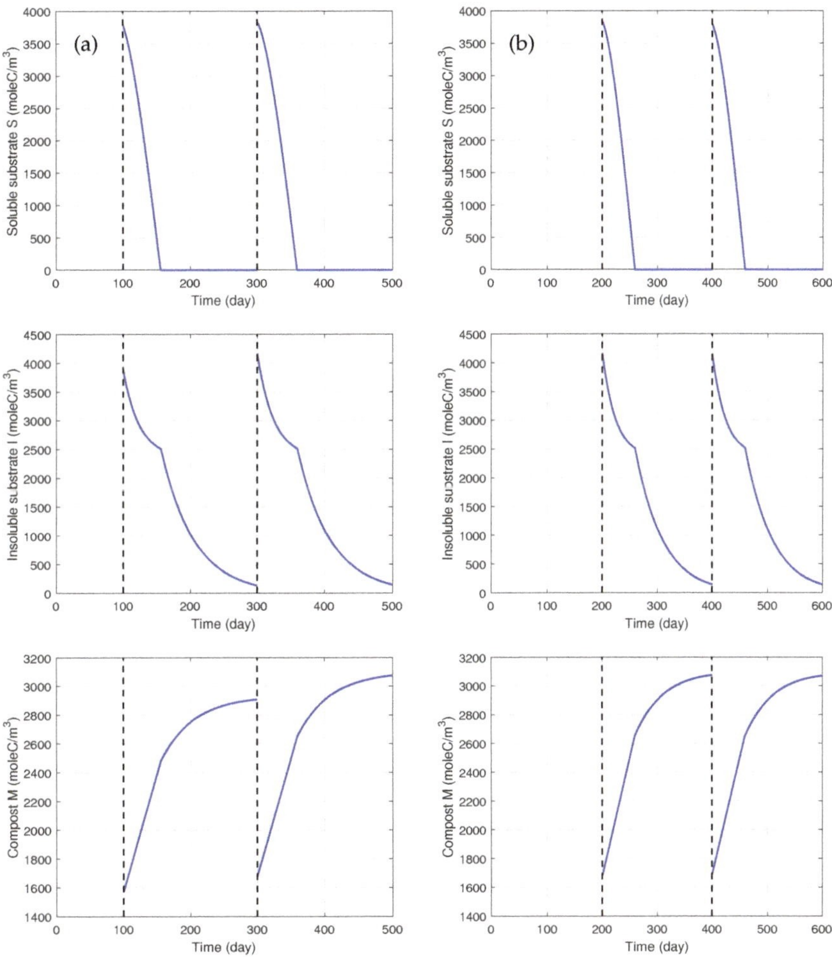

Figure 12. Composting process: Soluble substrate, insoluble substrate, inert material (compost) in (**a**) Cell 1 (left-hand-side column) and (**b**) Cell 2 (right-hand-side column).

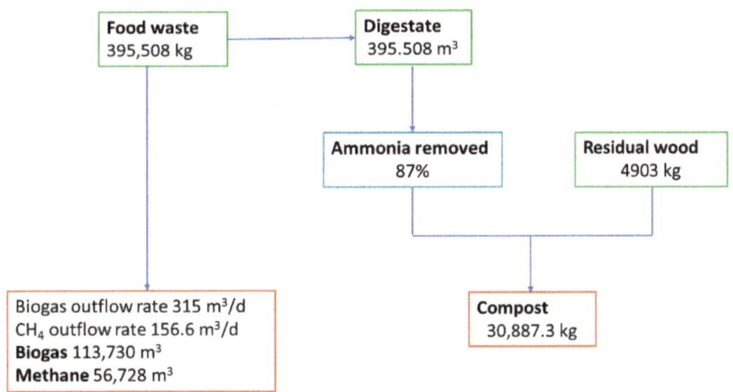

Figure 13. Products and their amounts obtained on the second branch of the biorefinery.

5. Conclusions and Perspectives

In this paper, the design of a small-scale biorefinery was presented and an evaluation based on data characterising the available biowaste on a yearly basis in a commune in Flanders was performed. The design started from the available feedstocks and was performed such that the side streams were minimised. The reported outcomes and the biorefinery operation represent a baseline scenario, which can be further improved through optimisation. The proposed biorefinery layout can be used for evaluation of feedstocks' potentials to produce high-valued products not only at the commune level but also at regional level. One of the main advantages of the proposed biorefinery is that it provides alternatives to the current practices at the local level, where food waste and landscaping waste are traditionally composted and burned.

The models employed in this biorefinery layout and the knowledge for their integration are the building blocks of the processing toolbox, which is one of the three core toolboxes in a decision support tool for the design of small-scale and flexi-feed biorefineries in a local setting. The processing toolbox will be linked with two additional toolboxes: *(i)* the bio-inventory toolbox [18] and *(ii)* the optimisation toolbox [33]. The former toolbox allows for drafting a survey of the locally available biowaste feedstocks and selecting one or multiple to be processed. Subsequently, this feedstock information will be employed by the processing toolbox to model and assess a suitable local and small-scale biorefinery layout, in a similar fashion as presented in this paper. To this end, the toolbox will be extended with new processes to account for the conversion of various feedstocks and the production of other high-value-added products [34,35]. Ultimately, as indicated above, the optimisation toolbox will employ the proposed biorefinery layout to further optimise the design and/or process settings. The decision support tool in its entirety will be detailed elsewhere.

Author Contributions: Conceptualization, M.S., V.D.B., S.A., S.B., M.P. and J.F.M.V.I.; software, M.S.; validation, M.S., V.D.B., S.A. and S.B.; writing—original draft preparation, M.S. and V.D.B.; writing—review and editing, M.S., V.D.B., S.A., S.B., M.P. and J.F.M.V.I.; supervision, J.F.M.V.I.; project administration, M.P. and J.F.M.V.I.; funding acquisition, S.A., M.P. and J.F.M.V.I. All authors have read and agreed to the published version of the manuscript.

Funding: This work was supported by the ERA-NET FACCE-SurPlus FLEXIBI Project, cofunded by VLAIO project HBC.2017.0176. V.D.B. is supported by FWO-SB Grant 1SC0922N. S.A. is supported by FWO grant 1224620N.

Institutional Review Board Statement: Not applicable.

Informed Consent Statement: Not applicable.

Acknowledgments: The authors would like to acknowledge the local council of De Pinte, Belgium, for their co-operation and the provision of the data necessary for this research.

Conflicts of Interest: The authors declare no conflict of interest. The funders had no role in the design of the study; in the collection, analyses or interpretation of data; in the writing of the manuscript, or in the decision to publish the results.

References

1. Cherubini, F.; Jungmeier, G.; Wellisch, M.; Willke, T.; Skiadas, I.; Van Ree, R.; de Jong, E. Toward a common classification approach for biorefinery systems. *Biofuels Bioprod. Bioref.* **2009**, *3*, 534–546. [CrossRef]
2. Naik, S.N.; Goud, V.V.; Rout, P.K.; Dalai, A.K. Production of first and second generation biofuels: A comprehensive review. *Renew. Sustain. Energy Rev.* **2010**, *14*, 578–597. [CrossRef]
3. De Buck, V.; Polanska, M.; Van Impe, J. Modeling Biowaste Biorefineries: A Review. *Front. Sustain. Food Syst.* **2020**, *4*, 11. [CrossRef]
4. Mohr, A.; Raman, S. Lessons from first generation biofuels and implications for the sustainability appraisal of second generation biofuels. *Energy Policy* **2013**, *63*, 114–122. [CrossRef]
5. Leong, H.Y.; Chang, C.K.; Khoo, K.S.; Chew, K.W.; Chia, S.R.; Lim, J.W.; Chang, J.S.; Show, P.L. Waste biorefinery towards a sustainable circular bioeconomy: A solution to global issues. *Biotechnol. Biofuels* **2021**, *14*, 87. [CrossRef]
6. Bruins, M.E.; Sanders, J.P.M. Small-scale processing of biomass for biorefinery. *Biofuels Bioprod. Biorefin.* **2012**, *6*, 135–145. [CrossRef]

7. Kolfschoten, R.C.; Bruins, M.E.; Sanders, J.P.M. Opportunities for small-scale biorefinery for production of sugar and ethanol in the Netherlands. *Biofuels Bioprod. Biorefin.* **2014**, *8*, 475–486. [CrossRef]

8. Jeevahan, J.; Anderson, A.; Sriram, V.; Durairaj, R.B.; Britto Joseph, G.; Mageshwaran, G. Waste into energy conversion technologies and conversion of food wastes into the potential products: A review. *Int. J. Ambient. Energy* **2021**, *42*, 1083–1101. [CrossRef]

9. Isikgor, F.; Becer, C. Lignocellulosic Biomass: A Sustainable Platform for Production of Bio-Based Chemicals and Polymers. *Polym. Chem.* **2015**, *6*, 4497–4559. [CrossRef]

10. Clark, J.H.; Deswarte, F.E.I. The Biorefinery Concept-An Integrated Approach. In *Introduction to Chemicals from Biomass*; Clark, J.H., Deswarte, F.E.I., Eds.; John Wiley & Sons, Ltd.: Chichester, UK, 2014; pp. 1–20. [CrossRef]

11. Sukumara, S.; Faulkner, W.; Amundson, J.; Badurdeen, F.; Seay, J. A multidisciplinary decision support tool for evaluating multiple biorefinery conversion technologies and supply chain performance. *Clean Technol. Environ. Policy* **2014**, *16*, 1027–1044. [CrossRef]

12. Martinkus, N.; Latta, G.; Rijkhoff, S.A.M.; Mueller, D.; Hoard, S.; Sasatani, D.; Pierobon, F.; Wolcott, M. A multi-criteria decision support tool for biorefinery siting: Using economic, environmental, and social metrics for a refined siting analysis. *Biomass Bioenergy* **2019**, *128*, 105330. [CrossRef]

13. Petre, E.; Selişteanu, D.; Roman, M. Control schemes for a complex biorefinery plant for bioenergy and biobased products. *Bioresour. Technol.* **2020**, *295*, 122245. [CrossRef] [PubMed]

14. Prunescu, R.M.; Blanke, M.; Jakobsen, J.G.; Sin, G. Model-based plantwide optimization of large scale lignocellulosic bioethanol plants. *Biochem. Eng. J.* **2017**, *124*, 13–25. [CrossRef]

15. Cheali, P.; Gernaey, K.V.; Sin, G. Synthesis and design of optimal biorefinery using an expanded network with thermochemical and biochemical biomass conversion platforms. *Comput. Aided Chem. Eng.* **2013**, *32*, 985–990. [CrossRef]

16. Cheali, P.; Quaglia, A.; Gernaey, K.V.; Sin, G. Effect of Market Price Uncertainties on the Design of Optimal Biorefinery Systems—A Systematic Approach. *Ind. Eng. Chem. Res.* **2014**, *53*, 6021–6032. [CrossRef]

17. Zondervan, E.; Nawaz, M.; de Haan, A.B.; Woodley, J.M.; Gani, R. Optimal design of a multi-product biorefinery system. *Comput. Chem. Eng.* **2011**, *35*, 1752–1766. [CrossRef]

18. De Buck, V.; Sbarciog, M.; Polanska, M.; Van Impe, J. Assessing the local biowaste potential of rural and developed areas using GIS-data and clustering techniques: Towards a decision support tool. *Front. Chem. Eng.* **2022**, *4*, 825045. [CrossRef]

19. European Biogas Association. *Annual Report 2020*; European Biogas Association: Brussels, Belgium, 2021. Available online: https://www.europeanbiogas.eu/eba-gie-biomethane-map/ (accessed on 31 March 2022).

20. Sbarciog, M.; Bhonsale, S.; De Buck, V.; Akkermans, S.; Polanska, M.; Van Impe, J. Modelling and simulation of co-digestion in anaerobic digestion systems. In Proceedings of the 10th Vienna International Conference on Mathematical Modelling (MATHMOD 2022), Vienna, Austria, 27–29 July 2022.

21. Batstone, D.J.; Keller, J.; Angelidaki, I.; Kalyuzhnyi, S.V.; Pavlostathis, S.G.; Rozzi, A.; Sanders, W.T.M.; Siegrist, H.; Vavilin, V.A. Anaerobic Digestion Model No. 1. In *IWA STR No. 13*; IWA Publishing: London, UK, 2002.

22. Borrega, M.; Nieminen, K.; Sixta, H. Effects of hot water extraction in a batch reactor on the delignification of birch wood. *BioResources* **2011**, *6*, 1890–1903.

23. Borrega, M.; Nieminen, K.; Sixta, H. Degradation kinetics of the main carbohydrates in birch wood during hot water extraction in a batch reactor at elevated temperatures. *Bioresour. Technol.* **2011**, *102*, 10724–10732. [CrossRef]

24. Nguyen, H.H. Modelling of Food Waste Digestion Using ADM1 Integrated with Aspen Plus. Ph.D. Thesis, University of Southampton, Southampton , UK, 2014.

25. Rosen, C.; Jeppsson, U. *Aspects on ADM1 Implementation within the BSM2 Framework*; Department of Industrial Electrical Engineering and Automation, Lund University: Lund, Sweden, 2006.

26. Weinrich, S.; Nelles, M. *Basics of Anaerobic Digestion—Biochemical Conversion and Process Modelling*; DBFZ: Leipzig, Germany, 2021.

27. Thamsiriroj, T.; Murphy, J. Modelling mono-digestion of grass silage in a 2-stage CSTR anaerobic digester using ADM1. *Bioresour. Technol.* **2011**, *102*, 948–959. [CrossRef]

28. Degermenci, N.; Nuri Ata, O.; Yildız, E. Ammonia removal by air stripping in a semi-batch jet loop reactor. *J. Ind. Eng. Chem.* **2012**, *18*, 399–404. [CrossRef]

29. Degermenci, N.; Yildız, E. Ammonia stripping using a continuous flow jet loop reactor: Mass transfer of ammonia and effect on stripping performance of influent ammonia concentration, hydraulic retention time, temperature, and air flow rate. *Environ. Sci. Pollut. Res.* **2021**, *28*, 31462–31469. [CrossRef] [PubMed]

30. Kofi, A.-W.; Martino, C.J.; Wilmarth, W.R.; Bennett, W.M.; Peters, R.S. Modeling air stripping of ammonia in an agitated vessel. In *Office of Scientific & Technical Information Technical Reports*; University of North Texas Libraries, UNT Digital Library: Aiken, SC, USA, 2005. Available online: https://digital.library.unt.edu/ark:/67531/metadc873356/m1/6/ (accessed on 20 July 2021).

31. Martalo, G.; Bianchi, C.; Buonomo, B.; Chiappini, M.; Vespri, V. Mathematical modeling of oxygen control in biocell composting plants. *Math. Comput. Simul.* **2020**, *177*, 105–119. [CrossRef]

32. Lopez, V.M.; De la Cruz, F.B.; Barlaz, M.A. Chemical composition and methane potential of commercial food wastes. *Waste Manag.* **2016**, *56*, 477–490. [CrossRef] [PubMed]

33. De Buck, V.; Nimmegeers, P.; Hashem, I.; Muñoz López, C.A.; Van Impe, J. Exploiting Trade-Off Criteria to Improve the Efficiency of Genetic Multi-Objective Optimisation Algorithms. *Front. Chem. Eng.* **2021**, *3*, 582123. [CrossRef]

34. Wagemann, K.; Tippkotter, N. (Eds.) *Biorefineries*; Springer: Cham, Switzerland, 2019.
35. Mohan, S.V.; Varjani, S.; Pandey, A. (Eds.) *Microbial Electrochemical Technology*; Elsevier: Amsterdam, The Netherlands, 2018.

Review

Application of Optimization and Modeling for the Enhancement of Composting Processes

Tea Sokač [1], Davor Valinger [2], Maja Benković [2], Tamara Jurina [2], Jasenka Gajdoš Kljusurić [2], Ivana Radojčić Redovniković [1] and Ana Jurinjak Tušek [2,*]

[1] Department of Biochemical Engineering, Faculty of Food Technology and Biotechnology, University of Zagreb, Pierottijeva 6, 10000 Zagreb, Croatia; tsokac@pbf.hr (T.S.); irredovnikovic@pbf.hr (I.R.R.)
[2] Department of Process Engineering, Faculty of Food Technology and Biotechnology, University of Zagreb, Pierottijeva 6, 10000 Zagreb, Croatia; davor.valinger@pbf.unizg.hr (D.V.); maja.benkovic@pbf.unizg.hr (M.B.); tamara.jurina@pbf.unizg.hr (T.J.); jasenka.gajdos@pbf.unizg.hr (J.G.K.)
* Correspondence: ana.tusek.jurinjak@pbf.unizg.hr

Abstract: Composting is a more environmentally friendly and cost-effective alternative to digesting organic waste and turning it into organic fertilizer. It is a biological process in which polymeric waste materials contained in organic waste are biodegraded by fungi and bacteria. Temperature, pH, moisture content, C/N ratio, particle size, nutrient content and oxygen supply all have an impact on the efficiency of the composting process. To achieve optimal composting efficiency, all of these variables and their interactions must be considered. To this end, statistical optimization techniques and mathematical modeling approaches have been developed over the years. In this paper, an overview of optimization and mathematical modeling approaches in the field of composting processes is presented. The advantages and limitations of optimization and mathematical modeling for improving composting processes are also addressed.

Keywords: composting; optimization; mathematical modeling

Citation: Sokač, T.; Valinger, D.; Benković, M.; Jurina, T.; Gajdoš Kljusurić, J.; Radojčić Redovniković, I.; Jurinjak Tušek, A. Application of Optimization and Modeling for the Enhancement of Composting Processes. *Processes* **2022**, *10*, 229. https://doi.org/10.3390/pr10020229

Academic Editors: Philippe Bogaerts and Alain Vande Wouwer

Received: 30 December 2021
Accepted: 24 January 2022
Published: 25 January 2022

Publisher's Note: MDPI stays neutral with regard to jurisdictional claims in published maps and institutional affiliations.

1. Introduction

The steady growth of industrial production and trade in many countries of the world has led to a rapid increase in the generation of municipal and industrial waste in the last decade [1]. About 50% of the waste generated worldwide consists of organic matter, generally from food, human and animal waste, garden and wood products [2]. A significant portion of this waste ends up in landfills and, if not properly treated, can pose a significant threat to the environment and human health [3]. The main producers are the agriculture and food sectors. Their waste can be used as a raw material to produce high value-added products, opening up a range of opportunities for sustainable production [4].

The treatment and disposal of waste is a very important and urgent issue, especially for local authorities who have to deal with this problem within their jurisdiction. Of course, many different strategies have been proposed to try to counteract and reduce waste production, such as composting [5] or anaerobic digestion. Anaerobic digestion has been suggested as an alternative method for high organic content waste. This method involves the degradation of organic matter without the presence of oxygen producing a potential energy source such as power generation or fuel gas [6]. Composting is an environmentally friendly and cost-effective substitute for processing organic waste and converting it into organic fertilizer. It is a biological process in which the polymeric waste materials contained in organic waste are degraded by the accelerated growth of fungi and bacteria [2,7]. In fact, it is a very complicated mechanism involving a variety of processes (microbiological, physicochemical and thermodynamic), all of which seem to be interrelated [8]. The products of the composting process are carbon dioxide and stable carbon forms that lead to the decomposition and mineralization of organic matter and the production of humic substances [7]. During the composting process, microorganisms release heat and energy as they decompose material. The heat generated increases the temperature of the compost pile, which ensures the inactivation of pathogenic microorganisms. For this reason, measuring the temperature of the pile is very important in evaluating the composting process [9]. The performance of the composting process is influenced by factors such as temperature, pH, moisture content, C/N ratio, particle size, nutrient content and oxygen supply [7,10,11]. The listed process variables can change frequently during the composting process. The largest and most significant temperature variation is observed during the thermophilic phase of the process. The changes in pH during composting are related to the proliferation of microorganisms. In addition, moisture content has a significant effect on the physical and chemical properties of the composting substrate. The C/N ratio and aeration are also very important for the multiplication of microorganisms [10]. Thus, to achieve maximum efficiency in composting, all these factors and their interactions must be considered. To obtain a respectable and highly stabilized compost, the working conditions must be optimized, which is crucial for setting up processes and improving their efficiency. Optimization is very important to ensure good quality of the final product by performing the process under experimental conditions that are the most suitable for specific phenomena taking place during the composting. For example, the optimization of the compost maturity (expressed as pH, or electrical conductivity, final C/N ration, germination index or ash content) requires appropriate set-up design of C/N ratio, moisture content, and aeration rate to ensure the conditions suitable for the microbial population growth which controls the organic matter degradation and, respectively, compost maturity [11,12]. Different optimization approaches have been described in the literature. For example, single-factor optimization has been used to determine the best process conditions [13,14], but this approach is imprecise and may lead to misinterpretation of data; it also does not illustrate the interacting effects of variables and does not ensure the selection of optimal conditions. On the other hand, statistics-based experimental designs allows simultaneous analysis of all variables affecting the process as well as their interactions. This approach saves time and reduces errors in detecting the interaction of process parameters and has recently been published in the literature [15–20].

A poorly performed composting process results in an insufficiently stabilized organic matter or immature compost, which can affect the soil environment and plant growth, be a source of disease, and cause damage to crops through phytotoxicity [21]. The quality of compost is related to its stability and maturity. Stability is a term that refers to the resistance of a product's organic matter to extensive degradation or to greater microbiological activity, while maturity describes the ability of a product to be used effectively in agriculture and is related to plant growth and phytotoxicity aspects. Many physical, chemical and biological tests have been proposed to evaluate the stability and maturity of compost, but there is still no single accepted test to evaluate both parameters [22] that would provide the information necessary for process control.

Composting is a dynamic process characterized by a large number of interrelated effects. The main phases of composting are shown in a following graph (Figure 1). In the first phase (mesophilic phase), energy-rich, easily degradable compounds such as sugars and proteins are degraded by fungi, actinobacteria and bacteria. The increase in a temperature higher than 45 °C leads to another composting phase called the thermophilic phase which is important because of the elimination of pathogens, parasites and ensures maximal sanitary conditions. The mesophilic microorganisms are replaced by thermophilic because they are adapted to higher temperatures and they are degrading more complex compounds. Compost temperature must not exceed 65 °C as this would destroy almost all microorganisms and cause enzyme denaturation which can lead to the end of the composting process. When the activity of the thermophilic microorganisms decreases due to the exhaustion of the substrate, the temperature starts to decrease. That is the beginning of the second mesophilic phase which is characterized by an increasing number of organisms that degrade starch or cellulose. In the maturation phase, the compost pile is stabilized for plant use and the proportion of fungi increases, while bacterial numbers decline. Also, during the maturation phase are formed compounds that are not further degradable, such as lignin-humus complexes [12,23]. Bacteria, fungi, and actinomycetes degrade organic materials, resulting in compositional change and release of energy and water during the composting process.

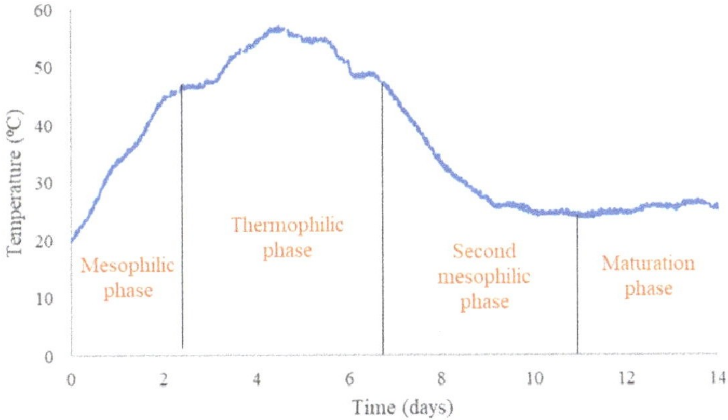

Figure 1. The main phases during the composting process (according to [24]).

As mentioned, composting is a highly complex process including numerous interconnected physical, chemical and biological phenomena. The particular connections are frequently non-linear and, therefore, numerous effects have to studied booth experimentally and theoretically. Therefore, mathematical modeling tools can be useful for interpretation of the complex dynamic interactions and for the development of the logical process design framework [25]. Mathematical models have been extensively used to gain valuable insights into the composting process [26–29]. Mathematical models of the composting process assist getting information on how different process variables and conditions like substrate composition, oxygen concentration, pollutants concentrations, composting duration, temperature and etc. effect the compost quality during the process [8]. Furthermore, as noted by Mason [30], mathematical models can reduce or even replace the need for experimental work when investigating novel processes. Considering the difficulties and costs associated with conducting laboratory and pilot experiments, it is desirable to improve the ability to study novel processes using models. To describe the composition process, the model must include the kinetics of the process as well as the mass and heat balances [2].

In this paper, an overview of optimization and mathematical modeling approaches in the field of composting processes is presented.

2. Literature Search

A comprehensive systematic review of the important scientific articles was conducted using the core collection in the Web of Science database for the period of the last 21 years (period from 2000 to 2021). The keywords "composting" and "optimization" and "composting" and "mathematical modeling" were used to search the title and abstract of the articles. Only indexed papers were selected for further analysis. For the field of optimization there were 77 articles, for the field of modeling a total of 222 articles.

3. Optimization of Composting Process

The estimation of optimal process parameters using practical and reliable models is still difficult and challenging and, therefore, multi-response optimization approaches have seen substantial growth and extension over the years [31]. According to Šibalija and Majstorović [32], optimization methods can be divided as follows:

a.   Conventional (statistical or mathematical based) method:

    i.   Iterative search techniques;

    ii.   Experimental design:

        1.   Method based on response surface methodology;

        2.   Methods based on factorial design;

        3.   Methods based on Taguchi design.

b.   Non-conventional (artificial intelligence based) methods:

    i.   Method based on fuzzy logic;

    ii.   Methods based on artificial neural networks;

    iii.   Methods based on metaheuristic algorithms:

        1.   Genetic algorithm;

        2.   Simulated annealing;

        3.   Particle swarm optimization;

        4.   Ant colony optimization;

        5.   Tabu search;

        6.   Artificial bee colony algorithm;

        7.   Biogeography-based optimization;

        8.   Teaching-learning based optimization.

    iv.   Methods based on expert systems.

In the literature, there are several optimization strategies for the composting process. Statistical optimization methods based on design of experiments and artificial neural network modeling are the mostly used [11].

3.1. Statistical Design of Experiments

When designing or developing complex products or processes, multiple responses have to be considered simultaneously [33] and the term "optimization" in this context refers to increasing process productivity or improving product characteristics [34]. In order to achieve process optimization, detailed knowledge of process performance is required, which is why precise procedures are needed for experiments. The process of preparing and conducting an experiment so that appropriate data can be collected and analyzed using statistical methods to derive relevant conclusions from the experimental data while reducing the number of experiments is referred to as statistical design of experiments (DOE) [35]. Multivariate experimental design is an effective tool to improve the quality of information obtained from an investigation while reducing the number of experiments to be performed [36]. According to Said and Amin [36], process optimization using statistical design of experiments can be divided into three steps:

(i)   experimental design;

(ii)   model development to describe the experimental data using statistics and regression analysis; and

(iii)   process optimization.

DOE optimization efficiently explores the space of the system under study at different sample points, which reduces the computational cost and facilitates the analysis.

Design of experiments is an effective analytical method for modeling and analyzing the effects of control elements on performance outcomes. The typical experimental design is challenging, especially when you are working with a large number of trials and an increasing number of process variables. Selecting the most important process variables is the most important step in experimental design [37]. DOEs come in a variety of forms, including full factorial, fractional factorial, Placket–Burman, central composite design, Box–Behnken, Taguchi, and others [38]. The type of DOE is chosen based on the objectives of the experiments and the set of variables to be studied [39,40]. For example, the randomized complete block design (RCBD) approach is used to optimize only one fundamental element that is expected to be the most influential parameter for variation in a process [39,40]. A full factorial experimental design ensures that all factorial effects can be determined separately by conducting 2^k trials (k is the number of factors studied). However, it is not ideal in practice that the number of experiments required increases with increasing number of factors [41]. Full factorial experimental designs can be simplified by using fractional factorial experimental designs, where only the factor-specific effects are examined. An example of a fractional factorial design is $2^{(k-p)}$ (p is the proportion of the design to be created) [42]. There are two types of fractional factorial experimental designs: regular and non-regular. Regular fractional factorial designs are generated by detecting associations between variables, although non-regular fractional factorial designs, such as Placket–Burmann designs, are occasionally used in investigative studies because they have a small run [43]. Moreover, response surface designs are limited to eight variables and are only accessible to continuous factors, while the Taguchi robust experiment design is based on the fractional factorial design and includes orthogonal designs [39,40]. By definition, the Taguchi method is an experimental methodology that reduces the number of experiments by using orthogonal arrangements and limiting the influence of control variables. The Taguchi method is an approach that involves a plan of experiments with the aim of obtaining data in a controlled manner, performing these tests, and analyzing the data to gain knowledge about the behavior of the given process [37,44].

3.2. Application of Optimization Methodology in Composting Processes

The application of the optimization methodology in the composting process over the last 20 years is shown in Figure 2 and Table 1. It can be seen that the interest in this field has increased especially in the last 6 years probably due to growing interest in sustainable waste treatment

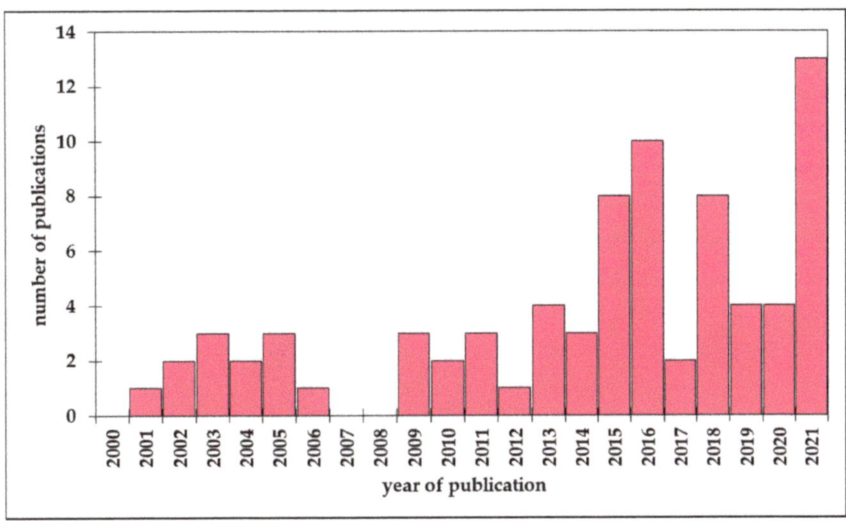

Figure 2. Number of publications on composting optimization in Web of Science database.

Table 1. Examples of optimization of waste composting.

Composting Substrate	Process Variables	Optimization Method	Major Results	Reference
Municipal solid waste	Pollutant concentration (pyrene), soil/compost mixing ratio, compost stability expressed as respiration index	Response surface based on central composite design (CCD)	Experimental design ensures to obtain reliable values of optimal process conditions for pyrene biodegradation. High values of coefficient of correlations ($R^2 > 0.69$) confirmed that second-order model was suitable for description of composting process.	[45]
Municipal solid waste	Temperature, moisture content, oxygen flow rate, free air space	One-factor-at-a-time	Optimisation of the process conditions based on data gathered in the composting plant increases the performance of biological treatment plants form 53% to 107%.	[46]
Municipal solid waste	Aeration, moisture, C/N ratio, time	Second order polynomials based on Box-Behnken central composite experimental design	Modelling results showed that composting time and C/N ratio are the most important variables influencing compost stability expressed as organic matter, chemical oxygen demand, nitrate concentration and biogedradability coefficient. R^2 where gather than 0.8 for all analysed output variables.	[47]
Pulp and paper mill sludge	Composting time, moisture, addition of hazelnut kernels	ANOVA test based on full factorial experimental design	60% removal of ammonium from compost can achieved by performing the composting process under estimated optimal conditions (5 weeks, moisture of 50% and addition of hazelnut kernel of 25%). ANOVA showed that all input variables have significant effect on the process output variable.	[48]
Palm oil mill effluent and empty fruit branches	Particle size, pH, initial carbon dosage, mix ratio of substrates	One-factor-at-a-time	The experimental conditions, the most suitable for effective composting process (electrical conductivity, protein content, organic matter content and C/N ratio), were estimated based on individual independent experiments.	[49]
Cattle sultry and cattle manure with maize/silage and peach-juice pulp	pH, C/N ration and moisture content	Factorial optimization	Dewar autothermal assay was efficiently used to describe the relationship between process input variables ($R^2 = 0.8$). The developed model gives information about the optimal composting conditions.	[50]
Medlar pruning waste with and without cattle manure	Cattle manure addition	One-factor-at-a-time	Results showed that addition of cattle manure increased electrical conductivity, the total nitrogen content and organic-matter stability of the compost.	[51]
Kitchen waste	Flay ash and bulking agent for moisture optimization Lime, temperature and inoculum size for C/N optimization	Response surface methodology based on Box–Benkhen design	Interaction between bulking agent and fly ash was significant for the compost moisture. Temperature, amount of lime addition and inoculum size have positive effect on C/N ratio of the compost. Second order polynomial equations described the experimental data with high precision; for moisture content $R^2 = 0.9975$ and for C/N ratio $R^2 = 0.9947$.	[15]

Table 1. *Cont.*

Composting Substrate	Process Variables	Optimization Method	Major Results	Reference
Municipal solid waste	Pollutant concentration (pyrene), soil/compost mixing ratio, compost stability expressed as respiration index	Response surface based on central composite design (CCD)	Experimental design ensures to obtain reliable values of optimal process conditions for pyrene biodegradation. High values of coefficient of correlations ($R^2 > 0.69$) confirmed that second-order model was suitable for description of composting process.	[45]
Paper mill sludge and corn waste	Materia type (corn cob and corn husk), materia ratio, moisture, and process duration	Statistical optimization based on full factorial design	Additive dosage and composting duration had a positive effect on the removal of ammonia during co-composting of pulp/paper mill sludge and corn wastes. On the other hand, additive type and moisture content had negative effects on the ammonia removal. 91.84% of ammonia was removed when composting process was performed under optimal process conditions.	[52]
Biodegradable solid waste	pH, moisture content, composting method	One-factor-at-a-time	When the moisture level was kept at 60% and composting was done in heaps and piles rather than pits and earthen pots, a long thermophilic phase was evident.	[53]
Food waste	Natural zeoilte and biochar from agricultural waste addition	One-factor-at-a-time	Because of its strong affinity for cations, adding zeolites to composting materials lowers the concentration of heavy metals in the final products. The addition of biochar to the composting process increases nitrogen retention.	[54]
Agricultural waste	Volatile solids, soluble BOD and CO_2 evolution during composting process	Radial basis functional (RBF) neural network model	RBF neural network model was successfully implemented ($R^2 = 0.99$) for planning of optimal agricultural waste mixture composition for composting process. Furthermore, the optimization based on the developed RBF model revealed that a waste combination of 70% vegetable waste, 20% cow manure, 10% sawdust, and 10 kg of dry leaves, coupled with 10 kg of dry leaves, give compost of preferable properties.	[55]

The analysis of the optimization methods presented in Table 1 shows that different experimental designs (central composite experimental design, full factorial experimental design, Box–Behnken experimental design, etc.) have been used for the analysis of the composting process during the last 15 years. However, it should be emphasized that most of the optimization procedures presented were carried out using the one-factor-at-a-time method. The one-factor-at-a-time method was used by Waqas et al. [54] and Chaher et al. [56] for the composting of food waste using in-vessel compost bioreactor by adding biochar prepared from lawn waste. The results showed that the addition of biochar improved the composting process and also the physiochemical properties of the final compost. The quality of the compost was evaluated using stabilization indicators such as moisture content, electrical conductivity, organic matter decomposition, change in total carbon and concentration of mineral nitrogen expressed as ammonium and nitrate. The same approach was described by Waqas et al. [57] for optimizing food waste composting by adding zeolites. The authors investigated the effects of natural zeolite and modified zeolite on the stability of compost produced in a bioreactor. Their results showed that modified zeolite significantly affected the compost properties (moisture content, electrical conductivity, organic matter, total carbon, mineral nitrogen, nitrification index, germination index). Bian et al. [58] also applied the one-factor-at-a-time method for agricultural waste composting with chicken manure, vegetable leaves and rice husks. The duration of thermal phase and conversion time on moisture, nutrients, carbon content and C/N ratio of compost were analyzed by seven composting experiments in a batch composter. Zhang et al. [59] investigated the influence of turning frequency, straw and microbial inoculum on the efficiency of dairy manure composting in a field composting process. Data variance analysis (ANOVA) and non-parametric Spearman correlations were applied to analyze the effects of each process variable on compost temperature, water content, and pH. Their results showed that to improve composting, the frequency of turning, the addition of straw and microbial inoculum at low temperatures (T < 0 °C, winter period of composting) must be adjusted. The analyzed results showed that the ratio of dairy manure to straw 2:1 (*v:v*), 4 days of turning and the addition of 1.3 L/t of inoculum were ideal for rapid heating and long duration of the thermophilic stage. The same approach, one-factor-at-a-time optimization, was used by Tabrika et al. [60] to optimize the composition of compost mixtures (tomato waste, olive pumice, sheep manure, chicken manure and sawdust). Their results indicate that the use of sheep manure and olive pumice improved the composting process of tomato waste. Lew et al. [61] optimized the temperature, moisture content, humidity and volume of liquid phase during composting of food waste in a smart bokashi composting system by also using a factor-by-time approach. Li et al. [62] estimated the optimal concentration of fulvic acid to improve the composting of straw and fungal residues based on the ANOVA analyzes of the experimental data, while Song et al. [63] used ANOVA analyses of the experimental data for optimization of the composting of food waste.

As described in the literature, statistical and mathematical optimization methods enable a statistically designed experiment that is cost-effective and allows analysis of the interactions between process variables. Sharma et al. [64] used a central composite design and response surface methodology to optimize floral waste composting. The floral wastes were composted using the stirred pile method with the addition of cow dung as inoculum origin and the stability and maturity of the compost was evaluated by pH, total organic carbon, electrical conductivity, ammonium nitrogen. C/N ratio, phosphorus, sodium, germination index, potassium and carbon dioxide production. The results obtained showed that second order polynomial equations can adequately describe the experimental data based on the input variables of the model (mass of floral waste and mass of cow dung). In the work of Calabi-Floody et al. [65], a three-level factorial design was used to evaluate the effects of three process variables (particle size, nitrogen addition, and *Trichoderma harzianum* inoculum concentration) on wheat straw composting. According to the results presented with the use of ANOVA, all three input variables and their interaction had a significant effect on the composting process. Multi-response optimization was used to investigate the effectiveness of the composting process to determine the technique most suitable to modify the wheat straw. It was found that the decrease in particle size has positive effect on the water holding capacity. Ajmal et al. [40] used Taguchi design method to optimize the temperature and time for composting process of agricultural wastes (poultry manure, vegetable wastes and rice straw) in vessels. They showed that performing the composting process under optimal conditions reduces the production of undesirable by-products. Moreover, Ajmal et al. [66] used Taguchi experimental design to optimize C/N ratio in agricultural waste composting based on composting temperature, composting time and initial inoculum amount in nine independent composting experiments.

Roman et al. [67] applied a Doehlert matrix experimental design (three variables at three levels) to evaluate the influence of hydrotreatment time, initial compost mass and hydrotreatment temperature on the municipal solid waste (MSW) composting process (total organic carbon, conductivity, liquid fraction volume and solid fraction carbon were the initial variables of the process) in 15 independent experiments simultaneously. From the experiments conducted, it was concluded that compost volume was the most important variable for the composting process. Gao et al. [19] applied simplex centroid design and response surface modeling followed by analysis of variance to optimize the composition of composting mixtures (pig manure, human feces, rice straw, and kitchen waste). In addition, Sokač et al. [24], presented the efficient use of the Box–Behnken experimental design (four variables at three levels) and response surface modelling (RSM) to optimize the composting process of biowaste in an adiabatic reactor with and without bioaugmentation. In addition to the optimization of the composting process, there are also examples of the application of the response surface methodology to optimize soil nutrient quality after compost addition (4–45% of compost based on soil weight) [60] and the application of RSM and artificial neural network (ANN) modeling to optimize the adsorption efficiency of municipal solid waste compost for the removal of reactive dyes from aqueous mixtures [61]. The results presented by Mazumder et al. [68] showed that higher compost addition (>15%) increased the amount of nutrients in the soil over time, while Dehghani et al. [69] stated that both RSM modeling and ANN modeling have a significant ability to describe and optimize the adsorption mechanisms.

Considering that the relationships between variables affecting the composting process are mostly highly non-linear, it is necessary to use a non-linear modeling approach such as artificial neural networks. This methodology has been described by Soto-Paz et al. [70] for optimizing the composting of a mixture of biowaste and sugarcane filter cake. The effects of mixing ratio and rotation frequency were modeled using the ANN model and the optimal process settings were estimated using the particle swarm optimization algorithm. Aycan Duemenci et al. [20] modeled and optimized the composting of olive mill waste with the addition of five natural minerals using the ANN method with high precision. An efficient application of machine learning based optimization was also reported by Yamawaki et al. [71] for the degradation of bio-based plastics in the compost pile, where the developed models predicted with high accuracy the molecular weight of biopolymers based on the moisture content of compost, composting time, degree of crystallinity of biopolymers in the pretreatment phase, and nuclear magnetic resonance spectra of the solid phase of composting, and by Golbaz et al. [72] for the development of an optimal primary composite for composting based on the wet weight of the sludge cake, the concentration of volatile solids in the dry solids, the sludge density, the Kjeldahl nitrogen amount of the dry solids in the sludge cake, the Kjeldahl nitrogen amount of the material for mixture modification, the Kjeldahl nitrogen concentration of the filler component, and the Kjeldahl nitrogen concentration of the recycled components in the sludge as input variables of the model. Another example of the efficient use of ANN modeling to optimize the composting process was given by Dragoi et al. [73]. They presented a method for developing the ANN model for predicting the change in total concentration of petroleum hydrocarbons and organic carbon during the composting process of oily sludge.

Taking into account listed results it can be noticed that there is potential of usage optimization methodology in the composting process. But there are still some limitations especially regarding to optimization based on design of experiments. According to Deaconu and Coleman [74] the major limitations of DOE are its degree of difficulty and number of experiments that must be undertaken to achieve statistical differences. Other issues include inability to manage specific variables (running the experiments with variables that will never yield an optimum value), difficulties recognizing all of the variables' impact on the process, and non-linearity between the variables which makes the interpretation of the interactions difficulty. So mentioned disadvantages tent to restrict how often DOE is employed in process optimization.

4. Mathematical Modeling of Composting Process

4.1. Some Basic Principles of Composting Process Modeling

Composting is a complicated bioprocess that involves several physical and biological processes. These interrelated and often highly non-linear processes produce a variety of phenomena that are difficult to study experimentally and analytically. Therefore, mathematical modeling can be useful for approximating the complex dynamic interplay that occurs during composting [75]. Mathematical modeling allows us to improve our knowledge of the process, reduce the time and energy required for experimental optimization, and enable computational simulations and development of process improvements [2]. According to Hangos and Cameron [76], mathematical models can be classified:

(i)   based on variables properties as deterministic (model variables are well known) and stochastic (model variables are random);
(ii)   based on deepened variable and their dependence on spatial position as lumped and distributed model;
(iii)   based on mathematical description of the process as continuous and discrete;
(iv)   based on mathematical structure as linear and non-linear.

Furthermore, due to their complexity mathematical model describing environmental process can be classified as:

(i)   mechanistic (white box) models are developed when all required data about process mechanisms are gathered;
(ii)   empirical (black box) models are developed when only experimental data are available and without understanding the mechanism in the process; and
(iii)   combined (gray box) models [77].

Systematically, composting is a three-phase process: solid phase (solid composting substrate), liquid phase and gas (air) [40]. Thus, it is a great challenge to describe all the changes that take place in each phase and also between the different phases. As described in detail by Mason [30], mathematical models of the composting process usually include dynamic heat and mass balances and the microorganisms' growth. Growth of microorganisms during the composting process is usually described by first-order kinetics, Monod kinetics or empirical expressions developed for the specific set of experimental data. In addition to Mason [30], Li et al. [10] reviewed the key variables affecting the food waste composting process and modeling approaches for describing the phenomena occurring during composting. In addition, Walling et al. [2], Ajmal et al. [8] and Yang et al. [13] also reviewed the application of mathematical modeling in the composting process. Walling et al. [2] discussed in detail the current limitations of the developed mathematical model and which aspects should be further researched and developed. On the other hand, Ajmal et al. [8] reviewed with special emphasis on the models describing the heat exchange during the process and Yang et al. [13] with special emphasis on the changes of carbon, nitrogen, phosphorus and potassium during the composting process.

4.2. Application of Mathematical Modeling in Composting Process

The growing interest in using of mathematical modeling in the field of composting is evident from the analysis of the data presented in Figure 3. It can be seen that the number of papers published using this method has increased steadily over the last 20 years. The increase in publications is mainly observed from 2016 to date.

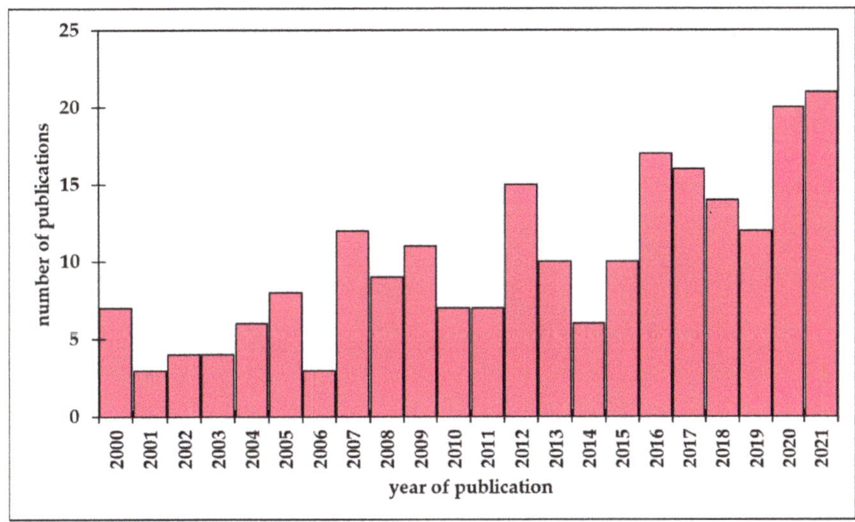

Figure 3. Number of publications on composting modeling in Web of Science.

Some specific characteristics and a brief description of some of the models used to describe the composting process from 2010 to 2021 can be found in Table 2. Based on the models presented, it can be seen that different modeling approaches have been used over the years.

There are examples of the use of simple empirical models [101], dynamic models [89–95] and artificial neural network models [88]. In general, simple empirical models are insufficient for simulating complex nonlinear dynamic processes. This is especially expressed when the analyzed process includes many performance modes with very different characteristics. Therefore, the fuzzy logic modeling approach was implemented to overcome the aforementioned limitations. As presented by Giusti and Marsili-Libelli [102], fuzzy logic modeling is very efficient for describing heat exchanges during the composting process. Moreover, Alavi et al. [103] applied kinetic modeling (first-order kinetic model was used for tetracycline degradation) and highly non-linear modeling with machine learning (artificial neural network modeling) to describe tetracycline elimination from chicken manure during co-composting with bagasse. The applicability of modeling from ANN was also confirmed by Roohi et al. [104] for planning phytoremediation of chromium- and zinc-contaminated soils by *Bromus tomentellus* with household waste compost and by Alavi et al. [105] for modeling the Elector–Fenton processing of compost leachate.

Table 2. Examples of the application of the modeling for the composting processes.

Composting Substrate	Model Formulation	Model Description	Reference
Industrial waste	Mass balance for oxygen and heat balance	Partial differential equations for description of temperature and oxygen concentration distribution within the composting pile at steady state. Energy balance includes energy change due to air follow though composting pile, heat generated by oxidation and heat generated by biological activity. Oxygen mass balance includes oxygen concentration change due to air follow though composting pile and due to oxidation.	[78]
Livestock manure	Hydrological model	Model describes the drainage of the waste water. In includes drainage from the composting pad area, direct precipitation falling and soil infiltration. Model uses weather data, data on vegetative filter strip buffer area data, soil infiltration rate, soil depth, water table depth, soli slope and plant type.	[79]
Vegetable waste	Mass balances for energy, oxygen, vapor and liquid water concentration	Partial differential equations for description of temperature, oxygen concentration and minimum liquid water-to-compost weight ratio during the composting process. Energy balance includes energy change due to air flow through composting pile, heat generated by oxidation and heat generated by biological activity. Oxygen mass balance includes oxygen concentration change due to air flow through composting pile and due to oxidation.	[80]
Bovine manure and sewage sludge	Organic N balance	Nitrogen mineralization kinetics in soils treated by different composts was analysed using five mathematical models: simple exponential model, double exponential model, special exponential model, hyperbolic model and parabolic model. Selected models allowed analysis of the mineralisation rate.	[81]
Switchgrass and dog food mixture	Balances for substrate and oxygen concentration and energy transfer balance	Model describes the change of substrate concentration (sugars and starches) and oxygen concentration in a form of differential equations. Energy balance was expressed as partial differential equation. Model was able to predict the microbial populations (bacteria, fungi and yeast) growth reliably.	[82]
Wood chips	Mathematical model of vertical moisture content movement	Mathematical model in form of differential equations, describes the dynamic of the moisture content change in a static composting pile. The amount of water in different composting layer dependent on input water flux, and output water flux by evaporation, diffusion, and percolation.	[83]
Fresh cow manure, organic faction of municipal solid waste, non-digested sludge from municipal wastewater treatment	Exponential equations for description of organic matter settlement during composting	Two-parameter and three-parameter models were used. Two-parameter model includes parameter that represent the time when settlement reaches 50% and parameter that present maximum possible settlement. Three-parameter model additionally includes parameter that presents the inflexion point in the settlement curve.	[84]

Table 2. *Cont.*

Composting Substrate	Model Formulation	Model Description	Reference
Green waste, biowaste and paper cardboard	Mass balances for organic matter change during composting	Mathematical model in form of differential equations describes the dynamics of organic matter change during the composting. Organic matter was analysed in five fractions (easily degradable soluble, slowly degradable soluble, hemicelluloses, cellulose, and lignin fractions). Microbial population growth rate was described by Monod kinetics.	[85]
Household waste	Energy transfer balance	Mathematical model in a form of differential equations and empirical algebraic equations describes the heat transfer during composting. Model includes heat production that is proportional to oxygen consumption rate and heat transfers by evaporation, convection between material and gas crossing the material, conduction and surface convection between gas and material in bottom and upper parts of the reactor,	[86]
Tobacco waste	Mass balance for substrate and biomass	Mathematical model in form of differential equations for description of biodegradation of organic matter from leachate which is produced during the composting process of tobacco waste. Applicability of Monod model, modified Monod model, Haldane model and expanded Haldane model for description of microbial growth was analysed.	[87]
Sewage sludge, straw and sawdust	Artificial neural network model	Artificial neural network model for description of ammonia emission during composting based on selected input variables (sampling time, composting mixture temperature, composting mixture pH, composting mixture conductivity, weight of dry mass in composting mixture, C/N ratio of the composting mixture and ammonia nitrate content in composting mixture). Selected ANNs predicted the ammonia emission during composting with high precision $R^2 > 0.970$.	[88]
Sewage sludge, branches, grass clippings and ground leaves	Mass balances for organic carbon and mass balances for organic pollutants	Mathematical model in a form of differential equations describes organic pollutants and organic carbon dynamics during composting. Model includes two modules that can be used separately or coupled. Microbial population growth rate was described by Monod kinetics.	[89]
Straw, yard waste, paper, wood fibers, screening residues and sewage sludge	Mass balances for organic micro-pollutants dynamics during composting	First-order (using a single degradation coefficient) and dual first-order (using two separate degradation coefficients) kinetic models were used to describe the degradation of organic micro-pollutants (representing the most common groups of organic chemicals present in sewage sludge) during composting.	[90]
Olive mill waste	Heat balance equation, water mass balance, oxygen consumption and carbon dioxide production balances, compost volume change balance, biomass growt balance, nitrogen and phosphorous mass balances	Integrated model in a form of differential equations for decryption and prediction physicochemical and biological mechanism during composting. Insoluble organic matter hydrolysis was described using first-order kinetics. Microbial biomass growth was modelled with a double-substrate limitation (by hydrolyzed available organic substrate and oxygen) using Monod kinetics. The inhibitory factors of temperature and moisture content were included in the system.	[91]

Table 2. *Cont.*

Composting Substrate	Model Formulation	Model Description	Reference
Sewage sludge and straw	Heat balance equation, organic matter balance, inert substrate balance, oxygen and carbon dioxide balances, water mass balance, biomass growt balance, ammonia mass balances	Integrated model in a form of 11 differential equations for decryption of aerobic composting dynamics. Model describes the growth of mesophilic and thermophilic microorganisms by Contois kinetics equation. Model also refers to easily hydrolysable organic matter and inert organic matter of the used composting substrate. According to model changes of carbon dioxide concentrations and ammonia concentrations are inverse to changes in the oxygen concentration.	[92]
Poultry manure and wheat straw	Heat balance, substrate mass balance, oxygen mass balance, carbon dioxide mass balance, water mass balance	Integrated model composed of eight ordinary differential equations that take into account microbial kinetics, mass balance, heat balance and stoichiometry. The Monod kinetic model described the microbial growth and its dependency on moisture content, oxygen and temperature.	[93]
Pig manure, poultry manure, wheat straw	Heat balance and oxygen mass balance	Oxygen uptake rate model in form of differential equations. Model takes into account oxygen consumption diffusion form gas phase to solid phase and oxygen consumption. Model also includes microorganisms growth according to Monod model and substrate hydrolysis in anaerobic conditions.	[94]
Animal manure	Heat balance, carbon dioxide mass balance, water mass balance and ammonia mass balances	Model for simulation of physicochemical and biological mechanism during composting in pile based on four modules; biodegradation, nitrogen transformation and volatilization, thermal exchanges and free air space evolution. Microbial growth was described by first order kinetic.	[95]
Sewage sludge and wheat straw	Heat balance, mass balances for moisture content and total mass	Integrated model for simulation of physicochemical and biological mechanism during composting. The model comprises first-order kinetics for organic matter (expressed as volatile solid) degradation, energy balance equation with two-dimensional heat transfer; and mass balance equation with one-dimensional mass transfer. Organic matter degradation includes adjusted functions of temperature, oxygen, moisture content and free air space.	[96]
Wood chips and dog food	Mass balances for: aerobic biomass, oxygen, soluble substrate, insoluble substrate, water, temperature and inert matter	Integrated biomass-dependant dynamic model that takes into account temperature and biological growth under moisture, oxygen and substrate content. The aerobic biomass growth was described by the Monod model with dependence on oxygen concentration, soluble substrate concentration, moisture content and temperature. Model differs slowly and rapidly degradable substrates.	[97]
Corn silage and cattle manure	Heat balance and substrate mass balance	Heat balance included biological heat production, heating of reactor system, heat of input and output gas, conductive and convective heat losses through reactor wall and, latent heat of water evaporation. Substrate degradation was described by first-order kinetics.	[98]

23. Martinez, M.M.; Ortega Blu, R.; Janssens, M.; Fincheira, P. Grape pomace compost as a source of organic matter: Evolution of quality parameters to evaluate maturity and stability. *J. Clean. Prod.* **2019**, *216*, 56–63. [CrossRef]
24. Sokač, T.; Šalić, A.; Kučić Grgić, D.; Šabić Runjevac, M.; Vidaković, M.; Jurinjak Tušek, A.; Horvat, Đ.; Juras Krnjak, J.; Vuković Domonac, M.; Zelić, B. An enhanced composting process with bioaugmentation: Mathematical modelling and process optimization. *Waste Manag. Res.* **2021**, *2021*, 1–9. [CrossRef]
25. Papračnin, E.; Pertic, I. Mathematical modeling and simulation of the composting process in a pilot reactor. *Glas. Hem. Tehnol. Bosne Herceg.* **2017**, *47*, 39–48.
26. He, X.; Han, L.; Ge, J.; Huang, G. Modelling for reactor-style aerobic composting based on coupling theory of mass-heat-momentum transport and Contois equation. *Bioresour. Technol.* **2018**, *253*, 165–174. [CrossRef]
27. Koolivand, A.; Godini, K.; Saeedi, R.; Abtahi, H.; Ghamari, F. Oily sludge biodegradation using a new two-phase composting method: Kinetics studies and effect of aeration rate and mode. *Process Biochem.* **2019**, *79*, 127–134. [CrossRef]
28. Loan, L.T.T.; Takahashi, Y.; Nomura, H.; Yabe, M. Modeling home composting behavior toward sustainable municipal organic waste management at the source in developing countries. *Resour. Conserv. Recycl.* **2019**, *140*, 65–71. [CrossRef]
29. Jain, M.S.; Paul, S.; Kalamdhad, A.S. Kinetics and physics during composting of various organic wastes: Statistical approach to interpret compost application feasibility. *J. Clean. Prod.* **2020**, *255*, 120324. [CrossRef]
30. Mason, I.G. Mathematical modelling of the composting process: A review. *Waste Manag.* **2006**, *26*, 3–21. [CrossRef]
31. Rothlauf, F. Optimization Methods. In *Design of Modern Heuristics*; Springer: Berlin/Heidelberg, Germany, 2011; pp. 45–102. [CrossRef]
32. Šibalija, T.V.; Majstorović, V.D. Optimisation methods. In *Advanced Multiresponse Process Optimization*; Springer: Berlin/Heidelberg, Germany, 2016; pp. 21–63. [CrossRef]
33. Tsai, C.W.; Tong, L.I.; Wang, C.H. Optimization of multiple responses using data envelopment analysis and response surface methodology. *J. Appl. Sci. Eng.* **2010**, *13*, 197–203.
34. Vera Candioti, L.; De Zan, M.M.; Cámara, M.S.; Goicoechea, H.C. Experimental design and multiple response optimization. Using the desirability function in analytical methods development. *Talanta* **2014**, *124*, 123–138. [CrossRef] [PubMed]
35. Uy, M.; Telford, J.K. Optimization by design of experiment techniques. In Proceedings of the 2009 IEEE Aerospace Conference, Big Sky, MT, USA, 7–14 March 2009; pp. 1–10. Available online: https://ieeexplore.ieee.org/document/4839625 (accessed on 23 July 2021).
36. Said, K.A.M.; Amin, M.A.M. Overview on the response surface methodology (RSM) in extraction processes. *JASPE* **2015**, *2*, 8–17.
37. Pang, J.S.; Ansari, M.N.M.; Zaroog, O.S.; Ali, M.H.; Sapuan, S.M. Taguchi design optimization of machining parameters on the CNC end milling process of halloysite nanotube with aluminium reinforced epoxy matrix (HNT/Al/Ep) hybrid composite. *HBRC J.* **2014**, *10*, 138–144. [CrossRef]
38. Saleem, M.M.; Somá, A. Design of experiments based factorial design and response surface methodology for MEMS optimization. *Microsyst. Technol.* **2015**, *21*, 263–276. [CrossRef]
39. Raykundaliya, D.; Shanubhogue, A. Comparison study: Taguchi methodology vis.-a-vis. response surface methodology through a case study of accelerated failure in spin-on-filter. *Int. Adv. Res. J. Sci. Eng. Technol.* **2015**, *2*, 1–5.
40. Ajmal, M.; Aiping, S.; Awais, M.; Ullah, M.S.; Saeed, R.; Uddin, S.; Ahmad, I.; Zhou, B.; Zihao, X. Optimization of pilot-scale in-vessel composting process for various agricultural wastes on elevated temperature by using Taguchi technique and compost quality assessment. *Process Saf. Environ. Prot.* **2020**, *140*, 34–45. [CrossRef]
41. Gunst, R.F.; Mason, R.L. Fractional factorial design. *Rev. Compt. Stat.* **2009**, *1*, 234–244. [CrossRef]
42. Rezende, C.A.; Atta, B.W.; Breitkreitz, M.C.; Simister, R.; Gomez, L.D.; McQueen-Mason, S.J. Optimization of biomass pretreatments using fractional factorial experimental design. *Biotechnol. Biofuels.* **2018**, *11*, 206. [CrossRef]
43. Phoa, F.K.H.; Xu, H.; Wong, W.K. The use of nonregular fractional factorial designs in combination toxicity studies. *Food Chem. Toxicol.* **2019**, *47*, 2183–2188. [CrossRef]
44. Aslitürk, I.; Akkus, H. Determining the effect of cutting parameters on surface roughness in hard turning using the Taguchi method. *Measurement* **2011**, *44*, 1697–1704. [CrossRef]
45. Sayara, T.; Sarra, M.; Sánchez, A. Optimization and enhancement of soil bioremediation by composting using the experimental design technique. *Biodegradation* **2010**, *21*, 345–356. [CrossRef] [PubMed]
46. Baptista, M.; Antunes, F.; Silveira, A. Diagnosis and optimization of the composting process in full-scale mechanical-biological treatment plants. *Waste Manag. Res.* **2011**, *29*, 565–573. [CrossRef] [PubMed]
47. Cabeza, I.O.; López, R.; Ruiz-Montoya, M.; Díaz, M.J. Maximising municipal solid waste—Legume trimming residue mixture degradation in composting by control parameters optimization. *J. Environ. Manag.* **2013**, *128*, 266–273. [CrossRef] [PubMed]
48. Aycan, N.; Turan, N.G.; Orgonenel, O. Optimization of process parameters for composting of pulp/paper mill sludge with hazelnut kernel using a statistical method. *Environ. Prot. Eng.* **2014**, *40*, 127–138. [CrossRef]
49. Mohammad, N.; Alam, M.Z.; Kabashi, N.A. Optimization of effective composting process of oil palm industrial waste by lignocellulolytic fungi. *J. Mater. Cycles Waste Manag.* **2015**, *17*, 91–98. [CrossRef]
50. Torres-Climent, A.; Martin-Mata, J.; Marhuenda-Egea, F.; Moral, R.; Barber, X.; Perez-Murcia, M.D.; Paredes, C. Composting of the solid phase of digestate from biogas production: Optimization of the moisture, C/N ratio, and pH conditions. Commun. *Soil Sci. Plant Anal.* **2015**, *46*, 197–207. [CrossRef]

51. Parades, C.; Moreno-Caselles, J.; Agullo, F.; Andreu-Rodriguez, J.; Torres Climent, A.; Bustamante, M.A. Optimization of medlar pruning waste composting process by cattle manure addition. *Commun. Soil Sci. Plant Anal.* **2015**, *6*, 228–237. [CrossRef]

52. Aycan, N.; Turan, N.G. Statistical optimization to model ammonia removal during co-composting of pulp/paper mill sludge and corn wastes. *Clean Soil Air Water* **2016**, *44*, 1572–1577. [CrossRef]

53. Sarkar, S.; Pal, S.; Chanda, S. Optimization of a vegetable waste composting process with a significant thermophilic phase. *Procedia Environ. Sci.* **2016**, *35*, 435–440. [CrossRef]

54. Waqas, M.; Nizami, A.S.; Aburiazaiza, A.S.; Barakat, M.A.; Rashid, I.M.; Ismail, I.M.I. Optimizing the process of food waste compost and valorizing its applications: A case study of Saudi Arabia. *J. Clean. Prod.* **2017**, *176*, 426–438. [CrossRef]

55. Varma, V.S.; Kalamdhad, A.S.; Kumar, B. Optimization of waste combinations during in-vessel composting of agricultural waste. *Waste Manag. Res.* **2017**, *35*, 101–109. [CrossRef] [PubMed]

56. Chaher, N.H.; Chakchouk, M.; Engler, N.; Nassour, A.; Nelles, M.; Hamdi, M. Optimization of food waste and biochar in-vessel co-composting. *Sustainability* **2020**, *12*, 1356. [CrossRef]

57. Waqas, M.; Nizami, A.S.; Aburiazaiza, A.S.; Barakat, M.A.; Asam, Z.Z.; Khattak, B.; Rashid, M.I. Untapped potential of zeolites in optimization of food waste composting. *J. Environ. Manag.* **2019**, *241*, 99–112. [CrossRef] [PubMed]

58. Bian, B.; Hu, X.; Zhang, S.; Lv, C.; Yang, Z.; Yang, W.; Zhang, L. Pilot-scale composting of typical multiple agricultural wastes: Parameter optimization and mechanisms. *Bioresour. Technol.* **2019**, *28*, 121482. [CrossRef] [PubMed]

59. Zhang, Q.; Liu, J.; Guo, H.; Li, E.; Yan, Y. Characteristics and optimization of dairy manure composting for reuse as a dairy mattress in areas with large temperature differences. *J. Clean. Product.* **2019**, *232*, 1053–1061. [CrossRef]

60. Tabrika, I.; Mayad, E.H.; Furze, J.N.; Zaafrani, M.; Azim, K. Optimization of tomato waste composting with integration of organic feedstock. *Environ. Sci. Pollut. Res. Int.* **2021**, *28*, 64140–64149. [CrossRef]

61. Lew, P.S.; Nik Ibrahim, N.N.L.; Kamarudin, S.; Thamrin, N.M.; Misnan, M.F. Optimization of bokashi-composting process using effective microorganisms-1 in smart composting bin. *Sensors* **2021**, *21*, 2847. [CrossRef]

62. Li, F.; Yu, H.; Li, Y.; Wang, Y.; Shen Resource, J.; Hu, D.; Han, Y. The quality of compost was improved by low concentrations of fulvic acid owing to its optimization of the exceptional microbial structure. *Bioresour. Technol.* **2021**, *342*, 125843. [CrossRef]

63. Song, B.; Manu, M.K.; Li, D.; Wang, C.; Varjani, S.; Ladumor, N.; Wong, J.W.C. Food waste digestate composting: Feedstock optimization with sawdust and mature compost. *Bioresour. Technol.* **2021**, *341*, 125759. [CrossRef]

64. Sharma, D.; Yadav, K.D.; Kumar, S. Biotransformation of flower waste composting: Optimization of waste combinations using Response Surface Methodology. *Bioresour. Technol.* **2018**, *270*, 198–207. [CrossRef] [PubMed]

65. Calabi-Floody, M.; Medina, J.; Suazo, J.; Ordiqueo, M.; Aponte, H.; Mora, M.; Rumpel, C. Optimization of wheat straw co-composting for carrier material development. *Waste Manag.* **2019**, *98*, 37–49. [CrossRef] [PubMed]

66. Ajmal, M.; Shi, A.; Awais, M.; Mengqi, Z.; Zihao, X.; Shabbir, A.; Ye, L. Ultra-high temperature aerobic fermentation pretreatment composting: Parameters optimization, mechanisms and compost quality assessment. *J. Environ. Chem. Eng.* **2021**, *9*, 105453. [CrossRef]

67. Roman, F.F.; de Tuesta, J.L.D.; Praca, P.; Silva, A.M.T.; Faria, J.L.; Gomes, H.T. Hydrochars from compost derived from municipal solid waste: Production process optimization and catalytic applications. *J. Environ. Chem. Eng.* **2021**, *9*, 104888. [CrossRef]

68. Mazumder, P.; Akhil, P.M.; Khwairakpam, M.; Mishra, U.; Kalamdhad, A.S. Enhancement of soil physico-chemical properties post compost application: Optimization using Response Surface Methodology comprehending Central Composite Design. *J. Environ. Manag.* **2021**, *289*, 112461. [CrossRef]

69. Dehghani, M.H.; Salari, M.; Karri, R.R.; Hamidi, F.; Bahadori, R. Process modeling of municipal solid waste compost ash for reactive red 198 dye adsorption from wastewater using data driven approaches. *Sci. Rep.* **2021**, *11*, 11613. [CrossRef] [PubMed]

70. Soto-Paz, J.; Alfonso-Morales, W.; Caicedo-Bravo, E.; Oviedo-Ocaña, E.R.; Torres-Lozada, P.; Manyoma, P.C.; Sanchez, A.; Komilis, D. A new approach for the optimization of biowaste composting using artificial neural networks and particle swarm optimization. *Waste Biomass Valori.* **2019**, *11*, 3937–3951. [CrossRef]

71. Yamawaki, R.; Tei, A.; Ito, K.; Kikuchi, J. Decomposition factor analysis based on virtual experiments throughout Bayesian optimization for compost-degradable polymers. *Appl. Sci.* **2021**, *11*, 2820. [CrossRef]

72. Golbaz, S.; Zamanzadeh, M.Z.; Pasalari, H.; Farzadkia, M. Assessment of co-composting of sewage sludge, woodchips, and sawdust: Feedstock quality and design and compilation of computational model. *Environ. Sci. Pollut. Res. Int.* **2021**, *28*, 12414–12427. [CrossRef]

73. Dragoi, E.-N.; Godini, K.; Koolivand, A. Modeling of oily sludge composting process by using artificial neural networks and differential evolution: Prediction of removal of petroleum hydrocarbons and organic carbon. *Environ. Technol. Innov.* **2021**, *21*, 101338. [CrossRef]

74. Deaconu, S.; Coleman, H.W. Limitations of statistical design of experiments approaches in engineering testing. *J. Fluid. Eng.* **2000**, *122*, 254. [CrossRef]

75. Petrić, I.; Selimbašić, V. Development and validation of mathematical model for aerobic composting process. *Chem. Eng. J.* **2008**, *139*, 304–317. [CrossRef]

76. Hangos, K.; Cameron, I. *Process Modeling and Model Analysis*; Academic Press: Cambridge, MA, USA, 2001; pp. 1–33.

77. Šalić, A.; Jurinjak Tušek, A.; Zelić, B. Modeling of environmental processes. In *Environmental Engineering*; Springer: Berlin/Heidelberg, Germany, 2018; pp. 317–356.

78. Luangwilai, T.; Sidhu, H.S.; Nelson, M.I.; Chen, X.D. Modelling air flow and ambient temperature effects on the biological self-heating of compost piles. *Asia-Pac. J. Chem. Eng.* **2010**, *5*, 609–618. [CrossRef]
79. Webber, D.F.; Mickelson, S.K.; Wulf, L.W.; Richard, T.L.; Ahn, H.K. Hydrologic modeling of runoff from a livestock manure windrow composting site with a fly ash pad surface and vegetative filter strip buffers. *J. Soil Water Conserv.* **2010**, *65*, 252–260. [CrossRef]
80. Luangwilai, T.; Sidhu, H.S.; Nelson, M.I. A two dimensional reaction-diffusion model of compost piles. *Anziam J.* **2010**, *53*, 34–52. [CrossRef]
81. Gil, M.V.; Carballo, M.T.; Calvo, L.F. Modelling N mineralization from bovine manure and sewage sludge composts. *Bioresour. Technol.* **2011**, *102*, 863–871. [CrossRef] [PubMed]
82. Fontenelle, L.T.; Corgié, S.C.; Walker, L.P. Integrating mixed microbial population dynamics into modeling energy transport during the initial stages of the aerobic composting of a switchgrass mixture. *Bioresour. Technol.* **2011**, *102*, 5162–5168. [CrossRef] [PubMed]
83. Seng, B.; Kaneko, H.; Hirayama, K.; Katayama-Hirayama, K. Development of water movement model as a module of moisture content simulation in static pile composting. *Environ. Technol.* **2012**, *33*, 1685–1694. [CrossRef]
84. Illa, J.; Prenafeta-Boldú, F.X.; Bonmatí, A.; Flotats, X. Empirical characterisation and mathematical modelling of settlement in composting batch reactors. *Bioresour. Technol.* **2012**, *104*, 451–458. [CrossRef]
85. Zhang, Y.; Lashermes, G.; Houot, S.; Doublet, J.; Steyer, J.P.; Zhu, Y.G.; Barriuso, E.; Garnier, P. Modelling of organic matter dynamics during the composting process. *Waste Manag.* **2012**, *32*, 9–30. [CrossRef]
86. de Guardia, A.; Petiot, C.; Benoist, J.C.; Druilhe, C. Characterization and modelling of the heat transfers in a pilot-scale reactor during composting under forced aeration. *Waste Manag.* **2012**, *32*, 1091–1105. [CrossRef] [PubMed]
87. Ćosić, I.; Vuković, M.; Gomzi, Z.; Briški, F. Comparison of various kinetic models for batch biodegradation of leachate from tobacco waste composting. *Rev. Chim.* **2012**, *63*, 967–971.
88. Boniecki, P.; Dach, J.; Pilarski, K.; Piekarska-Boniecka, H. Artificial neural networks for modeling ammonia emissions released from sewage sludge composting. *Atmos. Environ.* **2012**, *57*, 49–54. [CrossRef]
89. Lashermes, G.; Zhang, Y.; Houot, S.; Steyer, J.P.; Patureau, D.; Barriuso, E.; Garnier, P. Simulation of organic matter and pollutant evolution during composting: The COP-compost model. *J. Environ. Qual.* **2013**, *42*, 361–372. [CrossRef]
90. Sadef, Y.; Poulsen, T.G.; Bester, K. Modeling organic micro pollutant degradation kinetics during sewage sludge composting. *Waste Manag.* **2014**, *34*, 2007–2013. [CrossRef]
91. Vasiliadou, I.A.; Chowdhury, A.K.M.M.B.; Akratos, C.S.; Tekerlekopoulou, A.G.; Pavlou, S.; Vayenas, D.V. Mathematical modeling of olive mill waste composting process. *Waste Manag.* **2015**, *43*, 61–71. [CrossRef]
92. Białobrzewski, I.; Mikš-Krajnik, M.; Dach, J.; Markowski, M.; Czekała, W.; Głuchowska, K. Model of the sewage sludge-straw composting process integrating different heat generation capacities of mesophilic and thermophilic microorganisms. *Waste Manag.* **2015**, *43*, 72–83. [CrossRef]
93. Petric, I.; Mustafić, N. Dynamic modeling the composting process of the mixture of poultry manure and wheat straw. *J. Environ. Manag.* **2015**, *161*, 392–401. [CrossRef]
94. Ge, J.; Huang, G.; Huang, J.; Zeng, J.; Han, L. Modeling of oxygen uptake rate evolution in pig manure–wheat straw aerobic composting process. *Chem. Eng. J.* **2015**, *276*, 29–36. [CrossRef]
95. Oudart, D.; Robin, P.; Paillat, J.M.; Paul, E. Modelling nitrogen and carbon interactions in composting of animal manure in naturally aerated piles. *Waste Manag.* **2015**, *46*, 588–598. [CrossRef]
96. Zhang, J.; Chen, T.-B.; Gao, D. Simulation of the mathematical model of composting process of sewage sludge. *Compost Sci. Util.* **2016**, *24*, 73–85. [CrossRef]
97. Seng, B.; Kristanti, R.A.; Hadibarata, T.; Hirayama, K.; Katayama-Hirayama, K.; Kaneko, H. Mathematical model of organic substrate degradation in solid waste windrow composting. *Bioproc. Biosyst. Eng.* **2016**, *39*, 81–94. [CrossRef] [PubMed]
98. Wang, Y.; Pang, L.; Liu, X.; Wang, Y.; Zhou, K.; Luo, F. Using thermal balance model to determine optimal reactor volume and insulation material needed in a laboratory-scale composting reactor. *Bioresour. Technol.* **2016**, *206*, 164–172. [CrossRef]
99. Külcü, R. New kinetic modelling parameters for composting process. *J. Mater. Cycles Waste Manag.* **2016**, *18*, 734–741. [CrossRef]
100. Wang, Y.; Witarsa, F. Application of Contois, Tessier, and first-order kinetics for modeling and simulation of a composting decomposition process. *Bioresour. Technol.* **2016**, *220*, 384–393. [CrossRef] [PubMed]
101. Gutiérrez, M.C.; Siles, J.A.; Diz, J.; Chica, A.F.; Martín, M.A. Modelling of composting process of different organic waste at pilot scale: Biodegradability and odor emissions. *Waste Manag.* **2016**, *59*, 48–58. [CrossRef]
102. Giusti, E.; Marsili-Libelli, S. Fuzzy modelling of the composting process. *Environ. Model Softw.* **2010**, *25*, 641–647. [CrossRef]
103. Alavi, N.; Sarmadi, K.; Goudarzi, G.; Babaei, A.A.; Bakhshoodeh, R.; Paydary, P. Attenuation of tetracyclines during chicken manure and bagasse co-composting: Degradation, kinetics, and artificial neural network modeling. *J. Environ. Manag.* **2019**, *23*, 1203–1210. [CrossRef] [PubMed]
104. Roohi, R.; Jafari, M.; Jahantab, E.; Aman, M.S.; Moameri, M.; Zare, S. Application of artificial neural network model for the identification the effect of municipal waste compost and biochar on phytoremediation of contaminated soils. *J. Geochem. Explor.* **2019**, *39*, 1058–1068. [CrossRef]

105. Alavi, N.; Dehvari, M.; Alekhamis, G.; Goudarzi, G.; Neisi, A.; Babaei, A.A. Application of electro-Fenton process for treatment of composting plant leachate: Kinetics, operational parameters and modeling. *J. Environ. Health Sci. Eng.* **2019**, *17*, 417–431. [CrossRef]
106. Vidriales-Escobar, G.; Rentería-Tamayo, R.; Alatriste-Mondragón, F.; González-Ortega, O. Mathematical modeling of a composting process in a small-scale tubular bioreactor. *Chem. Eng. Res. Des.* **2017**, *120*, 360–371. [CrossRef]
107. Rentería-Tamayo, R.; Vidriales-Escobar, G.; González-Ortega, O.; Alatriste-Mondragón, F. Mathematical modeling of the mesophilic and thermophilic stages of a composting tubular reactor for sewage sludge sanitization. *Waste Biomass Valori.* **2018**, *11*, 955–966. [CrossRef]
108. Luangwilai, T.; Sidhu, H.S.; Nelson, M.I. One-dimensional spatial model for self-heating in compost piles: Investigating effects of moisture and air flow. *Food Bioprod. Proc.* **2018**, *108*, 18–26. [CrossRef]
109. Martalo, G.; Bianchi, C.; Buonomo, B.; Chiappini, M.; Vespri, V. Mathematical modeling of oxygen control in biocell composting plants. *Math. Comp. Simul.* **2020**, *177*, 105–119. [CrossRef]
110. Walling, E.; Vaneeckhaute, C. Novel simple approaches to modeling composting kinetics. *J. Environ. Chem. Eng.* **2021**, *9*, 105243. [CrossRef]
111. Tsiodra, C.; Stathakis, E.; Vlysidis, A.; Vlyssides, A. Development of a dynamic model for the degradation of fats, oils and greases during co-composting of olive mill solid and liquid wastes. *Fresenius Environ. Bull.* **2018**, *27*, 4900–4903.
112. Sable, S.; Mandal, D.K.; Ahuja, S.; Bhunia, H. Biodegradation kinetic modeling of oxo-biodegradable polypropylene/polylactide/nanoclay blends and composites under controlled composting conditions. *J. Environ. Manag.* **2019**, *249*, 109186. [CrossRef]
113. Sable, S.; Ahuja, S.; Bhunia, H. Biodegradation kinetic modeling of acrylic acid-grafted polypropylene during thermophilic phase of composting. *Iran. Polym. J.* **2020**, *29*, 735–747. [CrossRef]
114. Ebrahimzadeh, R.; Ghazanfari Moghaddam, A.; Sarcheshmehpour, M.; Mortezapour, H. A novel kinetic modeling method for the stabilization phase of the composting process for biodegradation of solid wastes. *Waste Manag. Res.* **2017**, *35*, 1226–1236. [CrossRef]
115. Samaei, M.R.; Jalili, M.; Abbasi, F.; Mortazavi, S.B.; Jonidi Jafari, A.; Bakhshi, B. Isolation and kinetic modeling of new culture from compost with high capability of degrading n-hexadecane, focused on Ochrobactrum oryzae and Paenibacillus lautus. *Soil Sediment Contam. Int. J.* **2020**, *29*, 384–396. [CrossRef]
116. Rafiee, R.; Obersky, L.; Xie, S.; Clarke, W.P. A mass balance model to estimate the rate of composting, methane oxidation and anaerobic digestion in soil covers and shallow waste layers. *Waste Manag.* **2017**, *63*, 196–202. [CrossRef] [PubMed]
117. Toledo, M.; Gutiérrez, M.C.; Siles, J.A.; Martín, M.A. Full-scale composting of sewage sludge and market waste: Stability monitoring and odor dispersion modeling. *Environ. Res.* **2018**, *167*, 739–750. [CrossRef] [PubMed]
118. Ghinea, C.; Apostol, L.C.; Prisacaru, A.E.; Leahu, A. Development of a model for food waste composting. *Environ. Sci. Pollut. Res. Int.* **2018**, *26*, 4056–4069. [CrossRef] [PubMed]
119. Soto-Paz, J.; Oviedo-Ocaña, E.R.; Manyoma, P.C.; Gaviría-Cuevas, J.F.; Marmolejo-Rebellón, L.F.; Torres-Lozada, P.; Sánchez, A.; Komilis, D. A multi-criteria decision analysis of co-substrate selection to improve biowaste composting: A mathematical model applied to Colombia. *Environ. Process.* **2019**, *6*, 673–694. [CrossRef]
120. Wang, H.; Bridges, W.; Chen, Z.; Gong, C.; Jiang, X. Comparing and modeling the thermal inactivation of bacteriophages as pathogenic viruses surrogates in chicken litter compost. *Compost Sci. Util.* **2020**, *28*, 87–99. [CrossRef]
121. Calisti, R.; Regini, L.; Proietti, P. Compost-recipe: A new calculation model and a novel software tool to make the composting mixture. *J. Clean. Prod.* **2020**, *270*, 122427. [CrossRef]
122. Sobieraj, K.; Stegenta-Dabrowska, S.; Koziel, J.A.; Białowiec, A. Modeling of CO accumulation in the headspace of the bioreactor during organic waste composting. *Energies* **2021**, *14*, 1367. [CrossRef]
123. Wojcieszak, D.; Zaborowicz, M.; Przybył, J.; Boniecki, P.; Jędruś, A. Assessment of the content of dry matter and dry organic matter in compost with neural modelling methods. *Agriculture* **2021**, *11*, 307. [CrossRef]

processes

Article

Optimal Darwinian Selection of Microorganisms with Internal Storage

Walid Djema [1,*], Térence Bayen [2] and Olivier Bernard [1,3]

[1] Université Côte d'Azur, Inria, INRAE, CNRS, Sorbonne Université, Biocore Team, 06902 Sophia Antipolis—Valbonne, France; olivier.bernard@inria.fr

[2] Laboratoire de Mathématiques d'Avignon (EA 2151), Avignon Université, 84018 Avignon, France; terence.bayen@univ-avignon.fr

[3] Laboratoire d'Océanographie de Villefranche (LOV), CNRS, Sorbonne University, IMEV, 06230 Villefranche-sur-Mer, France

* Correspondence: walid.djema@inria.fr

Abstract: In this paper, we investigate the problem of species separation in minimal time. Droop model is considered to describe the evolution of two distinct populations of microorganisms that are in competition for the same resource in a photobioreactor. We focus on an optimal control problem (OCP) subject to a five-dimensional controlled system in which the control represents the dilution rate of the chemostat. The objective is to select the desired species in minimal-time and to synthesize an optimal feedback control. This is a very challenging issue, since we are are dealing with a ten-dimensional optimality system. We provide properties of optimal controls allowing the strain of interest to dominate the population. Our analysis is based on the Pontryagin Maximum Principle (PMP), along with a thorough study of singular arcs that is crucial in the synthesis of optimal controls. These theoretical results are also extensively illustrated and validated using a direct method in optimal control (via the Bocop software for numerically solving optimal control problems). The approach is illustrated with numerical examples with microalgae, reflecting the complexity of the optimal control structure and the richness of the dynamical behavior.

Keywords: optimal control; modelling; microalgae; chemostat; nonlinear control; Pontryagin's principle; singular control; Droop model; photobioreactor

Citation: Djema, W.; Bayen, T.; Bernard, O. Optimal Darwinian Selection of Microorganisms with Internal Storage. *Processes* **2022**, *10*, 461. https://doi.org/10.3390/pr10030461

Academic Editors: Philippe Bogaerts and Alain Vande Wouwer

Received: 19 January 2022
Accepted: 14 February 2022
Published: 24 February 2022

Publisher's Note: MDPI stays neutral with regard to jurisdictional claims in published maps and institutional affiliations.

1. Introduction

The interaction between species coexisting in an ecosystem is complex and affected by external factors. Depending on their environment, some species will dominate, while others, less adapted, will progressively decline. This Darwinian pressure, when it can be manipulated [1], provides the opportunity of guiding the evolution of species of interest. This concept can be applied to artificial ecosystems to select individuals with a desired trait. Here, we focus on microalgae, unicellular photosynthetic microorganisms with promising potential for industrial applications [2,3]. The great biodiversity of microalgae opens the door for a large range of applications [4]. They are grown for their pigments, antioxidants or essential fatty acids [5], and, over the longer term, their efficient way of producing proteins, bricks for green chemistry, biofuel and CO_2 mitigation [2,6,7]. To date, microalgae do not have the place they deserve in biotechnology (see, e.g., [8–10]) and many optimization steps must be carried out to improve the economic and environmental performances of these processes at a large scale [6,11]. Currently, only wild organisms sampled in nature are used on an industrial scale. One of the key challenges is to improve the productivity of these strains.

Species in agriculture have been improved after centuries of selection and hybridization. The objective of this work is to develop an alternative approach adapted to microorganisms to select, on a shorter time scale, more productive microalgae strains, by Darwinian pressure. The idea is based on the competitive exclusion principle in a continuous reactor [12] stating that the species which more efficiently uses the available resources will win the competition. The conditions for a bacterial or microalgal species to win the competition for a limiting substrate have been well established, and the outcome of the competition is known to depend on the minimum requested substrate to support a growth rate equal to the dilution rate [12,13].

Experimental works for maintaining a long-term selection process to favour individuals of interest have already been carried out [14,15]. Since the experiments can last several months or several years, these approaches are costly in time. There is a margin of improvement by applying optimal control theory [16] to enhance the selection process for N strains competing for the same resource (the control parameter being the dilution rate).

One main issue is to decrease the operating time when the species of interest starts to dominate. Several works addressed the question of improving the selection process in minimal time in the case of the chemostat system with Monod's laws [17,18]. Microalgae are more complicated microorganisms better represented by the Droop model, taking into account the internal accumulation of the limiting nutrient [19,20]. Such a model for two strains in competition leads to a five-dimensional problem. The minimal time selection problem with this model is the main focus of the paper. It has been tackled in [21,22] after a simplification allowing for reducing the model dimension. This necessitated to oversimplify the initial dynamics which can play a role in the minimal time selection.

Optimal control [23] strategies ensuring the domination of the strain of interest are derived using the Pontryagin maximum principle [24]. Since the system is affine w.r.t. the control, we obtain various possible structures for an optimal control, namely, the concatenation of several bang arcs or of a bang arc with a singular arc of first order satisfying Legendre–Clebsch's condition. The paper is structured as follows: in Section 2, we introduce the model and present the optimal control problem. We also prove the reachability of the target set. In Section 3, we make explicit the necessary conditions provided by the Pontryagin Maximum Principle and we introduce properties of the switching function. A thorough study of singular arcs is provided in Section 4 thanks to geometrical control theory. The paper is concluded with numerical simulations of optimal strategies using a direct method in Section 5.

2. The Optimal Control Problem (OCP)

2.1. Droop Model and Main Assumptions

We consider the Droop model [19]. This emblematic variable yield model represents the growth rate of microorganims which can intracellularly store nutrients. When two strains are in competition, it results in a five-dimensional system. The growth of each strain depends on the intracellular quota-storage (q-variable) of the limiting nutrient (s-variable). More precisely, when two species/strains, of biomass concentrations x_1 and x_2 are competing for one limiting nutrient s in the bioreactor, the Droop model reads as follows:

$$\begin{cases} \dot{s} &= (s_{in} - s)D(t) - \rho_1(s)x_1 - \rho_2(s)x_2, \\ \dot{q}_1 &= \rho_1(s) - \mu_1(q_1)q_1, \\ \dot{x}_1 &= [\mu_1(q_1) - D(t)]x_1, \\ \dot{q}_2 &= \rho_2(s) - \mu_2(q_2)q_2, \\ \dot{x}_2 &= [\mu_2(q_2) - D(t)]x_2, \end{cases} \tag{1}$$

where q_i is the quota storage of the i-th species and s_{in} is the input substrate concentration. The dilution rate $D(\cdot)$ is a bounded non-negative control function such that $D(t) \in [0, D_{\max}]$, where $D_{\max} > 0$ is the maximal admissible value of the dilution rate, above the maximum actual growth rates [22] (this will be made more precise in Section 2.3) of the two species as shown in Figure 1. In addition, for $i = 1, 2$, ρ_i is a non-negative function representing the rate of substrate absorption, i.e., the uptake rate of the free nutrient

s, and μ_i is also a non-negative function representing the growth rate of the i-th species (see [25]).

Following for instance [25], we suppose that the uptake rates $\rho_i(s)$ are expressed as,

$$\rho_i(s) = \frac{\rho_{m_i} s}{K_i + s},\tag{2}$$

which corresponds to Michaelis–Menten's kinetics. Here, the parameters K_i and ρ_{m_i} are positive, $i = 1, 2$. In addition, we assume that, for $i = 1, 2$, there is $k_i > 0$ such that cell division does not occur if $k < k_i$. Concerning the Droop model, the kinetics μ_i are defined by

$$\left| \begin{array}{ll} \mu_i(q_i) = 0, & 0 \le q_i \le k_i, \\ \mu_i(q_i) = \mu_{i\infty}\left(1 - \frac{k_i}{q_i}\right), & k_i \le q_i. \end{array} \right.\tag{3}$$

In addition, for $i = 1, 2$, let M_i stand for,

$$M_i := \sup_{s \in [0, s_{in}]} \rho_i(s) = \rho_i(s_{in}) < \rho_{m_i},$$

and let $\bar{q}_1, \bar{q}_2$ be such that $\mu_i(\bar{q}_i)\bar{q}_i = M_i$, $i = 1, 2$ (observe that $\bar{q}_1, \bar{q}_2$ are uniquely defined). Thus, one has,

$$\rho_i(s_{in}) = \mu_i(\bar{q}_i)\bar{q}_i.$$

System (1) satisfies the following invariance property.

Proposition 1. *For every $q_{m_1} \ge \bar{q}_1$, and for every $q_{m_2} \ge \bar{q}_2$, the set*

$$\Omega := (0, s_{in}) \times [k_1, q_{m_1}] \times \mathbb{R}_+^* \times [k_2, q_{m_2}] \times \mathbb{R}_+^*,\tag{4}$$

is forward invariant by (1).

Proof. First, observe that, for $i = 1, 2$, x_i never vanishes whenever $x_i^0 = x_i(0) > 0$. Now, $(0, s_{in})$ is clearly invariant by the dynamics of $s(\cdot)$ since $\dot{s} \ge 0$ (resp. $\dot{s} \le 0$) whenever $s = 0$ (resp. $s = s_{in}$). Similarly, for $i = 1, 2$:

$$\begin{array}{lll} q_i = k_i & \Rightarrow & \dot{q}_i = \rho_i(s) > 0, \\ q_i = q_{m_i} & \Rightarrow & \dot{q}_i \le M_i - \mu_i(q_{m_i})q_{m_i} \le 0, \end{array}$$

where the last inequality follows from the choice of M_i and the fact that $q_{m_i} \ge \bar{q}_i$, $i = 1, 2$. This ends the proof. $\square$

The parameter q_{m_i} represents the *maximum internal storage quota*. Since Ω is invariant by (1) (Proposition 1), we notice that $\bar{q}_i$ stands for the *effective maximum internal storage quota* for $s \in (0, s_{in})$. Thus, in the sequel, we consider without loss of generality that $q_{m_i} = \bar{q}_i$ for $i = 1, 2$. In the sequel, we also assume that ρ_1, ρ_2 fulfill the following hypothesis:

Assumption 1. *The affinity of species 1 for the substrate is higher than the one of species 2, i.e., $K_{s_2} > K_{s_1}$, or equivalently:*

$$\frac{\rho_2''}{\rho_2'} > \frac{\rho_1''}{\rho_1'}.\tag{5}$$

From (2), we deduce that, for $s \ge 0$,

$$\rho_2''(s)\rho_1'(s) - \rho_1''(s)\rho_2'(s) = \frac{2K_1 K_2 \rho_{m_1} \rho_{m_2}(K_2 - K_1)}{(K_1 + s)^3(K_2 + s)^3}.$$

It means that species 1 with the lower K_i absorbs nutrients slightly faster.

We are now in a position to formulate the OCP of interest.

2.2. Statement of the Optimal Control Problem (OCP)

In this work, we suppose that the first species (with biomass concentration x_1) is the one of interest. Our aim is to compute the best feeding strategy, that is, the optimal (dilution rate) control function $D(\cdot)$, such that x_1 becomes predominant in the photobioreactor in minimal-time. This can be formulated and quantified in terms of the ratio between the two competing species. Intuitively, we wish to find an adequate control strategy (if possible optimal) $D(\cdot)$ for which, at the end of the process, we have $\frac{x_1}{x_2} \gg 1$.

Firstly, the set of admissible controls is defined as,

$$\mathcal{D} := \{D : [0, +\infty) \to [0, D_{\max}] \;;\; D(\cdot) \in \mathcal{L}^{\infty}_{loc}(\mathbb{R}_+)\},$$

where $\mathcal{L}^{\infty}_{loc}(\mathbb{R}_+)$ is the space of locally integrable functions on every compact on $\mathbb{R}_+$ and $D_{\max} > 0$ is the maximum pump feeding capacity. In practice, $D_{\max}$ is designed above the maximum growth rates of the coexisting species (see Section 2.3).

To handle the selection process between the two species, let us define a subset $\mathcal{T}$ of Ω as,

$$\mathcal{T} := \{X := (s, q_1, x_1, q_2, x_2) \in \Omega \;;\; x_2 \leq \varepsilon x_1\}.$$

We choose the parameter $\varepsilon > 0$ such that $\varepsilon \ll 1$ in such a way to quantify the contamination rate of the interesting strain x_1. Whenever a trajectory reaches the target set $\mathcal{T}$, this means that the biomass of the first species is significantly greater than the other one when reaching the target $\mathcal{T}$ at the terminal time (if possible).

Objective 1. *The optimal control problem (OCP) can then be stated: determine a dilution-based control strategy $D(\cdot)$ in such a way that trajectories of* (1) *starting from an initial condition within the set Ω reach the target set $\mathcal{T}$ in minimal-time, i.e.,*

$$\inf_{D \in \mathcal{D}} t_f^D \quad s.t. \quad X(t_f^D) \in \mathcal{T} \text{ and } X^0 \in \Omega, \tag{6}$$

where $X(\cdot)$ is the unique solution of (1) *associated with $D(\cdot) \in \mathcal{D}$ such that $X(0) = X^0 \in \Omega$ and $t_f^D \in (0, +\infty]$ is the first entry time of $X(\cdot)$ into the target set. In the sequel, we will use the simpler notation t_f instead of t_f^D.*

In other words, for every positive initial conditions $X^0 = (s^0, q_1^0, x_1^0, q_2^0, x_2^0)$ such that $q_i^0 > k_i$, we are seeking an admissible control strategy $D = D_{X_0} \in \mathcal{D}$, steering the trajectory $X(t)$ of the system (1) from X^0 to the target set $\mathcal{T}$ in minimal-time, for a fixed $D_{\max}$ (Figure 1) and a given contamination rate $\varepsilon \ll 1$. Note that, if one is able to synthesize such an optimal control for every $X^0 \in \Omega$, then one is able to construct an optimal feedback control over Ω as $X_0 \mapsto D_{X_0}(0)$. Such an optimal control problem falls into the class of minimal-time control problems governed by a mono-input affine controlled system, for which the synthesis of an optimal feedback control, thanks to geometric control theory, is a crucial (but also delicate) issue. In particular, handling the high dimension of the Droop model in competition and its resulting optimality system is challenging. Note also that the linearity of the problem w.r.t. D (in contrast, for instance, with strictly convex cost functionals) leads to technicalities because singular arcs usually occur in this setting, see Section 4.

2.3. Basic Properties

We now introduce the so-called *actual growth rates*. These key functions will have an important role in the optimal separation strategy. For that, let us firstly start with the following observations:

- The mapping $\rho_i : [0, s_{in}] \to [0, M_i]$ is one-to-one with $M_i < \rho_{m_i}$;

- For every $q_i \in [k_i, \bar{q}_i]$, one has $q_i \mu_i(q_i) \in [0, M_i]$ and $q_i \mapsto q_i \mu_i(q_i)$ is one-to-one from $[k_i, \bar{q}_i]$ into $[0, M_i]$;
- It follows that the composition $\delta_i : [k_i, \bar{q}_i] \to [0, s_{in}]$ such that

$$\delta_i(q_i) := \rho_i^{-1}(q_i \mu_i(q_i)) = \frac{K_i q_i \mu_i(q_i)}{\rho_{m_i} - q_i \mu_i(q_i)}, \quad q_i \in [k_i, \bar{q}_i],$$

is well-defined and is also one-to-one;
- Hence, the mapping $\delta_i^{-1} : [0, s_{in}] \to [k_i, \bar{q}_i]$ such that

$$\delta_i^{-1}(s) := \frac{k_i K_i \mu_{i\infty} + (\rho_{m_i} + \mu_{i\infty} k_i)s}{\mu_{i\infty}(K_i + s)}, \quad s \in [0, s_{in}],$$

is well-defined over $[0, s_{in})$ with values in $[k_i, \bar{q}_i]$ and is one-to-one.

From these observations, one can immediately check that the mappings δ_i, $i = 1, 2$ and δ_i^{-1} are increasing.

Indeed, for $i = 1$, one can write $\delta_1^{-1}(s) = c_1 + \frac{c_2}{s + c_3}$ with $c_1, c_3 > 0$ and

$$c_2 := \delta_i^{-1}(s) = C + K_{s_1} \left[\frac{k_{k_1} \mu_{1\infty}}{k_1 \mu_{1\infty} + \rho_{m_1}} - 1 \right] < 0,$$

which implies that δ_i^{-1} is increasing.

The *actual growth rate* of species i is then defined as the mapping $\mu_i \circ \delta_i^{-1}$,

$$\mu_i(\delta_i^{-1}(s)) = \frac{\rho_{m_i} \mu_{i\infty} s}{k_i K_i \mu_{i\infty} + (\rho_{m_i} + \mu_{i\infty} k_i)s}, \quad s \in [0, s_{in}], \quad i = 1, 2.$$

The resulting generic functions are illustrated in Figure 1: Let us now define,

$$\Delta(s) := \mu_1(\delta_1^{-1}(s)) - \mu_2(\delta_2^{-1}(s)), \quad s \in [0, s_{in}]. \tag{7}$$

Throughout the paper, we suppose that Δ satisfies the following assumption.

Assumption 2. *There is a unique $\hat{s} \in (0, s_{in})$ such that $\Delta(s) > 0$ for every $s \in (0, \hat{s})$ and $\Delta(s) < 0$ for every $s \in (\hat{s}, s_{in})$. In addition, Δ has a unique maximum $s_c \in [0, \hat{s}]$.*

Taking into account that $\Delta(\hat{s}) = 0$, the inequalities satisfied by Δ according to Assumption 2 can also be written:

$$\Delta(s)(s - \hat{s}) < 0, \quad s \in (0, s_{in}) \setminus \{\hat{s}\}.$$

If we assume that s is regulated to $s^*(t) = s_c$ using an appropriate control D, we notice that the q-variables are regulated to some unique $q_{ic} \in [k_i, q_{mi}]$, for $i = 1, 2$. The unique point (s_c, q_{1c}, q_{2c}), where $s_c \in [0, s_{in}]$, plays a crucial role in the optimal control strategy of (OCP) as discussed in Section 5. Finally, $D_{\max}$ (the maximum dilution rate) is assumed to be large enough in order to drive competition between the two species. More precisely, we assume that $D_{\max}$ satisfies the hypothesis:

Assumption 3. *The maximal value of the dilution rate $D_{\max}$ satisfies*

$$\forall s \in [0, s_{in}], \quad D_{\max} > \max\left(\mu_1(\delta_1^{-1}(s)), \mu_2(\delta_2^{-1}(s)) \right).$$

These assumptions will ensure reachability (as detailed in the next section) of the target $\mathcal{T}$, and establishes a generic framework where both species may win the competition for sufficiently large time (considering for instance various constant control parameters D that favor species 1 or 2). Thus, under these considerations, we ensure the well-posedness of the optimal control problem of interest. In this general framework, the objective is then to determine the optimal D that steers the trajectories in minimal-time to the desired target.

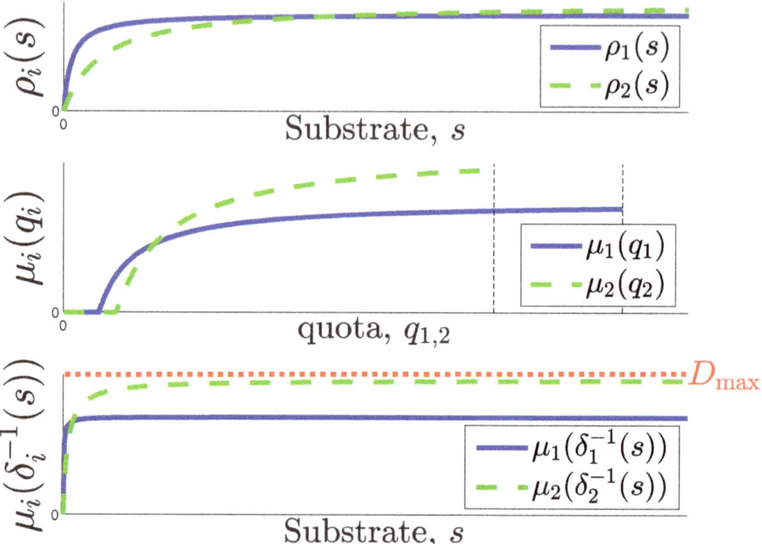

Figure 1. The illustrated absorption functions ρ_i and growth rates μ_i lead to a generic form of the actual growth rates $s \mapsto \mu_i(\delta_i(s))$ where both species may win the competition. The maximum dilution rate $D_{\max}$ is a fixed constant value above the maximum of the functions $s \mapsto \mu_i(\delta_i(s))$ for $i = 1, 2$, as stated in Assumption 3.

2.4. Reachability of the Target

Our next aim is to show that the target is reachable from every initial condition. First, let us recall that the set

$$\mathcal{M} := \{X \in \Omega \, ; \, x_1 q_1 + x_2 q_2 + s = s_{in}\} \tag{8}$$

is an invariant and attractive manifold for (1) for a given persistently exciting control (i.e., an admissible control function $D(\cdot)$ such that $\int_0^{+\infty} D(t) \, dt = +\infty$).

Proposition 2. *For every initial condition $X^0 \in \Omega$, there exists an admissible control $D(\cdot)$ and a time $t_e \geq 0$ such that $X(t_e) \in \mathcal{T}$, where $X(\cdot)$ is the unique solution of (1), starting from X^0, associated with $D(\cdot)$.*

Proof. Let $s^\dagger \in (0, \hat{s})$ and $X^0 \in \Omega$. Without any loss of generality, we may assume that $s(0) = s^\dagger$. Indeed, observe that $s = s^\dagger$ is not a steady-state of $\dot{s}$ whenever $D = 0$ over $\mathbb{R}_+$ or $D = D_{max}$ over $\mathbb{R}_+$. Thus, if we apply $D = 0$ (in that case $\dot{s} < 0$) or $D = D_{max}$ (in that case $\dot{s} > 0$), then $s(t) = s^\dagger$ is reached in a finite horizon. Consider the feedback control function,

$$D^\dagger(x_1, x_2) := \frac{\rho_1(s^\dagger)x_1 + \rho_2(s^\dagger)x_2}{s_{in} - s^\dagger},$$

in such a way that the unique solution of (1) associated with this control satisfies $s(t) = s^\dagger$ for every time $t \geq 0$ (Cauchy–Lipschitz's Theorem). We claim that there exists $t_1 \geq 0$ large enough such that $D^+(x_1(t), x_2(t)) \in [0, D_{max}]$ for every time $t \geq t_1$. Indeed, when $t \to +\infty$, taking into account (8), it follows:

$$D^+(x_1(t), x_2(t)) \sim \bar{D}(t) := \frac{\rho_1(s^+)x_1(t) + \rho_2(s^+)x_2(t)}{x_1(t)q_1(t) + x_2(t)q_2(t)}.$$

Now, when $t \to +\infty$, it is easy to see that $q_1(t)$ converges to $\frac{\rho_1(s^+)}{\mu_1^\infty} + k_1$, since q_1 satisfies the linear ODE $\dot{q}_1 + \mu_1^\infty q_1 = \rho_1(s^+) + \mu_1^\infty k_1)$. At steady-state, we thus have

$$\rho_1(s^+) = q_1 \mu_1(q_1).$$

In conclusion, when $t \to +\infty$,

$$\bar{D}(t) \sim \frac{q_1 \mu_1(q_1((t))x_1(t) + q_2(t)\mu_2(q_2(t))x_2(t)}{x_1(t)q_1(t) + x_2(t)q_2(t)},$$

or, equivalently, using that, at steady state, $s^+ = \rho_1^{-1}(q_1\mu_1(q_1))$ that is, $q_1 = \delta_i^{-1}(s^+)$, we end up with

$$\bar{D}(t) \sim \frac{\mu_1(\delta_1^{-1}(s^+))q_1(t)x_1(t) + \mu_2(\delta_2^{-1}(s^+))q_2(t)x_2(t)}{x_1(t)q_1(t) + x_2(t)q_2(t)}.$$

Thanks to Assumption 3, this last expression is upper bounded by D_{max} for t large enough, which proves our claim. Finally, posit $y_i := \ln(x_i)$ and observe that

$$\dot{y}_1 - \dot{y}_2 = \Delta(s^+) > 0.$$

It follows that $y_1(t) - y_2(t) \to +\infty$ when $t \to +\infty$, which implies that $\lim_{t \to +\infty} \frac{x_2(t)}{x_1(t)} = 0$. This ends the proof. $\square$

2.5. Motivation of Studying the OCP

Thanks to Proposition 2, the target set is reachable from any initial condition, thus the existence of an optimal control of (6) is standard (namely because the dynamics is affine w.r.t. the control): it is an application of the Fillipov Theorem, see, e.g., [26–28]. Considering such a control as in the proof of Proposition 2 then indeed allows the system to let the species of interest dominate the reactor, but this process can be long (see, for instance, Example 1). Another possible strategy is to use a constant control D. Following [12], depending on the value of D, species 1 may win the competition, i.e.,

$$\lim_{t \to +\infty} x_1(t) > 0 \quad ; \quad \lim_{t \to +\infty} x_2(t) = 0.$$

In that case, this (simple) strategy indeed allows for reaching the target. However, this convergence is asymptotic and depends on the value of D as in the competitive exclusion principle. Roughly speaking, if $D > \mu_1(\tilde{q}_1)$ (where $\tilde{q}_1$ and s^+ are such that $\tilde{q}_1 := \delta_1^{-1}(s^+)$ and $(\mu_1 \circ \delta_1^{-1})(s^+) = (\mu_2 \circ \delta_2^{-1})(s^+)$), then species 2 wins the competition, whereas, if $D < \mu_1(\tilde{q}_1)$, species 1 wins the competition. We refer to [12] for more details about the asymptotic behavior of (1) for a constant control D. Thus, the target set may not always be reachable with a constant control D. Our objective in this paper is precisely to propose a methodology to compute a control strategy to reach the target set $\mathcal{T}$ faster, playing on the control $D(\cdot)$ as illustrated in Figure 2.

Example 1. *Let us consider the following parameters: $\rho_{1m} = 0.8$, $\rho_{2m} = 0.95$, $K_{s_1} = 1$, $K_{s_2} = 1.4$, $k_1 = 1.1$, $k_2 = 1.4$, $\mu_{1\infty} = 1.8$, $\mu_{2\infty} = 1.7$, $s_{in} = 10$. The contamination rate is fixed to $\varepsilon = 0.05$. The initial conditions are given by: $s^0 = 2$, $q_i^0 = 2.5$ and $x_i^0 = 1$. As illustrated in Figure 2, the target $\mathcal{T}$ is reached after $t_f = 46.774$ days using the control $D(t) = D_{opt}$, while x_1 dominates the culture after $t_f = 62.68$ days using the constant control $D = 0.48$. Let us also point out* *that, using an arbitrary constant control $D \in [0, D_{\max}]$, we are not even sure that x_1 wins the competition (trajectories do not reach the target in that case).*

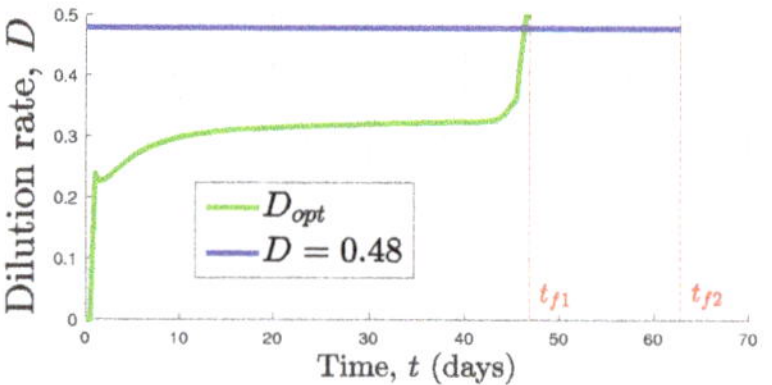
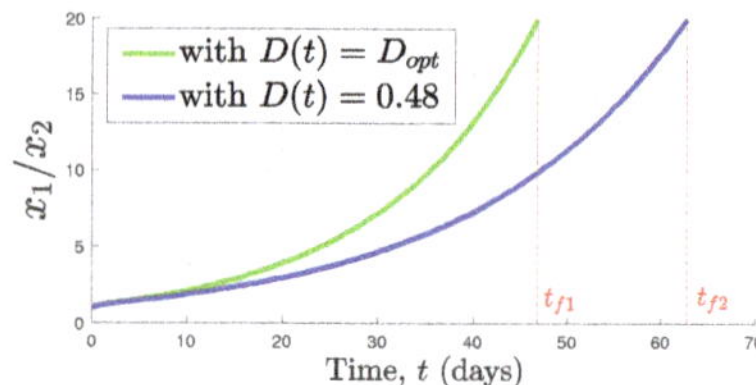

Figure 2. It is possible to find a constant feeding strategy $D \in [0, D_{\max}]$ that could favor one species or the other. However, this is a risky time-consuming process, while the target is reached much faster using an optimized control D_{opt}, thus saving several days of costly cultures.

3. Necessary Conditions on Optimal Controls

We start this section by generalizing previously obtained results [22,29] characterizing optimal solutions of (6). For that, we apply the PMP which allows for obtaining necessary conditions satisfied by optimal controls of (6). We denote by $X = (s, q_1, x_1, q_2, x_2)$ and $\lambda = (\lambda_s, \lambda_{q_1}, \lambda_{x_1}, \lambda_{q_2}, \lambda_{x_2})$, respectively, the state and adjoint variables (also called co-state or covector). The Hamiltonian associated with the optimal control problem

$$H = H(s, q_1, x_1, q_2, x_2, \lambda_s, \lambda_{q_1}, \lambda_{x_1}, \lambda_{q_2}, \lambda_{x_2}, \lambda^0, D),$$

is given by

$$
\begin{aligned}
H = \quad & \lambda^0 - (\rho_1(s)x_1 + \rho_2(s)x_2)\lambda_s + (\rho_1(s) - \mu_1(q_1)q_1)\lambda_{q_1} + \mu_1(q_1)x_1\lambda_{x_1} \\
& + (\rho_2(s) - \mu_2(q_2)q_2)\lambda_{q_2} + \mu_2(q_2)x_2\lambda_{x_2} + D[(s_{in} - s)\lambda_s - x_1\lambda_{x_1} - x_2\lambda_{x_2}].
\end{aligned}
$$

Let $X^0 \in \Omega \backslash \mathcal{T}$ and let $(X(\cdot), u(\cdot))$ be an optimal pair such that $X(\cdot)$ reaches the set $\mathcal{T}$ in a time $t_f \geq 0$. Thanks to the PMP, there exist an absolutely-continuous map $\lambda : [0, t_f] \to \mathbb{R}^5$ and $\lambda^0 \leq 0$ such that:

- The pair $(\lambda(\cdot), \lambda^0)$ is non-trivial, i.e., $(\lambda(\cdot), \lambda^0) \neq (0, 0)$.
- The covector satisfies

$$\dot{\lambda}(t) = -\nabla_X H(X(t), \lambda(t), \lambda^0, D(t)) \quad \text{a.e. } t \in [0, t_f]. \tag{9}$$

- The Hamiltonian maximization condition writes

$$D(t) \in \arg\max_{\xi \in [0, D_{\max}]} H(X(t), \lambda(t), \lambda^0, \xi) \quad \text{a.e. } t \in [0, t_f]. \tag{10}$$

- At the terminal time, the transversality condition writes:

$$\lambda(t_f) \in -N_{\mathcal{T}}(X(t_f)). \tag{11}$$

Here, $N_{\mathcal{T}}(X) = \{p \in \mathbb{R}^5 ; \forall Y \in \mathcal{T}, p \cdot (Y - X) \leq 0\}$ stands for the normal cone to $\mathcal{T}$ at some point $X \in \mathcal{T}$, see [28]. The adjoint Equation (9) is equivalent to:

$$
\begin{cases}
\dot{\lambda}_s = & (\rho_1'(s)x_1 + \rho_2'(s)x_2)\lambda_s - \rho_1'(s)\lambda_{q_1} - \rho_2'(s)\lambda_{q_2} + D\lambda_s, \\
\dot{\lambda}_{q_1} = & \mu_{1\infty}\lambda_{q_1} - \mu_1'(q_1)x_1\lambda_{x_1}, \\
\dot{\lambda}_{x_1} = & \rho_1(s)\lambda_s - \mu_1(q_1)\lambda_{x_1} + D\lambda_{x_1}, \\
\dot{\lambda}_{q_2} = & \mu_{2\infty}\lambda_{q_2} - \mu_2'(q_2)x_2\lambda_{x_2}, \\
\dot{\lambda}_{x_2} = & \rho_2(s)\lambda_s - \mu_2(q_2)\lambda_{x_2} + D\lambda_{x_2}.
\end{cases}
\tag{12}
$$

We define an extremal as a quadruplet $(X(\cdot), \lambda(\cdot), \lambda^0, D(\cdot))$ such that $(\lambda(\cdot), \lambda^0)$ is non zero and such that (1) and (9)–(11) is verified. Whenever $\lambda^0 = 0$, we say that the extremal is *abnormal* ; if $\lambda^0 \neq 0$, then we say that the extremal is *normal*. Because (6) is autonomous (i.e., the system does not depend explicitly on time), the Hamiltonian computed along an extremal is constant. In addition, since the terminal time is not fixed, we classically obtain, following optimal control theory, that $H = 0$.

The transversality condition is crucial for obtaining properties on optimal controls by reasoning backward in time from the terminal time $t = t_f$. We shall next extend earlier results [22] by taking into account explicitly the fact that $\mathcal{T}$ is a half-space of $\mathbb{R}^5$ and exploiting that $X(t_f)$ belongs to the set $E := \{X \in \mathbb{R}^5 \; ; \; x_2 - \varepsilon x_1 = 0\}$ (the boundary of the target set). Then, condition (11) can be transformed more explicitly as follows. At $X(t_f)$, the normal cone to $\mathcal{T}$ writes

$$N_{\mathcal{T}}(X(t_f)) = \mathbb{R}_+(0, 0, -\varepsilon, 0, 1).$$

Therefore, inclusion (11) is then equivalent to

$$\lambda_s(t_f) = \lambda_{q_1}(t_f) = \lambda_{q_2}(t_f) = 0, \tag{13}$$

together with

$$\lambda_{x_1}(t_f) + \varepsilon \lambda_{x_2}(t_f) = 0, \tag{14}$$

and the inequalities $\lambda_{x_1}(t_f) \geq 0$ and $\lambda_{x_2}(t_f) \leq 0$. Actually, one has $\lambda_{x_1}(t_f) > 0$ and $\lambda_{x_2}(t_f) < 0$. Suppose indeed that $\lambda_{x_1}(t_f) = 0$. Then, (14) would imply $\lambda_{x_2}(t_f) = 0$, and, since (9) is linear w.r.t. λ, we would obtain $\lambda \equiv 0$ over $[0, t_f]$. Using the constancy of H, we would also obtain $\lambda^0 = 0$ contradicting the PMP. We can then conclude that

$$\lambda_{x_1}(t_f) > 0 \quad \text{and} \quad \lambda_{x_2}(t_f) < 0. \tag{15}$$

Throughout the paper, we suppose that only normal extremals occur, i.e., $\lambda^0 < 0$, and, without any loss of generality, we may assume that $\lambda^0 = -1$ (up to a renormalization of the necessary conditions that are linear w.r.t. λ).

Remark 1. *Abnormal extremals are not generic. They correspond to the optimal path reaching the target set in some particular subset of the target set and are such that $\lambda^0 = 0$ (and thus $\lambda(\cdot) \neq 0$). From the conservation of H, this implies that $\mu_1(q_1(t_f)) = \mu_2(q_2(t_f))$ in such a way that $\lambda(t_f)$ is not uniquely defined in contrast with the normal case (see below). At such a singular point, the value function (the minimal time as a function of X^0) is also non-differentiable.*

Going back to the normal case, i.e., $\lambda^0 = -1$, the covector λ at $t = t_f$ can be completely determined (thanks to the conservation of H) as follows:

$$\lambda_s(t_f) = \lambda_{q_1}(t_f) = \lambda_{q_2}(t_f) = 0,$$
$$\lambda_{x_1}(t_f) = \frac{1}{x_1(t_f)(\mu_1(q_1(t_f)) - \mu_2(q_2(t_f)))} > 0, \tag{16}$$
$$\lambda_{x_2}(t_f) = -\frac{\lambda_{x_1}(t_f)}{\varepsilon} < 0.$$

Notice that the quantity $\mu_1(q_1(t_f)) - \mu_2(q_2(t_f))$ is non-zero along a normal extremal. As a consequence, the transversality condition (11) coupled with the conservation of H are equivalent to (16). The computation of λ at $t = t_f$ is useful to integrate the state adjoint system backward in time from the target set.

We now wish to exploit the Hamiltonian condition (10). It is of particular interest to introduce the *switching function*, which allows us to determine the optimal control D

according to the sign of the switching function. For that, let us denote by $\tilde{\Phi}$ the switching function:

$$\tilde{\Phi} := -(x_1\lambda_{x_1} + x_2\lambda_{x_2}) + (s_{in} - s)\lambda_s,$$

associated with the control function $D(\cdot)$.

$$\begin{aligned} \tilde{\Phi}(t) > 0 &\Rightarrow D(t) = D_{max}, \\ \tilde{\Phi}(t) < 0 &\Rightarrow D(t) = 0, \end{aligned} \tag{17}$$

for a.e. $t \in [0, t_f]$.

In the case where $\tilde{\Phi} > 0$, resp. $\tilde{\Phi} < 0$ over a time interval $[t_1, t_2]$, we say that the optimal control u is of bang type (denoted by B_+, resp. B_-). When the control $D(\cdot)$ is non-constant in every neighborhood of a time $t_c \in (0, t_c)$, we say that t_c is a *switching time*, and one must have $\tilde{\Phi}(t_c) = 0$. Next, when the switching function $\tilde{\Phi}$ vanishes over a time-interval $[t_1, t_2]$, we state that a *singular arc* occurs. In this case, the corresponding trajectory is *singular* over $[t_1, t_2]$, and such an arc will be denoted by $\mathcal{S}$. Singular arcs are essential to optimize the time to steer an initial condition to the target set. Now, we are ready to state some main features of the switching function $\tilde{\Phi}$, and then investigate properties of the singular paths.

Lemma 1.

(i)   The function $\tilde{\Phi}$ is continuously differentiable over $[0, t_f]$ and, moreover,

$$\dot{\tilde{\Phi}} = (s_{in} - s)(\rho_1'(s)[x_1\lambda_s - \lambda_{q_1}] + \rho_2'(s)[x_2\lambda_s - \lambda_{q_2}]). \tag{18}$$

(ii)   At the terminal time $t = t_f$, it holds:

$$\tilde{\Phi}(t_f) = \dot{\tilde{\Phi}}(t_f) = 0. \tag{19}$$

Proof. By differentiating $\tilde{\Phi}$ w.r.t. t, we find that

$$\dot{\tilde{\Phi}} = -\dot{s}\lambda_s + (s_{in} - s)\dot{\lambda}_s - [\dot{x}_1\lambda_{x_1} + \dot{x}_2\lambda_{x_2}].$$

Using (1) and (9), we obtain (18), which proves (i). For proving (ii), note that $x(t_f) \in E$ and $\lambda(t_f) \in E^\perp$, thus $\tilde{\Phi}(t_f) = 0$. Using (13), we also obtain $\dot{\tilde{\Phi}}(t_f) = 0$, which ends the proof. $\square$

Remark 2. *Following the formalism of geometric control theory, $\dot{\tilde{\Phi}}$ never involves D explicitly, but D is present in the expression of $\tilde{\Phi}^{(2k)}$, $k \geq 1$, see [27]. If $k \geq 1$ is the first integer for which the control is present in the expression of $\tilde{\Phi}^{(2k)}$, we usually say that the singular arc is of order k.*

At this step, an optimal control is a concatenation of bang and singular arcs:

$$\mathcal{B}_\pm, \mathcal{B}_\pm\mathcal{B}_\mp, \mathcal{B}_\pm\mathcal{S}, \mathcal{B}_\pm\mathcal{B}_\mp\mathcal{S}, \mathcal{B}_\pm\mathcal{B}_\mp\mathcal{S}\mathcal{B}_\pm, \dots$$

with possibly infinitely many crossing times (in particular, if there is a singular arc of order 2 [27]). The occurrence and properties of singular arcs as well as the various (possible) structure for an optimal control of (6) will be precisely the matter of the next section. The goal is to reduce (if possible) the number of possible structures for an optimal control.

4. Singular Arcs and Insights into Optimal Solutions

4.1. Legendre–Clsebsch's Necessary Condition and Computation of the Singular Control

The analysis of singular arcs requires to compute $\ddot{\tilde{\Phi}}$. Indeed, the Hamiltonian condition does not give any information about an optimal control during a singular phase. Thanks to the next computations, we shall also be able to deduce an expression of the singular control during a singular arc.

Doing so, let us define $\theta_1, \theta_2 : [0, T] \to \mathbb{R}$ as

$$
\begin{aligned}
\theta_1 &:= \rho_1'(s)[x_1\lambda_s - \lambda_{q_1}] + \rho_2'(s)[x_2\lambda_s - \lambda_{q_2}], \\
\theta_2 &:= \rho_1''(s)[x_1\lambda_s - \lambda_{q_1}] + \rho_2''(s)[x_2\lambda_s - \lambda_{q_2}].
\end{aligned}
$$

Hereafter, to simplify the layout, we do not write the dependency of θ_i, s, x_i, λ_s, λ_{q_i} w.r.t. the time. To shorten the notation, we also do not write explicitly the dependency of certain functions w.r.t. some variables. Using (12)–(18), one can write

$$
\dot{\Phi} = (s_{in} - s)\theta_1 \quad \text{and} \quad \dot{\lambda}_s = \theta_1 + D\lambda_s. \tag{20}
$$

Lemma 2.

(i)   *The derivative of θ_1 can be expressed as:*

$$
\begin{aligned}
\dot{\theta}_1 ={}& \theta_2 \dot{s} + [\rho_1' x_1 + \rho_2' x_2]\theta_1 + \lambda_s[x_1\mu_1\rho_1' + x_2\mu_2\rho_2'] \\
& - \rho_1'[\mu_{1\infty}\lambda_{q_1} - \mu_1'(q_1)x_1\lambda_{x_1}] - \rho_2'[\mu_{2\infty}\lambda_{q_2} - \mu_2'(q_2)x_2\lambda_{x_2}].
\end{aligned} \tag{21}
$$

(ii)   *The second derivative of $\tilde{\Phi}$ fulfills the equality:*

$$
\begin{aligned}
\ddot{\Phi} ={}& [-\theta_1 + (s_{in} - s)\theta_2]\dot{s} + (s_{in} - s)[\rho_1' x_1 + \rho_2' x_2]\theta_1 + (s_{in} - s)\lambda_s[x_1\mu_1\rho_1' + x_2\mu_2\rho_2'] \\
& - (s_{in} - s)\rho_1'[\mu_{1\infty}\lambda_{q_1} - \mu_1'(q_1)x_1\lambda_{x_1}] - (s_{in} - s)\rho_2'[\mu_{2\infty}\lambda_{q_2} - \mu_2'(q_2)x_2\lambda_{x_2}].
\end{aligned} \tag{22}
$$

Proof. By differentiating θ_1 w.r.t. t, we have

$$
\dot{\theta}_1 = \theta_2 \dot{s} + \rho_1'[\dot{x}_1\lambda_s + x_1\dot{\lambda}_s - \dot{\lambda}_{q_1}] + \rho_2'[\dot{x}_2\lambda_s + x_2\dot{\lambda}_s - \dot{\lambda}_{q_2}].
$$

Using (20) and (12), we obtain (21). Using that $\ddot{\Phi} = -\dot{s}\theta_1 + (s_{in} - s)\dot{\theta}_1$, we obtain (22). $\square$

Note that these computations have been verified thanks to a symbolic computation software. The next step is to establish whether Legendre–Clebsch's condition is verified or not along a singular arc. Recall that this condition is necessary for optimality and that it can be stated as follows (see, e.g., [27,30,31]).

Theorem 1 (Legendre–Clebsch's condition [27]). *Let $I = [t_1, t_2]$ be such that the trajectory is singular over $[t_1, t_2]$. Then, one has:*

$$
\ddot{\Phi}_{|_D} \geq 0, \tag{23}
$$

which is fulfilled over $I = [t_1, t_2]$.

Using the expression of the derivative $\ddot{\Phi}$ given in (22), we provide in the next lemma the expression of the second derivative, $\ddot{\Phi}_{|_D}$.

Lemma 3. *Let $I = [t_1, t_2]$ be such that the trajectory is singular over $[t_1, t_2]$. Then, one has:*

$$
\ddot{\Phi}_{|_D} = (s_{in} - s)^2\theta_2 = (x_1\lambda_s - \lambda_{q_1})\frac{(\rho_1''\rho_2' - \rho_2''\rho_1')}{\rho_2'}. \tag{24}
$$

Proof. In (22), the only term involving the control D is related to $\dot{s}$. We obtain (24) using that $\theta_1 \equiv 0$ over $[t_1, t_2]$. $\square$

Proposition 3. *Along a singular arc that occurs over a time interval $[t_c, t_f]$, it holds that:*

$$
\lambda_s = 0,
$$

over $[t_c, t_f]$ and Legendre–Clebsch's condition (with a strict inequality) is fulfilled over $[t_c, t_f)$.

Proof. Because the trajectory is singular over $[t_c, t_f]$, one has $\theta_1 \equiv 0$ over this interval; thus, $\lambda_s(\cdot)$ satisfies:

$$\dot{\lambda}_s = D\lambda_s, \quad \lambda_s(t_f) = 0.$$

It follows that $\lambda_s \equiv 0$ over $[t_c, t_f]$. In a left neighborhood of $t = t_f$, one has from (12) $\dot{\lambda}_{q_1} < 0$; thus, since λ_{q_1} vanishes at $t = t_f$, we necessarily have $\lambda_{q_1} > 0$ in a left neighborhood of t_f. Because λ_s is zero over $[t_c, t_f]$, we deduce that $x_1\lambda_s - \lambda_{q_1} = -\lambda_{q_1} < 0$ over $[t_c, t_f)$ (at $t = t_f$, λ_{q_1} vanishes at t_f, as well as $\ddot{\Phi}_{|D}$). Over $[t_c, t_f]$, we note that

$$\dot{\lambda}_{x_1} = \lambda_{x_1}(D - \mu_1(q_1)),$$

hence λ_{x_1} does not vanish over $[t_c, t_f]$. Suppose that λ_{q_1} vanishes over $[t_c, t_f]$ at a time $t' \in [t_c, t_f]$. Then, one must have $\dot{\lambda}_{q_1}(t') \geq 0$ since $\lambda_{q_1} > 0$ over (t', t_f). However, at $t = t'$, the adjoint equation implies that

$$\dot{\lambda}_{q_1}(t') = -\mu_1'(q_1(t'))x_1(t')\lambda_{x_1}(t') < 0$$

because $q_1(t') > k_1$, $\mu_1' > 0$ over $(k_1, +\infty)$, $x_1 > 0$, and $\lambda_{x_1} > 0$ over $[t_c, t_f]$. This is a contradiction and thus λ_{q_1} does not vanish over t_c, t_f. Assumption 1 implies that $\rho_1''\rho_2' - \rho_2''\rho_1' < 0$, thus $\ddot{\Phi}_{|D} > 0$ over $[t_c, t_f)$ as desired. We can then conclude that Legendre–Clebsch's condition (with a strict inequality) is fulfilled over the whole interval $[t_c, t_f)$. $\square$

A consequence of the previous proposition is that, when a singular arc occurs over some time interval $[t_c, t_f]$, then it is of order 1. Based on this proposition, we shall only consider singular arcs of first order in the remaining of the paper. If Legendre–Clebsch's condition holds true, the singular arc is said to be of turnpike type [26]. The expression defining the singular control can then be derived using (22). Next, let $\varsigma(X, \lambda)$ be defined by:

$$\varsigma(X, \lambda) := \theta_2[\rho_1 x_1 + \rho_2 x_2] - \lambda_s(x_1\mu_1\rho_1' + x_2\mu_2\rho_2') - \rho_1'[\mu_{1\infty}\lambda_{q_1} - \mu_1'(q_1)x_1\lambda_{x_1}] \\ - \rho_2'[\mu_{2\infty}\lambda_{q_2} - \mu_2'(q_2)x_2\lambda_{x_2}]. \tag{25}$$

We now give an expression of the singular control as a feedback of the state and covector.

Proposition 4. *Suppose that an extremal is singular over $[t_1, t_2]$ and that (23) is verified over $[t_1, t_2]$ with a strict inequality. Then, the singular control D_s is given by*

$$D_s(X, \lambda) := \frac{\varsigma(X, \lambda)}{(s_{in} - s)\theta_2}, \tag{26}$$

where we recall that $\theta_2 = \rho_1''(s)[x_1\lambda_s - \lambda_{q_1}] + \rho_2''(s)[x_2\lambda_s - \lambda_{q_2}]$ and ς is given by (25).

Proof. This expression follows from (22) in which $\dot{s}$ is replaced by $(s_{in} - s)D - \rho_1 x_1 - \rho_2 x_2$ and $\theta_1 \equiv 0$ (since $\dot{\Phi} = \ddot{\Phi} = 0$). $\square$

Corollary 1. *If the singular arc occurs over some time interval $[t_c, t_f)$, expression (26) simplifies into*

$$D_s(X, \lambda) := \frac{\tilde{\varsigma}(X, \lambda)}{(s_{in} - s)\theta_2}, \tag{27}$$

where $\tilde{\varsigma}$ is given by

$$\tilde{\varsigma}(X, \lambda) := \theta_2[\rho_1 x_1 + \rho_2 x_2] - \rho_1'[\mu_{1\infty}\lambda_{q_1} - \mu_1'(q_1)x_1\lambda_{x_1}] - \rho_2'[\mu_{2\infty}\lambda_{q_2} - \mu_2'(q_2)x_2\lambda_{x_2}],$$

and, in this case, θ_2 simplifies also into $\theta_2 = -\rho_1''\lambda_{q_1} - \rho_2''\lambda_{q_2}$ because $\lambda_s \equiv 0$.

Remark 3.

(i) *Note that Legendre–Clebsch's condition (with a strict inequality) is equivalent to $\theta_2 > 0$ (and that this condition is always verified over $[t_c, t_f)$ with t_c close to t_f.*

(ii) *In view of the general expression giving the singular control, see (26), there is no guarantee a priori that the singular control D_s is always with values in $[0, D_{max}]$, i.e., that the singular arc is always admissible (even if Legendre–Clebsch's condition is verified). This can bring additional difficulties; however, we may discard this point by choosing D_{max} large enough.*

(iii) *Notice that (27) is at least active at $t = t_f$ in the case where a singular arc D_s steers the model trajectories towards the target $\mathcal{T}$, since the transversality conditions ensure that $\lambda_s(t_f) = 0$.*

4.2. About the Occurrence of a Terminal Singular Arc at the Terminal Time

The aim of this section is to discuss the possibility of having a singular arc over some time interval $[t_f - \tau, t_f]$ (with $\tau > 0$) and the structure of optimal controls. Our main questioning is as follows:

Does any optimal trajectory contain a singular arc over some time interval $[t_f - \tau, t_f]$?

To analyze this point, let us summarize properties of the switching function at $t = t_f$ (that are consequences of transversality conditions associated with the codimension 1 target):

- The switching function and its derivative vanish at $t = t_f$:

$$\dot{\Phi}(t_f) = \Phi(t_f) = 0. \tag{28}$$

- The second derivative of the switching function satisfies:

$$\ddot{\Phi}_{|_D}(t_f) = 0. \tag{29}$$

- In addition, Legendre–Clebsch's condition (23) is always satisfied along a singular arc defined in a left neighborhood of the terminal time $t = t_f$.

The necessary conditions (28) and (29) are a very good indication for the occurrence of a singular arc and are thus strong arguments to answer positively to the above question. Thus, we could now wonder whether or not conditions (28) and (29) are sufficient to ensure the occurrence of a singular arc in some time interval $[t_f - \tau, t_f]$. It appears that this question is complex and falls into the setting of geometric optimal control theory. As far as we know, such conditions are not equivalent to the occurrence of a singular arc over some time interval $[t_f - \tau, t_f]$ (this may depend, in particular, on the initial condition). It is, however, worth mentioning that these conditions (in particular (28)) are commonly used numerically to implement a singular arc in shooting methods [32]. In our context of Droop model, it is very interesting to notice that singular arcs are the cornerstone of the optimal control, in particular for a large set of initial conditions that are biologically meaningful (typically for heterogeneous cultures where $x_1^0/x_2^0 \approx 1$). However, the answer to the above question is not always true and depends on the initial condition (as it has been confirmed using direct optimization methods, see Section 5). Indeed, as illustrated in Example 2—Section 5, when the initial conditions x_i^0 are taken very close to the target $(x_2(t_f)/x_1(t_f) = \varepsilon)$, and $\mu_1(q_1^0) - \mu_2(q_2^0) > 0$, the singular arc does not appear or appears marginally at $t = t_f$ to satisfy the transversality conditions.

Recall that $\Phi(t_f) = \dot{\Phi}(t_f) = 0$. Hence, the sign of Φ depends on $\ddot{\Phi}(t_f)$ that is computed in the next lemma.

Lemma 4. *At the terminal time, one has*

$$\mu_1(q_1(t_f)) - \mu_2(q_2(t_f)) > 0. \tag{30}$$

In addition, the second derivative of the switching function exists at $t = t_f$ and:

$$\overset{\cdots}{\Phi}(t_f) = \frac{[s_{in} - s(t_f)][\rho_2'(s(t_f))\mu_2'(q_2(t_f)) - \rho_1'(s(t_f))\mu_1'(q_1(t_f))]}{\mu_2(q_2(t_f)) - \mu_1(q_1(t_f))}. \tag{31}$$

Proof. Inequality (30) follows from the expression of $\lambda(t_f)$ and the transversality condition (16). The expression of $\overset{\cdots}{\Phi}(t_f)$ in (31) follows from (22). $\square$

We can now define the function:

$$t \mapsto \xi(t) := \rho_2'(s(t))\mu_2'(q_2(t)) - \rho_1'(s(t))\mu_1'(q_1(t)).$$

From the previous lemma, we deduce the behavior of an optimal path near the terminal time:

- First, the target set can only be reached at some point $X(t_f)$ such that (30) is fulfilled.
- In addition, if $\xi(t_f) \neq 0$, then, in a left neighborhood of $t = t_f$, the optimal control $D(\cdot)$ is of bang type and satisfies

$$D(t) = \text{sign}(\xi(t)).$$

- If a singular arc occurs in a left neighborhood of $t = t_f$, then one must have $\xi(t_f) = 0$, i.e., a singular arc reaches the target in the subset of $\mathcal{T}$ defined as:

$$\mathcal{T}' := \{X = (s, q_1, x_1, q_2, x_2) \in \mathcal{T} \; ; \; \mu_1(q_1) - \mu_2(q_2) > 0 \text{ and}$$
$$\rho_2'(s)\mu_2'(q_2) - \rho_1'(s)\mu_1'(q_1) = 0\}.$$

4.3. Toward an Optimal Synthesis Characterizing the Optimal Solutions

Reducing the number of switching times is in general non-tractable for nonlinear optimal control problems governed by a system in dimension greater than three. Nevertheless, thanks to the properties of singular arcs, we obtained previously and of the switching function at $t = t_f$, we can expect a limited number of possible structures for an optimal control as we formulate in the next conjecture.

Conjecture 1. *Every initial condition in Ω is steered optimally to the target set via a control D that has a finite number of switching times. In addition, for almost every initial condition, an optimal control presents the following structure:*

$$\mathcal{B}_\pm \mathcal{B}_\mp \mathcal{S}.$$

For a large set of initial conditions in some subset $\Omega^\dagger \subset \Omega$ (far from the target), there is a single bang arc and a terminal singular arc, whereas, for some initial conditions close to the target set, no singular arc occurs (i.e., $\mathcal{S}$ is of zero duration).

This conjecture has been verified numerically for a large number of initial conditions (see Section 5). Our argumentation to confirm this conjecture is as follows.

- From the PMP, we have seen that, for every initial condition in Ω, an optimal control is a concatenation of bang arcs $\mathcal{B}_\pm$ and singular arcs $\mathcal{S}$.

  Moreover, since the switching function $\overset{\cdot}{\Phi}$ satisfies the strong requirements $\overset{\cdot}{\Phi}(t_f) = \overset{\cdot}{\Phi}(t_f) = 0$, $\overset{\cdots}{\Phi}_{|_D}(t_f) = 0$

  (from the transversality conditions) as well as Legendre–Clebsch's condition, we conjecture that, for almost all initial conditions, an optimal control is singular in a left neighborhood of the terminal time. This implies in particular that the number of switchings is finite since we proved that any terminal singular arc is of first order. In addition, the number of switchings is minimal in general (apart when chattering occurs, see [27], which is not the case here). Thus, an optimal control should be of type $B_\pm B_\mp S$ with one or two bang arcs before the terminal singular arc.

As we have seen, we also must have $\xi(t_f) = 0$, which only involves state variables at the terminal time. This surprising condition mixing the first derivative of the basic functions in the Droop model, with the s variable on one side and the q_i variables on the other side, is, however, hard to interpret biologically.

- For some marginal—-but admissible—initial conditions outside of $\Omega^\dagger$, the structure of an optimal control of (6) may be of bang type for almost all $t \in [0, t_f)$, or $[0, t_f - \tau]$, with very small $\tau > 0$ (see, e.g., Example 2 in the next section). This is the case when typically $x_1^0 \gg x_2^0$, i.e., the initial condition is very close to the target set $\mathcal{T}$, with in addition $\mu_1(q_1^0) - \mu_2(q_2^0) > 0$. Thus, the requirement $\mu_1(q_1(t_f)) - \mu_2(q_2(t_f)) > 0$ is easily satisfied. Thus, in this particular situation, it comes as no surprise that the fastest path to reach the target $\mathcal{T}$ is the one exploiting the fact that $\dot{x}_1(0) \gg \dot{x}_2(0)$ (since $x_1^0 \gg x_2^0$) along with $D(t) = 0$, since it maximizes $\dot{x}_1(t)$ (we recall that $\dot{x}_i = (\mu_i(q_i) - D(t))x_i$).

It is worth noticing that, when no singular arc occurs, the strategy mainly consists of "pushing" x_1 and x_2 as quickly as possible towards the target $\mathcal{T}$ using $D = 0$, when the initial conditions x_1^0 and x_2^0 are very close to $\mathcal{T}$. Nevertheless, this strategy may not be the optimal one whenever q_1^0 and q_2^0 are "far" from satisfying $\mu_1(q_1(t_f)) - \mu_2(q_2(t_f)) > 0$. This is typically the case illustrated in Example 3 in the next section.

For an optimal control of type $B_\pm S$, the occurrence of a singular arc is related to the so-called turnpike phenomenon that we now explain in this framework. For a large subset of initial conditions $\mathcal{S}^\dagger \subset \Omega$ that are biologically the most relevant, the structure of the optimal control is *bang-singular* $B_\pm S$. The singular arc is the control D_s given in (27) that reaches the target $\mathcal{T}$. Moreover, this singular phase coincides with optimal trajectories $(s(t), q_1(t), q_2(t))$ that stay most of the time close to the critical point (s_c, q_{1c}, q_{2c}) defined in Section 2.3 (related to the *actual* growth rates and the function $\Delta(s) = \mu_1(\delta_1^{-1}(s)) - \mu_2(\delta_2^{-1}(s))$). Observe, for instance, the trajectories s, q_1 and q_2 in Figure 3a. We also believe that the concatenation of bang arcs before the major singular phase exclusively aims at moving (s^0, q_1^0, q_2^0) towards (s_c, q_{1c}, q_{2c}). Then, the singular arc D_s takes over at a switching-time instant $t = t_s$ and ensures that the associated singular trajectory, denoted $(s_s(t), q_{1s}(t), q_{2s}(t))$, satisfies the so-called *turnpike inequality* (see, e.g., [33]),

$$\|s_s(t) - s_c\| + \|q_{1s}(t) - q_{1c}\| + \|q_{2s}(t) - q_{2c}\| \leq a_1 \left(e^{-a_2 t} + e^{a_2(t - t_f)} \right),$$

$$a_1, a_2 > 0, \quad \text{for all } t \in [t_s, t_f]. \tag{32}$$

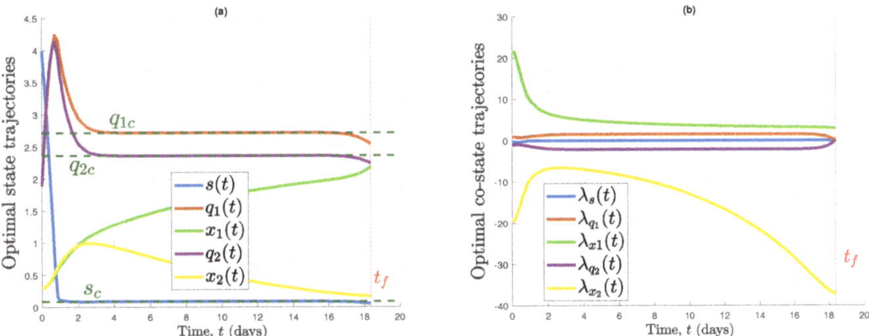

Figure 3. (**a**) Optimal state in Droop model, (**b**) the resulting optimal co-state trajectories, which satisfy in particular all the transversality conditions of the PMP.

This is, for instance, the case for optimal controls illustrated in Figure 4 (of type $\mathcal{B}_- S$) and in Figure 5a (of type $\mathcal{B}_- \mathcal{B}_+ S$). The inequality (32) usually holds when the time interval

$[0, t_f]$ is not excessively short [33–35], which is the generic case in the Droop model (1) associated with (6). Indeed, in practice, the most significant biological experiments aim to separate species and select x_1 starting from an homogeneous culture (a well-balanced initial culture with $x_1^0/x_2^0 \approx 1$) or even from $x_2^0 \gg x_1^0$ with the challenging issue of selecting the minority species (x_1), which is not naturally promoted. In these cases, Droop's kinetics ensure that the minimum selection time t_f cannot be excessively short and therefore singular arcs as well as the *turnpike*-type behavior appear systematically in the optimal strategy of (6).

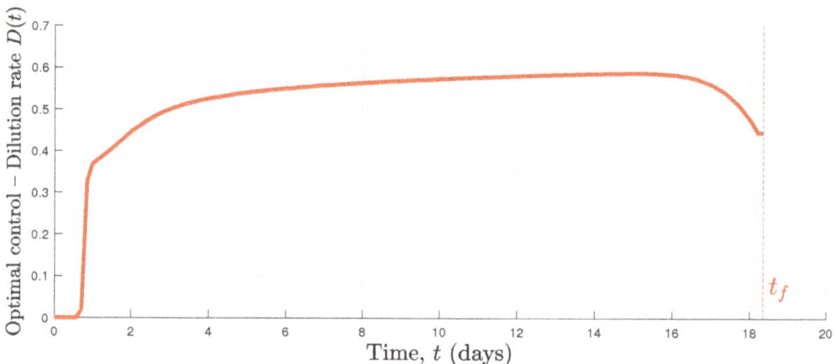

Figure 4. The optimal control in Example 2 ($s^0 = 4$, $q_i^0 = 1.9$, $x_i^0 = 0.3$) is of *bang(0)-singular* type.

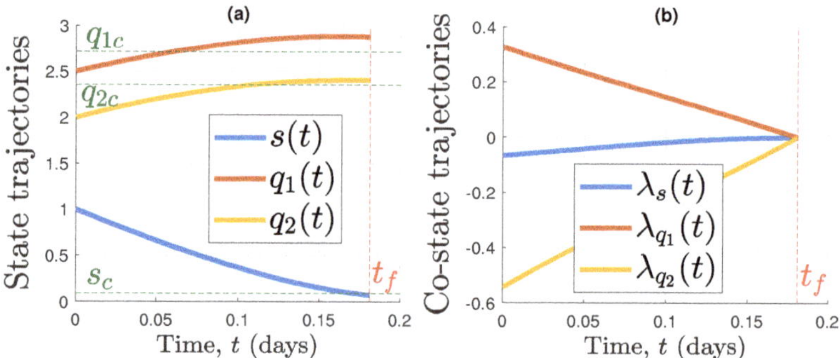

Figure 5. (a) The optimal state trajectories s, q_1 and q_2 (associated with the optimal control D in Figure 6) get closer over time to the critical point (s_c, q_{1c}, q_{2c}). The target $\mathcal{T}$, with $\varepsilon = 0.08$, is reached quickly, without resorting to the singular arc. The transversality conditions are satisfied, and in particular $\lambda_s(t_f) = \lambda_{q_1}(t_f) = \lambda_{q_1}(t_f) = 0$, as illustrated in (b).

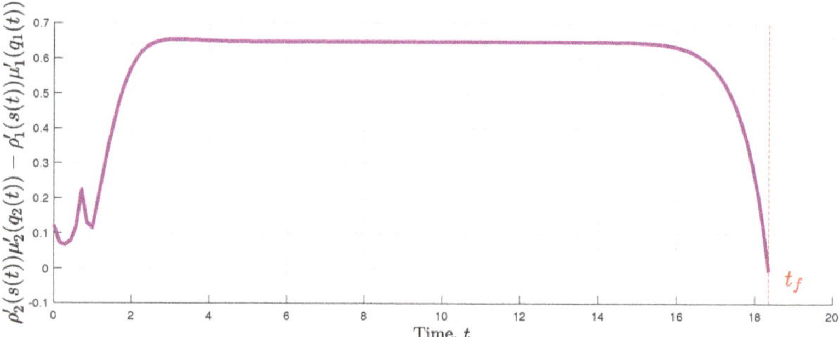

Figure 6. Evolution of the quantity $\rho_2'(s(t))\mu_2'(q_2(t)) - \rho_1'(s(t))\mu_1'(q_1(t))$ along the optimal trajectories given in Figure 3a.

5. Direct Optimization and Numerical Results

In this section, a direct-optimization approach is performed in order to solve (6) and illustrate the different cases discussed in Section 4.3. The numerical direct methods that we use throughout this paper are implemented in Bocop [36] (an optimal control toolbox), and they have the characteristic to transform the studied (6) into a nonlinear programming problem (NLP) in finite-dimension [37], through the discretization step of the control and the state variables [38]. Numerical results are organized as follows:

- An optimal control of type $B_-\mathcal{S}$ is developed throughout Example 1.
- An optimal control of type B_- is developed throughout Example 2.
- An optimal control of type $B_-B_+\mathcal{S}$ is developed throughout Example 3.

In all the numerical examples, we consider the model parameters given in Table 1, with the settings in Table 2.

Table 1. The model parameters used in Section 5.

	ρ_{m_i}	K_i	$\mu_{i\infty}$	k_i
Species/strain 1	7	0.3	1.7	1.75
Species/strain 2	8	0.6	1.8	1.80

Table 2. Model and OCP settings. The contamination rate ε characterizes the target set $\mathcal{T}$.

s_{in}	Control D	Contamination Rate ε
6	$[0, D_{max} = 1.5]$	0.08

Assumption 3 is verified when $D_{\max} = 1.5$, namely because $D_{\max}$ is precisely chosen above the maximum actual growth rates of the species. The contamination rate in all the examples is fixed to a significantly small value, $\varepsilon = 0.08$.

In Bocop, the state variables (and even the time, in minimal-time OCPs) of the Droop model (1) are discretized with a `Lobatto` scheme based on Runge–Kutta methods of type `Lobatto-IIIC` of order 6, which uses an implicit trapezoidal rule. The main settings used in Bocop are given in Table 3.

Table 3. Bocop settings used in Section 5.

Discretization Method	Lobatto IIIC (Implicit, 4-Stage, Order 6)
Time steps	130
NLP tolerance	$<10^{-14}$

Example 2. *In the first example, we consider the Droop model resulting from the parameters in Table 1 and Figure 7, associated with the settings in Tables 2 and 3, and the initial conditions given in Table 4.*

Table 4. The initial conditions used in Example 2.

s^0	q_1^0	x_1^0	q_2^0	x_2^0
4	1.9	0.3	1.9	0.3

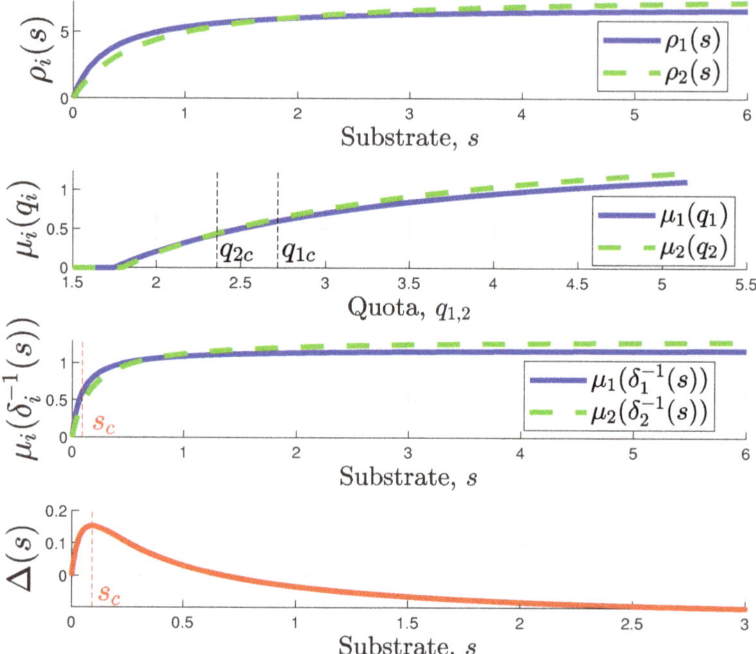

Figure 7. Growth and absorption (uptake) functions, respectively μ_i and ρ_i, $i = 1, 2$, produced using the model parameters given in Table 1, and used throughout the Examples 1–3. We note that the value of s that maximizes the function $\Delta(s)$ is given $s_c = 0.092$. This point maximizes the difference between the actual growth rates (as discussed in Section 2.3) and defines the unique *static* solution (s_c, q_{1c}, q_{2c}). All the assumptions that ensure the well-posedness of the generic (6) are satisfied in this case. In particular, we highlight that $\Delta(s)$ can be positive and negative, and both species may win the competition using an appropriate control D (see Section 2.4).

In this example, we have a well-balanced initial culture since $x_1^0/x_2^0 = 1$ (see Section 4.3).

The direct optimization method allows us to determine the optimal control D, given in Figure 4, that steers the model trajectories towards $\mathcal{T}$ (with $\varepsilon = 0.08$) in minimal-time $t_f = 18.3526$ days.

We check and analyze the evolution of the switching function $\tilde{\Phi}$, its derivatives, and the co-state of the substrate s (Figure 8) in order to characterize the switching instant $t_s \in (0, t_f)$. This time-instant coincides with $\lambda_s = 0$ (since the singular arc is the one reaching the target $\mathcal{T}$), $\tilde{\Phi} = \dot{\tilde{\Phi}} = 0$ (thus activating D_s, according to the PMP). We also notice that the condition $\ddot{\tilde{\Phi}}(t_f) = 0$ is also satisfied. The optimal state and co-state trajectories are depicted in Figure 3, where we notice that $s(t)$, $q_1(t)$ and $q_2(t)$ evolve around the *static* critical point (s_c, q_{1c}, q_{2c}) for almost all $t \in [t_s, t_f]$, see the *turnpike*-like property discussed in Section 4.3.

The optimal control strategy aims to maximize the difference between the *actual* growth rates (the function $\Delta(s)$) as illustrated in Figure 9. The initial arc *bang(0)* drives s^0 towards s_c (we recall that $\dot{s} = (s_{in} - s)D - \rho_1(s)x_1 - \rho_2(x_2)$, $s^0 = 4$, $s_{in} = 6$, and, $s_c = 0.092$). The quantity $\mu_1(q_1(t)) - \mu_2(q_2(t))$ is maximized, with a delayed-dynamics, as a consequence of maximizing $\Delta(s)$. At the final time of $t_f = 18.3526$ days, we have $\mu_1(q_1(t_f)) - \mu_2(q_2(t_f)) > 0$. Finally, we check in Figure 6 that, at $t = t_f$, we have

$$\frac{\rho_2'(s)}{\rho_1'(s)}(t_f) = \frac{\mu_1'(q_1)}{\mu_2'(q_2)}(t_f).$$

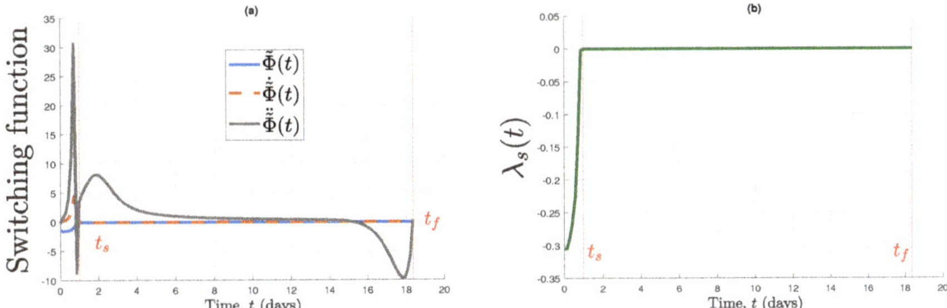

Figure 8. Characterization of the switching instant and of the singular arc in Example 2. (**a**) The switching function and its derivatives. (**b**) The co-state of the substrate s.

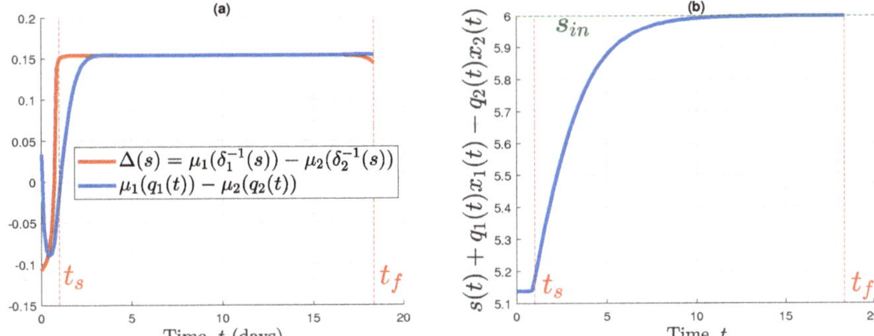

Figure 9. (**a**) Evolution of $\Delta(s(t))$ and $\mu_1(q_1(t)) - \mu_2(q_2(t))$ along the optimal trajectories. The optimal control aims to maximize the function Δ. We also check in (**b**), using the optimal model trajectories given in Figure 3, that $s(t) + q_1(t)x_1(t) + q_2(t)x_2(t)$ converges towards s_{in} (see Section 2.4).

The behavior of the optimal control and optimal trajectories described in Example 2 is definitively the most compelling one (with *bang(0)-singular* or *bang($D_{\max}$)-singular* arcs) from a biological standpoint, since it is the one that systematically appears when the final time is not extremely short. Indeed, in practice, initial conditions start more commonly sufficiently "far" from the target $\mathcal{T}$, leading to a final time that allows the singular arc and the *turnpike*-like behaviors to hold.

Example 3. *Now, let us consider the initial conditions in Table 5.*

Table 5. The initial conditions used in Example 3.

s^0	q_1^0	x_1^0	q_2^0	x_2^0
1	2.5	1.2	2	0.1

It is worth noticing that the initial conditions in Example 3 are intuitively favourable for reaching the target $\mathcal{T}$ in a very short time, since $\mu_1(q_1^0) - \mu_2(q_2^0) > 0$ and $x_2^0/x_1^0 = 0.083$, while $\varepsilon = 0.08$ (very close to the target). The optimal control in this case is mainly a *bang(0)* over time, as illustrated in Figure 10 (see Section 4.3 for more details). The optimal state trajectories and co-state trajectories (that satisfy the transversality conditions) are illustrated in Figure 5.

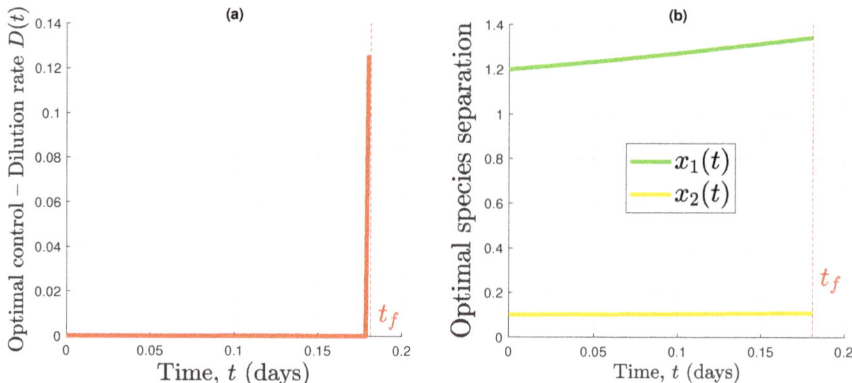

Figure 10. The optimal control given in (**a**) is *bang(0)* almost everywhere over $[0, t_f)$. It achieves species separation (**b**) and selects the species x_1 at $t_f = 1.1810$ days, since $x_1(t_f) = 1.3425$ and $x_2(t_f)/x_1(t_f) = \varepsilon = 0.08$.

Example 4. *In the last example, let us consider the initial conditions in Table 6.*

Table 6. The initial conditions used in Example 4.

s^0	q_1^0	x_1^0	q_2^0	x_2^0
5	1.8	2.2	5	0.18

In this situation, we notice that initial conditions are still favourable for reaching the target $\mathcal{T}$ in a very short time because $x_2^0/x_1^0 = 0.0818$ (with $\varepsilon = 0.08$, so x_i^0 are very close to the target). However, we also note that $\mu_1(q_1(0)) - \mu_2(q_2(0)) < 0$, while $\mu_1(q_1(t_f)) - \mu_2(q_2(t_f))$ should be positive at $t = t_f$ (see Section 3). Thus, the issue of minimal-time separation is slightly more complex than Example 3, and we obtain a structure for the optimal control of *bang(0)-bang(1)-singular* type, as illustrated in Figure 11. The corresponding model trajectories and optimal co-states are provided in Figure 12.

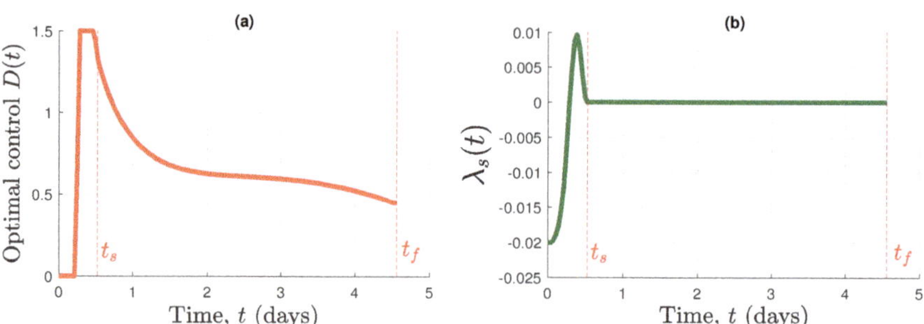

Figure 11. The optimal control given in (**a**) is *bang(0)-bang($D_{\max}$)-singular* over $[0, t_f)$, where $t_f = 4.5541$ days. A first switching instant from *bang(0)* to *bang($D_{\max}$)* occurs around 0.3 days, then t_s occurs when $\lambda_s = 0$, as illustrated in (**b**), starting the singular arc that steers the model trajectories towards $\mathcal{T}$, with $\varepsilon = 0.08$.

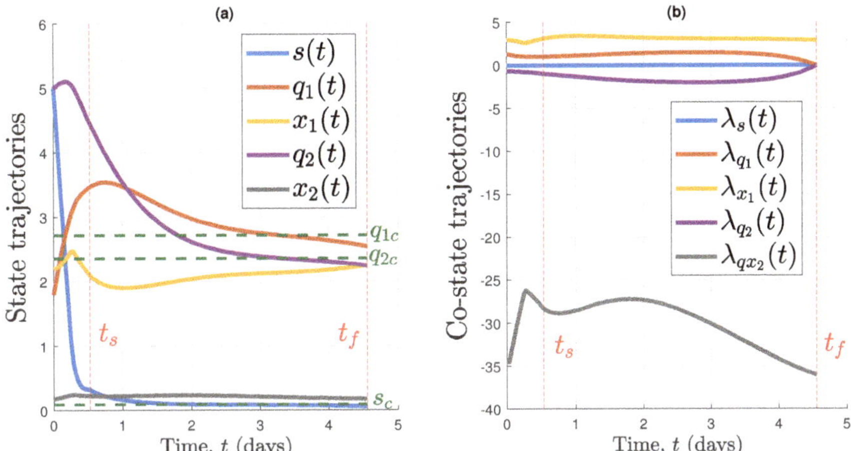

Figure 12. The optimal model trajectories (**a**), and the optimal co-state trajectories (**b**) that satisfy the transversality conditions.

6. Conclusions

The minimal-time OCP for the competition between two microbial species accumulating nutrients is a key issue. Progressing along this problem will definitely help for experiments that nowadays last more than 6 months [14,15]. However, the competition described by the Droop model turns out to be significantly more complicated than for the Monod model in dimension 2 [17]. We have improved the preliminary results about this problem to be found in [22,29] in order to provide an optimal synthesis depending on the initial condition. In particular, we applied the PMP, we discussed the structure of the optimal control, and we identified the singular arc steering optimal paths to the desired target set in a minimal amount of time. This study also highlights the *turnpike* behavior [33] although an exact verification in our case is not possible in our setting since the problem is affine w.r.t. entries in contrast with [35] that, in general, requires coercive hypotheses on the Hamiltonian w.r.t. entries. As usual in optimal control problems that are affine in the control, the study of singular arcs via geometric methods is a crucial issue.

This study also raised a mathematical (open) question outside the scope of this paper on the existence or not of a terminal singular arc whenever the switching function, its derivative, and second derivative w.r.t. the input vanish on a terminal manifold of codimension 1 (see, e.g., [39]).

Future work will focus on the determination of closed loop (sub-optimal) controllers to be applied for bioreactor control subject to uncertainties that are inherently present in biological systems. Finally, the proposed strategy must now be tested experimentally to assess the gain in experimental time it can offer.

Author Contributions: Conceptualization, W.D., T.B. and O.B.; methodology, W.D., T.B. and O.B.; software, W.D.; validation, W.D., T.B. and O.B.; formal analysis, W.D., T.B. and O.B.; investigation, W.D. and T.B.; resources, W.D., T.B. and O.B.; data curation, W.D., T.B. and O.B.; writing—original draft preparation, W.D., T.B. and O.B.; writing—review and editing, W.D., T.B. and O.B.; visualization, W.D., T.B. and O.B.; supervision, W.D., T.B. and O.B.; project administration, W.D., T.B. and O.B.; funding acquisition, W.D., T.B. and O.B. All authors have read and agreed to the published version of the manuscript.

Funding: The authors would like to thank the *Bio-MSA* (Graine-ADEME) project, funded by the *Agence Nationale de la Recherche (ANR)* in France, for partially supporting this work.

Institutional Review Board Statement: Not applicable.

Informed Consent Statement: Not applicable.

Data Availability Statement: Not applicable.

Acknowledgments: The authors are grateful to J.-B. Caillau and J.-B. Pomet for fruitful exchanges about singular arcs.

Conflicts of Interest: The authors declare no conflict of interest.

References

1. Masci, P.; Bernard, O.; Grognard, F. Continuous selection of the fastest growing species in the chemostat. *IFAC Proc. Vol.* **2008**, *41*, 9707–9712. [CrossRef]
2. Angermayr, S.A.; Hellingwerf, K.J.; Lindblad, P.; de Mattos, M.J.T. Energy biotechnology with cyanobacteria. *Curr. Opin. Biotechnol.* **2009**, *20*, 257–263. [CrossRef] [PubMed]
3. Ducat, D.C.; Way, J.C.; Silver, P.A. Engineering cyanobacteria to generate high-value products. *Trends Biotechnol.* **2011**, *29*, 95–103. [CrossRef] [PubMed]
4. Wijffels, R.H.; Barbosa, M.J.; Eppink, M.H. Microalgae for the production of bulk chemicals and biofuels. *Biofuels Bioprod. Biorefin. Innov. Sustain. Econ.* **2010**, *4*, 287–295. [CrossRef]
5. Mata, T.M.; Martins, A.A.; Caetano, N.S. Microalgae for biodiesel production and other applications: A review. *Renew. Sustain. Energy Rev.* **2010**, *14*, 217–232. [CrossRef]
6. Thangavel, M.; Pugazhendhi, A. Utilization of algae for biofuel, bio-products and bio-remediation. *Biocatal. Agric. Biotechnol.* **2019**, *17*, 326–330.
7. Morales, M.; Collet, P.; Lardon, L.; Hélias, A.; Steyer, J.P.; Bernard, O. Life-cycle assessment of microalgal-based biofuel. In *Biofuels from Algae*; Elsevier: Amsterdam, The Netherlands, 2019; Chapter 20, pp. 507–550. ISBN 9780444641922.
8. Cervantes-Gaxiola, M.E.; Hernández-Calderón, O.M.; Rubio-Castro, E.; Ortiz-del-Castillo, J.R.; González-Llanes, M.D.; Rios-Iribe, E.Y. In silico study of the microalgae-bacteria symbiotic system in a stagnant pond. *Comput. Chem. Eng.* **2020**, *135*, 106740. [CrossRef]
9. De-Luca, R.; Trabuio, M.; Barolo, M.; Bezzo, F. Microalgae growth optimization in open ponds with uncertain weather data. *Comput. Chem. Eng.* **2018**, *117*, 410–419. [CrossRef]
10. Malek, A.; Alamoodi, N.; Almansoori, A.S.; Daoutidis, P. Optimal dynamic operation of microalgae cultivation coupled with recovery of flue gas CO_2 and waste heat. *Comput. Chem. Eng.* **2017**, *105*, 317–327. [CrossRef]
11. Nhat, P.V.H.; Ngo, H.H.; Guo, W.S.; Chang, S.W.; Nguyen, D.D.; Nguyen, P.D.; Bui, X.T.; Zhang, X.B.; Guo, J.B. Can algae-based technologies be an affordable green process for biofuel production and wastewater remediation? *Bioresour. Technol.* **2018**, *256*, 491–501. [CrossRef]
12. Smith, H.L.; Waltman, P. *The Theory of the Chemostat, Dynamics of Microbial Competition*; Press of Cambridge University: Cambridge, UK, 1995.
13. Grognard, F.; Masci, P.; Benoît, E.; Bernard, O. Competition between phytoplankton and bacteria: Exclusion and coexistence. *J. Math. Biol.* **2015**, *70*, 959–1006. [CrossRef] [PubMed]
14. Bonnefond, H.; Grimaud, G.; Rumin, J.; Bougaran, G.; Talec, A.; Gachelin, M.; Boutoute, M.; Pruvost, E.; Bernard, O.; Sciandra, A. Continuous selection pressure to improve temperature acclimation of Tisochrysis lutea. *PLoS ONE* **2017**, *12*, e0183547.
15. Gachelin, M.; Boutoute, M.; Carrier, G.; Talec, A.; Pruvost, E.; Guihéneuf, F.; Bernard, O.; Sciandra, A. Enhancing PUFA-rich polar lipids in Tisochrysis lutea using adaptive laboratory evolution (ALE) with oscillating thermal stress. *Appl. Microbiol. Biotechnol.* **2021**, *105*, 301–312. [CrossRef]
16. Nolasco, E.; Vassiliadis, V.S.; Kähm, W.; Adloor, S.D.; Al Ismaili, R.; Conejeros, R.; Espaas, T.; Gangadharan, N.; Mappas, V.; Scott, F.; et al. Optimal control in chemical engineering: Past, present and future. *Comput. Chem. Eng.* **2021**, *155*, 107528. [CrossRef]
17. Bayen, T.; Mairet, F. Optimization of the separation of two species in a chemostat. *Automatica* **2014**, *50*, 1243–1248. [CrossRef]
18. Bayen, T.; Mairet, F. Optimization of strain selection in evolutionary continuous culture. *Int. J. Control* **2017**, *90*, 2748–2759. [CrossRef]
19. Droop, M.R. Vitamin B_{12} and marine ecology. (IV). The kinetics of uptake growth and inhibition in Monochrysis lutheri. *J. Mar. Biol. Assoc.* **1968**, *48*, 689–733. [CrossRef]
20. Droop, M.R. Some thoughts on nutrient limitation in algae. *J. Phycol.* **1973**, *9*, 264–272. [CrossRef]
21. Djema, W.; Giraldi, L.; Bernard, O. An Optimal Control Strategy Separating Two Species of Microalgae in Photobioreactors. In Proceedings of the Conference on Dynamics and Control of Process Systems, including Biosystems—12th DYCOPS, Florianopolis, Brazil, 23–26 April 2019; Volume 52, pp. 910–915.
22. Djema, W.; Bernard, O.; Giraldi, L. Separating Two Species of Microalgae in Photobioreactors in Minimal Time. *J. Process Control (JPC)* **2020**, *87*, 120–129. [CrossRef]
23. Andrés-Martínez, O.; Biegler, L.T.; Flores-Tlacuahuac, A. An indirect approach for singular optimal control problems. *Comput. Chem. Eng.* **2020**, *139*, 106923. [CrossRef]
24. Pontryagin, L.S. *Mathematical Theory of Optimal Processes*; Springer: Berlin/Heidelberg, Germany, 1964.
25. Caperon, J.; Meyer, J. Nitrogen-limited growth of marine phytoplankton—II. Uptake kinetics and their role in nutrient limited growth of phytoplankton. *Deep Sea Res. Oceanogr. Abstr.* **1972**, *19*, 619–632. [CrossRef]

26. Boscain, U.; Piccoli, B. *Optimal Syntheses for Control Systems on 2D Manifolds*; Springer: Berlin/Heidelberg, Germany, 2004; Volume 43.
27. Schättler, H.; Ledzewicz, U. *Geometric Optimal Control*; Springer: New York, NY, USA, 2012.
28. Vinter, R. *Optimal Control, Systems and Control: Foundations and Applications*; Birkhauser: Basel, Switzerland, 2000.
29. Djema, W.; Bernard, O.; Bayen, T. Optimal control separating two microalgae species competing in a chemostat. In Proceedings of the 59th IEEE Conference on Decision and Control (CDC), IEEE Proceedings 2020, Jeju, Korea, 14–18 December 2020; pp. 2368–2373.
30. Bayen, T.; Cots, O. Tangency Property and Prior-Saturation Points in Minimal Time Problems in the Plane. *Acta Appl. Math.* **2020**, *170*, 515–537. [CrossRef]
31. Bonnard, B.; Chyba, M. *Singular Trajectories and Their Role in Control Theory*; Springer: Berlin/Heidelberg, Germany, 2003; Volume 40.
32. Aronna, M.S.; Bonnans, J.F.; Martinon, P. A Shooting Algorithm for Optimal Control Problems with Singular Arcs. *J. Optim. Theory Appl.* **2013**, *158*, 419–459. [CrossRef]
33. Djema, W.; Giraldi, L.; Maslovskaya, S.; Bernard, O. Turnpike features in optimal selection of species represented by quota models. *Automatica* **2021**, *132*, 109804. [CrossRef]
34. Porretta, A.; Zuazua, E. Long time versus steady state optimal control. *SIAM J. Cont. Opti.* **2013**, *51*, 4242–4273. [CrossRef]
35. Trélat, E.; Zuazua, E. The turnpike property in finite-dimensional nonlinear optimal control. *J. Differ. Equ.* **2015**, *258*, 81–114. [CrossRef]
36. Bonnans, F.J.; Giorgi, D.; Grelard, V.; Heymann, B.; Maindrault, S.; Martinon, P.; Tissot, O.; Liu, J. *BOCOP: An Open Source Toolbox for Optimal Control—A Collection of Examples*; Technical Reports, Project-Team Commands; Inria, Saclay: Palaiseau, France, 2017.
37. Biegler, L.T. *Nonlinear Programming: Concepts, Algorithms, and Applications to Chemical Processes*; Series MPS-SIAM on Optimization (Book Number 10); SIAM: Philadelphia, PA, USA, 2010; p. 415.
38. Betts, J.T. *Practical Methods for Optimal Control and Estimation Using Nonlinear Programming*, 2nd ed.; Advances in Design & Control; SIAM: Philadelphia, PA, USA, 2010; Volume 19, p. 427.
39. Bonnard, B.; Rouot, J. *Towards Geometric Time Minimal Control without Legendre Condition and with Multiple Singular Extremals for Chemical Networks*; Applied Mathematics, Advances in Nonlinear Biological Systems: Modeling and Optimal Control; American Institute of Mathematical Sciences: San Jose, CA, USA, 2020; pp. 1–34.

Article

Best Operating Conditions for Biogas Production in Some Simple Anaerobic Digestion Models

Tewfik Sari

ITAP, INRAE, Institut Agro, University of Montpellier, 34196 Montpellier, France; tewfik.sari@inrae.fr

Abstract: We consider one-step and two-step simple models of anaerobic digestion that are able to adequately capture the main dynamical behaviour of the full anaerobic digestion model ADM1. We do not consider specific growth functions. We only require them to satisfy certain qualitative assumptions. These assumptions are satisfied for concave growth functions, but they are also satisfied for a large class of growth functions found in many applications. We consider the maximisation of the biogas production with respect to the operating parameters of the model, which are the dilution rate and the substrate input concentration. We give the best operating conditions and we describe them as a subset of the set of operating parameters. Our models incorporate biomass decay terms, corresponding to maintenance. Numerical plots with specified growth functions and biological parameters illustrate the obtained results.

Keywords: anaerobic digestion; biogas; chemostat; maintenance; operating diagram; optimization; productivity; stability

MSC: 34D20; 34H20; 65K10; 92C75

Citation: Sari, T. Best Operating Conditions for Biogas Production in Some Simple Anaerobic Digestion Models. *Processes* **2022**, *10*, 258. https://doi.org/10.3390/pr10020258

Academic Editors: Philippe Bogaerts and Alain Vande Wouwer

Received: 22 December 2021
Accepted: 24 January 2022
Published: 28 January 2022

Publisher's Note: MDPI stays neutral with regard to jurisdictional claims in published maps and institutional affiliations.

1. Introduction

Anaerobic digestion (AD) is a well known and established technology for treating waste in the methanisation of sewage sludge from wastewater treatment plants. AD enables the water industry to treat waste water as a resource for generating energy and recovering valuable by-products. In the context of renewable energy, it has now become an attractive alternative to fossil carbon [1]. AD is a complex biological process in which organic material is converted into biogas (methane) in an environment without oxygen [2–6]. One of its main disadvantages is its sensitivity to disturbances, which can lead to instability problems, in addition to a decrease in the biogas production rate [7]. Indeed, the conditions and technological parameters characterising the methane fermentation process include many parameters: hydraulic retention time, organic loading rate, anaerobic sludge concentration in the bioreactor, substrate dewatering, organic matter content, substrate dosage, mixing method and frequency, temperature, and many others.

When the experimenter does not have a mechanistic mathematical model of the process being studied, one method for selecting the best conditions for biogas production is to carry out multi-variant tests and select the most efficient variants and then optimise them and develop a mathematical model. This first approach is presented in [8–10]. On the other hand, when the experimenter has a model of the process being studied and knows or has identified its biological parameters, a good way to optimise biogas production is to look for the optimal flow rate of the bioreactor that produces the most biogas. This approach is presented in [11–21]. Therefore, mechanistic mathematical models are a good basis for monitoring and developing control strategies to optimise the operation of such processes. The present paper is a contribution to this second approach to the problem: we assume that we know a mechanistic mathematical model of the process and that we have already identified its biological parameters, and then we look to the best operating conditions for biogas production.

However, having a model of AD is not that easy. Indeed, the complexity of AD has motivated the development of mechanistic mathematical models, such as the widely used Anaerobic Digestion Model Model No. 1 (ADM1) [2]. This model has a large number of state variables and parameters. It is impossible to obtain an analytical characterization of the steady states and to describe the operating diagram (OD), that is to say, to identify the asymptotic behaviour of existing steady states as a function of the operating parameters (substrate inflow concentrations and dilution rate). To the author's knowledge, only numerical investigations are available [3]. Therefore, although ADM1 is a complex model that is widely accepted as a common platform for AD process modelling and simulation, it has a large number of parameters and states that hinder its analytic study. Due to the analytic intractability of the full ADM1, progress has been made towards the construction of simpler models that preserve biological meaning. The simplest model of the chemostat with only one biological reaction, where one substrate is consumed by one microorganism, is well understood [22–24]. However a one-step model is too simple to encapsulate the essence of AD.

More realistic models of AD are two-step models. An important contribution to the modelling of AD as a two-step is the model presented in [25], hereafter denoted as the AM2 model and studied in [15,26]. It has been shown that under some circumstances, this very simple two-step model is able to adequately capture the main dynamical behaviour of the full ADM1 [27,28]. AM2 is a four-dimensional system of ordinary differential equations and takes acidogenesis and methanogenesis into consideration. In the first step, the organic substrate is consumed by the acidogenic bacteria and produces a substrate, the Volatile Fatty Acids (VFA), while in the second step, the methanogenic population consumes VFA and produces biogas.

Another interesting simple AD model, with eight state variables, was considered in [21,29,30]. This model takes into consideration acidogenesis, acetogenesis, and methanogenesis. We also mention the mathematical model considered in [31], which also added the hydrolysis step in the model. It is also worth mentioning the models of AD that include the evolution of biogas and hydrogen [32–34].

The problem of optimising biogas production for one-step AD models is studied in [13,14] and for the AM2 model in [11,15–18]. This problem is also analysed in [21,31,35], where models with more steps for AD are considered.

The OD of a model has operating parameters as its coordinates, and the various regions defined within it correspond to qualitatively different asymptotic behaviours. The operating parameters are the input concentrations of substrates and the flow rate. We call them operating parameters, although they are not always under the control of the experimenter. Indeed, in most practical cases, one can at best store material upstream and control the flow rate. The concentration of the input substrate is rarely a control parameter. However, this parameter is known to the experimenter and is not of the same nature as the biological parameters, on which the experimenter can only act with great difficulty. In most of the results, we will assume that the input concentration of substrate is fixed, and we want to determine the corresponding optimal flow rate. Apart from the operating parameters, which can vary, all other parameters have biological meaning and are fitted using experimental data from ecological and/or biological observations of organisms and substrates. When the biological parameters are determined, it is then easy to plot the operating diagram and thus have a prediction of the behaviour of the system as a function of the operating parameters.

The OD is then the bifurcation diagram that shows how the system behaves when we vary the operating (control) parameters. This diagram shows how extensive the parameter region is, where some asymptotic behaviours occur. This bifurcation diagram is very useful to understand the model from both the mathematical and biological points of view. Its importance for bioreactors was emphasized in [36]. This diagram is often constructed both in the biological literature [15,29,35–38] and the mathematical literature [3,21,30,39–44].

In the present work, we consider the one-step model and the AM2 model, and we give the best operating conditions for biogas production, that is to say, we give the subset of the OD corresponding to the maximal flow rate of the biogas. This set of the best operating conditions in the OD indicates to the experimenter how to choose the operating condition such that the system produces the maximum of biogas. The surprising result for AM2 is that the optimal steady state can involve the extinction of the acidogenic bacteria [11]. This property was also observed for more complex models [21,31]. We address this problem and fully describe the operating conditions under which this situation is encountered. Another very important phenomenon, which was observed in [35], is that the best biogas produced is sometimes obtained for operating parameters for which the system has bistability. This issue is also addressed, and the set of operating parameters for which the system may be in such a situation is fully described.

The paper is organized as follows. In Section 2, we describe the one-step and two-step models of AD that are studied in this paper. We give the steady states of the models and their biogas flow rate or productivity. We state the problems of optimisation that will be considered later. The results for one-step models are given in Section 3.1. The particular case when the biomass mortality is neglected is considered in Section 3.1.8, and applications to various growth functions that were considered in the literature are given in Appendix A.5. The results for two-step models are listed in Section 3.2, and the applications of our theory to the classical AM2 model are emphasized in Section 3.2.4. We discuss and compare our results with the results of the existing literature in Section 3.3. Finally, Sections 4 and 5 draw some discussions, conclusions, and perspectives. The proofs and supplementary information are given in Appendixes A and B.

2. Materials and Methods

We consider a continuous stirred-tank reactor (CSTR), also called a bioreactor or a chemostat, where a single population of micro-organisms is growing on a single limiting substrate. We also consider the more complex situation where this population produces a substrate which is itself consumed by a second population. The limiting substrate is fed into the culture vessel with a constant concentration at flow rate Q. The culture medium is withdrawn at the same flow rate Q so that the culture volume V in the vessel is kept constant.

The dilution rate D is defined as $D = Q/V$ and is the inverse of the residence time. We will take into account that the residence time of the liquid (culture medium) in the bioreactor may be shorter than that of the solids (micro-organisms), which is common in bioreactors.

We also take maintenance into account. Consumption of energy for all processes other than growth is called maintenance. In situations where microbial cells are located in a favourable environment, maintenance can often be neglected. In other situations, however, a significant portion of the energy-yielding substrate that could be used for growth is consumed for maintenance [45]. In the ADM1 model and also in some simple models of AD, maintenance is taken into account as decay [2,37,38,43,44].

It is assumed that the other required substrates are provided in excess, that the culture medium is perfectly mixed and that the environmental conditions (temperature and pH) are regulated at appropriate constant values.

2.1. One-Step Models

Although the one-step model is too simple to encapsulate the essence of AD, it is useful for the understanding of some basic facts concerning optimization of biogas in bioreactors. Consider a one-step model of the form:

$$kS \xrightarrow{r} X + k_1 CH_4 \tag{1}$$

where one substrate S is consumed by one micro-organism X and produces biogas with reaction rate $r = \mu(S)X$, where μ is the growth function and k and k_1 are pseudo-stoichiometric coefficients. Let D be the dilution rate and S^{in} the concentrations of input substrate. The dynamical equations of the model are [22–24,46,47]

$$\begin{aligned} \dot{S} &= D(S^{in} - S) - k\mu(S)X \\ \dot{X} &= (\mu(S) - D_1)X \end{aligned} \tag{2}$$

where D_1, the removal rate of the micro-organisms, takes the form

$$D_1 = \alpha D + a, \tag{3}$$

where a is the decay term corresponding to maintenance effects and $\alpha \in (0,1]$ is a parameter allowing us to decouple the Hydraulic Retention Time, $\text{HRT} = 1/D$ and the Solid Retention Time $\text{SRT} = 1/(\alpha D)$. The stoichiometric coefficient k_1 in (1) appears in the mathematical equations of the model when we consider the biogas flow rate; see Section 2.1.2. The stoichiometric coefficient k can be reduced to 1; see Appendix A.1. However, since the stoichiometric coefficient has its own importance for the biologist, and since our aim is to give the biologist a useful tool for the best operating conditions of the chemostat model, we do not make this reduction and we present the results in the original model (2). The mathematical analysis of (2) is well-known [22,24]. For the convenience of the reader, we recall in this paper the main results and state them using the OD; see Appendix A.2.

2.1.1. Steady States

We assume that μ is not necessarily monotonic, i.e., that the inhibition by substrate S can be taken into account in the model. We make now the following hypothesis.

Hypothesis 1. *The function μ is C^1 and satisfies $\mu(0) = 0$, and there exists $S^m \in (0, +\infty]$, such that $\mu'(S) > 0$ for $0 < S < S^m$. If $S^m < +\infty$, then, in addition, $\mu'(S) < 0$ for $S > S^m$.*

The case $S^m = +\infty$ corresponds to an increasing function. This case is called the *Monod case*, since it is satisfied by the usual Monod growth function

$$\mu(S) = \frac{mS}{K+S}. \tag{4}$$

The case $S^m < +\infty$ corresponds to an increasing and then decreasing function and models the inhibition by the substrate at high concentrations. This case is called the *Haldane case*, since it is satisfied by the usual Haldane growth function

$$\mu(S) = \frac{mS}{K+S+S^2/K_i}. \tag{5}$$

We need to define the break-even concentrations:

Definition 1. *When $S^m = +\infty$, the break-even concentration $\lambda(D)$ is the unique solution of equation $\mu(S) = D$. It is defined for $D < \mu(+\infty)$. When $S^m < +\infty$, there can be two break-even concentrations $\lambda(D)$ and $\bar{\lambda}(D)$. They are the solutions of equation $\mu(S) = D$, such that $\lambda(D) < S^m < \bar{\lambda}(D)$. The first one is defined for $0 < D < \mu(S^m)$. The second one is defined for $\mu(+\infty) < D < \mu(S^m)$. They have the same limit value $\lambda(D^m) = \bar{\lambda}(D^m)$ for $D^m = \mu(S^m)$. If $D > \mu(S^m)$, by convention we let $\lambda(D) = +\infty$ and $\bar{\lambda}(D) = +\infty$.*

Besides the washout steady state $F_0 = (S^{in}, 0)$, (2) has the positive steady states

$$F_1 = \left(\lambda(D_1), \frac{D}{kD_1}\left(S^{in} - \lambda(D_1)\right)\right), \qquad F_2 = \left(\bar{\lambda}(D_1), \frac{D}{kD_1}\left(S^{in} - \bar{\lambda}(D_1)\right)\right). \tag{6}$$

When $S^m = +\infty$, only F_1 exists. The conditions of existence an stability of the steady state, together with the OD of (2), are given in Appendix A.2. Note that F_1 is stable whenever its exists, while F_2 is unstable whenever its exists.

2.1.2. Steady State Optimization of Biogas Production

The biogas is simply a product of the biological reactions and it has no feedback on the dynamical Equation (2). The biogas flow rate, denoted by G_{CH_4}, is proportional to the microbial activity, as proposed in [46,48–50]:

$$G_{CH_4} = k_1 \mu(S^*) X^* \tag{7}$$

where (S^*, X^*) is a steady state of (2). Let us denote by G_i, the rate of production of biogas, defined by (7), and evaluated at steady state F_i, $i = 0, 1, 2$. One has $G_0 = 0$, and using the components of the steady states F_1 and F_2 given in (6), G_1 and G_2 are given by

$$
\begin{aligned}
G_1(D, S^{in}) &= \tfrac{k_1}{k} D\big(S^{in} - \lambda(\alpha D + a)\big) \quad \text{for} \quad S^{in} \geq \lambda(\alpha D + a), \\
G_2(D, S^{in}) &= \tfrac{k_1}{k} D\big(S^{in} - \bar{\lambda}(\alpha D + a)\big) \quad \text{for} \quad S^{in} \geq \bar{\lambda}(\alpha D + a).
\end{aligned}
\tag{8}
$$

Our aim is to determine the set of operating conditions for which the biogas production is maximal. We consider the biogas flow rate G_2 corresponding to the unstable equilibrium F_2 because we do not know if this flow rate is always lower than that of the stable equilibrium F_1. If it was possible that, for some operating condition D and S^{in}, $G_2(D, S^{in}) > G_1(D, S^{in})$, then the problem of the stabilization of the reactor at its unstable steady state F_2 by using some feedback control would have been an interesting challenge. However, this possibility is excluded, as stated in the following remark.

Remark 1. *Note that G_2 is defined if and only if $S^{in} \geq \bar{\lambda}(\alpha D + a)$. Since $\bar{\lambda}(\alpha D + a) > \lambda(\alpha D + a)$, G_1 is also defined and we have $G_1(D, S^{in}) > G_2(D, S^{in})$.*

Hence, the operating conditions D and S^{in} which produce the maximum of biogas are obtained by the maximization of $G_1(D, S^{in})$.

Problem 1. *Determine the set of operating conditions for which G_1 is maximal.*

2.1.3. Steady State Optimization of Biomass Production

AD is used because it allows material to be degraded without producing too much biomass, which is a good thing because in the environmental field we do not really know what to do with the sludge produced. If we want to produce biomass, it is rather in biotechnologies such as pharmaceuticals or food processing that we should be looking. Let us forget about AD for a moment and assume that the industrial goal of the process is the production of micro-organisms. When a continuous culture system is viewed as a production process, its performance may be judged by the quantity of bacteria produced, which is called the productivity of biomass. The total output from a continuous culture unit in the steady state is equal to the product of flow rate and concentration of organisms. Therefore, the productivity of (2) at steady state (S^*, X^*) is given by [20,47]

$$P = Q X^* \tag{9}$$

where $Q = VD$ is the flow rate, and V is the volume of the CSTR. Let us denote by P_i, the productivity evaluated at steady state F_i, $i = 0, 1, 2$. One has $P_0 = 0$ and using the components of the steady states F_1 and F_2, given in (6), P_1 and P_2 are given by

$$
\begin{aligned}
P_1(D, S^{in}) &= \tfrac{VD^2}{k(\alpha D + a)}\big(S^{in} - \lambda(\alpha D + a)\big) \quad \text{for} \quad S^{in} \geq \lambda(\alpha D + a), \\
P_2(D, S^{in}) &= \tfrac{VD^2}{k(\alpha D + a)}\big(S^{in} - \bar{\lambda}(\alpha D + a)\big) \quad \text{for} \quad S^{in} \geq \bar{\lambda}(\alpha D + a).
\end{aligned}
\tag{10}
$$

Our aim is to determine the set of operating conditions for which the productivity is maximal. Note that, as for the biogas flow rate, the productivity at F_1 is greater the the productivity at F_2: $P_1(D, S^{in}) > P_2(D, S^{in})$. Hence, the operating conditions D and S^{in} that maximize productivity are obtained by maximizing $P_1(D, S^{in})$.

Problem 2. *Determine the set of operating conditions for which P_1 is maximal.*

2.1.4. The Case without Mortality

Note that when $a = 0$, we have

$$G_1(D, S^{in}) = \tfrac{k_1}{k} D(S^{in} - \lambda(\alpha D)), \qquad G_2(D, S^{in}) = \tfrac{k_1}{k} D(S^{in} - \bar{\lambda}(\alpha D)),$$
$$P_1(D, S^{in}) = \tfrac{V}{k\alpha} D(S^{in} - \lambda(\alpha D)), \qquad P_2(D, S^{in}) = \tfrac{V}{k\alpha} D(S^{in} - \bar{\lambda}(\alpha D)).$$

Therefore, G_i and P_i, $i = 1, 2$ are proportional. Hence, we can make the following remark.

Remark 2. *When $a = 0$, optimizing P_1, given by (10), is the same as optimizing G_1, given by (8); that is, Problems 1 and 2 have the same solution. However, this is no longer true when $a > 0$.*

For increasing functions (i.e., $S^m = +\infty$), in the case $a = 0$, the equivalent Problems 1 and 2 have been solved in [51]; in the case $a > 0$, Problem 1 has been solved in [52] and Problem 2 in [53]. In Sections 3.1.1 and 3.1.5, we will give the solutions to these problems in the more general case where the growth function μ satisfies the Hypothesis 1 and is not necessarily monotonic.

2.2. Two-Step Models

We consider the general two-step model with a cascade of two biological reactions, where one substrate S_1 is consumed by one microorganism X_1 (*acidogenic* bacteria, in the AM2 model), to produce a product S_2 that serves as the main limiting substrate for a second microorganism X_2 (*methanogenic* bacteria in the AM2 model) as schematically represented by the following reaction scheme (see [25]):

$$k_1 S_1 \xrightarrow{r_1} X_1 + k_2 S_2, \quad k_3 S_2 \xrightarrow{r_2} X_2 + k_4 CH_4 \tag{11}$$

where $r_1 = \mu_1(S_1)X_1$ and $r_2 = \mu_2(S_2)X_2$ are the kinetics of the reactions and k_i, $i = 1, \ldots, 4$ are pseudo-stoichiometric coefficients. In fact, biological reactions also produce CO_2; see Equations (1) and (2) in [25]. However, since in this section we are only interested in the biogas production, we do not focus on the CO_2 production. Let D be the dilution rate and S_1^{in} and S_2^{in} the concentrations of input substrates S_1 and S_2, respectively. The dynamical equations of the model take the form:

$$
\begin{aligned}
\dot{S}_1 &= D(S_1^{in} - S_1) - k_1 \mu_1(S_1) X_1, \\
\dot{X}_1 &= (\mu_1(S_1) - D_1) X_1, \\
\dot{S}_2 &= D(S_2^{in} - S_2) + k_2 \mu_1(S_1) X_1 - k_3 \mu_2(S_2) X_2, \\
\dot{X}_2 &= (\mu_2(S_2) - D_2) X_2,
\end{aligned}
\tag{12}
$$

where, as in (3), the removal rates of the micro-organisms D_1 and D_2 take the form

$$D_i = \alpha_i D + a_i, \quad i = 1, 2, \tag{13}$$

where $\alpha_i \in (0, 1]$, $i = 1, 2$, is a parameter allowing us to decouple the HRT and the SRT. This decoupling is necessary when considering technology such as systems where biomass is fixed onto supports (as in fixed or fluidized bed reactors) or still retained in the system by membranes such as in MBRs (Membrane Bioreactors); see [54,55]. The model (12) is an extension of the AM2 model presented in [25], with $\alpha_1 = \alpha_2$, $a_1 = a_2 = 0$, and kinetics μ_1 and μ_2 of Monod and Haldane types, respectively.

The pseudo-stoichiometric coefficients k_i in (12) can be reduced to 1; see Appendix B.1. However, since these coefficients have their own importance for the biologist and since our aim is to discuss the best operating conditions, we do not make this reduction and we present the results in the original model (12). The model has a cascade structure which renders its analysis easy. We give in Appendix B.3 the main results on the existence and stability of the steady states of (12), and we express them using the OD.

2.2.1. Steady States

We consider (12) with general kinetics functions μ_1 and μ_2, satisfying the following qualitative properties:

Hypothesis 2. *The function μ_1 is C^1, $\mu_1(0) = 0$, $\mu_1'(S_1) > 0$ for $S_1 > 0$. Let $m_1 = \mu_1(+\infty)$.*

Hypothesis 3. *The function μ_2 is C^1, $\mu_2(0) = 0$, $\mu_2(+\infty) = 0$, and there exists $S_2^m > 0$ such that $\mu_2'(S_2) > 0$ for $0 < S_2 < S_2^m$, and $\mu_2'(S_2) < 0$ for $S_2 > S_2^m$.*

We consider the break-even concentrations as stated in Definition 1. The growth function μ_1 admits only one break-even concentration, denoted λ_1, while the growth function μ_2 admits two break-even concentrations, which will be denoted λ_2 and $\bar{\lambda}_2$. We summarize in Table 1 the definitions of these break-even concentrations, together with two auxiliary functions that are used in the description of the biogas flow-rates at steady states of (12).

Table 1. Break-even concentrations and auxiliary functions.

$\lambda_1(D)$ is the unique solution of equation $\mu_1(S_1) = D$, for $D < m_1$
$\lambda_2(D) < \bar{\lambda}_2(D)$ are the solutions of equation $\mu_2(S_2) = D$, for $D < \mu_2(S_2^m)$ $\lambda(0) = 0$, $\bar{\lambda}_2(0) = +\infty$ and $\lambda(D) = \bar{\lambda}_2(D)$ for $D = \mu_2(S_2^m)$ $H_1(D) = \lambda_2(D_2) + \frac{k_2}{k_1}\lambda_1(D_1)$, $H_2(D) = \bar{\lambda}_2(D_2) + \frac{k_2}{k_1}\lambda_1(D_1)$

The system (12) can have up to six steady states, denoted E_{ij}, where $i = 0, 1$ and $j = 0, 1, 2$. The components of the steady states are given in Table A3. The existence and stability conditions of the steady states of (12) are given in Appendix B.3. Note that E_{11} is stable whenever it exists, while E_{01} is stable if and only if it exists and E_{11} does not exist. Moreover the steady states E_{02} and E_{12} are unstable whenever they exist.

2.2.2. Steady State Optimization of Biogas Production

As in the one-step model, the biogas is simply a product of the biological reactions and it has no feedback on the dynamical Equation (12). As we noticed in (7), the mass flow of the methane production, denoted by G_{CH_4}, is proportional to the microbial activity (see Equation (12) in [25]):

$$G_{CH_4} = k_4 \mu_2(S_2) X_2.$$

Let us denote by G_{ij} the production of biogas at steady states E_{ij} for $i = 0, 1$ and $j = 0, 1, 2$. Using the components of the steady states given in Table A3, it is seen that $G_{00} = G_{10} = 0$ and G_{ij} for $i = 0, 1$ and $j = 1, 2$ are defined as in Table 2.

Table 2. The biogas production at steady state E_{ij}, $i = 0, 1$, $j = 1, 2$; $\lambda_2(D)$, $\bar{\lambda}_2(D)$ and $H_j(D)$, $j = 1, 2$, are defined in Table 1.

Biogas Production	Domain of Definition
$G_{01}\left(D, S_2^{in}\right) = \frac{k_4}{k_3} D\left(S_2^{in} - \lambda_2(D_2)\right)$	$\lambda_2(D_2) \leq S_2^{in}$
$G_{02}\left(D, S_2^{in}\right) = \frac{k_4}{k_3} D\left(S_2^{in} - \bar{\lambda}_2(D_2)\right)$	$\bar{\lambda}_2(D_2) \leq S_2^{in}$
$G_{11}\left(D, S_1^{in}, S_2^{in}\right) = \frac{k_4}{k_3} D\left(S_2^{in} + \frac{k_2}{k_1} S_1^{in} - H_1(D)\right)$	$\lambda_1(D_1) \leq S_1^{in}, H_1(D) \leq S_2^{in} + \frac{k_2}{k_1} S_1^{in}$
$G_{12}\left(D, S_1^{in}, S_2^{in}\right) = \frac{k_4}{k_3} D\left(S_2^{in} + \frac{k_2}{k_1} S_1^{in} - H_2(D)\right)$	$\lambda_1(D_1) \leq S_1^{in}, H_2(D) \leq S_2^{in} + \frac{k_2}{k_1} S_1^{in}$

Our aim is to find set of operating conditions for which the flow rate of biogas is maximal.

Remark 3. *We always have $G_{01} > G_{02}$ and $G_{11} > G_{12}$; see Section 3.2.1.*

Hence, the operating conditions D, S_1^{in}, and S_2^{in}, which produce the maximum of biogas, are obtained by the maximization of $G_{01}(D, S_2^{in})$ or $G_{11}(D, S_1^{in}, S_2^{in})$. The main problem is then to compare the maximum of biogas production G_{11} at E_{11}, where both species are present, with the maximum of biogas production G_{01} at E_{01} where species X_1 is extinct and species X_2 is present. Surprisingly, the optimal biogas production does not always occur at E_{11}, as was noticed by [11,21,31]. Therefore we have to solve the following problem.

Problem 3. *Determine the sets of operating conditions, for which G_{01} and G_{11} are maximal. Compare the maximum of G_{01} to that of G_{11}.*

3. Results

3.1. One-Step Models

The OD of the one-step model (2) is described in Appendix A.2.

3.1.1. Best Operating Conditions for Biogas Production

Let G_1 defined by (8) and S^{in} fixed. Our aim is to maximize the function $D \mapsto G_1(D, S^{in})$. Note that this function is proportional to the function G defined by

$$G(D) = D(S^{in} - \lambda(\alpha D + a)). \tag{14}$$

The function G is depending on the parameter S^{in}. It is defined for $D \in I(S^{in})$, where the interval $I(S^{in})$ is given by

$$I(S^{in}) = \begin{cases} [0, \delta(S^{in})] & \text{if } S^{in} < S^m \\ [0, \delta(S^m)] & \text{if } S^{in} \geq S^m \end{cases} \quad \text{with} \quad \delta(S) = \frac{\mu(S) - a}{\alpha} \tag{15}$$

The function G_1 has an absolute maximum if G has one and this maximum is reached at the same point where G reaches its maximum. By the *Extreme Value Theorem*, since G is continuous on the closed interval $I(S^{in})$, it must attain a maximum. Let us consider the set of arguments of the maximum of G, denoted by $g(S^{in})$ and defined by

$$g(S^{in}) = \underset{D \in I(S^{in})}{\text{argmax}}\, G := \left\{ D^* \in I(S^{in}) : G(D) \leq G(D^*) \text{ for all } D \in I(S^{in}) \right\}. \tag{16}$$

To obtain the maximum value of $G(D)$, we differentiate (14) with respect to D, and we solve the equation $G'(D) = 0$. The derivative of G is given by

$$G'(D) = S^{in} - \gamma(D)$$

where γ is defined by

$$\gamma(D) = \lambda(\alpha D + a) + \alpha D \lambda'(\alpha D + a). \tag{17}$$

Remark 4. *Since $\mu(\lambda(D)) = D$, we have $\lambda'(D) = 1/\mu'(\lambda(D))$. Therefore the function γ is written*

$$\gamma(D) = \lambda(\alpha D + a) + \frac{\alpha D}{\mu'(\lambda(\alpha D + a))}.$$

We have the following result

Proposition 1. *Let $D^* \in g(S^{in})$. We have $S^{in} = \gamma(D^*)$, where γ is defined by* (17).

Proof. The proof is given in Appendix A.3.1. $\square$

Therefore, the curve

$$\Gamma = \left\{ (D, S^{in}) : S^{in} = \gamma(D) \right\} \tag{18}$$

of SOP contains the operating conditions for which G_1 is maximal.

In Figure 1, we plot the Γ curve in the OD of (2). We have shown a curve Γ, which is the graph of an increasing function. However, this does not always happen; see Appendix A.5.5. When Γ is not increasing, there may be several maxima of the biogas flow. In Section 3.1.3, we give sufficient conditions for the maximum to be unique. Since $\lambda'(D) > 0$, we deduce that $\gamma(D) > \lambda(\alpha D + a)$ for $D > 0$. On the other hand,

$$\gamma(0) = \lambda(a), \quad \text{and} \quad \lim_{D \to \delta(S^m)} \gamma(D) = +\infty.$$

From these properties we deduce the following remark.

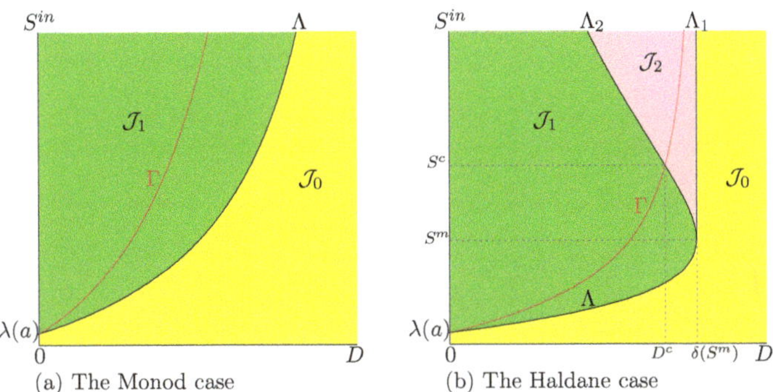

(a) The Monod case   (b) The Haldane case

Figure 1. The OD of (2). The curve Γ is the set of best operating conditions.

Remark 5. *If $S^m = +\infty$, the curve Γ is contained in the region $\mathcal{J}_1$ (the green region) of the OD, see Figure 1a. If $S^m < +\infty$, Γ is contained in $\mathcal{J}_1 \cup \mathcal{J}_2$ (the green and pink regions), and, since $\mu'(S^m) = 0$, the vertical line Λ_1 is an asymptote of Γ, see Figure 1b. Note that in the Haldane case $(S^m < \infty)$, the curve Γ enters in the bistability region $\mathcal{J}_2$ at point (D^c, S^c).*

3.1.2. How to Determine the Maximum of Biogas Production

From Proposition 1, to obtain $g(S^{in})$, we must solve the equation $S^{in} = \gamma(D)$. However, this equation can be complicated to solve because $\gamma(D)$ is itself defined by $\lambda(D)$, which is the solution of the equation $\mu(S) = D$. We have at our disposal another description of $g(S^{in})$. Indeed, we can write

$$G(D) = \tfrac{1}{\alpha} H(\lambda(\alpha D + a)), \tag{19}$$

where H is defined by

$$H(S) = (\mu(S) - a)(S^{in} - S), \quad \text{for} \quad \lambda(a) \le S \le S^{in} \tag{20}$$

From (19), it is deduced that the absolute maximum of G corresponds to the absolute maximum of H and vice versa. To obtain the maximum value of $H(S)$, we differentiate H with respect to S and we solve the equation $H'(S) = 0$. The derivative of H is given by

$$H'(S) = \mu'(S)(S^{in} - S) - \mu(S) + a.$$

Hence, $H'(S) = 0$ if and only if $S^{in} = \eta(S)$, where $\eta(S)$ is defined by

$$\eta(S) = S + \frac{\mu(S) - a}{\mu'(S)} \quad \text{for} \quad S \ge \lambda(a). \tag{21}$$

We have the following result.

Proposition 2. *Let S^* be the maximum of H on $(\lambda(a), S^{in})$. Let $D^* = \frac{\mu(S^*) - a}{\alpha}$. Then $D^* \in g(S^{in})$. Moreover, we have $S^{in} = \eta(S^*)$, where η is defined by (21).*

Proof. The proof is given in Appendix A.3.2. $\square$

Remark 6. *With the first method, we must first solve the equation $\mu(S) = D$ to obtain $\lambda(D)$ and then solve the equation $\gamma(D) = S^{in}$ to obtain the optimal $D^* \in g(S^{in})$. With the second method, we simply solve the equation $\eta(S) = S^{in}$ to get the maximum S^* and then take $D^* = \frac{\mu(S^*) - a}{\alpha} \in g(S^{in})$.*

3.1.3. Uniqueness of the Maximum

Hypothesis 1 is not enough to guarantee that the biogas flow rate admits a unique global maximum; see Appendix A.5.5. We make the following hypothesis.

Hypothesis 4. *For all $S^{in} > 0$, $g(S^{in})$, defined by (16), has a unique element, which is denoted by $D_G^*(S^{in})$.*

From Proposition 1 we deduce then the answer to Problem 1: assume that Hypotheses 1 and 4 are satisfied. Then, the set of best operating conditions for biogas production of (2) is the curve Γ of SOP defined by:

$$\Gamma = \left\{ (D, S^{in}) : S^{in} = \gamma(D) \right\} = \left\{ (D, S^{in}) : D = D_G^*(S^{in}) \right\}. \tag{22}$$

From Propositions 1 and 2, it is deduced that Hypothesis 4 is satisfied when the equations

$$S^{in} = \gamma(D) \quad \text{or} \quad S^{in} = \eta(S)$$

have a unique solution. A sufficient condition for this is that the functions $\gamma(D)$ and $\eta(S)$ are increasing. The following result gives sufficient conditions for Hypothesis 4 to be valid.

Lemma 1. *Assume that Hypothesis 1 is satisfied and, in addition, μ is C^2. The following conditions are equivalent*

1. *$\gamma' > 0$ on $\left(0, \frac{\mu(S^m) - a}{\alpha}\right)$.*
2. *$(\mu - a)\mu'' < 2(\mu')^2$ on $(\lambda(a), S^m)$.*
3. *$\left(\frac{1}{\mu - a}\right)'' > 0$ on $(\lambda(a), S^m)$.*
4. *$\eta' > 0$ on $(\lambda(a), S^m)$.*

If these equivalent conditions are satisfied, then Hypothesis 4 is satisfied. If $\mu'' < 0$ on $(\lambda(a), S^m)$, then the conditions are satisfied.

Proof. The proof is given in Appendix A.3.3. $\quad\square$

3.1.4. Best Operating Conditions

We first analyse the Monod case ($S^m = \infty$). We show in Figure 2 the set Γ of best operating conditions and we describe how to use this set to obtain practically the maximum of biogas production. Let S^{in} be fixed. The intersections of Γ and Λ with the horizontal line where S^{in} is kept constant define the values $D_G^*(S^{in})$, defined in Hypothesis 4, and $\delta(S^{in}) = \frac{\mu(S^{in})-a}{\alpha}$, defined by (15), see Figure 2a. The function $D \mapsto G_1(D, S^{in})$ is defined on $[0, \delta(S^{in})]$ and attains its maximum $G^*(S^{in})$ for $D = D_G^*(S^{in})$; see Figure 2b.

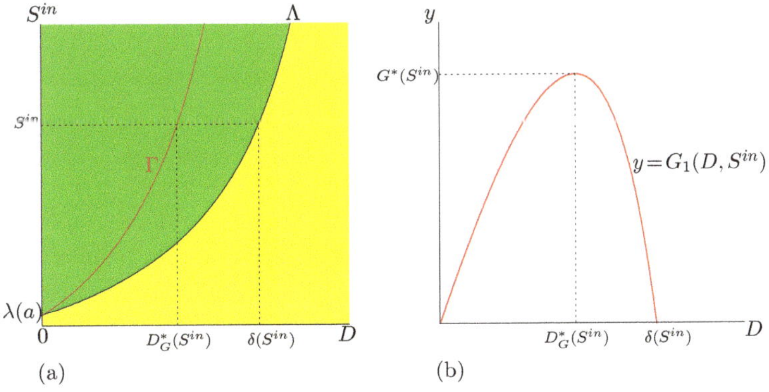

Figure 2. The best operating conditions of biogas flow rate for the Monod case. (**a**): The curve Γ in SOP shows the optimal value $D_G^*(S^{in})$. (**b**): The function $D \mapsto G_1(D, S^{in})$ is defined on $[0, \delta(S^{in})]$, and attains its maximum, $G^*(S^{in})$, for $D = D_G^*(S^{in})$.

In the Haldane case ($S^m < \infty$), the description is a little more complicated. If S^{in} is fixed, the function $D \mapsto G_1(D, S^{in})$ attains its maximum $G_1^*(S^{in})$ for $D = D_G^*(S^{in})$, obtained by taking the intersection of Γ with the horizontal line where S^{in} is kept constant, as it is seen in Figure 3. However, there exist two threshold values S^c and S^m, depicted in Figure 1b. If $S^{in} \leq S^m$, only G_1 is defined (see Figure 3a) while G_1 and G_2 are both defined when $S^{in} > S^m$ (see Figure 3b,c). On the other hand, if $S^{in} > S^c$, then the dilution rate $D_G^*(S^{in})$, which maximises biogas production, corresponds to the bistability mode of the chemostat; see Figure 3c. More precisely, we make the following remark.

Remark 7. *Assume that Hypotheses 1 and 4 hold. Let $D = D^c$ be the unique solution to equation $\gamma(D) = \bar{\lambda}(\alpha D + a)$. Let $S^c = \gamma(D^c)$.*

- *If $S^{in} < S^c$ then for the operating parameters S^{in} and $D = D_G^*(S^{in})$, F_1 is GAS.*
- *If $S^{in} > S^c$ then for the operating parameters S^{in} and $D = D_G^*(S^{in})$, F_0 and F_1 are both stable.*

Indeed, since γ is increasing and $\bar{\lambda}$ is decreasing, curves Γ and Λ_2 have a unique intersection point (D^c, S^c); see Figure 1b. The OD shows that if $S^{in} < S^c$ then $\left(D_G^*(S^{in}), S^{in}\right) \in \mathcal{J}_1$, that is to say, the best operating conditions are in the green region $\mathcal{J}_1$, where F_1 is GAS and if $S^{in} > S^c$, then $\left(D_G^*(S^{in}), S^{in}\right) \in \mathcal{J}_2$; that is to say, the best operating conditions are in the pink region $\mathcal{J}_2$ of bistability of F_0 and F_1.

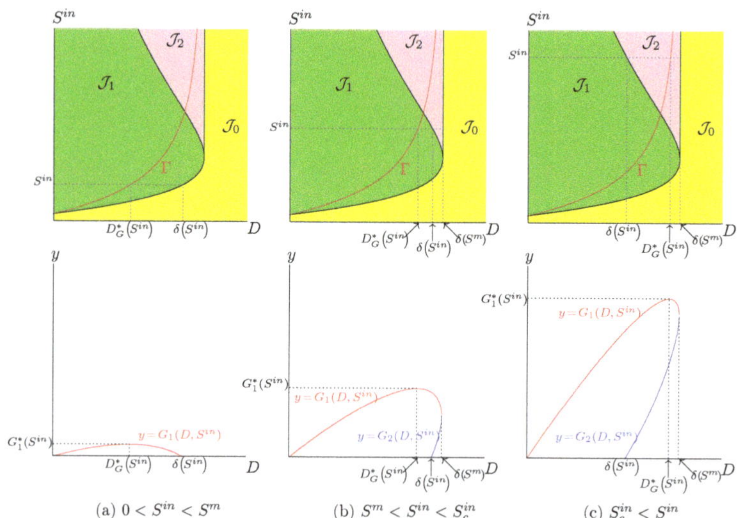

Figure 3. The set of best operating conditions Γ (in red) shows the optimal dilution rate $D_G^*(S^{in})$ corresponding to three typical values of S^{in}.

Figure 3 shows three typical values of S^{in} and the corresponding optimal dilution rates $D_G^*(S^{in})$. The corresponding biogas productions are depicted in the same figure. The main results are summarized as follows:

- If $S^{in} < S^m$, the biogas production $G_1(D, S^{in})$ is defined for $D \in [0, \delta(S^{in})]$; see Figure 3a.
- If $S^{in} > S^m$, the biogas production $G_1(D, S^{in})$ is defined for $D \in [0, \delta(S^m)]$, and the biogas production $G_2(D, S^{in})$ is defined for $D \in [\delta(S^{in}), \delta(S^m)]$; see Figure 3b,c.
- If $S^{in} < S^c$, and the chemostat is operated at the optimal dilution rate $D_G^*(S^{in})$, then the system converges towards the positive steady state F_1 giving the maximum of biogas; see Figure 3a,b.
- If $S^{in} > S^c$ and the chemostat is operated at the optimal dilution rate $D_G^*(S^{in})$, then, according to the initial condition, the system converges either to the positive steady state F_1, giving maximum biogas, or the washout steady state F_0, with no biogas production; see Figure 3c.

3.1.5. Best Operating Conditions for Biomass Production

Let P_1 be defined by (10) and S^{in} fixed. Our aim is to maximise the function $D \mapsto P_1(D, S^{in})$. Note that this function is proportional to the function $P : D \mapsto p(D)$ defined by

$$P(D) = \frac{D^2}{\alpha D + a}\left(S^{in} - \lambda(\alpha D + a)\right), \quad \text{for} \quad D \in I(S^{in}) \tag{23}$$

where $I(S^{in})$ is defined by (15). Therefore P_1 has an absolute maximum if P has one and this maximum is reached at the same point where P reaches its maximum. As in the case of the biogas flow rate, we consider the arguments of the maximum of P

$$p(S^{in}) = \underset{D \in I(S^{in})}{\operatorname{argmax}} p := \left\{ D^* \in I(S^{in}) : P(D) \leq P(D^*) \text{ for all } D \in I(S^{in}) \right\}. \tag{24}$$

To obtain the maximum value of $P(D)$, we differentiate (23) with respect to D, and we solve the equation $P'(D) = 0$. The derivative of P is given by

$$P'(D) = \frac{D(\alpha D + 2a)}{(\alpha D + a)^2}\left(S^{in} - \pi(D)\right)$$

where π is defined by

$$\pi(D) = \lambda(\alpha D + a) + \frac{\alpha D(\alpha D + a)}{\alpha D + 2a}\lambda'(\alpha D + a). \tag{25}$$

Remark 8. *Using* $\lambda'(D) = 1/\mu'(\lambda(D))$, *the function* π *can be written*

$$\pi(D) = \lambda(\alpha D + a) + \frac{\alpha D(\alpha D + a)}{(\alpha D + 2a)\mu'(\lambda(\alpha D + a))}.$$

We have the following result

Proposition 3. *Let* $D^* \in p(S^{in})$. *We have* $S^{in} = \pi(D^*)$, *where* π *is defined by* (25).

Proof. The proof is given in Appendix A.4.1. $\square$

Therefore, the curve

$$\Pi = \left\{ (D, S^{in}) : S^{in} = \pi(D) \right\} \tag{26}$$

of SOP contains the operating conditions for which P_1 is maximal. In Figure 4, this set is shown in the OD depicted in Figure 1, together with the set Γ. Note that if $a > 0$, then

$$\lambda(\alpha D + a) < \pi(D) < \gamma(D). \tag{27}$$

Therefore, curve Π is above curve Λ and below curve Γ; see Figure 4.

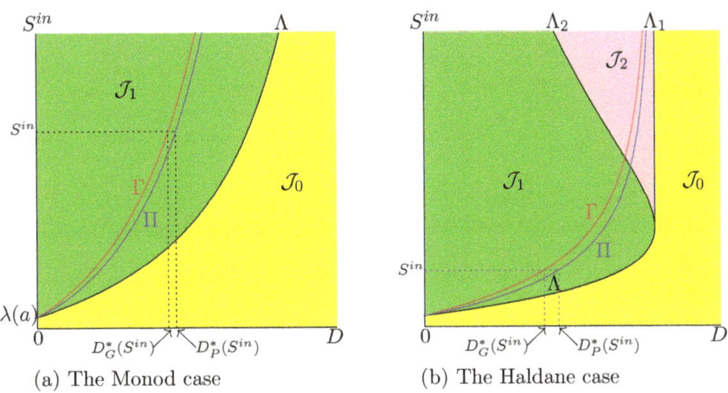

(a) The Monod case   (b) The Haldane case

Figure 4. The curves Γ (in red) and Π (in blue).

3.1.6. How to Determine the Maximum of Biomass Production?

From Proposition 3, to obtain $p(S^{in})$ we must solve the equation $S^{in} = \pi(D)$, which can be difficult to solve. We have at our disposal another description of $p(S^{in})$. We can write

$$P(D) = \frac{1}{\alpha^2}Q(\lambda(\alpha D + a)), \tag{28}$$

where Q is defined by

$$Q(S) = \frac{(\mu(S) - a)^2}{\mu(S)}(S^{in} - S), \quad \text{for} \quad \lambda(a) \leq S \leq S^{in} \tag{29}$$

From (28), it is deduced that the absolute maximum of P corresponds to the absolute maximum of Q and vice versa. To obtain the maximum value of $Q(S)$, we differentiate Q with respect to S, and we solve the equation $Q'(S) = 0$. The derivative of Q is given by

$$Q'(S) = \frac{(\mu(S) - a)(\mu(S) + a)\mu'(S)}{(\mu(S))^2}(S^{in} - S) - \frac{(\mu(S) - a)^2}{\mu(S)}.$$

Hence, $Q'(S) = 0$ if and only if $S^{in} = \rho(S)$, where $\rho(S)$ is defined by

$$\rho(S) = S + \frac{(\mu(S) - a)\mu(S)}{(\mu(S) + a)\mu'(S)}, \quad \text{for} \quad S \geq \lambda(a). \tag{30}$$

More precisely, we have the following result.

Proposition 4. *Let S^* be the maximum of p on $(\lambda(a), S^{in})$. Then $D^* = \frac{\mu(S^*) - a}{\alpha}$. We have $D^* \in p(S^{in})$. Moreover, we have $S^{in} = \rho(S^*)$, where ρ is defined by* (30).

Proof. The proof is given in Appendix A.4.2. $\square$

Remark 9. *With the first method we must first solve the equation $\mu(S) = D$ to obtain $\lambda(D)$, and then solve the equation $\pi(D) = S^{in}$ to obtain the optimal $D^* \in p(S^{in})$. With the second method, we simply solve the equation $\rho(S) = S^{in}$ to get the maximum S^* and then take $D^* = \frac{\mu(S^*) - a}{\alpha} \in p(S^{in})$.*

3.1.7. Uniqueness of the Maximum

Hypothesis 1 is not enough to guarantee that the biomass productivity admits a unique global maximum; see Appendix A.5.5. We make the following hypothesis.

Hypothesis 5. *For all $S^{in} > 0$, $p(S^{in})$, defined by* (24)*, has a unique element, which is denoted by $D_P^*(S^{in})$.*

From Proposition 3, we obtain the answer to Problem 2: Assume that Hypotheses 1 and 5 hold. Then, the set of best operating conditions for the productivity of (2) is the curve Π of SOP defined by:

$$\Pi = \left\{ (D, S^{in}) : S^{in} = \pi(D) \right\} = \left\{ (D, S^{in}) : D = D_P^*(S^{in}) \right\}. \tag{31}$$

From (27), we deduce that if $a > 0$, then $D_G^*(S^{in}) < D_P^*(S^{in})$; see Figure 4.

From Propositions 3 and 4, it is deduced that the uniqueness of $D_P^*(S^{in})$ is guaranteed when the equations

$$S^{in} = \pi(D) \quad \text{or} \quad S^{in} = \rho(S)$$

have a unique solution. A sufficient condition for this is that the functions $\pi(D)$ and $\rho(S)$ are increasing. The following result gives sufficient conditions for Hypothesis 5 to be valid.

Lemma 2. *Assume that Hypothesis 1 is satisfied and, in addition, μ is C^2. The following conditions are equivalent*

1. $\pi' > 0$ on $\left(0, \frac{\mu(S^m) - a}{\alpha} \right)$.
2. $\frac{(\mu - a)(\mu + a)}{\mu + 2a}\mu'' < 2(\mu')^2$ on $(\lambda(a), S^m)$.
3. $\rho' > 0$ on $(\lambda(a), S^m)$.

If these equivalent conditions are satisfied, then Hypothesis 5 is satisfied. If $\mu'' < 0$ on $(\lambda(a), S^m)$ or $\left(\frac{1}{\mu - a} \right)'' > 0$ on $(\lambda(a), S^m)$, then the conditions are satisfied.

Proof. The proof is given in Appendix A.4.3. $\square$

3.1.8. The Case without Mortality

The functions γ, H, and η, defined by (17), (20), and (21), respectively, that were used for the optimization of the biogas flow rate G are summarized in Table 3. Note that the functions G and H are related by formula (19). Similarly, the functions π, Q, and ρ, defined by (25), (29) and (30), respectively, which were used for the optimization of the productivity P, are summarized in Table 3. Note that the functions P and Q are related by formula (28).

Table 3. The functions γ, H and η used for the optimization of the biogas flow rate G. The functions π, Q and ρ used for the optimization of the productivity P. Note that $G(D) = \frac{1}{\alpha} H(\lambda(\alpha D + a))$ and $P(D) = \frac{1}{\alpha} Q(\lambda(\alpha D + a))$.

Biogas Production	**Biomass Productivity**
$G(D) = D(S^{in} - \lambda(\alpha D + a))$	$P(D) = \frac{D^2}{\alpha D + a}(S^{in} - \lambda(\alpha D + a))$
$\gamma(D) = \lambda(\alpha D + a) + \alpha D \lambda'(\alpha D + a)$	$\pi(D) = \lambda(\alpha D + a) + \frac{\alpha D(\alpha D + a)}{\alpha D + 2a} \lambda'(\alpha D + a)$
$H(S) = (\mu(S) - a)(S^{in} - S)$	$Q(S) = \frac{(\mu(S) - a)^2}{\mu(S)}(S^{in} - S)$
$\eta(S) = S + \frac{\mu(S) - a}{\mu'(S)}$	$\rho(S) = S + \frac{(\mu(S) - a)\mu(S)}{(\mu(S) + a)\mu'(S)}$

Biogas Production = Biomass Productivity ($a = 0$)	
$G(D) = \alpha P(D) = D(S^{in} - \lambda(\alpha D))$	
$\gamma(D) = \pi(D) = \lambda(\alpha D) + \alpha D \lambda'(\alpha D)$	
$H(S) = Q(S) = \mu(S)(S^{in} - S)$	
$\eta(S) = \rho(S) = S + \frac{\mu(S)}{\mu'(S)}$	

Table 3 shows that in the case without mortality, one has $G = \alpha P$, $\gamma = \pi$, $H = Q$, and $\eta = \rho$. Hence, if $a = 0$, we have $D_G^*(S^{in}) = D_P^*(S^{in})$. In the following, this value is referred to as $D^*(S^{in})$. Therefore, for the optimization of the biogas flow rate or the productivity of the biomass, a first method consists in solving the equation

$$\lambda(\alpha D) + \alpha D \lambda'(\alpha D) = S^{in}.$$

to obtain the optimal value of the dilution rate $D^*(S^{in})$. The second method consists in solving the equation

$$\eta(S) = S^{in}, \quad \text{where} \quad \eta(S) := S + \frac{\mu(S)}{\mu'(S)} \tag{32}$$

to get the maximum $S^*(S^{in})$ and then take $D^*(S^{in}) = \frac{1}{\alpha}\mu(S^*(S^{in}))$. Hence, without loss of generality, one can put $\alpha = 1$ and solve Equation (32) or equation

$$\gamma(D) = S^{in}, \quad \text{where} \quad \gamma(D) := \lambda(D) + D\lambda'(D). \tag{33}$$

The results of Lemmas 1 and 2 become the same in the case that $a = 0$. We summarize them below, in this special case.

Lemma 3. *Assume that Hypothesis 1 is satisfied and, in addition μ is C^2. The following conditions are equivalent*

1. $\gamma' > 0$ *on* $(0, \mu(S^m))$, *where γ is defined in* (33).
2. $\mu\mu'' < 2(\mu')^2$ *on* $(0, S^m)$.
3. $\left(\frac{1}{\mu}\right)'' > 0$ *on* $(0, S^m)$.
4. $\eta' > 0$ *on* $(0, S^m)$, *where η is defined in* (32).

If these equivalent conditions are satisfied, then each of Equations (32) *and* (33) *has a unique solution; i.e., Hypothesis 4 is satisfied. If $\mu'' < 0$ on $(0, S^m)$, then the conditions are satisfied.*

In Appendix A.5, we apply the preceding results to various growth functions that were considered in the literature.

3.2. Two-Step Models

The steady state and their stability of the two-step model (12) are given in Appendix B.2, and the OD is described in Appendix B.3.

3.2.1. Comparison of Biogas Flow Rates

Recall that E_{11} is stable whenever it exists; E_{01} can be stable but is unstable whenever E_{11} exists, and E_{02} and E_{12} are unstable whenever they exist. Is it possible that for some operating condition D, S_1^{in}, and S_2^{in}, the biogas production at an unstable steady state is greater than at a stable one? This possibility is excluded, as is stated in the following result.

Proposition 5.

- *For all operating conditions D and S_2^{in} where G_{02} is defined, then G_{01} is also defined, and $G_{01}\left(D, S_2^{in}\right) > G_{02}\left(D, S_2^{in}\right)$.*
- *For all operating conditions D, S_1^{in} and S_2^{in} where G_{12} is defined, then G_{11} is also defined, and $G_{11}\left(D, S_1^{in}, S_2^{in}\right) > G_{12}\left(D, S_1^{in}, S_2^{in}\right)$.*
- *For all operating conditions D, S_1^{in} and S_2^{in} where G_{01} and G_{11} are both defined, we have $G_{11}\left(D, S_1^{in}, S_2^{in}\right) > G_{01}\left(D, S_2^{in}\right)$.*

Proof. The proof is given in Appendix B.5.1. $\square$

This result shows that $G_{01} > G_{02}$ and $G_{11} > G_{12}$, which justifies Remark 3. Therefore, in Problem 3, we can restrict our attention to the maximisation of G_{01} and G_{11}. The result also shows that when E_{11} and E_{01} are both defined, then we have $G_{11} > G_{01}$. Table A6 shows that both E_{11} and E_{01} exist simultaneously only in regions $\mathcal{I}_6$, $\mathcal{I}_7$, and $\mathcal{I}_8$, and that in this case E_{11} is stable while E_{01} is unstable. However, it is possible for one to be defined without the other being defined, as shown in Table A6. Indeed, in the regions $\mathcal{I}_1$ and $\mathcal{I}_2$, E_{01} exists and is stable, while E_{11} does not exist and in the regions $\mathcal{I}_4$ and $\mathcal{I}_5$, E_{11} exists and is stable, while E_{01} does not exist. Therefore, the maximum of G_{11} and G_{01} can be obtained for different values of the dilution rate D, and the last part of Problem 3 is to fix S_1^{in} and S_2^{in} and compare

$$\max_{D} G_{01}\left(D, S_2^{in}\right) \quad \text{and} \quad \max_{D} G_{11}\left(D, S_1^{in}, S_2^{in}\right).$$

3.2.2. Best Operating Conditions for G_{01} and G_{11}

Let us fix the operating parameters S_1^{in} and S_2^{in}. We restrict our attention to the case $a_1 = a_2 = 0$ and $\alpha_1 = \alpha_2 = \alpha$, which was considered in [11]. The general case can be considered without added difficulty. Our aim is to compute the values of D for which the functions

$$D \mapsto G_{01}\left(D, S_2^{in}\right) \quad \text{and} \quad D \mapsto G_{11}\left(D, S_1^{in}, S_2^{in}\right)$$

reach their maxima. These functions are proportional to the functions

$$G_0(D) = D\left(S_2^{in} - \lambda_2(\alpha D)\right) \tag{34}$$

$$G_1(D) = D\left(S_2^{in} + \tfrac{k_2}{k_1} S_1^{in} - \lambda_2(\alpha D) - \tfrac{k_2}{k_1}\lambda_1(\alpha D)\right) \tag{35}$$

respectively, where λ_1 and λ_2 are defined in Table 1. Therefore, G_{01} has an absolute maximum if G_0 has one, and this maximum is reached at the same point where G_0 reaches its maximum. Similarly, G_{11} has an absolute maximum if G_1 has one, and this maximum is reached at the same point where G_1 reaches its maximum. To obtain the maximum of G_0, we differentiate G_0 with respect to D. The derivative is given by

$$G_0'(D) = S_2^{in} - \gamma_2(\alpha D)$$

where γ_2 is defined by

$$\gamma_2(D) = \lambda_2(D) + D\lambda_2'(D). \tag{36}$$

Similarly, the derivative of G_1 is given by

$$G_1'(D) = S_2^{in} - \gamma_2(\alpha D) + \tfrac{k_2}{k_1}\left(S_1^{in} - \gamma_1(\alpha D)\right)$$

where γ_1 is defined by

$$\gamma_1(D) = \lambda_1(D) + D\lambda_1'(D). \tag{37}$$

Remark 10. *Using $\lambda_1'(D) = 1/\mu_1'(\lambda_1(D))$ and $\lambda_2'(D) = 1/\mu_2'(\lambda_2(D))$, the functions γ_2 and γ_1 can be written*

$$\gamma_2(D) = \lambda_2(D) + \frac{D}{\mu_2'(\lambda_2(D))}, \quad \gamma_1(D) = \lambda_1(D) + \frac{D}{\mu_1'(\lambda_1(D))}.$$

We make the following assumptions:

Hypothesis 6. *The function $\gamma_2 : I_2 \to (0 + \infty)$, defined on $I_2 = (0, \mu_2(S_2^m))$ by (36), is C^1, and for all $D \in I_2$ we have $\gamma_2'(D) > 0$.*

Hypothesis 7. *The function $\mu_1 : I_1 \to (0 + \infty)$, defined on $I_1 = (0, m_1)$ by (37), is C^1 and for all $D \in I_1$ we have $\gamma_1'(D) > 0$.*

If Hypothesis 6 is satisfied, then the function γ_2 is invertible, and for each S_2^{in}, the equation

$$S_2^{in} = \gamma_2(\alpha D) \tag{38}$$

has a unique solution, denoted

$$D_0^*(S_2^{in}) = \tfrac{1}{\alpha}\gamma_2^{-1}(S_2^{in}), \tag{39}$$

where γ_2^{-1} is the inverse function of γ_2. On the other hand, if Hypotheses 6 and 7 are satisfied, the function $\gamma_2 + \frac{k_2}{k_1}\gamma_1$ is C^1 and increasing, since it is the sum of two increasing functions. Therefore, for each S_1^{in} and S_2^{in}, the equation

$$S_2^{in} + \tfrac{k_2}{k_1}S_1^{in} = \gamma_2(\alpha D) + \tfrac{k_2}{k_1}\gamma_1(\alpha D) \tag{40}$$

has a unique solution, denoted

$$D_1^*(S_1^{in}, S_2^{in}) = \tfrac{1}{\alpha}\gamma^{-1}\left(S_2^{in} + \tfrac{k_2}{k_1}S_1^{in}\right), \tag{41}$$

where γ^{-1} is the inverse function of $\gamma := \gamma_2 + \frac{k_2}{k_1}\gamma_1$.

The following result gives the answer to the first part of Problem 3.

Proposition 6. *Assume that Hypotheses 2, 3, 6, and 7 are satisfied. Then $G_{01}(D, S_2^{in})$ reaches its maximum at $D_0^*(S_2^{in})$, defined by (39) and $G_{11}(D, S_2^{in}, S_2^{in})$ reaches its maximum at the right-hand end of its defining interval, or at $D_1^*(S_1^{in}, S_2^{in})$, defined by (41).*

Proof. The proof is given in Appendix B.5.2. $\square$

The set of best operating conditions for biogas production at E_{01} is the surface Γ_0 of SOP, defined by:

$$\Gamma_0 = \left\{(D, S_1^{in}, S_2^{in}) : S_2^{in} = \gamma_2(\alpha D)\right\} = \left\{(D, S_1^{in}, S_2^{in}) : D = D_0^*(S_2^{in})\right\} \tag{42}$$

It is the set of operating conditions that produce the maximum of G_{01}. The set of best operating conditions for biogas production at E_{11} is the surface Γ_1 of SOP, defined by:

$$\Gamma_1 = \left\{(D, S_1^{in}, S_2^{in}) : S_2^{in} + \tfrac{k_2}{k_1}S_1^{in} = \gamma(\alpha D)\right\} = \left\{(D, S_1^{in}, S_2^{in}) : D = D_1^*(S_1^{in}, S_2^{in})\right\} \tag{43}$$

This is the set of operating conditions which produce the maximum of G_{11}.

We plot the sets Γ_0 and Γ_1 in the 2-dimensional ODs in the (D, S_1^{in})-plane shown in Figure 5. Since S_2^{in} is fixed, the set Γ_0, in blue in the figures, is the vertical line $D = D_0^*(S_2^{in})$, while Γ_1, in red in the figures, is the curve of equation $S_1^{in} = \frac{k_1}{k_2}\left(\gamma(\alpha D) - S_2^{in}\right)$. Let S_1^{in} and S_2^{in} be fixed. Consider the OD for which S_2^{in} is equal to the fixed value considered and look for the intersections of Γ_0 and Γ_1 with the horizontal line where S_1^{in} is kept constant at the fixed value considered. The abscissas of these intersections are the optimal dilution rates $D_0^*\left(S_2^{in}\right)$ and $D_1^*\left(S_2^{in}\right)$ defined by (39) and (41), respectively.

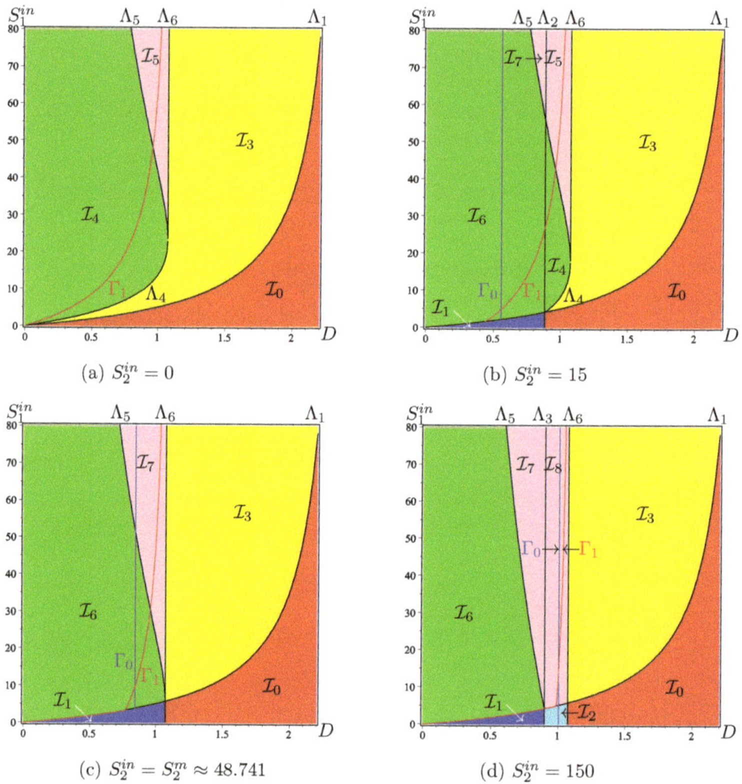

Figure 5. The 2-dimensional OD in $\left(D, S_1^{in}\right)$, obtained by cuts at S_2^{in} constant of the 3-dimensional OD shown in Figure A6. The curve Γ_1, in red, is the set of maximisation of G_{11}. The vertical line Γ_0, in blue, is the set of maximisation of G_{01}.

Remark 11. *As for the one-step model with a Haldane type growth function, shown in Figure 1b, there exists a threshold value S_1^c corresponding to the intersection point $\left(D^c, S_1^c\right)$ of curves Γ_1 and Λ_5, such that, if $S_1^{in} > S_1^c$, then the best operating point lies in the bistability pink region; see Figure 6a. The value $D = D^c$ is the solution of equation*

$$\bar{\lambda}_2(\alpha D) = \lambda_2(\alpha D) + \alpha D \lambda_2'(\alpha D) + \frac{k_2}{k_1}\alpha D \lambda_1'(\alpha D), \tag{44}$$

which gives the abscissa of the point of intersection of Γ_1 and Λ_5, and S_1^c is given by

$$S_1^c = \lambda_1(\alpha D^c) + \frac{k_1}{k_2}\left(\bar{\lambda}_2(\alpha D^c) - S_2^{in}\right).$$

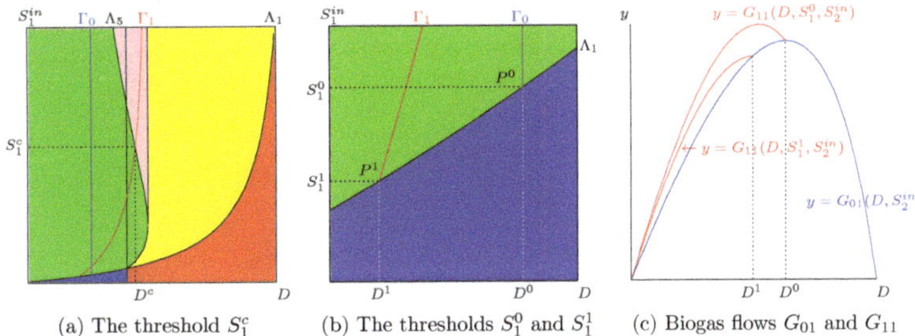

(a) The threshold S_1^c (b) The thresholds S_1^0 and S_1^1 (c) Biogas flows G_{01} and G_{11}

Figure 6. (**a**): The point $(D^c, S_1^c) = \Gamma_1 \cap \Lambda_5$. (**b**): A zoom showing the points $P^0 = \Gamma_0 \cap \Lambda_1$ and $P^1 = \Gamma_1 \cap \Lambda_1$. (**c**): The function $D \mapsto G_{01}(D, S_2^{in})$ in blue and the functions $D \mapsto G_{11}(D, S_1^{in}, S_2^{in})$, in red, for $S_1^{in} = S_1^0$ and $S_1^{in} = S_1^1$.

3.2.3. The Maximum of G_{01} Can Be Larger than the Maximum of G_{11}

In addition to the threshold S_1^c, Figures 5 and 6 show two other thresholds obtained by considering the intersection of the Γ_0 and Γ_1 curves with the Λ_1 curve. We depict in Figure 6 a typical situation and show in a zoom the points of intersection. Let $P^0 = (D^0, S_1^0)$ be the point of intersection of Γ_0 with Λ_1; see Figure 6b. If $S_1^{in} = S_1^0$, then the productivity G_{11} is defined for $0 \le D \le D^0$ and reaches its maximum for some $D_1^*(S_1^0, S_2^{in}) < D^0$. Moreover, we have

$$\max_D G_{01}(D, S_2^{in}) = G_{01}(D^0, S_2^{in}) = G_{11}(D^0, S_1^0, S_2^{in}).$$

Therefore, see Figure 6c, we have

$$\max_D G_{11}(D, S_1^0, S_2^{in}) > \max_D G_{01}(D, S_2^{in}).$$

Since the function $S_1^{in} \mapsto G_{11}(D, S_1^{in}, S_2^{in})$ is increasing, the same result is true for any $S_1^{in} > S_1^0$. Note that S_1^0 depends on S_2^{in} and is a solution of the set of equations

$$S_1^{in} = \lambda_1(\alpha D), \qquad S_2^{in} = \gamma_2(\alpha D)$$

which give the point of intersection of Λ_1 and Γ_0. Therefore, (S_1^0, S_2^{in}) belongs to the curve

$$\Sigma_0 = \left\{ (S_1^{in}, S_2^{in}) : S_2^{in} = \gamma_2\big(\mu_1(S_1^{in})\big) \right\}. \tag{45}$$

Similarly, let $P^1 = (D^1, S_1^1)$ be the point of intersection of the curves Γ_1 and Λ_1; see Figure 6b. If $S_1^{in} = S_1^1$ then the productivity G_{11} is defined for $0 \le D \le D^1$ and reaches its maximum for $D = D^1$. Since $D^1 < D^0$, we have (see Figure 6c),

$$G_{11}(D^1, S_1^1, S_2^{in}) = G_{01}(D^1, S_2^{in}) < G_{01}(D^0, S_2^{in}).$$

Therefore,

$$\max_D G_{11}(D, S_1^1, S_2^{in}) < \max_D G_{01}(D, S_2^{in}).$$

The same result is true for any $S_1^{in} < S_1^1$, because the function $S_1^{in} \mapsto G_{11}(D, S_1^{in}, S_2^{in})$ is increasing. Note that S_1^1 depends on S_2^{in} and is a solution to the set of equations

$$S_1^{in} = \lambda_1(\alpha D), \qquad S_2^{in} + \frac{k_2}{k_1} S_1^{in} = \gamma_2(\alpha D) + \frac{k_2}{k_1}\gamma_1(\alpha D),$$

which give the point of intersection of Λ_1 and Γ_1. Therefore (S_1^1, S_2^{in}) belongs to the curve

$$\Sigma_1 = \left\{ (S_1^{in}, S_2^{in}) : S_2^{in} = \sigma_1(S_1^{in}) \right\}, \quad \text{where} \quad \sigma_1(S_1^{in}) = \gamma_2\big(\mu_1(S_1^{in})\big) + \frac{k_2}{k_1}\frac{\mu_1(S_1^{in})}{\mu_1'(S_1^{in})}. \tag{46}$$

The curves Σ_0 and Σ_1 are illustrated in Figure 7b. We have the following result.

Proposition 7. *Let Σ_0 and Σ_1 be the curves of the (S_1^{in}, S_2^{in}) plane defined by (45) and (46), respectively. If (S_1^{in}, S_2^{in}) is at the right of Σ_0, then we have*

$$\max_D G_{11}(D, S_1^1, S_2^{in}) > \max_D G_{01}(D, S_2^{in}).$$

If the function μ_1 / μ_1' is increasing and (S_1^{in}, S_2^{in}) is at the left of Σ_1, then we have

$$\max_D G_{11}(D, S_1^1, S_2^{in}) < \max_D G_{01}(D, S_2^{in}).$$

Proof. The proof is given is Appendix B.5.3. $\square$

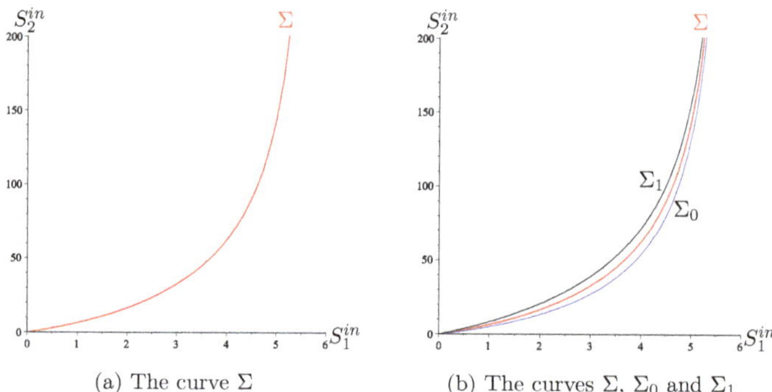

(a) The curve Σ

(b) The curves Σ, Σ_0 and Σ_1

Figure 7. To the left of the curve Σ we have $\max_D G_{01} > \max_D G_{11}$ and to its right we have $\max_D G_{01} < \max_D G_{11}$.

Now, we give the curve Σ lying between the Σ_0 and Σ_1 curves, such that the maximum of biogas flow rate is obtained for E_{01} at the left of Σ and for E_{11} at the right of Σ; see Figure 7a. We need the following hypothesis.

Hypothesis 8. *We assume that the function ϕ defined by $\phi(D) = D^2 \lambda_2'(D)$ is increasing.*

Therefore, ϕ has an inverse function ϕ^{-1} defined by $D = \phi^{-1}(B)$ if and only if D is the solution to equation $\phi(D) = B$. Consider the curve Σ defined by the parametric equations

$$S_2^{in} = \gamma_2(\Delta(D)), \quad S_1^{in} = \gamma_1(\alpha D) + \tfrac{k_1}{k_2}(\gamma_2(\alpha D) - \gamma_2(\Delta(D))) \tag{47}$$

where $\Delta(D)$ is defined by

$$\Delta(D) := \phi^{-1}\left(\alpha^2 D^2 \left(\lambda_2'(\alpha D) + \tfrac{k_2}{k_1}\lambda_1'(\alpha D)\right)\right). \tag{48}$$

The following result gives the answer to the second part of Problem 3.

Proposition 8. *Assume that Hypothesis 8 is satisfied and, in addition, the curve C defined by the parametric Equation (47) is the graph of an increasing function $S_2^{in} \mapsto S_1^{in}$. Then it is the subset of the (S_1^{in}, S_2^{in}) plane, where*

$$\max_D G_{01}(D, S_2^{in}) = \max_D G_{11}(D, S_1^{in}, S_2^{in}). \tag{49}$$

To the left of C, we have $\max_D G_{01} > \max_D G_{11}$ and to its right, we have $\max_D G_{01} < \max_D G_{11}$.

Proof. The proof is given in Appendix B.5.4. □

Remark 12. *By combining the result of Remark 11 with that of Proposition 8, we deduce that the curve Σ and the straight line $\mathcal{C}$ defined by*

$$\mathcal{C} := \left\{ (S_1^{in}, S_2^{in}) : S_2^{in} + \tfrac{k_2}{k_1} S_1^{in} < \gamma_2(\alpha D^c) + \tfrac{k_2}{k_1}\gamma_1(\alpha D^c) \right\},$$

where $D = D^c$ is the solution of Equation (44), divide the plane (S_1^{in}, S_2^{in}) into three regions:

$$\mathcal{R}_0 := \left\{ (S_1^{in}, S_2^{in}) \text{ lies to the left of } \Sigma \right\}$$
$$\mathcal{R}_1 := \left\{ (S_1^{in}, S_2^{in}) \text{ lies to the right of } \Sigma \text{ and to the left of } \mathcal{C} \right\}$$
$$\mathcal{R}_2 := \left\{ (S_1^{in}, S_2^{in}) \text{ lies to the right of } \Sigma \text{ and } \mathcal{C} \right\}.$$

In the region $\mathcal{R}_0$, we have $\max_D G_{10} > \max_D G_{11}$. In the region $\mathcal{R}_1$, we have $\max_D G_{10} < \max_D G_{11}$, and the optimal dilution rate corresponds to the global asymptotic stability of E_{11}. In the region $\mathcal{R}_2$, we also have $\max_D G_{10} < \max_D G_{11}$, but the optimal dilution rate corresponds to the bistability of E_{11} and E_{10}.

Since the steady state E_{10} does not produce biogas, if the bioreactor is operated in the $\mathcal{R}_2$ region, care should be taken to initialise it in the basin of attraction of E_{11} and not in the basin of E_{10}. The regions are illustrated in Figure 8a, obtained with the parameter values given in Table A8. Let us illustrate the behaviour of $G_{01}(D, S_2^{in})$ and $G_{11}(D, S_1^{in}, S_2^{in})$, as functions of D, for the operating points $o_k \in \mathcal{R}_k$, $k = 0, 1, 2$, shown in Figure 8a. Figure 8b shows the OD in the (D, S_1^{in}) plane and $S_2^{in} = 15$. The horizontal lines $S_1^{in} = 1.5, 10$, and 50, corresponding to the points $o_0 = (1.5, 15)$, $o_1 = (10, 15)$, and $o_2 = (50, 15)$, respectively, give the optimal dilution rates. For o_0, the maximum of the biogas flow is obtained for E_{01}; see Figure 8c. For o_1, the maximum of the biogas flow is obtained for E_{11}, and E_{11} is GAS; see Figure 8d. For o_2, the maximum of the biogas flow is obtained for E_{11}, but E_{11} is only LAS; see Figure 8e.

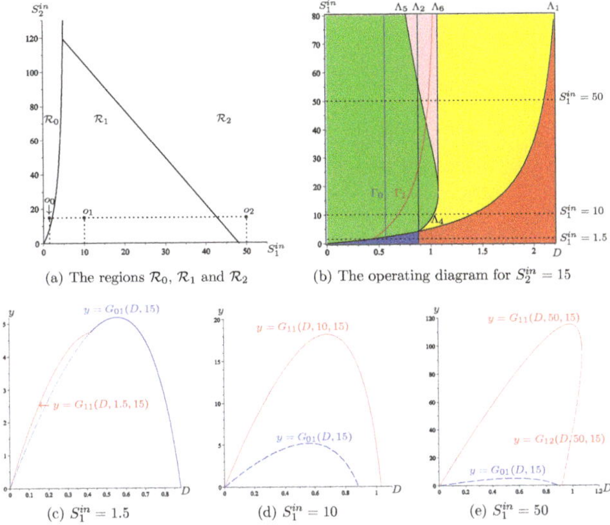

(a) The regions $\mathcal{R}_0$, $\mathcal{R}_1$ and $\mathcal{R}_2$ (b) The operating diagram for $S_2^{in} = 15$

(c) $S_1^{in} = 1.5$ (d) $S_1^{in} = 10$ (e) $S_1^{in} = 50$

Figure 8. The biogas flow rates $D \mapsto G_{01}(D, S_2^{in})$ in blue, $D \mapsto G_{11}(D, S_1^{in}, S_2^{in})$ in red, and $D \mapsto G_{12}(D, S_1^{in}, S_2^{in})$ in dashed red, corresponding to the operating points (**c**) $o_0 = (1.5, 15)$, (**d**) $o_1 = (10, 15)$ and (**e**) $o_2 = (50, 15)$. The flow rate biogas of a stable steady state is drawn in bold, while it is drawn in dashed line when the steady state is unstable.

3.2.4. Applications to the Classical AM2 Model

The dynamical equations of the model are

$$
\begin{aligned}
\dot{S}_1 &= D\big(S_1^{in} - S_1\big) - k_1\mu_1(S_1)X_1, \\
\dot{X}_1 &= \big(\mu_1(S_1) - \alpha D\big)X_1, \\
\dot{S}_2 &= D\big(S_2^{in} - S_2\big) + k_2\mu_1(S_1)X_1 - k_3\mu_2(S_2)X_2, \\
\dot{X}_2 &= \big(\mu_2(S_2) - \alpha D\big)X_2,
\end{aligned}
\tag{50}
$$

where the kinetics μ_1 and μ_2 are given by

$$
\mu_1(S_1) = \frac{m_1 S_1}{K_1 + S_1}, \qquad
\mu_2(S_2) = \frac{m_2 S_2}{K_2 + S_2 + S_2^2/K_i},
\tag{51}
$$

For the Monod and Haldane functions, Hypotheses 2 and 3 are satisfied and the break-even concentrations can be calculated explicitly. For the convenience of the reader we summarize in Table 4 the expressions of the break even concentrations and the auxiliary functions that are needed in the description of the results. The OD in the three dimensional SOP, corresponding to the biological value parameters given in Table A8 is shown in Figure A6 of the Appendix. The two-dimensional diagrams in the (D, S_1^{in}) plane, where S_2^{in} is kept constant, are depicted in Figure A7. The two-dimensional diagrams in the (S_1^{in}, S_2^{in}) plane, where D is kept constant, are depicted in Figure A8.

Table 4. Auxiliary function in the classical AM2 model.

$\mu_1(S_1) = \frac{m_1 S_1}{K_1 + S_1}$	
$\lambda_1(D) = \frac{DK_1}{m_1 - D}$,	Defined for $0 \le D < m_1 = \mu_1(+\infty)$
$\mu_2(S_2) = \frac{m_2 S_2}{K_2 + S_2 + S_2^2/K_i}$,	$S_2^m = \sqrt{K_2 K_i},\ \mu_2(S_2^m) = \frac{m_2}{1 + 2\sqrt{K_2/K_i}}$
$\lambda_2(D) = \frac{(m_2 - D) - \sqrt{(m_2 - D) - 4D^2 K_2/K_i}}{2D} K_i$,	Defined for $0 < D \le \mu_2(S_2^m)$
$\bar{\lambda}_2(D) = \frac{(m_2 - D) + \sqrt{(m_2 - D)^2 - 4D^2 K_2/K_i}}{2D} K_i$	Defined for $0 < D \le \mu_2(S_2^m)$
$\gamma_1(D) = \lambda_1(D) + D\lambda_1'(D)$,	Defined for $D < m_1$
$\gamma_2(D) = \lambda_2(D) + D\lambda_2'(D)$,	Defined for $D < \mu_2(S_2^m)$
$\gamma(D) = \gamma_2(D) + \frac{k_2}{k_1}\gamma_1(D)$	Defined for $D < \min(m_1, \mu_2(S_2^m))$

Since $\mu_1'' < 0$ and $\mu_2'' < 0$ on $(0, S_2^m)$, from Lemma 3 we deduce that $\gamma_1' > 0$ and $\gamma_2' > 0$. Therefore Hypotheses 6 and 7 are satisfied. From Proposition 6, we deduce that the curves Γ_0 and Γ_1, defined by (42) and (43) are the sets of best operating conditions for G_{01} and G_{11}, respectively. These sets are shown in Figure 5, for some of the ODs depicted in Figure A7.

On the other hand, since $\lambda_2'' > 0$, we deduce that $\phi' > 0$, where $\phi(D) = D^2\lambda_2'(D)$. Hence Hypothesis 8 is satisfied. The inverse function of ϕ can be computed explicitly. We have

$$
\phi^{-1}(B) = m_2 \frac{(m_2 K_i + 2B)\sqrt{BK_2 K_i(m_2 K_i + B)} - (m_2 K_i + B)K_i B}{K_2 m_2^2 K_i^2 + (4K_2 - K_i)(m_2 K_i + B)B}
$$

Note that the function μ_1/μ_1' is increasing. Therefore, the result of Proposition 7 is true. Straightforward computation shows that the curve Σ is increasing. Hence, the result of Proposition 8 is true. The curve Σ of the (S_1^{in}, S_2^{in})-plane,

$$
\max_D G_{01}(D, S_2^{in}) = \max_D G_{11}(D, S_1^{in}, S_2^{in}),
$$

and the curves Σ_0 and Σ_1 are shown in Figure 7. Finally the regions $\mathcal{R}_0$, $\mathcal{R}_1$, and $\mathcal{R}_2$ and the behaviour of the biogas flow rates $D \mapsto G_{01}(D, S_2^{in})$ and $D \mapsto G_{11}(D, S_1^{in}, S_2^{in})$ are depicted in Figure 8 for three operating points $o_j \in \mathcal{R}_j, j = 0, 1, 2$.

3.3. Relationship with Previous Results

The OD of the one-step model is well known in the existing literature [22,36]. In these references, the dilution rate is shown on the vertical axis, and the input substrate concentration is shown on the horizontal axis. In this paper, we have reversed the axes, because, as we then consider the biogas flow rate, or productivity, as a function of the dilution rate, it is interesting to have the dilution rate on the horizontal axis in all graphs.

In practical applications, when maximising biogas or biomass production, the substrate concentration S^{in} is given and the optimal dilution rate $D^*(S^{in})$, depending on S^{in}, that maximises biogas or biomass production must be determined. For the Monod function, the formula giving the optimal dilution appears in several reference books; see for example Formula (13.70) in [12] or Formula (6.83) in [19]. For the Monod and Haldane functions, it appears in [20,21] and were used for the optimization of bioreactors by extremum seeking. The approach used here is to try to directly exploit the equation of which the optimal D is a solution and to represent its solutions in the OD. To the best of our knowledge, the set of best operating conditions for biogas or bimass production have only recently been drawn in the OD [51–53]. In these papers the main problem is to consider the optimisation of biogas flow rate or biomass productivity in the serial chemostat and to compare the performances of the serial chemostat with a single chemostat of the same total volume.

In the case without biomass mortality, the mathematical analysis of the two-step model was given in [15], in the case $\alpha = 1$, and in [26] in the case $\alpha \leq 1$. The OD was given in [42]. Here we have extended these results to the case including mortality. The maximization of biogas flow for this model has been well studied in [11]. For example, the curves Σ_0 and Σ_1 were described(see Figure 4 in [11]), where the curves are called C_2 and C_3, respectively. The existence of the curve Σ was predicted; see Remark 7 in [11]. However, neither its analytical equation nor its numerical representation was given in [11]. Note that the curves Σ_0 and Σ_1 have vertical asymptotes; see Figure 7. We deduce that the curve Σ also has one. Therefore, the region $\mathcal{R}_0$ is not wide. That is to say, for an S_2^{in} as large as one wants, it is enough that S_1^{in} exceeds a certain threshold, corresponding to the vertical asymptote, for the system to be in the $\mathcal{R}_1$ or $\mathcal{R}_2$ region.

The representation of the set of optimal operating conditions in the OD, as well as its use to deduce the various properties of biogas production, is not found in the existing literature. In particular, the identification of the threshold at which the system will operate in a bistability regime is new and answers practical questions of great interest for bioreactors and their management. These questions are related to the so-called stability criteria named "overloading tolerance" or "destabilization risk index" [26,56]. This index alerts the experimenter as soon as the system approaches a regime of bistability. Bistability in the model occurs when the unstable steady states E_{02} or E_{12} exist. For example, although the steady state E_{12} is unstable, if it exists, its existence completely changes the functioning of the system. Indeed, in this case, the steady state E_{10}, of washout of the methanogenic bacteria (without biogas production), becomes stable, and the positive steady state E_{11} loses its global stability. This important issue is not addressed in [11], where the authors do not consider the steady states E_{02} and E_{12}. They justify their disregard by the fact that these steady states are unstable, that their biogas flow rate is lower than the biogas flow rate of the associated steady states E_{01} and E_{11}, and also because according to them their conditions of existence are the same as those of the steady states E_{01} and E_{11}; see Section 3 in [11]. The first two reasons are of course correct, but the third is not. Indeed, E_{11} can exist without E_{12} existing. On the other hand, when E_{12} exists, E_{11} must also exist, and we have the phenomenon of bistability of E_{10} and E_{11}. In this paper, we considered all steady states, which allowed us to highlight the important region of bistability (coloured in pink in the figures) and thus to provide a valuable tool for the experimenter to avoid monitoring the system in this region, or at least to be very careful if he should do so.

4. Discussion

In this paper, we have determined the set of operating parameters that optimise the biogas flow in simple AD models. We have represented these sets in the OD of the model. This representation allowed us to obtain a simple graphic visualisation of the optimal operating conditions. It also allows direct discovery of the properties of these optimal conditions.

To illustrate the simplicity with which the properties appear in the OD, let us consider the case with inhibition by the substrate when its concentration is high (Haldane function). It is well known that when the inflowing substrate concentration of the bioreactor is high, the system presents bistability, with a risk of convergence towards the washout steady state. It is natural then to ask whether operating conditions that maximise the biogas flow can lead to this bistability situation. This phenomenon was already observed, using the OD, in a more complex system [35]. The main result of this paper is to address this problem and to give a complete answer in one-step and two-steps models. Although we have an explicit formula for the optimal dilution rate as a function of the substrate input concentration, this formula does not allow us to easily determine whether or not the system is in the instability zone. On the other hand, drawing the set of optimal conditions in the OD immediately shows that this set enters the bistability zone and allows to find the critical threshold of the substrate input concentration at which the system will operate in the instability zone; see the threshold S^c in Figure 1b. This shows the value of the OD in understanding the model.

The contribution of the OD to the understanding of the system's behaviour is even more spectacular in the case of the AM2 model. In this case, there are three operating parameters, and the OD must be represented in the plane formed by two of them by fixing the third. The role of this third parameter is described by a series of diagrams. The sets of optimal operating conditions are surfaces in the space of the three operating parameters, whose traces in the two dimensional ODs are curves. It is immediately apparent whether these curves fall within the areas where the system behaviour may be at risk and the thresholds can be easily found. Three regions can then be determined in the plane of the concentrations of the two input substrates. In one of the regions, the maximum biogas flow rate of the steady state where both acidogenic and methanogenic bacteria are present is reached for a value of the dilution ratio for which the acidogenic bacteria are washed out. In a second region, the maximum is reached for a value of the dilution rate for which the positive steady state is GAS. In a third region, the maximum is reached for a value of the dilution rate for which the system presents à bistability behaviour; see Figure 8. These regions have not been identified in the existing literature.

Some figures in this paper (see Figures 1–4, 6, A4, and A5) are made without graduations on the axes because they represent generic situations where the growth functions verify our general hypothesis and the biological parameters are not specified. However, in practice, to construct an OD, one fixes the growth functions and biological parameters and then draws the curves separating the regions of the OD. Indeed, the OD is a tool for the experimenter who knows the biological parameter values of the model he is considering, and then plots its OD. We do that in Appendix A.5 for some classical growth functions; see Figures A1–A3. See also Figures 5, 7 and 8 in Section 3.2, for the AM2 model, whose biological parameters are given in Table A8. See also Figures A6–A8 in the Appendix B.

Another result obtained with the help of the OD of a two-step model is worth mentioning here. It was shown in [42,57] that under certain circumstances, increasing the dilution rate can globally stabilize two-step biological systems. This kind of surprising and unexpected result was obtained also for a two-step model where the first reaction has a Contois kinetics instead of a Monod one [58]. These studies have shown how unexpected properties can be discovered and studied by analysing the OD of the model. Our findings in this paper are a further illustration of the relevance of the OD in the study of one-step and two-step models.

The two-step models of the form (12) present a commensalistic relationship between microorganisms. For definitions and complementary information on commensalism, the reader may consult [59]. Methanogenic bacteria use for their growth the product of the acidogenic bacteria, but acidogenic bacteria are not affected by the growth of the methanogenic bacteria. More complex models are those studied in [38–40,43], which present a syntrophic relationship between the micro organisms: the first population is affected by the growth of the second population. For more details and information on commensalism and syntrophy, the reader is referred to [40,43,59–64] and the references therein. The ODs of some of these models are well understood; see [38–40,43]. Studying the biogas or biomass production for these more complex and more realistic models of AD is a challenging question. It is the subject for future research directions. The determination of the OD and the optimal productivity of synthetic microbial communities considered in [65] is also an interesting question that deserves further attention.

5. Conclusions

In this work, we considered one-step and two-step simple models of AD which are able to adequately capture the main dynamical behaviour of the full ADM1 and have the advantage that a complete analysis for the existence and local stability of their steady states is available. These models have been validated on real data. We considered that the biological parameters of the models have been calibrated on the data. Therefore, the OD of the model can be constructed, and the results can be illustrated in the OD. The best operating conditions for biogas production or biomass production are obtained as subsets of the OD.

For a one-step model, the set Γ of best operating conditions for biogas production is described as a curve of equation $S^{in} = \gamma(D)$; see Figure 2 for the Monod case and Figure 3 for the Haldane case. These curves permit the optimal dilution $D_G^*(S^{in})$ for which the biogas production is maximal to be obtained graphically and easily. The explicit expression for $D_G^*(S^{in})$ is not always available, and even when it is known, see Appendix A.5. On the other hand, the graphical visualisation of $D_G^*(S^{in})$ in the OD allows us to predict the behaviour of the system when it is operated at this optimal dilution rate, as illustrated in Figures 2 and 3 and A1–A3.

When there are no maintenance terms included in the model, it is known that biogas production and biomass production are given by the same expressions. Therefore, the maximum of these quantities is obtained for the same operating conditions. However, when maintenance is included in the model, the subsets of best operating conditions for biogas production and biomass production are not the same; see Figure 4.

For a two-step model, we obtain two subsets, Γ_0 and Γ_1, of maximal biogas production, corresponding to the steady states E_{01} and E_{11}, respectively; see Figure 5. The steady state E_{01} corresponds to the washout of the first biomass, while E_{11} corresponds to the persistence of the two populations. For certain operating conditions, the biogas production of E_{01} can be higher than that of E_{11}. We have determined the set of values for the input substrate concentrations for which this occurs; see Figure 7. We have identified two other subsets of operating conditions in which the system behaves in different ways; see Figure 8. In one set the optimal dilution rate corresponds to an operating regime where the system is functioning at a GAS steady state, while, in the second, there is bistability. It may be in the experimenter's interest to run the system with operating parameters that give rise to bistability, since the biogas flow rate is then greater. However, they must be careful to initialise it in the basin of attraction of the steady state E_{11}, because otherwise it may converge towards the steady state E_{10}, which does not produce biogas.

Our findings illustrate how the OD is a useful tool for the understanding of the behaviour of one-step and two-step models. The OD can be constructed once the biological parameters of the model are fixed. It can also be constructed qualitatively, without specifying the values of the biological parameters. It is therefore a powerful tool for the mathematical analysis of a model when the growth functions are not specified. It is also a tool that allows us to answer important and natural questions that we might not have asked ourselves without this tool. Therefore, the OD allows new interesting questions to be asked and answered about the model. When studying any problem concerning the chemostat, it is useful to represent the results obtained in the OD. This gives a very clear overview of the system and its operating modes. In this paper, we have illustrated the effectiveness of this approach in the study of the maximisation of the biogas flow rate and the productivity of the biomass.

Funding: This research received no external funding.

Institutional Review Board Statement: Not applicable.

Informed Consent Statement: Not applicable.

Data Availability Statement: Data sharing is not applicable to this article as no datasets were generated or analysed during the current study.

Acknowledgments: The author is grateful to Radhouane Fekih-Salem and Jérôme Harmand for insightful comments on this work. The author is also grateful to two anonymous reviewers for comments that greatly improved this manuscript. The author thanks the UNESCO ICIREWARD project ANUMAB and the Euro-Mediterranean research network Treasure (http://www.inrae.fr/treasure) (accessed on 20 January 2022).

Conflicts of Interest: The author declares no conflict of interest.

Abbreviations

The following abbreviations are used in this manuscript:

AD	Anaerobic Digestion
ADM1	The IWA Anaerobic Digestion Model No 1, see [2]
AM2	Anaerobic Digestion Model of [25]
CSTR	Continuous Stirred Tank Reactor or Bioreactor, or Chemostat
GAS	Globally Asymptotically Stable
HRT	Hydraulic Retention Time
LAS	Locally Asymptotically Stable
MBR	Membrane Bioreactor
OD	Operating Diagram
SOP	Set of Operating Parameters
SRT	Solid Retention Time
U	Unstable
VFA	Volatile Fatty Acids

Appendix A. One-Step Model

Appendix A.1. Model Reduction

We consider the chemostat model (2). It is usual in mathematical theory [22,24] to make the change of variable $x = kX$, which transforms (2) into

$$\begin{aligned} \dot{S} &= D(S^{in} - S) - \mu(S)x \\ \dot{x} &= (\mu(S) - D_1)x \end{aligned}$$

Therefore, the stoichiometric coefficient k can be reduced to 1 in (2).

Appendix A.2. The Operating Diagram of the One-Step Model

In order to construct the OD of (2), one needs to determine and compute the boundaries of the regions of the diagram, i.e., to compute the parameter values at which a qualitative change in the dynamic behaviour of (2) occurs. For (2), these boundaries are the curves

$$\begin{aligned} \Lambda &= \left\{ (D, S^{in}) : S^{in} = \lambda(\alpha D + a) \right\}, \\ \Lambda_2 &= \left\{ (D, S^{in}) : S^{in} = \bar{\lambda}(\alpha D + a) \right\}, \\ \Lambda_1 &= \left\{ (D, S^{in}) : \alpha D + a = \mu(S^m) \text{ and } S^{in} \geq S^m \right\} \end{aligned} \qquad (A1)$$

These curves separate the Set of Operating Parameters (SOP)

$$\text{SOP} = \left\{ (D, S^{in}) : D \geq 0 \text{ and } S^{in} \geq 0 \right\},$$

in three regions, denoted $\mathcal{J}_0$, $\mathcal{J}_1$, and $\mathcal{J}_2$, corresponding to different behaviours of (2), as depicted in Table A1.

Table A1. Existence and stability of steady states of (2) in the three regions of the operating space. The last column shows the color in which the region is depicted in the OD shown in Figures 1–4 and A1–A3.

Region		F_0	F_1	F_2	Color
$\mathcal{J}_0 = \left\{ \left(D, S^{in} \right) : S^{in} \leq \lambda(\alpha D + a) \right\}$		GAS			Yellow
$\mathcal{J}_1 = \left\{ \left(D, S^{in} \right) : \lambda(\alpha D + a) < S^{in} \leq \bar{\lambda}(\alpha D + a) \right\}$		U	GAS		Green
$\mathcal{J}_2 = \left\{ \left(D, S^{in} \right) : S^{in} > \bar{\lambda}(\alpha D + a) \right\}$		LAS	LAS	U	Pink

GAS, LAS, and U mean that the steady state is Globally Asymptotically Stable, Locally Asymptotically Stable, or Unstable, respectively. No letter means that the steady state does not exist in the region. Note that

$$\Lambda \cup \Lambda_2 = \left\{ (D, S^{in}) : D = \frac{\mu(S^{in}) - a}{\alpha} \right\}.$$

We plot in Figure 1 the curves Λ, Λ_1, and Λ_2 in SOP and the regions delimited by these curves. This figure, together with Table A1, is the OD of (2). This diagram is well known in the literature [22,36]. When $S^m = +\infty$, then only Λ exists ($\Lambda_1 = \Lambda_2 = \varnothing$). In this case, the OD contains only the regions $\mathcal{J}_0$ and $\mathcal{J}_1$. The main difference between Figure 1a, obtained for the Monod case ($S^m = +\infty$), and Figure 1b, obtained for the Haldane case ($S^m < +\infty$), is the appearance of the region of bistability $\mathcal{J}_2$. In this region, both steady states F_0 and F_1 are LAS and the asymptotic behaviour of a solution depends on its initial condition. If the initial condition belongs to the basin of attraction of F_0, then the species X is washed out from the chemostat. If the initial condition belongs to the basin of attraction of F_1, then, when $t \to +\infty$, the concentration $X(t)$ of the species tends to $X^* = \frac{D}{kD_1}\left(S^{in} - \lambda(D_1) \right)$. The green region $\mathcal{J}_1$ is the "target" operating regions, as it corresponds to the global stability of the steady state, where the species survive. The pink region $\mathcal{J}_2$ corresponds to the bistability of F_0 (no biogas production) and F_1 (with biogas production). If the chemostat is operated in the region $\mathcal{J}_2$, then, for a good operation of the system, its state at start up should correspond to the convergence toward F_1 rather than F_0.

Appendix A.3. Maximization of Biogas Production

Appendix A.3.1. Proof of Proposition 1

The function G defined by (14) is $\mathcal{C}^1$ on the interior of $I(S^{in})$ and its derivative is given by

$$G'(D) = S^{in} - \gamma(D),$$

where γ is defined by (17). Therefore, if $g(S^{in})$ is in the interior of $I(S^{in})$, by Fermat's theorem, any point $D^* \in g(S^{in})$ is a critical point of G; i.e., $G'(D^*) = 0$, which is equivalent to $S^{in} = \gamma(D^*)$. The proof of the proposition is complete if we prove that the set $g(S^{in})$ is in the interior of $I(S^{in})$. If $S^{in} < S^m$, then G is defined for $0 \leq D \leq \delta$, where $\delta = \frac{\mu(S^{in}) - a}{\alpha}$, is positive if $0 < D < \delta$ and satisfies $G(0) = 0$ and

$$G(\delta) = \delta(S^{in} - \lambda(\alpha\delta + a)) = \delta(S^{in} - \lambda(\mu(S^{in}))) = \delta(S^{in} - S^{in}) = 0.$$

Therefore, the maximum cannot be attained in 0 or δ. Similarly if $S^m < +\infty$ and $S^{in} \geq S^m$, then G is defined for $0 \leq D \leq \delta$, where $\delta = \frac{\mu(S^m)-a}{\alpha}$, is positive if $0 < D < \delta$ and satisfies $G(0) = 0$ and

$$G(\delta) = \delta(S^{in} - \lambda(\alpha\delta + a)) = \delta(S^{in} - \lambda(\mu(S^m))) = \delta(S^{in} - S^m) \geq 0.$$

Moreover, if $S^{in} > S^m$, we have

$$\lim_{D \to \mu(S^m)} \lambda'(D) = +\infty.$$

Hence,

$$\lim_{D \to \delta} G'(D) = -\infty.$$

Therefore, the maximum cannot be attained in 0 or δ and $g(S^{in})$ is in the interior of $I(S^{in})$.

Appendix A.3.2. Proof of Proposition 2

Since $H(\lambda(a)) = H(S^{in}) = 0$ and $H(S) > 0$ for $\lambda(a) < S < S^{in}$, the maximum of H is attained at a point $S^* \in (\lambda(a), S^{in})$. By Fermat's theorem, S^* is a critical point of H; i.e., $H'(S^*) = 0$. We have

$$G'(D) = H'(\lambda(\alpha D + a))\lambda'(\alpha D + a).$$

Hence, H has a maximum at S^* if and only if G has a maximum at $D^* = \frac{\mu(S^*)-a}{\alpha}$. The derivative of H is given by

$$H'(S) = \mu'(S)(S^{in} - S) - \mu(S) + a.$$

Hence, $H'(S) = 0$ if and only if $S^{in} = \eta(S)$, where η is defined by (21). From $H'(S^*) = 0$, it is deduced that $S^{in} = \eta(S^*)$.

Appendix A.3.3. Proof of Lemma 1

If μ is C^2, so is λ and the derivative of γ is given by

$$\gamma'(D) = 2\alpha\lambda'(\alpha D + a) + \alpha^2 D\lambda''(\alpha D + a).$$

Using $\mu(\lambda(D)) = D$, we have

$$\lambda'(D) = \frac{1}{\mu'(\lambda(D))} \quad \text{and} \quad \lambda''(D) = -\frac{\mu''(\lambda(D))\lambda'(D)}{(\mu'(\lambda(D)))^2}. \tag{A2}$$

Hence,

$$\gamma'(D) = \alpha\lambda'(\alpha D + a)\left(2 - \frac{\alpha D\mu''(\lambda(\alpha D+a))}{(\mu'(\lambda(\alpha D+a)))^2}\right).$$

Since $\lambda' > 0$ it is deduced that $\gamma'(D) > 0$ if and only if for all $D \in (0, \delta(S^m))$,

$$\alpha D\mu''(\lambda(\alpha D + a)) < 2(\mu'(\lambda(\alpha D + a)))^2.$$

Using the change of variable $S = \lambda(\alpha D + a)$, this condition is equivalent to: for all $S \in (\lambda(a), S^m)$,

$$(\mu(S) - a)\mu''(S) < 2(\mu'(S))^2.$$

Therefore (1) $\Leftrightarrow$ (2). Moreover, we have

$$\left(\frac{1}{\mu-a}\right)' = -\frac{\mu'}{(\mu-a)^2}, \qquad \left(\frac{1}{\mu-a}\right)'' = \frac{2(\mu')^2-(\mu-a)\mu''}{(\mu-a)^3}.$$

Hence, $(1/(\mu-a))'' > 0$ if and only if $(\mu-a)\mu'' < 2(\mu')^2$. Therefore (2) $\Leftrightarrow$ (3). The derivative of η is given by

$$\eta'(S) = 2 - \frac{(\mu(S)-a)\mu''(S)}{(\mu'(S))^2}.$$

Therefore (2) $\Leftrightarrow$ (4). If $\mu'' < 0$ on $(0, S^m)$, then since $\mu' > 0$ on $(\lambda(a), S^m)$, the condition $(\mu(S)-a)\mu''(S) < (\mu'(S))^2$ is obviously satisfied.

Appendix A.4. Maximization of Biomass Production

Appendix A.4.1. Proof of Proposition 3

The function P defined by (23) is C^1 on the interior of $I(S^{in})$, and its derivative is given by

$$P'(D) = \frac{D(\alpha D + 2a)}{(\alpha D + a)^2}\left(S^{in} - \pi(D)\right),$$

where π is defined by (25). Therefore, if the set $p(S^{in})$ is in the interior of $I(S^{in})$, by Fermat's theorem, any point $D^* \in p(S^{in})$ is a critical point of P; i.e., $P'(D^*) = 0$, which is equivalent to $S^{in} = \pi(D^*)$. The proof that $p(S^{in})$ is in the interior of $I(S^{in})$ is the same as the proof that $g(S^{in})$ is in the interior of $I(S^{in})$ given in Appendix A.3.1.

Appendix A.4.2. Proof of Proposition 4

Since $Q(\lambda(a)) = Q(S^{in}) = 0$ and $Q(S) > 0$ for $\lambda(a) < S < S^{in}$, the maximum of Q is attained at a point $S^* \in (\lambda(a), S^{in})$. By Fermat's theorem, S^* is a critical point of Q, i.e., $Q'(S^*) = 0$. We have

$$P'(D) = \tfrac{1}{\alpha}Q'(\lambda(\alpha D + a))\lambda'(\alpha D + a)$$

Hence, Q has a maximum at S^* if and only if P has a maximum at $D^* = \frac{\mu(S^*)-a}{\alpha}$. Moreover, $Q'(S) = 0$ if and only if $S^{in} = \rho(S)$, where η is defined by (30). From $Q'(S^*) = 0$, it is deduced that $S^{in} = \rho(S^*)$.

Appendix A.4.3. Proof of Lemma 2

If μ is C^2, so is λ, and the derivative of π is given by

$$\pi'(D) = \alpha\frac{\alpha D+a}{\alpha D+2a}\left(\frac{2(\alpha D+3a)}{\alpha D+2a}\lambda'(\alpha D+a) + \alpha D\lambda''(\alpha D+a)\right).$$

Using (A2), we have

$$\pi'(D) = \alpha\frac{\alpha D+a}{\alpha D+2a}\lambda'(\alpha D+a)\left(\frac{2(\alpha D+3a)}{\alpha D+2a} - \frac{\alpha D\mu''(\lambda(\alpha D+a))}{(\mu'(\lambda(\alpha D+a)))^2}\right).$$

Since $\lambda' > 0$ it is deduced that $\pi'(D) > 0$ if and only if for all $D \in (0, \delta(S^m))$,

$$\alpha D\frac{\alpha D+2a}{\alpha D+3a}\mu''(\lambda(\alpha D+a)) < 2(\mu'(\lambda(\alpha D+a)))^2.$$

Using the change of variable $S = \lambda(\alpha D + a)$, this condition is equivalent to the following: for all $S \in \lambda(a)0, S^m)$,

$$\frac{(\mu(S)-a)(\mu(S)+a)}{\mu(S)+2a}\mu''(S) < 2(\mu'(S))^2.$$

Therefore, (1) $\Leftrightarrow$ (2). The derivative of ρ is given by

$$\rho'(S) = \frac{\mu(S)}{(\mu(S)+a)^2(\mu'(S))^2}\left(2(\mu(S)+2a)(\mu'(S))^2 - (\mu(S)-a)(\mu(S)+a)\mu''(S)\right).$$

Therefore (2) $\Leftrightarrow$ (3). If $\mu'' < 0$ on $(\lambda(a), S^m)$, then, since $\mu' > 0$ on $(\lambda(a), S^m)$, the condition $(\mu(S) - a)\mu''(S) < (\mu'(S))^2$ is obviously satisfied. Moreover, we have seen in Lemma 1 that if the condition $\left(\frac{1}{\mu - a}\right)''(S) > 0$ holds, then we have

$$(\mu(S) - a)\mu''(S) < 2(\mu'(S))^2.$$

Therefore, we have

$$\frac{(\mu(S) - a)(\mu(S) + a)}{\mu(S) + 2a}\mu''(S) < (\mu(S) - a)\mu''(S) < 2(\mu'(S))^2,$$

which is the condition 2 in the lemma.

Appendix A.5. Applications to Some Usual Growth Functions

For simplicity, we restrict our attention to the case where $\alpha = 1$ and $a = 0$. In this case, $D^*(S^{in})$ is obtained by solving Equation (33). One can also solve Equation (32), to get the maximum $S^*(S^{in})$, and then take

$$D^*(S^{in}) = \mu(S^*(S^{in})). \tag{A3}$$

Appendix A.5.1. Monod Growth Rate

This growth function is given by (4). This function satisfies Hypothesis 1 with $S^m = +\infty$. Since $\mu'' < 0$, using Lemma 3, we obtain that Hypothesis 4 is satisfied. Straightforward computations show that

$$\lambda(D) = \frac{DK}{m - D}, \qquad \gamma(D) = \frac{DK(2m - D)}{(m - D)^2}, \qquad \eta(S) = S^2/K + 2S.$$

Hence, $S^*(S^{in})$, the (unique) solution of equation $S^{in} = \eta(S)$, and $D^*(S^{in})$ are given by

$$S^*(S^{in}) = \sqrt{K^2 + KS^{in}} - K, \quad D^*(S^{in}) = \mu(S^*(S^{in})) = m\left(1 - \sqrt{\frac{K}{K + S^{in}}}\right).$$

This formula for $D^*(S^{in})$ is well known in the literature; see for example [12,19,20]. In Figure A1a, we show the OD, together with the set of best operating conditions Γ and the biogas flow rate $G(D, S^{in})$, with $S^{in} = 10$, for the Monod growth function (4), with $m = 1$ and $K = 5$. This figure shows how the optimal dilution rate $D^*(S^{in})$ can be graphically determined. Although we have an explicit formula for $D^*(S^{in})$, this graphical construction can be very useful as it allows the dilution rate that the experimenter should choose to optimise the biogas flow rate to be visualised in the OD.

Appendix A.5.2. Hill Growth Rate

This growth function is given by

$$\mu(S) = \frac{mS^p}{K^p + S^p}, \qquad p \geq 1. \tag{A4}$$

This function satisfies Hypothesis 1 with $S^m = +\infty$. Moreover, we have

$$\left(\frac{1}{\mu}\right)''(S) = \frac{p(p+1)K^p}{mS^{p+2}}.$$

Hence, $(1/\mu)'' > 0$, and using Lemma 3, we obtain that Hypothesis 4 is satisfied. Notice that for $p > 1$, the Hill function (A4) is not concave on $(0, +\infty)$. Straightforward computations show that

$$\lambda(D) = \left(\frac{D}{m - D}\right)^{\frac{1}{p}}K, \qquad \gamma(D) = \left(\frac{D}{m - D}\right)^{1/p}\frac{(p+1)m - pD}{p(m - D)}K, \qquad \eta(S) = \frac{K^{-p}S^{p+1} + (p+1)S}{p}.$$

Hence, $S^*(S^{in})$, the (unique) solution of equation $S^{in} = \eta(S)$ is the positive solution of equation

$$K^{-p}S^{p+1} + (p+1)S - pS^{in} = 0.$$

One has explicit formulas for $S^*(S^{in})$ when $p = 1$ (the Monod case) and $p = 2$

$$S^*(S^{in}) = \left(K^2 S^{in} + \sqrt{K^6 + K^4(S^{in})^2}\right)^{1/3} - \frac{K^2}{\left(K^2 S^{in} + \sqrt{K^6 + K^4(S^{in})^2}\right)^{1/3}} \quad \text{if} \quad p = 2.$$

We can deduce also the explicit expression of $D^*(S^{in})$, the (unique) solution to equation $S^{in} = \gamma(D)$ by using (A3). This example illustrates the fact that the second method is much more practicable than the first one, since the direct resolution of equation $S^{in} = \gamma(D)$ is not easy.

In Figure A1b, we show the OD, together with the set of best operating conditions Γ and the biogas flow rate $G(D, S^{in})$, with $S^{in} = 10$, for the Hill growth function (A4), with $p = 2$, $m = 1$ and $K = 5$. This figure shows how the optimal dilution rate $D^*(S^{in})$ can be graphically determined. This graphical construction is very useful as it allows the dilution rate that the experimenter should choose to optimise the biogas flow rate to be visualised in the OD. Indeed, the above explicit formula for $S^*(S^{in})$, and hence for $D^*(S^{in})$, is not very informative. Moreover, for $p > 2$, we do not have an explicit formula for $D^*(S^{in})$, whereas the graphical construction can be done for any p.

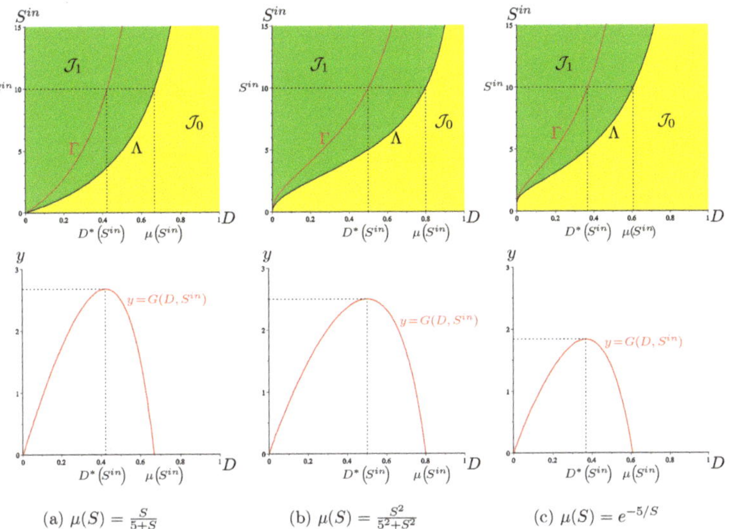

(a) $\mu(S) = \frac{S}{5+S}$ (b) $\mu(S) = \frac{S^2}{5^2+S^2}$ (c) $\mu(S) = e^{-5/S}$

Figure A1. The set of best operating conditions Γ (in Red) shows the optimal dilution rate $D^*(S^{in})$ for three increasing growth functions and $S^{in} = 10$, $a = 0$, $\alpha = 1$.

Appendix A.5.3. Desmond–Le Quéméner and Bouchez Growth Rate

This growth function is given by [66]

$$\mu(S) = me^{-k/S}. \tag{A5}$$

This function satisfies Hypothesis 1 with $S^m = +\infty$. Moreover, we have

$$\left(\frac{1}{\mu}\right)''(S) = \frac{k}{mS^3}\left(2 + \frac{k}{S}\right)e^{k/S}.$$

Hence, $(1/\mu)'' > 0$, and using Lemma 3, we obtain that Hypothesis 4 is satisfied. Notice that the function (A5) is not concave on $(0, +\infty)$. Straightforward computations show that

$$\lambda(D) = \frac{k}{\ln(m/D)}, \qquad \gamma(D) = \frac{k}{\ln(m/D)}\left(1 + \frac{1}{\ln(m/D)}\right), \qquad \eta(S) = S + \frac{S^2}{k}.$$

Therefore

$$S^*(S^{in}) = \frac{\sqrt{k^2 + 4kS^{in}} - k}{2} \quad \text{and} \quad D^*(S^{in}) = \mu\big(S^*(S^{in})\big) = me^{-\frac{\sqrt{k^2+4kS^{in}}+k}{2S^{in}}}.$$

In Figure A1c, we show the OD, together with the set of best operating conditions Γ and the biogas flow rate $G(D, S^{in})$, with $S^{in} = 10$, for the growth function (A5), with $m = 1$ and $k = 5$. This figure shows how the optimal dilution rate $D^*(S^{in})$ can be graphically determined. Although we have an explicit formula for $D^*(S^{in})$, this graphical construction can be very useful as it allows visualising in the OD the dilution rate that the experimenter chooses to optimise the biogas flow rate.

Appendix A.5.4. Haldane Growth Rate

This growth function is given by (5). It satisfies Hypothesis 1, with

$$S^m = \sqrt{KK_i} \quad \text{and} \quad \max_{S \geq 0} \mu(S) = \mu(S^m) = \frac{m}{1 + 2\sqrt{K/K_i}}.$$

Since $\mu''(S) < 0$ on $(0, S^m)$, using Lemma 3, we obtain that Hypothesis 4 is satisfied. We have

$$\lambda(D) = \frac{m - D - \sqrt{\Delta}}{2D} K_i = \frac{2D}{m - D + \sqrt{\Delta}} K, \qquad \bar{\lambda}(D) = \frac{m - D + \sqrt{\Delta}}{2D} K_i,$$

where $\Delta = (m - D)^2 - 4D^2 K/K_i$, defined for $0 \leq D \leq \mu(S^m)$. Note that Δ tends toward $(m - D)^2$ when $K_i \to +\infty$. Hence $\lambda(D) \to \frac{DK}{m - D}$ and $\bar{\lambda}(D) \to +\infty$. We find the case of Monod. Straightforward calculations show that

$$\gamma(D) = \frac{2DK(2m - D + 4DK/K_i)}{\Delta + (m - D + 4DK/K_i)\sqrt{\Delta}}, \qquad \eta(S) = \frac{(2K + S)K_i S}{KK_i - S^2}.$$

The solution of $S^{in} = \eta(S)$ is given by

$$S^*(S^{in}) = \frac{\sqrt{KK_i\big((K + S^{in})K_i + (S^{in})^2\big)} - KK_i}{K_i + S^{in}}.$$

Hence, $D^*(S^{in})$, the solution of $S^{in} = \gamma(D)$, is given by (A3), i.e.,

$$D^*(S^{in}) = \mu\big(S^*(S^{in})\big) = \frac{m(K_i + S^{in})\big(\sqrt{KK_i((K + S^{in})K_i + (S^{in})^2)} - KK_i\big)}{2K\big((K + S^{in})K_i + (S^{in})^2\big) + (K_i + S^{in} - 2K)\sqrt{KK_i((K + S^{in})K_i + (S^{in})^2)}}.$$

These formulas for $S^*(S^{in})$ and $D^*(S^{in})$ are known in the literature [20]. Note that the equation $S^{in} = \gamma(D)$ is equivalent to an algebraic quadratic equation of degree two which can be solved explicitly. We obtain the formula

$$D^*(S^{in}) = \begin{cases} m\left(\frac{K_i}{K_i - 4K} - \frac{K_i + 2S^{in}}{K_i - 4K}\sqrt{\frac{KK_i}{(K + S^{in})K_i + (S^{in})^2}}\right) & \text{if } K_i \neq 4K, \\ m\frac{S^{in}(4K + S^{in})}{2(2K + S^{in})^2} & \text{if } K_i = 4K. \end{cases}$$

Note that when $K_i \to +\infty$, then $D^*(S^{in}) \to m\left(1 - \sqrt{\frac{K}{K + S^{in}}}\right)$. We find the case of Monod.

On the other hand, equation $\gamma(D) = \bar{\lambda}(D)$ is equivalent to the third-degree polynomial equation:

$$(4K - K_i)^2 D^3 + 3mK_i(4K - K_i)D^2 + 3m^2 K_i(K_i - K)D - m^3 K_i^2 = 0.$$

Therefore D^c, considered in Remark 7, is the unique positive solution of this equation and can be computed explicitly. Let us illustrate the results of Section 3.1.4 in the particular case of the Haldane function given by $m = 1$, $K = 5$, and $K_i = 5$. The OD and the set Γ of best operating conditions are depicted in Figure A2. The biogas flow is shown for five values of S^{in}. The curves Γ and Λ_2 intersect at $(D^c, S^c) = (0.293, 9.397)$. If $S^{in} > S^c$, then the optimal dilution rate $D^*(S^{in})$ corresponds to the bistability region (pink region) $\mathcal{J}_2$. Depending on the initial condition, the system can go to the washout of the species with no biogas production, or its persistence, with maximal biogas production.

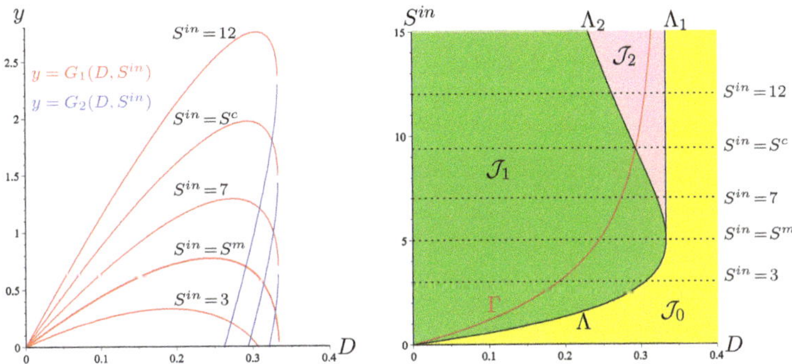

Figure A2. The set of optimal biogas production for the Haldane function (5), with $m = 1$, $K = 5$, $K_i = 5$. We have $S^m = 5$, $D^c = 0.293$, and $S^c = 9.397$.

Appendix A.5.5. An Example with Two Maxima

It is known that Hypothesis 1 is not enough to guarantee that the biogas flow rate admits a unique global maximum (Hypothesis 4); see Figure 5.1 in [14]. Even if the function f is increasing, it is possible that the biogas flow rates have two maxima. For example, consider the function

$$\mu(S) = \frac{mS^6 + S}{K^6 + S^6 + S}, \quad \text{with } m = 2, \quad K^6 = 0.1,$$

which is obtained from the Hill function (A4) (with $p = 6$) by adding S to the numerator and denominator. This function is increasing; see Figure A3a. However, for some values of S^{in}, the biogas flow rate has three local extrema; see Figure A3d. Numerical exploration shows that the the set of arguments of the maximum of G is as follows

$$g(S^{in}) = \begin{cases} 0.705 & \text{if } S^{in} = 1 \\ \{0.786, 1.277\} & \text{if } S^{in} = 1.7625 \\ 1.475 & \text{if } S^{in} = 2.1 \end{cases}$$

This behaviour is consistent with the plot of the curve Γ; see Figure A3c. The function η is given by:

$$\eta(S) = S + \frac{\mu(S)}{\mu'(S)} = S + \frac{(S + mS^6)(K^6 + S + S^6)}{K^6 + 5(m-1)S^6 + 6K^6 mS^5}.$$

The plot of this function shows that it is not increasing; see Figure A3b. Therefore, from Lemma 3, we can easily predict that the function γ is not increasing, as depicted in Figure A3c. Hence, Hypothesis 4 is not satisfied.

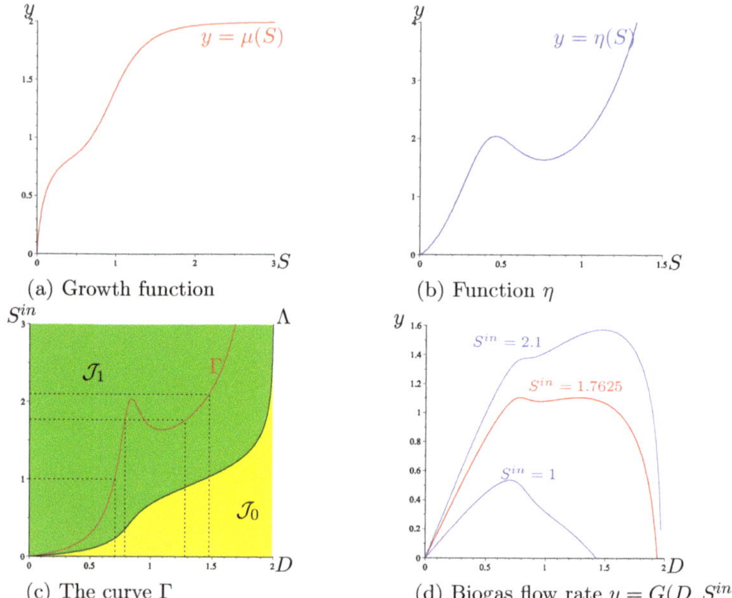

(a) Growth function
(b) Function η
(c) The curve Γ
(d) Biogas flow rate $y = G(D, S^{in})$

Figure A3. An increasing growth function with two maxima of the biogas flow rate.

Appendix B. Two-Step Models

Appendix B.1. Model Reduction

The linear change of variables

$$s_1 = \frac{k_2}{k_1}S_1, \quad x_1 = k_2 X_1, \quad s_2 = S_2, \quad x_2 = k_3 X_2$$

transforms (12) into

$$
\begin{aligned}
\dot{s}_1 &= D\left(s_1^{in} - s_1\right) - f_1(s_1)x_1, \\
\dot{x}_1 &= (f_1(s_1) - D_1)x_1, \\
\dot{s}_2 &= D\left(s_2^{in} - s_2\right) + f_1(s_1)x_1 - f_2(s_2)x_2, \\
\dot{x}_2 &= (f_2(s_2) - D_2)x_2,
\end{aligned}
\tag{A6}
$$

where

$$s_1^{in} = \frac{k_2}{k_1}S_1^{in}, \quad s_2^{in} = S_1^{in}, \quad f_1(s_1) = \mu_1\left(\frac{k_1}{k_2}s_1\right), \quad f_2(s_2) = \mu_2(s_2)$$

Therefore, the stoichiometric coefficients k_i, $i = 1, 2, 3$ are reduced to 1. However, as explained in Section 2.2, we do not work with the reduced model (A6) and we present the results in the original model (12).

Appendix B.2. The Steady States of a Two-Step Model

The model (12) has a cascade structure, which renders its mathematical analysis easy. There is no additional difficulty compared to the case considered in [26] in which $\alpha_1 = \alpha_2 = \alpha$ and $a_1 = a_2 = 0$. We recall that the break-even concentrations were defined in Table 1. We summarize in Table A2 the definitions of some additional auxiliary functions that are used in the description of the steady states of (12).

Table A2. Auxiliary functions. The functions λ_1, λ_2, and $\bar{\lambda}_2$ and H_i, $i = 1,2$ are defined in Table 1.

$$S_2^{in*}\left(D, S_1^{in}, S_2^{in}\right) = S_2^{in} + \frac{k_2}{k_1}\left(S_1^{in} - \lambda_1(D_1)\right)$$

$$X_1^*\left(D, S_1^{in}\right) = \frac{D}{k_1 D_1}\left(S_1^{in} - \lambda_1(\alpha D)\right)$$
$$X_{21}\left(D, S_2^{in}\right) = \frac{D}{k_3 D_2}\left(S_2^{in} - \lambda_2(D_2)\right)$$
$$X_{22}\left(D, S_2^{in}\right) = \frac{D}{k_3 D_2}\left(S_2^{in} - \bar{\lambda}_2(D_2)\right)$$
$$X_{2i}^*\left(D, S_1^{in}, S_2^{in}\right) = \frac{D}{k_3 D_2}\left(S_2^{in} + \frac{k_2}{k_1}S_1^{in} - H_i(D)\right), i = 1,2$$

The system (12) can have up to six steady states, denoted E_{ij}, where $i = 0,1$ and $j = 0,1,2$. The convention used is as follows: if $i = 0$, it means that $X_1 = 0$ and if $i = 1$, then $X_1 > 0$. Similarly, if $j = 0$, it means that $X_2 = 0$ and if $j = 1,2$, then $X_2 > 0$. It should be noticed that E_{00}, where $X_1 = 0$ and $X_2 = 0$, is the washout steady state where acidogenic and methanogenic bacteria are extinct; E_{0i}, $i = 1,2$, where $X_1 = 0$ and $X_2 > 0$, is the steady state of washout of acidogenic bacteria, while methanogenic bacteria are maintained; E_{10}, where $X_1 > 0$ and $X_2 = 0$ is the steady state of washout of methanogenic bacteria, while acidogenic bacteria are maintained; E_{1i}, $i = 1,2$, where $X_1 > 0$ and $X_2 > 0$ is the steady state of coexistence of acidogenic and methanogenic bacteria. The components of the steady states are given in Table A3.

Table A3. The steady states of (12). The functions λ_1, λ_2, and $\bar{\lambda}_2$ are defined in Table 1. The functions S_2^{in*}, X_1^*, X_{2i} and X_{2i}^*, $i = 1,2$ are defined in Table A2.

	S_1	S_2	X_1	X_2
E_{00}	S_1^{in}	S_2^{in}	0	0
E_{01}	S_1^{in}	$\lambda_2(D_2)$	0	$X_{21}\left(D, S_2^{in}\right)$
E_{02}	S_1^{in}	$\bar{\lambda}_2(D_2)$	0	$X_{22}\left(D, S_2^{in}\right)$
E_{10}	$\lambda_1(D_1)$	$S_2^{in*}\left(D, S_1^{in}, S_2^{in}\right)$	$X_1^*\left(D, S_1^{in}\right)$	0
E_{11}	$\lambda_1(D_1)$	$\lambda_2(D_2)$	$X_1^*\left(D, S_1^{in}\right)$	$X_{21}^*\left(D, S_1^{in}, S_2^{in}\right)$
E_{12}	$\lambda_1(D_1)$	$\bar{\lambda}_2(D_2)$	$X_1^*\left(D, S_1^{in}\right)$	$X_{22}^*\left(D, S_1^{in}, S_2^{in}\right)$

Table A4. Necessary and sufficient conditions for the existence and stability of steady states of (12). The functions λ_1, λ_2, $\bar{\lambda}_2$, and H_i, $i = 1,2$ are defined in Table 1.

	Existence Conditions	Stability Conditions
E_{00}	Always exists	$S_1^{in} < \lambda_1(D_1)$ and $S_2^{in} \notin \left[\lambda_2(D_2), \bar{\lambda}_2(D_2)\right]$
E_{01}	$S_2^{in} > \lambda_2(D_2)$	$S_1^{in} < \lambda_1(D_1)$
E_{02}	$S_2^{in} > \bar{\lambda}_2(D_2)$	Unstable if it exists
E_{10}	$S_1^{in} > \lambda_1(D_1)$	$S_2^{in} + \frac{k_2}{k_1}S_1^{in} \notin \left[H_1(D), H_2(D)\right]$
E_{11}	$S_1^{in} > \lambda_1(D_1)$ and $S_2^{in} + \frac{k_2}{k_1}S_1^{in} > H_1(D)$	Stable if it exists
E_{12}	$S_1^{in} > \lambda_1(\alpha D)$ and $S_2^{in} + \frac{k_2}{k_1}S_1^{in} > H_2(D)$	Unstable if it exists

Table A5. The surfaces Λ_i, $i = 1 \cdots 6$ and the regions $\mathcal{I}_k$, $k = 0 \cdots 8$.

$$\Lambda_1 = \left\{ (D, S_1^{in}, S_2^{in}) : S_1^{in} = \lambda_1(D_1) := \lambda_1(\alpha_1 D + a_1) \right\}$$
$$\Lambda_2 = \left\{ (D, S_1^{in}, S_2^{in}) : S_2^{in} = \lambda_2(D_2) := \lambda_2(\alpha_2 D + a_2) \right\}$$
$$\Lambda_3 = \left\{ (D, S_1^{in}, S_2^{in}) : S_2^{in} = \bar{\lambda}_2(D_2) := \bar{\lambda}_2(\alpha_2 D + a_2) \right\}$$
$$\Lambda_4 = \left\{ (D, S_1^{in}, S_2^{in}) : S_2^{in} + \tfrac{k_2}{k_1} S_1^{in} = H_1(D) \right\}$$
$$\Lambda_5 = \left\{ (D, S_1^{in}, S_2^{in}) : S_2^{in} + \tfrac{k_2}{k_1} S_1^{in} = H_2(D) \right\}$$
$$\Lambda_6 = \left\{ (D, S_1^{in}, S_2^{in}) : D = \delta_2 := \tfrac{\mu_2(S_2^m) - a_2}{\alpha_2} \right\},$$

$$\mathcal{I}_0 = \left\{ \left(\left(D, S_1^{in}, S_2^{in}\right) : S_1^{in} < \lambda_1(D_1) \text{ and } S_2^{in} < \lambda_2(D_2) \right\}$$
$$\mathcal{I}_1 = \left\{ \left(\left(D, S_1^{in}, S_2^{in}\right) : S_1^{in} < \lambda_1(D_1) \text{ and } \lambda_2(D_2) < S_2^{in} \leq \bar{\lambda}_2(D_2) \right\}$$
$$\mathcal{I}_2 = \left\{ \left(\left(D, S_1^{in}, S_2^{in}\right) : S_1^{in} < \lambda_1(D_2) \text{ and } S_2^{in} > \bar{\lambda}_2(D_2) \right\}$$
$$\mathcal{I}_3 = \left\{ \left(\left(D, S_1^{in}, S_2^{in}\right) : S_1^{in} > \lambda_1(D_1) \text{ and } S_2^{in} + \tfrac{k_2}{k_1} S_1^{in} < H_1(D) \right\}$$
$$\mathcal{I}_4 = \left\{ \left(\left(D, S_1^{in}, S_2^{in}\right) : S_1^{in} > \lambda_1(D_1), S_2^{in} \leq \lambda_2(D_2) \text{ and } H_1(D) < S_2^{in} + \tfrac{k_2}{k_1} S_1^{in} \leq H_2(D) \right\}$$
$$\mathcal{I}_5 = \left\{ \left(\left(D, S_1^{in}, S_2^{in}\right) : S_1^{in} > \lambda_1(D_1), S_2^{in} \leq \lambda_2(D_2) \text{ and } S_2^{in} + \tfrac{k_2}{k_1} S_1^{in} > H_2(D) \right\}$$
$$\mathcal{I}_6 = \left\{ \left(\left(D, S_1^{in}, S_2^{in}\right) : S_1^{in} > \lambda_1(D_1), S_2^{in} > \lambda_2(D_2) \text{ and } S_2^{in} + \tfrac{k_2}{k_1} S_1^{in} \leq H_2(D) \right\}$$
$$\mathcal{I}_7 = \left\{ \left(\left(D, S_1^{in}, S_2^{in}\right) : S_1^{in} > \lambda_1(D_1), \lambda_2(D_2) < S_2^{in} \leq \bar{\lambda}_2(D_2) \text{ and } S_2^{in} + \tfrac{k_2}{k_1} S_1^{in} > H_2(D) \right\}$$
$$\mathcal{I}_8 = \left\{ \left(\left(D, S_1^{in}, S_2^{in}\right) : S_1^{in} > \lambda_1(D_1) \text{ and } S_2^{in} > \bar{\lambda}_2(D_2) \right\}$$

Appendix B.3. Operating Diagram

In order to construct the OD of (12), one needs to determine and compute the boundaries of the regions of the diagram, i.e., to compute the parameter values at which a qualitative change in the dynamic behaviour of (12) occurs. These boundaries are six surfaces, denoted Λ_i, $k = 1 \ldots 6$, in the Set of Operating Parameters (SOP)

$$\text{SOP} = \left\{ (D, S_1^{in}, S_2^{in}) : D \geq 0, S_1^{in} \geq 0 \text{ and } S_2^{in} \geq 0 \right\}.$$

These surfaces separate SOP in nine regions, denoted $\mathcal{I}_k$, $k = 0, \ldots, 8$. These regions correspond to the system behaviour shown in Table A6.

Table A6. Existence and stability of steady states of (12) in the nine regions of the operating space. The last column shows the color in which the region is depicted in Figures 5, 6, 8, A4, A5, A7, and A8.

Region	E_{00}	E_{01}	E_{02}	E_{10}	E_{11}	E_{12}	Color
$\mathcal{I}_0$	GAS						Red
$\mathcal{I}_1$	U	GAS					Blue
$\mathcal{I}_2$	LAS	LAS	U				Cyan
$\mathcal{I}_3$	U			GAS			Yellow
$\mathcal{I}_4$	U			U	GAS		Green
$\mathcal{I}_5$	U			LAS	LAS	U	Pink
$\mathcal{I}_6$	U	U		U	GAS		Green
$\mathcal{I}_7$	U	U		LAS	LAS	U	Pink
$\mathcal{I}_8$	U	U	U	LAS	LAS	U	Pink

The definitions of the surfaces Λ_i and the regions $\mathcal{I}_k$ are given in Table A5. We plot in Figure A6 these surfaces with the biological parameters fixed as in Table A8. Since it is not easy to visualize regions in the three-dimensional operating parameters space, D and S_1^{in} are used as coordinates of the OD, while S_2^{in} is kept constant. The effects of S_2^{in} are shown in a series of operating diagrams; see Figures 5 and A7.

Remark A1. *In Figures 5, 6, 8, A4, A5, A7, and A8, presenting ODs, a region is coloured according to the colour in Table A6. Each colour corresponds to different asymptotic behaviour:*

- *Red for the washout of both species; that is, the steady state E_{00} is globally asymptotically stable (GAS), which occurs in region $\mathcal{I}_0$.*
- *Blue for the washout of acidogenic bacteria while methanogenic bacteria are maintained; that is, the steady state E_{01} is GAS, which occurs in region $\mathcal{I}_1$.*
- *Cyan for the bistability of E_{00} and E_{01}, which are both (locally) stable. This behaviour occurs in region $\mathcal{I}_1$. Depending on the initial condition, the system can go to the washout of both species or the washout of only the acidogenic bacteria.*
- *Yellow for the washout of methanogenic bacteria while acidogenic bacteria are maintained; that is the steady state E_{10} is GAS, which occurs in region $\mathcal{I}_3$.*
- *Green for the global asymptotic stability of the positive steady state E_{11}; which occurs in $\mathcal{I}_4$ and $\mathcal{I}_6$. These regions differ only by the existence, in the second region, of the unstable boundary steady state E_{01}.*
- *Pink for the bistability of E_{10} and E_{11}, which are both locally asymptotically stable. This behaviour occurs in regions $\mathcal{I}_5$, $\mathcal{I}_7$, and $\mathcal{I}_8$. These regions differ only by the possible existence of the unstable boundary steady states E_{01} or E_{02}. Depending on the initial condition, the system can go to the washout of methanogenic bacteria or the coexistence of both species.*

Appendix B.3.1. Operating Diagram in (S_1^{in}, S_2^{in}) Where D Is Kept Constant

The fact that there are nine regions is easily seen when considering the sections of SOP through a plane (S_1^{in}, S_2^{in}) where D is kept constant. Let us denote

$$\delta_1 = \tfrac{m_1 - a_1}{\alpha_1}, \quad \delta_2 = \tfrac{\mu_2(S_2^m) - a_2}{\alpha_2} \tag{A7}$$

The surface Λ_1 is defined for $D < \delta_1$, the surfaces Λ_2 and Λ_3 are defined for $D < \delta_2$, and the surfaces Λ_4 and Λ_5 are defined for $D < \min(\delta_1, \delta_2)$, where δ_1 and δ_2 are given by (A7). The intersections of the surfaces Λ_i, $i = 1 \ldots 5$, with a plane where D is kept constant are straight lines: vertical line for Λ_1, horizontal lines for Λ_2 and Λ_3, and oblique lines for Λ_4 and Λ_5; see Figure A4. We consider in this figure the case $\delta_1 > \delta_2$. This case corresponds to the situation where $\alpha_1 = \alpha_2$, $a_1 = a_2$, and

$$\mu_2(S^m) = \max_{S_2 \geq 0} \mu_2(S_2) < \max_{S_1 \geq 0} \mu_1(S_1) = \mu_1(+\infty),$$

which is most likely to occur in a real model. The case $\delta_1 \leq \delta_2$ is similar; see [42]. Since the curves are straight lines, the nine regions of the OD are easy to picture. The regions are coloured according to the colours in Table A6. This table gives the system behaviour in the nine regions.

Figure A4 shows the following features. For $0 < D < \delta_2$, all regions exist; see Figure A4a. For increasing D, the vertical line Λ_1 moves to the right and tends towards the vertical line defined by $S_1^{in} = \lambda_1(\alpha\delta_2 + a_1)$. At the same time, the horizontal lines Λ_2 and Λ_3 move towards each other and tend toward the horizontal line defined by $S_2^{in} = S_2^m$, so that the regions $\mathcal{I}_1$, $\mathcal{I}_4$, $\mathcal{I}_6$, and $\mathcal{I}_7$ shrink and disappear; see Figure A4b. For $D = \delta_2$, the OD changes dramatically, since regions $\mathcal{I}_1$, $\mathcal{I}_4$, $\mathcal{I}_6$, and $\mathcal{I}_7$ shrink and disappear; see Figure A4b, obtained for $D < \delta_2$ and $D \approx \delta_2$. For $D > \delta_2$ and $D \approx \delta_2$, regions $\mathcal{I}_0$, $\mathcal{I}_3$ invade the whole operating plane, so that regions $\mathcal{I}_2$, $\mathcal{I}_5$, and $\mathcal{I}_8$ also disappear; see Figure A4c. For $\delta_2 < D < \delta_1$, only regions $\mathcal{I}_0$ and $\mathcal{I}_3$ appear; see Figures A4d. For increasing D, the vertical line Λ_1 moves to the right and tends towards infinity, so that, for $D \geq \delta_1$, only region $\mathcal{I}_0$ appears.

In Figure A4, the axes are not graduated, because the figure corresponds to a general case where the growth functions μ_1 and μ_2 verify Hypotheses 2 and 3 and the biological parameters are not specified. The intersections of the OD with planes where D is constant provide an easy way to see that the OD contains nine regions. However, as we are interested in this paper in the biogas flow rate as a function of D, it is preferable to have ODs that include D as a coordinate and in which, for example, S_2^{in} is fixed. We describe these diagrams in the following section.

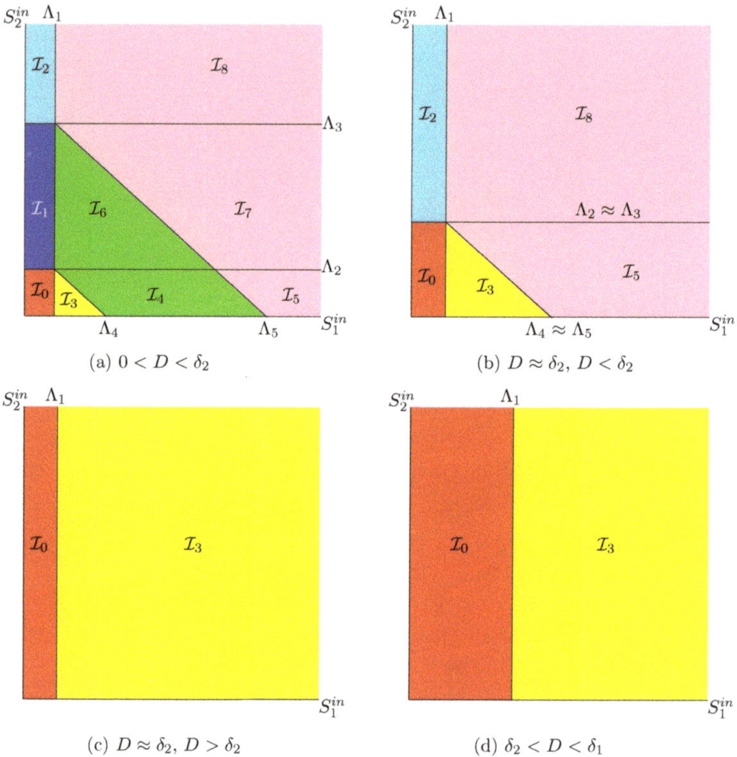

Figure A4. The 2-dimensional OD in $\left(S_1^{in}, S_2^{in}\right)$, obtained by cuts at the D constant of the 3-dimensional OD of (12), where δ_1 and δ_2 are given by (A7). If $D \geq \delta_1$, the region $\mathcal{I}_0$ invades the whole plane.

Appendix B.3.2. Operating Diagram in (D, S_1^{in}) Where S_2^{in} Is Kept Constant

Since we want to plot the intersections of the regions $\mathcal{J}_k$ with a (D, S_1^{in})-plane, where S_2^{in} is kept constant, we must determine the intersections of the surfaces Λ_i with this plane. These intersections are the curves whose equations are given in Table A7.

Table A7. Intersections of Λ_k with a $\left(D, S_1^{in}\right)$-plane, where S_2^{in} is kept constant.

Λ_1	Curve of function $S_1^{in} = \lambda_1(\alpha_1 D + a_1)$ or $D = \frac{\mu_1(S_1^{in}) - a_1}{\alpha_1}$
Λ_2	Vertical line $D = \frac{\mu_2(S_2^{in}) - a_2}{\alpha_2}$ or $S_2^{in} = \lambda_2(\alpha_2 D + a_2)$, if $S_2^{in} \leq S_2^m$
Λ_3	Vertical line $D = \frac{\mu_2(S_2^{in}) - a_2}{\alpha_2}$ or $S_2^{in} = \bar{\lambda}_2(\alpha_2 D + a_2)$, if $S_2^{in} \geq S_2^m$
Λ_4	Curve of function $S_1^{in} = \frac{k_1}{k_2}\left(H_1(D) - S_2^{in}\right)$ restricted to $S_1^{in} > \lambda_1(\alpha_1 D + a_1)$
Λ_5	Curve of function $S_1^{in} = \frac{k_1}{k_2}\left(H_2(D) - S_2^{in}\right)$ restricted to $S_1^{in} > \lambda_1(\alpha_1 D + a_1)$
Λ_6	Vertical line $D = \frac{\mu_2(S_2^m) - a_2}{\alpha_2}$ or $S_2^m = \lambda_2(\alpha_2 D + a_2) = \bar{\lambda}_2(\alpha_2 D + a_2)$

From the equations of curves Λ_4 and Λ_4 and using the $\lambda_2 < \bar{\lambda}_2$, we see that the curve Λ_5 is above the curve Λ_4, which is itself above the curve Λ_1. Note that Λ_1 and Λ_4 are increasing, while Λ_5 is not necessarily increasing, since $H_2(D)$ is the sum of the increasing function $\frac{k_2}{k_1}\lambda_1(\alpha_1 D + a_1)$ and the decreasing function $\bar{\lambda}_2(\alpha_2 D + a_2)$. In Figure A5, we have depicted the curves in the particular case where the curve Λ_5 is decreasing. The general case is left to the reader. It is similar to the case (B) and (C) considered in [42].

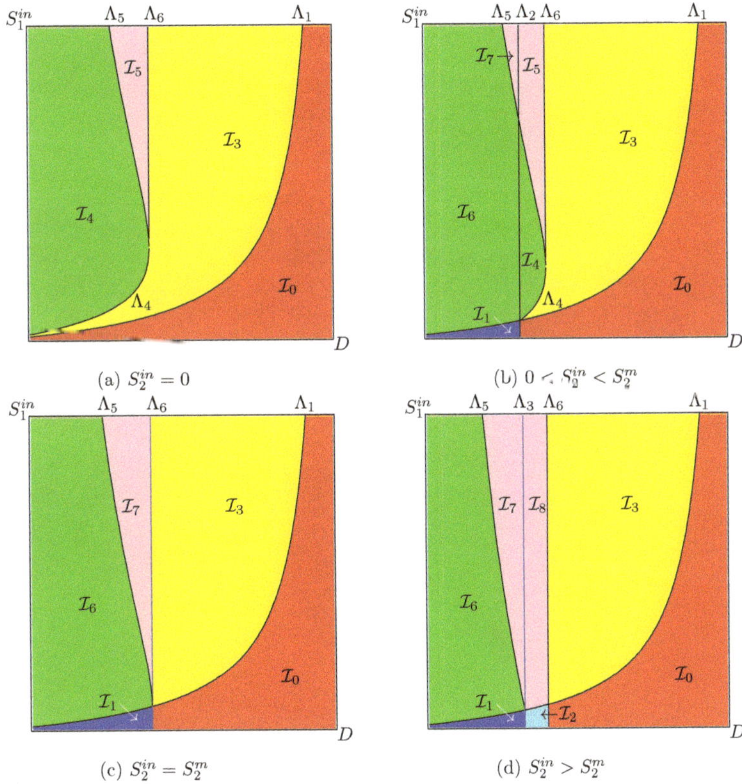

(a) $S_2^{in} = 0$ (b) $0 < S_2^{in} < S_2^m$

(c) $S_2^{in} = S_2^m$ (d) $S_2^{in} > S_2^m$

Figure A5. The 2-dimensional OD in $\left(D, S_1^{in}\right)$ obtained by cuts at the S_2^{in} constant of the 3-dimensional OD of (12).

From the equations of the curves given in Table A7, we deduce that if $0 \leq S_2^{in} \leq S_2^m$, then curves Λ_4, Λ_5 and Λ_6 intersect at point

$$\Lambda_4 \cap \Lambda_5 \cap \Lambda_6 = \left\{ \left(\delta_2, \frac{k_1}{k_2}(S_2^m - S_2^{in}) + \lambda_1(\alpha_1 \delta_2 + a_1) \right) \right\},$$

while curves Λ_1, Λ_2 and Λ_4 intersect at point

$$\Lambda_1 \cap \Lambda_2 \cap \Lambda_4 = \left\{ \left(\delta(S_2^{in}), \lambda_1(\alpha_1 \delta(S_2^{in}) + a_1) \right) \right\}, \quad \text{where} \quad \delta(S_2^{in}) = \frac{\mu_2(S_2^{in}) - a_2}{\alpha_2};$$

see Figure A5a,b. Similarly, if $S_2^{in} = S_2^m$, then

$$\Lambda_2 = \Lambda_3 = \Lambda_6 \quad \text{and} \quad \Lambda_1 \cap \Lambda_5 \cap \Lambda_6 = \left\{ (\delta_2, \lambda_1(\alpha_1 \delta_2 + a_1)) \right\};$$

see Figure A5c, and if $S_2^{in} > S_2^m$, then

$$\Lambda_1 \cap \Lambda_3 \cap \Lambda_5 = \left\{ \left(\delta(S_2^{in}), \lambda_1(\alpha_1 \delta(S_2^{in}) + a_1) \right) \right\}, \quad \Lambda_1 \cap \Lambda_6 = \left\{ (\delta_2, \lambda_1(\alpha_1 \delta_2 + a_1)) \right\};$$

see Figure A5d. Therefore, the curves intersect as depicted in Figure A5, where the regions are coloured according to the colours in Table A6. This figure shows the following features: For $S_2^{in} = 0$, only the regions $\mathcal{I}_0$, $\mathcal{I}_3$, $\mathcal{I}_4$, and $\mathcal{I}_5$ exist; see Figure A5a. For $0 < S_2^{in} < S_2^{m}$, the curve Λ_2 appears, giving birth to $\mathcal{I}_1$, $\mathcal{I}_6$, and $\mathcal{I}_7$ regions; see Figure A5b. For increasing S_2^{in}, Λ_4, and Λ_5 curves are translated downwards, while the vertical line Λ_2 moves to the right and tends towards the vertical line Λ_6, as S_2^{in} tends to S_2^{m}. For $S_2^{in} = S_2^{M}$, the curve Λ_4 disappears, while Λ_2 becomes equal to Λ_6, so that $\mathcal{I}_4$ and $\mathcal{I}_5$ regions have disappeared; see Figure A5c. For $S_2^{in} > S_2^{m}$, the curve Λ_3 appears, giving birth to $\mathcal{I}_2$ and $\mathcal{I}_8$ regions; see Figure A5d. For increasing S_2^{in}, the vertical line Λ_3 moves to the left, while the Λ_5 curve is translated downwards.

Appendix B.4. The Operating Diagram to the AM2 Model

In this section, we show the ODs of the model (50,51), with the biological parameter values given in Table A8. These parameter values can be found in Tables III and V of [25]. These values have been also used by [11]. The OD in the three-dimensional SOP is shown in Figure A6. The two-dimensional diagrams in the (D, S_1^{in}) planes where S_2^{in} is kept constant are depicted in Figure A7. The two-dimensional diagrams in the (S_1^{in}, S_2^{in}) planes where D is kept constant are depicted in Figure A8.

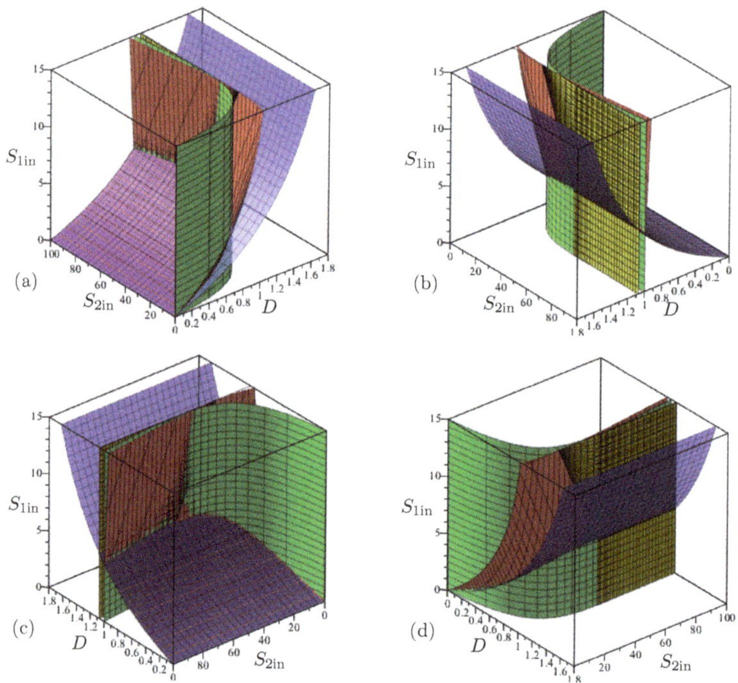

Figure A6. The surfaces Λ_1 (in Blue), Λ_2 and Λ_3 (in Green), Λ_4 and Λ_5 (in Red), and Λ_6 (in Yellow), defined in Table A5 separate the 3-dimensional operating space $\left(D, S_1^{in}, S_2^{in}\right)$ in 9 regions $\mathcal{I}_k$, $k = 0, \cdots, 8$. Front (**a**), rear (**b**), left (**c**), and right (**d**) view of the surfaces Λ_i. The biological parameter values are given in Table A8.

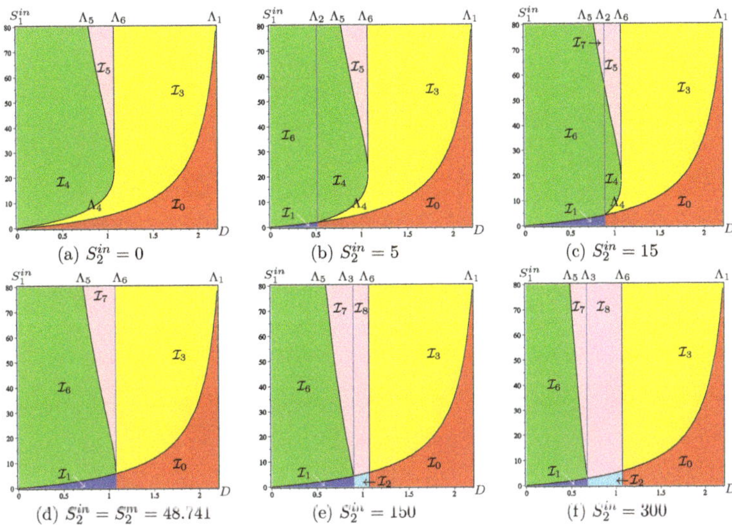

Figure A7. The 2-dimensional OD in $\left(D, S_1^{in}\right)$, obtained by cuts at S_2^{in} constant of the 3-dimensional OD shown in Figure A6.

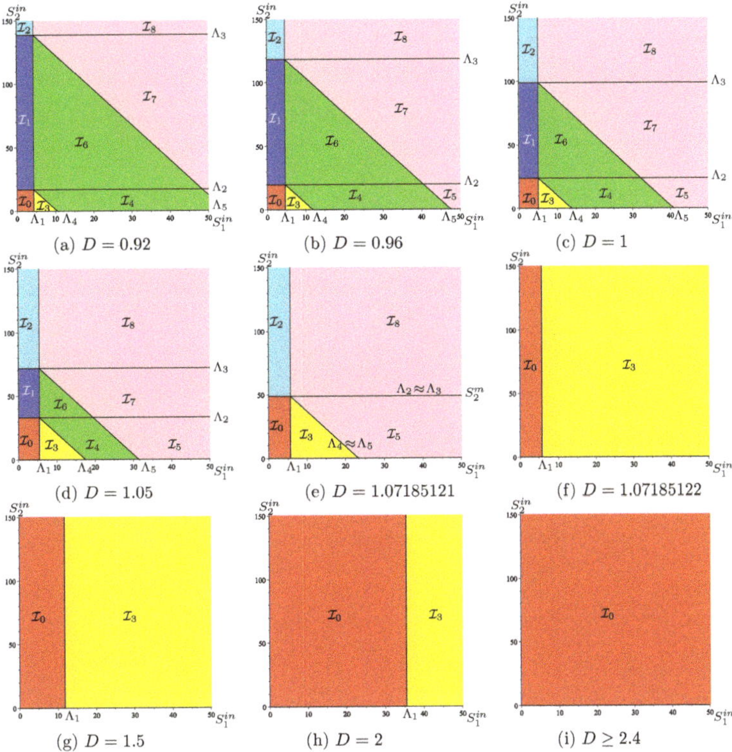

Figure A8. The 2-dimensional OD in $\left(S_1^{in}, S_2^{in}\right)$, obtained by cuts at D constant of the 3-dimensional OD shown in Figure A6.

Table A8. Nominal parameters values, and $\alpha = 0$, used in Figures 5, 7, 8, A6, A7, and A8.

Parameter	m_1	K_1	m_2	K_2	K_I	k_1	k_2	k_3
Unit	d^{-1}	g/L	d^{-1}	mmol/L	mmol/L		mmol/g	mmol/g
Value	1.2	7.1	0.74	9.28	256	42.14	116.5	268

Appendix B.5. Maximization of Biogas Production

Appendix B.5.1. Proof of Proposition 5

From Table 2, it is seen that G_{02} is defined if and only if $\bar{\lambda}_2(D_2) < S_2^{in}$. Since $\bar{\lambda}_2(D_2) > \lambda_2(D_2)$, the results show that G_{01} is also defined and $G_{01}(D, S_2^{in}) > G_{02}(D, S_2^{in})$. This proves the first item of the proposition.

From Table 2, it is seen that G_{12} is defined if and only if $H_2(D) < S_2^{in} + \frac{k_2}{k_1}S_1^{in}$. Since $H_2(D) > H_1(D)$, the results show that G_{11} is also defined and $G_{11}(D, S_1^{in}, S_2^{in}) > G_{12}(D, S_1^{in}, S_2^{in})$. This proves the second item of the proposition.

If G_{11} is defined, then $S_1^{in} > \lambda_1(D_1)$. Hence,

$$S_2^{in} + \frac{k_2}{k_1}S_1^{in} - H_1(D) = S_2^{in} - \lambda_2(D_2) + \frac{k_2}{k_1}\left(S_1^{in} - \lambda_1(D_1)\right) > S_2^{in} - \lambda_2(D_2).$$

Therefore, if G_{01} is defined, we have $G_{11}(D, S_1^{in}, S_2^{in}) > G_{01}(D, S_2^{in})$. This proves the third item of the proposition.

Appendix B.5.2. Proof of Proposition 6

The proof follows the same ideas and computations as the proof of Proposition 1. See Appendix A.3.1 for the details.

Appendix B.5.3. Proof of Proposition 7

Since the functions γ_2 and μ_1 are increasing, the function $S_1^{in} \mapsto \gamma_2\left(\mu_1\left(S_1^{in}\right)\right)$ is increasing. Therefore, the condition $S_1^{in} > S_1^0$ is equivalent to the fact that the point (S_1^{in}, S_2^{in}) lies to the right of the curve Σ_0. Similarly, if the function μ_i / μ_1' is increasing, then the function

$$S_1^{in} \mapsto \gamma_2\left(\mu_1\left(S_1^{in}\right)\right) + \frac{k_2}{k_1}\frac{\mu_1(S_1^{in})}{\mu_1'(S_1^{in})}$$

is increasing. Therefore the condition $S_1^{in} < S_1^1$ is equivalent to the fact that the point (S_1^{in}, S_2^{in}) lies to the left of the curve Σ_1.

Appendix B.5.4. Proof of Proposition 8

Equation (49) is equivalent to the equation

$$G_0(D_0^*(S_2^{in})) = G_1(D_1^*(S_1^{in}, S_2^{in}))$$

where $D_0^*(S_2^{in})$ is the solution to (38) and $D_1^*(S_1^{in}, S_2^{in})$ is the solution to (40). Therefore, using (34) and (35), we deduce that we need to solve the following system of three equations with four unknowns S_1^{in}, S_2^{in}, D_0, and D_1.

$$D_0\left(S_2^{in} - \lambda_2(\alpha D_0)\right) = D_1\left(S_2^{in} + \frac{k_2}{k_1}S_1^{in} - \lambda_2(\alpha D_1) - \frac{k_2}{k_1}\lambda_1(\alpha D_1)\right), \tag{A8}$$

$$S_2^{in} = \gamma_2(\alpha D_0), \tag{A9}$$

$$S_2^{in} + \frac{k_2}{k_1}S_1^{in} = \gamma_2(\alpha D_1) + \frac{k_2}{k_1}\gamma_1(\alpha D_1). \tag{A10}$$

Substituting (A9) and (A10) into (A8), we obtain

$$D_0(\gamma_2(\alpha D_0) - \lambda_2(\alpha D_0)) = D_1\left(\gamma_2(\alpha D_1) + \frac{k_2}{k_1}\gamma_1(\alpha D_1) - \lambda_2(\alpha D_1) - \frac{k_2}{k_1}\lambda_1(\alpha D_1)\right).$$

Replacing γ_2 and γ_1 by their expressions (36) and (37), respectively, we obtain

$$D_0^2 \lambda_2'(\alpha D_0) = D_1^2\left(\lambda_2'(\alpha D_1) + \tfrac{k_2}{k_1}\lambda_1'(\alpha D_1)\right).$$

Therefore, αD_0 is a solution to equation

$$\phi(\alpha D_0) = \alpha^2 D_1^2\left(\lambda_2'(\alpha D_1) + \tfrac{k_2}{k_1}\lambda_1'(\alpha D_1)\right),$$

where ϕ is as in Hypothesis (8). Using this hypothesis, we obtain $\alpha D_0 = \Delta(D_1)$, where Δ is given by (48). Substituting in (A9) and (A10), we obtain

$$S_2^{in} = \gamma_2(\Delta(D_1)), \qquad \gamma_2(\Delta(D_1)) + \tfrac{k_2}{k_1}S_1^{in} = \gamma_2(\alpha D_1) + \tfrac{k_2}{k_1}\gamma_1(\alpha D_1).$$

These equations show that the point (S_1^{in}, S_2^{in}) belongs to the curve $\mathcal{C}$, defined by equations (47). The system formed by the three Equations (A8)–(A10) shows that the reciprocal is also true, i.e., any point on curve $\mathcal{C}$ is a point where $\max_D G_0 = \max_D G_1$. Since the partial derivative of G_1 with respect to S_1^{in} is positive, we see that we have $\max_D G_1 > \max_D G_0$ to the right of curve $\mathcal{C}$.

References

1. Kelessidis, A.; Stasinakis, A.S. Comparative study of the methods used for treatment and final disposal of sewage sludge in European countries. *Waste Manag.* **2012**, *32*, 1186–1195. [CrossRef] [PubMed]
2. Batstone, D.J.; Keller, J.; Angelidaki, I.; Kalyuzhnyi, S.V.; Pavlostathis, S.G.; Rozzi, A.; Sanders, W.T.M.; Siegrist, H.; Vavilin, V.A. The IWA Anaerobic Digestion Model No 1 (ADM1). *Water Sci. Technol.* **2002**, *45*, 65–73. [CrossRef] [PubMed]
3. Bornhöft, A.; Hanke-Rauschenbach, R.; Sundmacher, K. Steady-state analysis of the Anaerobic Digestion Model No. 1 (ADM1). *Nonlinear Dyn.* **2013**, *73*, 535–549. [CrossRef]
4. Donoso-Bravo, A.; Mailier, J.; Martin, C.; Rodríguez, J.; Aceves-Lara, C.A.; Vande Wouwer, A. Model selection, identification and validation in anaerobic digestion: A review. *Water Res.* **2011**, *45*, 5347–5364. [CrossRef] [PubMed]
5. Shen, S.; Premier, G.; Guwy, A.; Dinsdale, R. Bifurcation and stability analysis of an anaerobic digestion model. *Nonlinear Dyn.* **2007**, *48*, 465–489. [CrossRef]
6. Wade, M.J. Not just numbers: Mathematical modelling and its contribution to anaerobic digestion processes. *Processes* **2020**, *8*, 888. [CrossRef]
7. Kim, M.; Ahn, Y.H.; Speece, R. Comparative process stability and efficiency of anaerobic digestion; mesophilic vs. thermophilic. *Water Res.* **2002**, *36*, 4369–4385. [CrossRef]
8. Abouelenien, F.; Miura, T.; Nakashimada, Y.; Elleboudy, N.S.; Al-Harbi, M.S.; Ali, E.F.; Shukry, M. Optimization of Biomethane Production via Fermentation of Chicken Manure Using Marine Sediment: A Modelling Approach Using Response Surface Methodology. *Int. J. Environ. Res. Public Health* **2021**, *18*, 1988. [CrossRef]
9. Kazimierowicz, J.; Dzienis, L.; Dębowski, M.; Zieliński, M. Optimisation of methane fermentation as a valorisation method for food waste products. *Biomass Bioenergy* **2021**, *144*, 105913. [CrossRef]
10. Lin, C.-Y.; Chai, W.S.; Lay, C.-H.; Chen, C.-C.; Lee, C.-Y.; Show, P.L. Optimization of Hydrolysis-Acidogenesis Phase of Swine Manure for Biogas Production Using Two-Stage Anaerobic Fermentation. *Processes* **2021**, *9*, 1324. [CrossRef]
11. Bayen, T.; Gajardo, P. On the steady state optimization of the biogas production in a two-stage anaerobic digestion model. *J. Math. Biol.* **2019**, *78*, 1067–1087. [CrossRef] [PubMed]
12. Doran, P.M. Reactor Engineering, In *Bioprocess Engineering Principles*, 2nd ed.; Academic Press: Cambridge, MA, USA, 2013; pp. 761–852. [CrossRef]
13. Ghouali, A.; Sari, T.; Harmand, J. Maximizing biogas production from the anaerobic digestion. *J. Process Control.* **2015**, *36*, 79–88. [CrossRef]
14. Harmand, J.; Lobry, C.; Rapaport, A.; Sari, T. *Optimal Control in Bioprocesses: Pontryagin's Maximum Principle in Practice*; Wiley ISTE, Ed.; Wiley: Hoboken, NJ, USA, 2019.
15. Sbarciog, M.; Loccufier, M.; Noldus, E. Determination of appropriate operating strategies for anaerobic digestion systems. *Biochem. Eng. J.* **2010**, *51*, 180–188. [CrossRef]
16. Sbarciog, M.; Loccufier, M.; Vande Wouwer, A. On the optimization of biogas production in anaerobic digestion systems. *IFAC Proc.* **2011**, *44*, 7150–7155. [CrossRef]
17. Sbarciog, M.; Loccufier, M.; Vande Wouwer, A. An optimizing start-up strategy for a bio-methanator. *Bioprocess Biosyst. Eng.* **2012**, *35*, 565–578. [CrossRef]
18. Sbarciog, M.; Moreno, J.A.; Vande Wouwer, A. A biogas-based switching control policy for anaerobic digestion systems. *IFAC Proc.* **2012**, *45*, 603–608.

19. Shuler, M.L.; Fikret Kargi, F.; DeLisa, M. *Bioprocess Engineering, Basic Concepts*, 3rd ed.; Prentice Hall: Hoboken, NJ, USA, 2017.

20. Wang, H.H.; Krstic, M.; Bastin, G. Optimizing Bioreactors by Extremum Seeking. *Int. J. Adapt. Control. Signal Process.* **1999**, *13*, 651–669. [CrossRef]

21. Weedermann, M.; Wolkowicz, G.; Sasara, J. Optimal biogas production in a model for anaerobic digestion. *Nonlinear Dyn.* **2015**, *81*, 1097–1112. [CrossRef]

22. Harmand, J.; Lobry, C.; Rapaport, A.; Sari, T. *The Chemostat: Mathematical Theory of Micro-organism Cultures*; Wiley ISTE, Ed.; Wiley: Hoboken, NJ, USA, 2017.

23. Monod, J. La technique de culture continue: Théorie et applications. *Ann. L'Institut. Pasteur.* **1950**, *79*, 390–410. [CrossRef]

24. Smith, H.L.; Waltman, P. *The Theory of the Chemostat: Dynamics of Microbial Competition*; Cambridge University Press: Cambridge, UK, 1995.

25. Bernard, O.; Hadj-Sadock, Z.; Dochain, D.; Genovesi, A.; Steyer, J.P. Dynamical model development and parameter identification for an anaerobic waste water treatment process. *Biotechnol Bioeng.* **2001**, *75*, 424–438. [CrossRef]

26. Benyahia, B.; Sari, T.; Cherki, C.; Harmand, J. Bifurcation and stability analysis of a two step model for monitoring anaerobic digestion processes. *J. Process Control* **2012**, *22*, 1008–1019. [CrossRef]

27. Arzate, J.A.; Kirstein, M.; Ertem, F.C.; Kielhorn, E.; Ramirez, M.H.; Neubauer, P.; Cruz-Bournazou, M.N.; Junne, S. Anaerobic digestion model (AM2) for the description of biogas processes at dynamic feedstock loading rates. *Chem. Ing. Tech.* **2017**, *89*, 686–695. [CrossRef]

28. García-Dièguez, C.; Bernard, O.; Roca, E. Reducing the anaerobic digestion model no.1 for its application to an industrial waste water treatment plant treating winery efluent waste water. *Bioresour. Technol.* **2013**, *132*, 244–253. [CrossRef] [PubMed]

29. Sbarciog, M.; Vande Wouwer, A. A constructive approach to assess the stability of anaerobic digestion systems. *Chem. Eng. Sci.* **2020**, *227*, 115931. [CrossRef]

30. Weedermann, M.; Seo, G.; Wolkowics, G. Mathematical Model of Anaerobic Digestion in a Chemostat: Effects of Syntrophy and Inhibition. *J. Biol. Dyn.* **2013**, *7*, 59–85. [CrossRef]

31. Daoud, Y.; Abdellatif, A.; Harmand, J. Modèles mathématiques de digestion anaérobie: Effet de l'hydrolyse sur la production du biogaz. *ARIMA J.* **2019**, *28*, 35–59. [CrossRef]

32. Borisov, M.; Dimitrova, N.; Simeonov, I. Mathematical Modelling and Stability Analysis of a Two-Phase Biosystem. *Processes* **2020**, *8*, 791. [CrossRef]

33. Chorukova, E.; Simeonov, I. Mathematical modelling of the anaerobic digestion in two-stage system with production of hydrogen and methane including three intermediate products. *Int. J. Hydrog. Energy* **2020**, *45*, 11550–11558. [CrossRef]

34. Giovannini, G.; Sbarciog, M.; Steyer, J.P.; Chamy, C.; Vande Wouwer, A. On the derivation of a simple dynamic model of anaerobic digestion including the evolution of hydrogen. *Water Res.* **2018**, *134*, 209–225. [CrossRef]

35. Khedim, K.; Benyahia, B.; Cherki, B.; Sari, T.; Harmand, J. Effect of control parameters on biogas production during the anaerobic digestion of protein-rich substrates. *Appl. Math. Model.* **2018**, *61*, 351–376. [CrossRef]

36. Pavlou, S. Computing operating diagrams of bioreactors. *J. Biotechnol.* **1999**, *71*, 7–16. [CrossRef]

37. Wade, M.; Pattinson, R.; Parker, N.; Dolfing, J. Emergent behaviour in a chlorophenol-mineralising three tiered microbial 'food web'. *J. Theor. Biol.* **2016**, *389*, 171–186. [CrossRef] [PubMed]

38. Xu, A.; Dolfing, J.; Curtis, T.; Montague, G.; Martin, E. Maintenance affects the stability of a two-tiered microbial 'food chain'? *J. Theor. Biol.* **2011**, *276*, 35–41. [CrossRef] [PubMed]

39. Daoud, Y.; Abdellatif, A.; Sari, T.; Harmand, J. Steady state analysis of a syntrophic model: The effect of a new input substrate concentration. *Math. Model. Nat. Phenom.* **2018**, *13*, 31. [CrossRef]

40. Fekih-Salem, R.; Daoud, Y.; Abdellatif, N.; Sari, T. A mathematical model of anaerobic digestion with syntrophic relationship, substrate inhibition and distinct removal rates. *SIAM J. Appl. Dyn. Syst.* **2021**, *20*, 621–1654. [CrossRef]

41. Mtar, T.; Fekih-Salem, R.; Sari, T. Interspecific density-dependent model of predator-prey relationship in the chemostat. *Int. J. Biomath.* **2021**, *14*, 2050086. [CrossRef]

42. Sari, T.; Benyahia, B. The operating diagram for a two-step anaerobic digestion model. *Nonlinear Dyn.* **2021**, *105*, 2711–2737. [CrossRef]

43. Sari, T.; Harmand, J. A model of a syntrophic relationship between two microbial species in a chemostat including maintenance. *Math. Biosci.* **2016**, *275*, 1–9. [CrossRef]

44. Sari, T.; Wade, M. Generalised approach to modelling a three-tiered microbial food-web. *Math. Biosci.* **2017**, *291*, 21–37. [CrossRef]

45. Fredrickson, A.G.; Tsuchiya, H.M. Microbial kinetics and dynamics. In *Chemical Reactor Theory, A Review*; Lapidus, L., Amundson, N.R., Eds.; Prentice-Hall: Englewood Cliffs, NJ, USA, 1977; pp. 405–483.

46. Bastin, G.; Dochain, D. *On-Line Estimation and Adaptive Control of Bioreactors*; Elsevier Science Publishers: Amsterdam, The Netherlands, 1990.

47. Herbert, D.R.; Elsworth, R.; Telling, R.C. The continuous culture of bacteria: A theoretical 686 and experimental study. *J. Gen. Microbiol.* 1956, *14*, 601–622. [CrossRef]

48. Dochain, D.; Bastin, G. Adaptive identification and control algorithms for nonlinear bacterial growth systems. *Automatica* **1984**, *20*, 621–634. [CrossRef]

49. Polihronakis, M.; Petrou, L.; Deligiannis, A. Parameter adaptive control techniques for anaerobic digesters; real-life experiments. *Comput. Chem. Eng.* **1993**, *17*, 1167–117. [CrossRef]

50. Renard, P.; Dochain, D.; Bastin, G.; Naveau, H.; Nyns, E.J. Adaptive control of anaerobic digestion processes—A pilot-Scale application. *Biotechnol. Bioeng.* **1988**, *31*, 287–294. [CrossRef] [PubMed]
51. Dali-Youcef, M.; Rapaport, A.; Sari, T. Study of performance criteria of serial configuration of two chemostats. *Math. Biosci. Eng.* **2020**, *17*, 6278–6309. [CrossRef] [PubMed]
52. Dali-Youcef, M.; Rapaport, A.; Sari, T. Performances Study Criteria of Two Interconnected Chemostats with Mortality. 2021. Available online: https://hal.inrae.fr/hal-03318978 (accessed on 20 January 2022).
53. Dali-Youcef, M.; Sari, T. The Productivity of Two Serial Chemostats. 2021. Available online: https://hal.inrae.fr/hal-03445797 (accessed on 20 January 2022).
54. Benyahia, B.; Sari, T. Effect of a new variable integration on steady states of a two-step Anaerobic Digestion Model. *Math. Biosci. Eng.* **2020**, *17*, 5504–5533. [CrossRef]
55. Benyahia, B.; Sari, T.; Cherki, C.; Harmand, J. Anaerobic membrane bioreactor modelling in the presence of Soluble Microbial Products (SMP)—The Anaerobic Model AM2b. *Chem. Eng. J.* **2013**, *228*, 1011–1022. [CrossRef]
56. Hess, J.; Bernard, O. Design and Study of a risk management criterion for an unstable anaerobic waste water treatment process. *J. Process Control* **2008**, *18*, 71–79. [CrossRef]
57. Harmand, J.; Rapaport, A.; Dochain, D. Increasing the dilution rate can globally stabilize two-step biological systems. *J. Process Control.* **2020**, *95*, 67–74. [CrossRef]
58. Hanaki, M.; Harmand, J.; Mghazli, Z.; Rapaport, A.; Sari, T.; Ugalde, P. Mathematical study of a two-stage anaerobic model when the hydrolysis is the limiting step. *Processes* **2021**, *9*, 2050. [CrossRef]
59. Stephanopoulos, G. The dynamics of commensalism. *Biotechnol. Bioeng.* **1981**, *23*, 2243–2255. [CrossRef]
60. Burchard, A. Substrate degradation by a mutualistic association of two species in the chemostat. *J. Math. Biol.* **1994**, *32*, 465–489. [CrossRef]
61. El-Hajji, M.; Mazenc, F.; Harmand, J. A mathematical study of a syntrophic relationship of a model of anaerobic digestion process. *Math. Biosci. Eng.* **2010**, *7*, 641–656. [PubMed]
62. Sari, T.; El-Hajji, M.; Harmand, J. The mathematical analysis of a syntrophic relationship between two microbial species in a chemostat. *Math. Biosci. Eng.* **2012**, *9*, 627–645. [PubMed]
63. Reilly, P.J. Stability of commensalistic systems. *Biotechnol. Bioeng.* **1974**, *16*, 1373–1392. [CrossRef]
64. Wade, M.J.; Harmand, J.; Benyahia, B.; Bouchez, T.; Chaillou, S.; Cloez, B.; Godon, J.J.; Moussa-Boudjemaa, B.; Rapaport, A.; Sari, T.; et al. Perspectives in mathematical modelling for microbial ecology. *Ecol. Model.* **2016**, *321*, 64–74. [CrossRef]
65. Di, S.; Yang, A. Analysis of productivity and stability of synthetic microbial communities. *J. R. Soc. Interface* **2019**, *16*, 20180859. [CrossRef]
66. Desmond-Le Quéméner, E.; Bouchez, T. A thermodynamic theory of microbial growth. *ISME J.* **2014**, *8*, 1747–1751. [CrossRef]

Article

Off-Gas-Based Soft Sensor for Real-Time Monitoring of Biomass and Metabolism in Chinese Hamster Ovary Cell Continuous Processes in Single-Use Bioreactors

Tobias Wallocha * and Oliver Popp

Roche Diagnostics GmbH, Pharma Research and Early Development, 82377 Penzberg, Germany; oliver.popp@roche.com
* Correspondence: tobias.wallocha@roche.com; Tel.: +49-8856-6013298

Abstract: In mammalian cell culture, especially in pharmaceutical manufacturing and research, biomass and metabolic monitoring are mandatory for various cell culture process steps to develop and, finally, control bioprocesses. As a common measure for biomass, the viable cell density (VCD) or the viable cell volume (VCV) is widely used. This study highlights, for the first time, the advantages of using VCV instead of VCD as a biomass depiction in combination with an oxygen-uptake- rate (OUR)-based soft sensor for real-time biomass estimation and process control in single-use bioreactor (SUBs) continuous processes with Chinese hamster ovary (CHO) cell lines. We investigated a series of 14 technically similar continuous SUB processes, where the same process conditions but different expressing CHO cell lines were used, with respect to biomass growth and oxygen demand to calibrate our model. In addition, we analyzed the key metabolism of the CHO cells in SUB perfusion processes by exometabolomic approaches, highlighting the importance of cell-specific substrate and metabolite consumption and production rate qS analysis to identify distinct metabolic phases. Cell-specific rates for classical mammalian cell culture key substrates and metabolites in CHO perfusion processes showed a good correlation to qOUR, yet, unexpectedly, not for qGluc. Here, we present the soft-sensing methodology we developed for qPyr to allow for the real-time approximation of cellular metabolism and usage for subsequent, in-depth process monitoring, characterization and optimization.

Keywords: process analytical technologies (PAT); off-gas analytic; real-time monitoring; viable cell biomass; perfusion process; continuous process; single-use bioreactor (SUB); oxygen uptake rate (OUR); soft sensor

Citation: Wallocha, T.; Popp, O. Off-Gas-Based Soft Sensor for Real-Time Monitoring of Biomass and Metabolism in Chinese Hamster Ovary Cell Continuous Processes in Single-Use Bioreactors. *Processes* **2021**, *9*, 2073. https://doi.org/10.3390/pr9112073

Academic Editor: Philippe Bogaerts

Received: 21 October 2021
Accepted: 16 November 2021
Published: 19 November 2021

Publisher's Note: MDPI stays neutral with regard to jurisdictional claims in published maps and institutional affiliations.

1. Introduction

Chinese hamster ovary cells (CHO) represent the backbone of commercial and research expression hosts used for the manufacture of monoclonal antibodies (mAB) for therapeutic purposes. In the last decades, a variety of different production strategies and processes have been developed to ensure high yields and product quality as well as operational efficiency and reproducibility [1]. In order to cope with these demands, a smart synergy of flexible, state-of-the-art bioreactor systems and up-to-date process analytical techniques (PAT) is essential and highly recommended by the FDA [2]. Single-use bioreactor (SUB) systems have gained tremendous interest in the biopharmaceutical industry as these bioreactors and their peripheral solutions can remarkably elevate the efficiency and, of course, the flexibility of modern manufacturing processes [3]. SUB systems are currently an often used alternative and cover a broad range of different applications despite lacking many conventional engineering parameters or availability of compatible single-use sensors [4,5]. In the context of modern bioprocess development and optimization, the online monitoring of crucial key performance indicators (KPIs) is necessary to ensure high process and product quality [6]. Therefore, monitoring sensors that can be applied to SUB systems and deliver high-grade, real-time information is a significant need, especially for one of the most valuable indicators, the biomass concentration [7,8]. Even if common sensor systems, e.g., hard-type sensors that have been adjusted to match single-use bioreactor requirements, are available [9], there is still a challenge to measure the biomass concentration online [10]. As hard-type sensor probes need to fulfill a variety of prerequisites, offline-based methods, such as image analysis with trypan blue staining and subsequent cell counting, lack limited amount of samples, are time consuming and require sample taking [11]. An appropriate alternative for online measuring of biomass is soft sensors. Soft sensors constitute the interoperation of hard-type sensors that can be implemented, or peripherally attached, to the bioreactor system in numerous ways using a software-based model as estimator [12,13]. Utilization of different soft-sensor approaches for online determination of biomass has been reported in many valuable contributions focusing on radio frequency impedance, Raman spectroscopy or off-gas analytic techniques [10,11,14–20]. Online biomass estimation via off-gas measurement generally relies on the oxygen uptake rate (OUR) as this variable is one of the best indicators for cell physiological activity and correlates very well with metabolic turnover rates and the concentration of viable biomass.

The knowledge of the OUR as a metabolic marker allows a deeper understanding of intrinsic physiological performance of the biomass and can be merged with other process variables to create meaningful new information in the sense of soft sensing [17,21–25]. The online measure of the OUR can be achieved by the global mass balance (GMB) approach which is easy to implement in SUB systems and enables a disturbance-free measure of the OUR. In perfusion and continuous processes, process interventions or other perturbations that might affect the steady-state mode are undesirable [24]. Only the knowledge of volumetric flow, composition of the gas in the inlet and outlet and the fermenter volume is required, constituting an advantage when compared to liquid-phase measurements, such as the dynamic method [23]. Perfusion processes require an optimized perfusion strategy to allow VCD and cellular biomass generation and continuous product formation and, at the same time, avoid unintentionally high economical demands of perfusion media. The exchange of unconditioned media is highly dependent upon the applied perfusion rate D, which ensures a sufficient supply of substrates to keep, in the best case, the substrate concentrations in a steady state. In parallel, the constant removal of conditioned media preserves the level of known metabolites, such as lactate, ammonia and metabolic and abiotic break-down products of amino acids, nucleotides and lipoid metabolism [26], as well as unknown cytotoxic/cytostatic metabolites at uncritical levels. It has been shown previously that the cell-specific nutrient uptake rates in perfused cell cultures correlated with the cell-specific growth rate for hybridomas and recombinant CHO cell lines [27]. This knowledge strengthens the goal to analyze and use cell- and biomass-specific rates and further metabolic KPIs for process development and optimization. In this work, we present a real-time off-gas-based biomass soft sensor that can be applied for the perfusion-based biomass growth phase of continuous processes. We used a proof-of-concept data set of 14 similar SUB continuous fermentation processes including 14 different mAB expressing CHO cell lines for model calibration. We aimed for this heterologous set-up of cell lines in order to cover a broad spectrum of different metabolically active cells as the oxygen demand between cell lines can differ significantly [24]. The soft sensor consists of two different models to predict the biomass in terms of viable cell density (VCD) and viable cell volume concentration (VCV) using a multilinear regression approach. The OUR was measured as a major input variable for both using the GBM technique and observed noise distortions were minimized by data preprocessing to improve model accuracy. Model prediction quality assessment was done by RMSE as well as MAPE and MdAPE calculation, enabling an in-depth analysis of errors and their distribution. Real-time biomass prediction was then applied on three different and unknown cell lines to the prediction models by utilization of two moving average methods to remove the OUR signal noise. Furthermore, we elaborate on the question of how biomass can be described most properly in modern bioprocesses. VCD, as a commonly used measure, lacks the information of the cellular volume and refers solely to the number of cells.

Our utilization of VCV as a measure for biomass delivers more information, taking into account cell volume, which can lead to more precise correlations with the OUR [14]. In addition, cell size can have a direct impact on oxygen demand, leading to higher oxygen requirements from larger cells compared to smaller cells, reflecting a positive correlation between OUR and cell size [22]. We also highlight the advantage of the shown off-gas-based biomass soft sensor in SUB continuous processes and illustrate how the biomass can be described best when VCD or VCV are applied as descriptive measures. Process variables, such as specific oxygen uptake rate per single cell or per viable cell volume, can raise a different picture when the volume per cell is not constant.

2. Materials and Methods

2.1. Cell Lines

For this study, 17 different in-house-generated Chinese hamster ovary clonal cell lines (CHO-K1), engineered to produce 14 different mAbs, were used (Table 1). All clones were cultivated using a proprietary, chemically-defined (CD), serum-free, in-house base and perfusion medium. In general, cells can also be cultivated in commercially chemical-defined media, such as CD-CHO (Thermo Fisher Scientific, Waltham, MA, USA).

Table 1. Overview of all processes with their relation to the expressed antibody and clone that were used to create the biomass prediction models and for validation.

No.	Process	Expressed Antibody	Clone	Purpose
1	P-01	A-01	C-01	Training
2	P-02	A-02	C-02	Training
3	P-03	A-03	C-03	Training
4	P-04	A-04	C-04	Training
5	P-05	A-01	C-05	Training
6	P-06	A-05	C-06	Training
7	P-07	A-06	C-07	Training
8	P-08	A-07	C-08	Training
9	P-09	A-05	C-09	Training
10	P-10	A-08	C-10	Training
11	P-11	A-09	C-11	Training
12	P-12	A-05	C-12	Training
13	P-13	A-10	C-13	Training
14	P-14	A-11	C-14	Training
15	P-15	A-12	C-15	Validation
16	P-16	A-13	C-16	Validation
17	P-17	A-14	C-17	Validation

2.2. Cell Cultivation and in Process Control

Cells were thawed in a shake flask and maintained in a humidified shaking incubator (Multitron Cell, Infors AG, Headoffice, Switzerland) at 36.5 °C under 7% (v/v) carbon dioxide (CO_2), applying a constant shaking rate and relative humidity of 70%. Cell passage took place every 3–4 days for scale-up purposes. After 10 days, cells were transferred into a wave-mixed SUB (Biostat® RM, Sartorius Stedim Biotech GmbH, Göttingen, Germany) for a 4 day long inoculation train as the batch phase for further cell propagation. During this step, the temperature was controlled at 36.5 °C and device-internal optical probes were used to control pH at 7.00 and dissolve oxygen (DO) to 30% saturation by gassing with a mixture of process air, nitrogen (N_2), carbon dioxide (CO_2) and pure oxygen (O_2). Rocking motion was held constant at 15 rocks/min at an angle of 9 degrees.

After the inoculation train process was finished, cells were transferred to a stirred SUB (HyPerforma™ SUB, Thermo Fisher Scientific, Waltham, MA, USA) with an appropriate seeding cell density depending on the doubling time in the preceding inoculation train process. Temperature was controlled to 36.5 °C and stirrer speed was set constant at 140 rpm. DO concentration was measured using a stainless steel optical probe (VisiFerm™

DO ECS 225, Hamilton, Switzerland) and controlled to 30% saturation analogous to the wave-mixed process. pH monitoring was done by pH probe (Inpro[®] 3253/225/PT100, Mettler Toledo, Columbus, OH, USA) and regulated by CO_2 gassing and 1 M sodium carbonate addition to pH 7.00.

Cell retention was enabled using a hollow fiber module (KrosFlo[®] MBT[®], Repligen, Waltham, MA, USA) with 0.2 µm pore size. Perfusion was started 24 h after inoculation and modified stepwise according to the following protocol: 24 h after inoculation, perfusion mode was started with normalized fermentation volumes per day (vvd_n) of 1 and increased to 2 vvd_n after another 24 h. Further increase was done after 72 h to a vvd_n of 3 until the last raise was applied 120 h after inoculation up to a vvd_n of 4.55 towards process end. During perfusion mode, the fermentation volume was kept constant by weight-controlled addition of fresh perfusion media and no cell bleed took place. After the dynamic state with altering perfusion rates, the steady-state process would start with a constant normalized perfusion rate of 4.55 vvd_n with parallel cell bleed to keep biomass concentration and product titer stable while product yield rises (Figure 1A). The biomass soft sensor presented in this work is proposed for automatically controlling cell bleeding during the steady-state perfusion process.

Offline samples were drawn at least once per day using a sterile syringe (Omnifix Luer Lock Solo, B. Braun Melsungen AG, München, Germany) and aliquoted for further analytic purposes.

2.3. Oxygen Balancing and OUR Calculation

To quantify the OUR, the well-known global mass balance approach is used as shown in Equation (1):

$$\frac{dc_{O2}}{dt} = OTR(t) - OUR(t) \tag{1}$$

The rate of oxygen that is transferred from the gas phase into the liquid phase (OTR) is influenced by several factors, such as the fluid-side oxygen mass transfer coefficient ($k_L a$), maximum possible oxygen solubility (c_{o2}^*) and the current dissolved oxygen concentration ($c_{O2,L}$). Therefore, Equation (1) can be written as:

$$\frac{dc_{O2}}{dt} = k_L a(t) \cdot (c_{O2}^*(t) - c_{O2,L}(t)) - OUR(t) \tag{2}$$

Since the oxygen saturation during all fermentation processes in this work was controlled to 30%, steady-state conditions can be assumed, thus OTR equals OUR, leading the temporal change of soluble oxygen concentration to 0. Within this condition, the OUR calculation is possible using a sensitive off-gas analyzer and the further application of a mass balance approach that is based on the mass of oxygen that enters ($O_{2,in}$) and leaves ($O_{2,out}$) the bioreactor system [5]. Since the temperature of the gas mixture (consisting of air, N_2, CO_2, O_2) that flows into the system is known, the oxygen mass intake can be calculated using the ideal gas law (R = gas constant, M_{O2} = molar mass of oxygen). The same principle applies to the calculation of the oxygen mass leaving the system where the measured oxygen volume fraction and the gas flow rate are used. As all fermentation processes were operated without any overpressure, the inlet gassing flow rate equals the outlet flow rate ($F_{in} = F_{out}$). Furthermore, the gas inlet and outlet temperature is presumed to be equal due to the length of the tubing ($T_{in} = T_{out}$). Finally, the current liquid fermentation volume (V_L) and the ambient pressure (P_{amb}) are needed to calculate the OUR as shown in Equation (3):

$$OUR(t) = (O_{2,in}(t) - O_{2,out}(t)) \cdot \frac{1}{V_L(t)} = \frac{(V_{O2,in}(t) - V_{O2,out}(t)) \cdot M_{O2} \cdot P_{amb}}{R \cdot T \cdot V_L(t)} \tag{3}$$

For the graphical representation in this study, the volumetric OUR was normalized to the maximum value in the training data set.

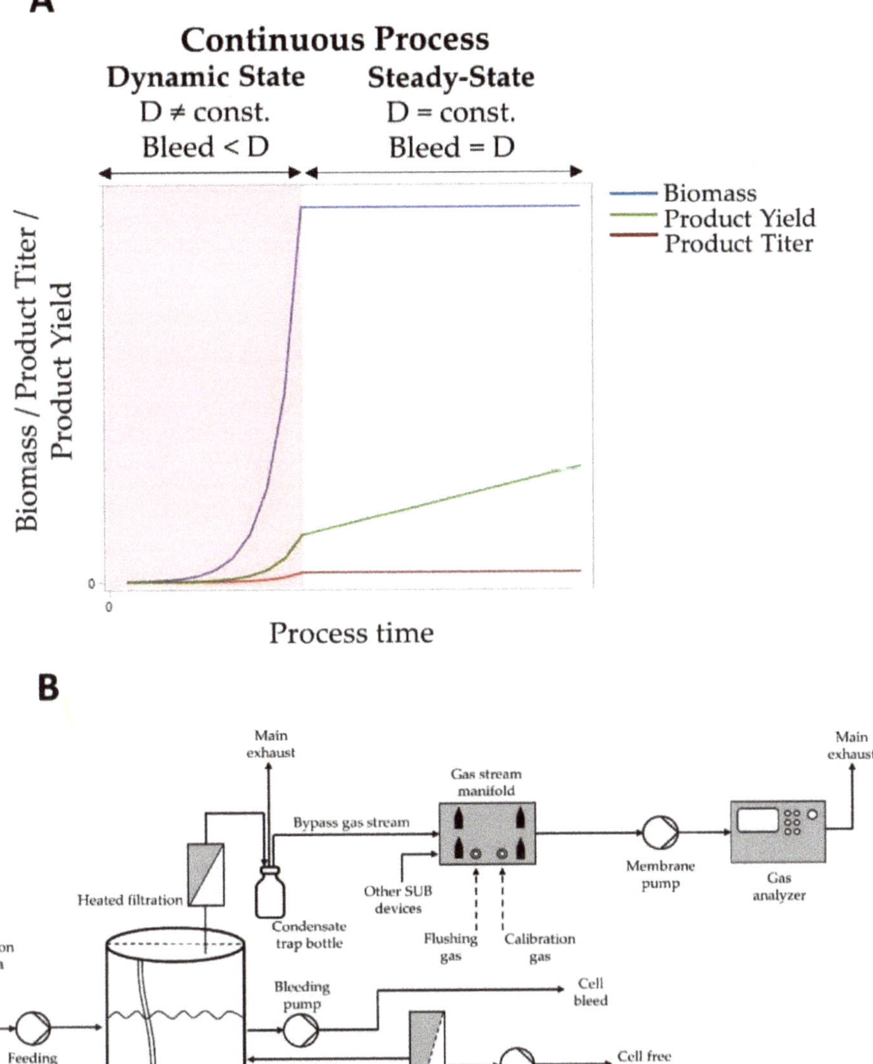

Figure 1. Schematic overview of (**A**) continuous process consisting of perfusion-based dynamic state (red marked area) and cell-bleed-based steady-state phase and (**B**) the off-gas measurement set-up for continuous processes in single-use bioreactors.

2.4. Off-Gas Measurement Set-Up

To perform off-gas analytic measurements in single-use bioreactors, a bypass solution was applied. The gas mixture enters the SUB at the bottom via a microsparger or open pipe. After the gas leaves the bioreactor, it passes heated filters and is then transported into a bottle that serves as a divider as well as a condensate trap. Here the gas stream is separated into two parts where the major part leaves the system via main exhaust whereas a smaller portion is distributed by a self-made gas manifold chamber and actively drawn using a membrane pump (Laboport® N96, KNF Neuberger GmbH, Freiburg, Germany) to the gas analyzer (DasGip® GA4, Eppendorf AG, Hamburg, Germany). The gas manifold and the multi-channel gas analyzer allow simultaneous off-gas measurements on up to four fermentation systems in parallel. Therefore, no multiplexing or flushing steps were necessary. We used gas-tight Teflon tubing for the whole transportation of the gas stream after it leaves the heated sterile filters as silicone tubing tends to be permeable for gases (Figure 1B). The gas analyzer was two-point calibrated before each process with air and a defined gas mixture (Linde AG, Höllriegelskreuth, Germany) containing 10% CO_2 and 2% O_2. Unused gas analyzer channels were flushed with humidified air to preserve sensor lifetime.

2.5. Data Collection and Preprocessing

All used offline and online data points for model generation were taken from perfusion-based biomass growth phases of continuous processes (P-01 to P-14). These processes were carried out in two identical SUB fermentation systems. Processes were technically identical in terms of used media, cultivation conditions and process operating conditions, as described above. Table 1 provides an overview on the respective cell lines, expressed antibody and data used for model creation and validation. In order to build a biomass prediction model, the collected OUR data gathered from off-gas analytics needed to be preprocessed to remove signal noise and measurement distortions.

All OUR raw data were fitted by higher order polynomials using corresponding regressions. The degree of the applied polynomials was chosen by the highest correlation coefficient (R^2) in order to select the most descriptive regression model for each process. Since the OUR is the main factor affecting model quality, this pretreatment step was mandatory for achieving high model prediction accuracy.

2.6. Model Generation and Assessment

The prediction model was built and evaluated using the statistical software JMP® 15.2.0 (SAS Institute, Cary, NC, USA). For the modeling procedure, treated OUR variables (OUR_{fitted}), offline VCV and VCD and process time were used to fit a multilinear regression model. To address model performance, the root mean square error (RMSE) with Bessel's correction was calculated as shown in Equation (4):

$$\text{RMSE} = \sqrt{\frac{\sum_{i=1}^{n}(y_i - y_i')^2}{n-1}} \tag{4}$$

In order to compare the performance of both prediction models, the mean absolute percentage error (MAPE) was calculated (Equation (5)):

$$\text{MAPE} = \frac{100}{n}\sum_{i=1}^{n}\left|\frac{y_i - y_i'}{y_i}\right| \tag{5}$$

In both equations, y_i represents the observed values, y_i' the corresponding predicted values and n the number of fitted points in total. Since the RMSE and MAPE are based on averages, outliers can negatively distort their predication [28]. Therefore, the MAPE was also calculated using the median of absolute percentage errors (MdAPE) to paint a more robust view on model accuracy.

2.7. Real-Time Prediction and Validation

The models were implemented in data visualization and analyzer software SEEQ (SEEQ Corporation, Seattle, WA, USA) for calculation of VCV and VCD predictions in real time. Therefore, three new data sets from technical replicate processes (P-15, 16 and

17) with unknown cell lines to both models (C-15, 16 and 17) were used to validate the prediction models. Because online OUR raw data have low signal-to-noise ratios, as described above, two moving average smoothing algorithms were applied to assess their impact on final prediction accuracy. The used algorithm was either a Savitzky–Golay (SG) or a locally estimated scatterplot smoothing (LOESS) algorithm with equivalent analytical design regarding the investigation and sample output time range. Through the described analysis, real-time signal cleansing of OUR raw data was possible, leading to more accurate predictions. Figure 2 gives an overview of the performed model generation and validation workflow.

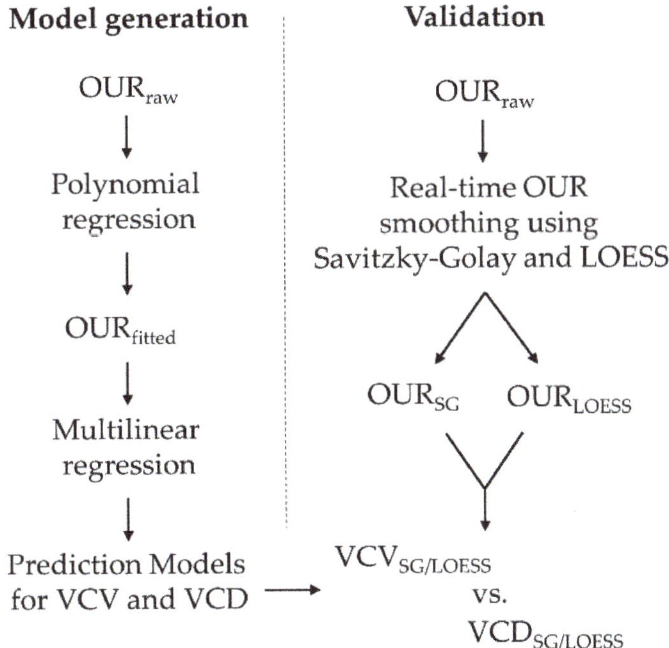

Figure 2. Workflow overview for biomass model generation and consequent validation.

2.8. Off-Line Measurements

All cell physiological measures, such as viable cell density (VCD), average cell diameter (ACD), average volume per cell (AVC) and culture viability, were determined using an automatic cell counting device (Cedex HiRes®, Roche Diagnostics, Mannheim, Germany). In this work, the shape of a cell is assumed to be spherical, hence the AVC is calculated as follows:

$$AVC = \frac{4}{3}\pi\left(\frac{ACD}{2}\right)^3 \tag{6}$$

All samples were immediately processed after the sample was taken, as described in Section 2.3. Accordingly, 300 µL of the cell containing sample was transferred into a Cedex HiRes sample cup and measured directly to avoid long-term storage. Depending on VCD concentration, the sample was diluted properly using 3% (m/v) Pluronic F68 dissolved in PBS. Furthermore, the measured VCD was used to calculate VCV according to Equation (7):

$$VCV = VCD \cdot AVC \tag{7}$$

For the graphical representation in this study, the biomass measures VCD and VCV were normalized to the maximum value in the training data set.

A biochemical analyzer (Cedex Bio HT®, Roche Diagnostics, Mannheim, Germany) was used to determine the metabolites glucose, lactate, pyruvate and ammonium. Therefore, cell suspension was centrifuged (Heraeus Multifuge 1S-R, Thermo Fisher Scientific, Waltham, MA, USA) at $3500 \times g$ for 10 min. The cell pellet was discarded, and the supernatant was used subsequently.

Amino acid analysis was performed using an in-house LC-MS (Ultivo Triple Quadrupole LC/MS System, Agilent Technologies Inc., Santa Clara, CA, USA) procedure with stable isotope-labeled internal standards for calibration.

2.9. Cell-Specific Substrate and Metabolite Consumption and Production Rate, Product Formation Rate and Yield Calculation

The cell-specific substrate consumption and metabolite production rates in the dynamic state of continuous process were calculated, as recently described by Bausch et al., [29] according to the following balancing Equation (8):

$$\frac{dS}{dt} = D(S_{in} - S) + q_S X \tag{8}$$

where S represents the molar concentration of the substrate or metabolite, D is the perfusion rate, S_{in} is the substrate molar concentration in the perfusion medium, X is the cell number and qS is the molar cell-specific substrate/metabolite production rate. In a simplified approach neglecting abiotic degradation of instable compounds, the cell-specific qS at discrete process time points, i, are calculated as described in Equation (9):

$$qS_i = \frac{1}{X_i}\left(\frac{(S_i - S_{i-1})}{(t_i - t_{i-1})} - D(S_{in} - S_i) \right) \tag{9}$$

A negative and positive value for qS represent consumption and production of a compound, respectively. The product formation rate qP can be calculated analogous to Equation (9). The metabolic yield coefficients $Y_{Lac/Glc}$ and $Y_{NH4/Gln}$ for the assessment of the metabolic state and efficiency were calculated as follows in Equations (10) and (11), respectively:

$$Y_{Lac,i/Glc,i} = \frac{q_{Lac,i}}{q_{Glc,i}} \tag{10}$$

$$Y_{NH4,i/Gln,i} = \frac{q_{NH4,i}}{q_{Gln,i}} \tag{11}$$

3. Results and Discussion

3.1. Online Parameter Evaluation and Preprocessing

To monitor the biomass formation of 14 different CHO cell lines expressing different target proteins, we used an at-line-based viable cell density assessment, as described above. Although all CHO lines were derived from the same native CHO host cell line, we detected, as shown before by others, significant, process time-dependent differences in cell growth characteristics, such as viable cell density formation and timing for cell doubling, as well as in volume per cell among all tested clones (Figure 3A,C,E). Usually, biomass formation analysis is performed only once per day, resulting in an erroneous, discrete monitoring of this critical KPI, in conflict with the continuous use of this variable for dynamic calculations of important bioprocess key performance indicators, such as the cell-specific product formation rate and feedback control process strategies. Continuous assessment of cell biomass formation is a prerequisite for efficient bioprocess development and economic target protein production.

For the biomass soft-sensor model, general assumptions were made. Process variations occur from: Media lot-to-lot differences, pH and DO probe behavior, mass flow divergences, off-gas sensor response time and the metabolic performance of the used clone. These variations can influence the oxygen transfer and/or its solubility and, therefore, the oxygen level becomes a sum parameter. Specifically, the different metabolic behavior of the clones (that leads to different controller responses and correction agent additions) might have the greatest impact on the fermentation broth and its physicochemical characteristics in terms of oxygen transfer. In addition, ambient conditions may vary during the course of a fermentation that affect the off-gas measurement. Residence time of the off-gas in the headspace as well as in the tubing and condensate trap bottles may further influence the proper calculation of the OUR [30]. Differences in the gas inlet and outlet temperature also have an impact on measured volume fractions, especially in cases where an off-gas cooler is used [16]. Since we did not use any off-gas cooling, we assumed the inlet and outlet gas temperature difference to be negligible. It has been shown that correction functions or description models generated from perturbation experiments can be applied to enhance accuracy of off-gas measurements for OUR calculations [30–32]. These approaches require considerable effort and profound knowledge about the characteristics of O_2 transport kinetics within the whole system. However, we took none of the mentioned factors into consideration as we wanted to create a robust and relatively simple model that allows for easy implementation and good prediction quality in contrast to alternative, soft-sensing approaches. As described most recently by Tuveri et al., precise estimation of bioprocess variables such as biomass can be realized by comprehensive yet more complex approaches, such as the utilization of Kalman filters [33]. However, in our study, we were aiming for a biomass soft-sensor model that can handle the metabolic diversity and its effects on process properties that are caused by the varying clonal behavior of not yet in-depth, characterized CHO cell lines.

All available online variables were investigated regarding their ability to predict the biomass in terms of VCD and VCV. We found the OUR and process time (PT) to be the most predictive variables using a JMP predictor screening algorithm, which confirmed the known, high correlation of cellular biomass and respective volumetric OUR in cell culture (Figure 3B). The assembled OUR data (OUR_{raw}) showed a low signal-to-noise ratio at the beginning of all processes up to several process days. The ratio was heavily influenced by low biomass concentrations as well as DO and pH controller response. Once the biomass reached a critical level accompanied by higher oxygen demand, the measured OUR signal became stable (Figure 4A).

Due to the use of different cell lines with diverse growth behavior as part of the data set, this condition differed clearly with respect to process time (Figure 4B) and mainly influenced the choice of the used higher order polynomials to describe the data set of each process as accurate as possible. We found polynomials of the third to fifth degree to fit best to the observed OUR raw data sets. Table 2 provides an overview of used polynomial degree and their related R^2.

Table 2. Overview of applied polynomial degree to describe the OUR raw data.

Process	Polynomial Degree	R^2
P-01	5	0.863
P-02	4	0.983
P-03	4	0.878
P-04	4	0.888
P-05	4	0.712
P-06	4	0.971
P-07	4	0.991
P-08	3	0.892
P-09	3	0.982
P-10	3	0.977
P-11	3	0.918
P-12	4	0.956
P-13	4	0.977
P-14	4	0.954

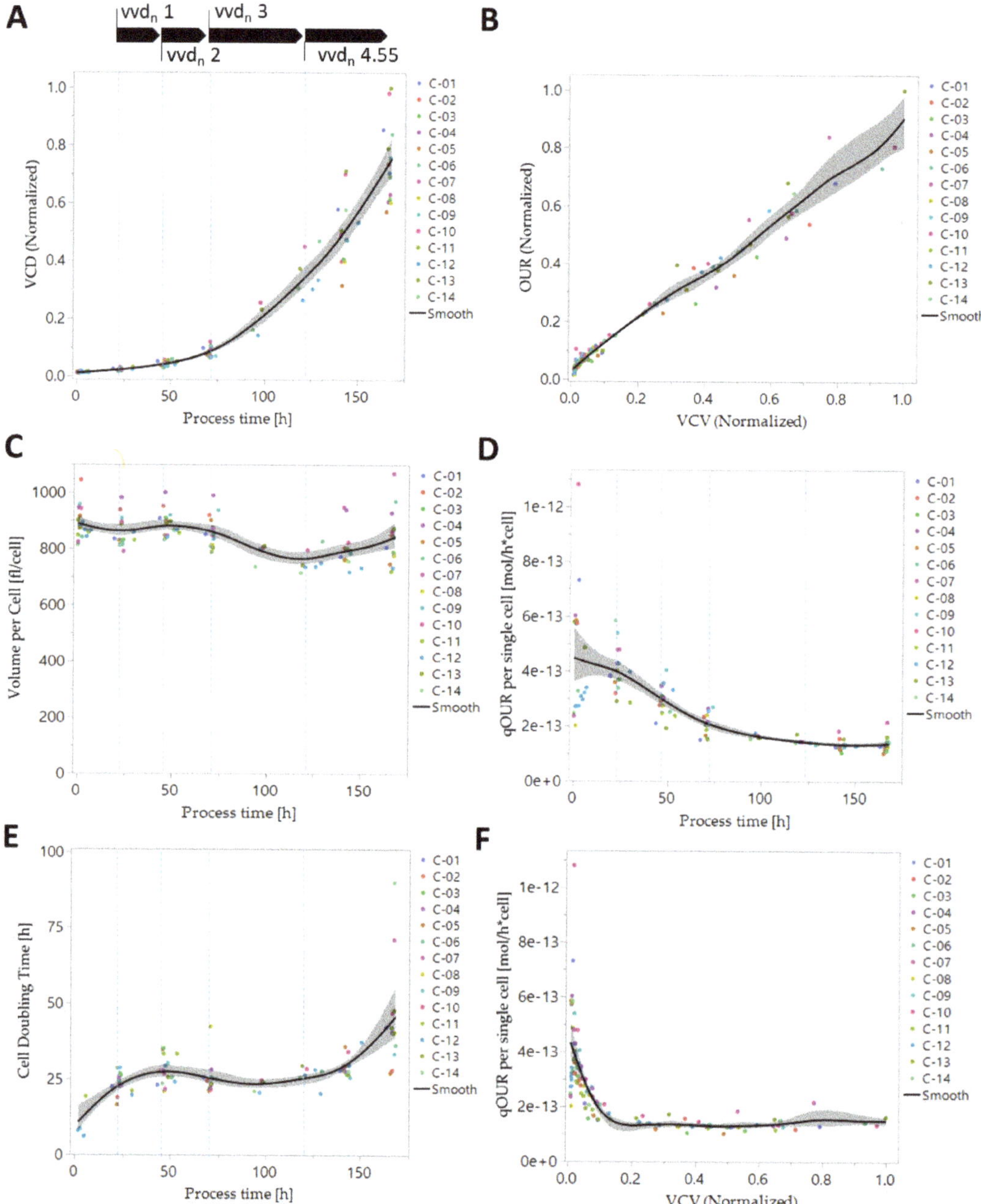

Figure 3. Time course of (**A**) normalized viable cell density (VCD), (**B**) normalized viable cell volume (VCV) vs. normalized volumetric oxygen uptake rate (OUR), (**C**) cell volume, (**D**) cell-specific OUR, (**E**) cell doubling time and (**F**) cell-specific OUR vs. normalized VCV of 14 different CHO cell lines (training data set, see Table 1) expressing different target proteins in a seven-day perfusion process. Black arrows and blue dotted lines show the perfusion rate protocol with respective normalized perfusion rate (in volume media per volume fermenter and day, vvd$_n$) and timing strategy. The black lines represent the fit among all tested clones and runs and the grey area highlights the confidence of the fit with $\alpha = 0.05$.

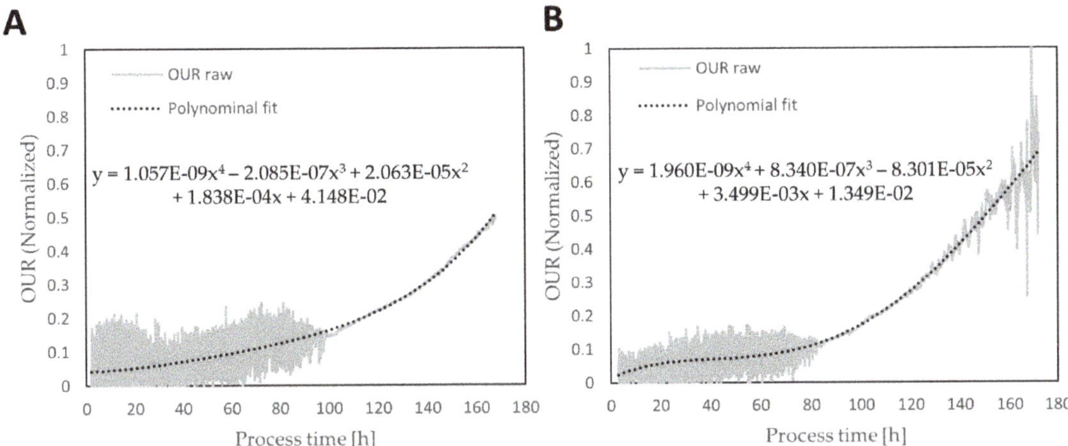

Figure 4. (**A**) Example of normalized OUR raw data fit from process P-04 with high signal–noise ratio in the first 100 h and following stable signal towards end of fermentation. A polynomial fit of fourth grade was used to describe the OUR with an R^2 of 0.89. (**B**) Example of normalized OUR raw data fit from process P-09 with high signal–noise ratio in the first 85 h and also towards end of fermentation. A polynomial fit of fourth grade was used to describe the OUR with an R^2 of 0.95.

3.2. Biomass Model Generation and Assessment

Two descriptive models were built using the preprocessed OUR and process time (PT) as input variables to predict the VCV and the VCD, respectively. Both regression models allow a good description of the biomass for each variable (Figure 5A,B). The VCV model has a normalized prediction error of RMSE = 0.0339, whereas the VCD model reaches 0.0469. Referring to relative model performance evaluation, the accuracy for VCV prediction was calculated as $MAPE_{VCV}$ = 31.79% and $MdAPE_{VCV}$ = 13.19%. Lower forecast performance values were obtained from the VCD model with $MAPE_{VCD}$ = 56.59% and $MdAPE_{VCD}$ = 19.78%. The differences between MAPE- and MdAPE-derived values can be explained by the nature of the observed errors and their distribution during the fermentations that were used to create these models. Despite the similarity from the observed residuals to the normal distribution (Shapiro–Wilk for VCV residuals is 0.89 and 0.82 for VCD), it is noticeable that, within both models, the difference between actual and predicted values begins to scatter with progressing process time and biomass concentration (Figure 5C,D). Small dimension residuals were observed up to 60–80 h after the process start and were highest towards the end of processes. However, the lack of prediction performance of both models is located at the beginning of the processes, as the magnitude of absolute percentage errors (APE) reveals (Figure 5E,F). Both prediction models show comparable behavior regarding the APE distribution but, significantly, lower APE magnitudes were found from the VCV model. This local APE density is mainly influenced by the high signal noise produced by the OUR raw data combined with comparably low biomass concentrations and, therefore, low oxygen demands. Despite the fact that the OUR data is preprocessed as described, the impact of the low signal-to-noise ratio heavily reduces the accuracy of both models. Furthermore, this is the leading cause for the described differences between MAPE and MdAPE values as the median is not as affected as the mean is by high APE occurrence, as mentioned. Additionally, a likely explanation for an increase in scattering residuals might be related to the necessary dilution to stay within the manufacturer's specifications and calibration ranges for the Cedex HiRes® cell density assay.

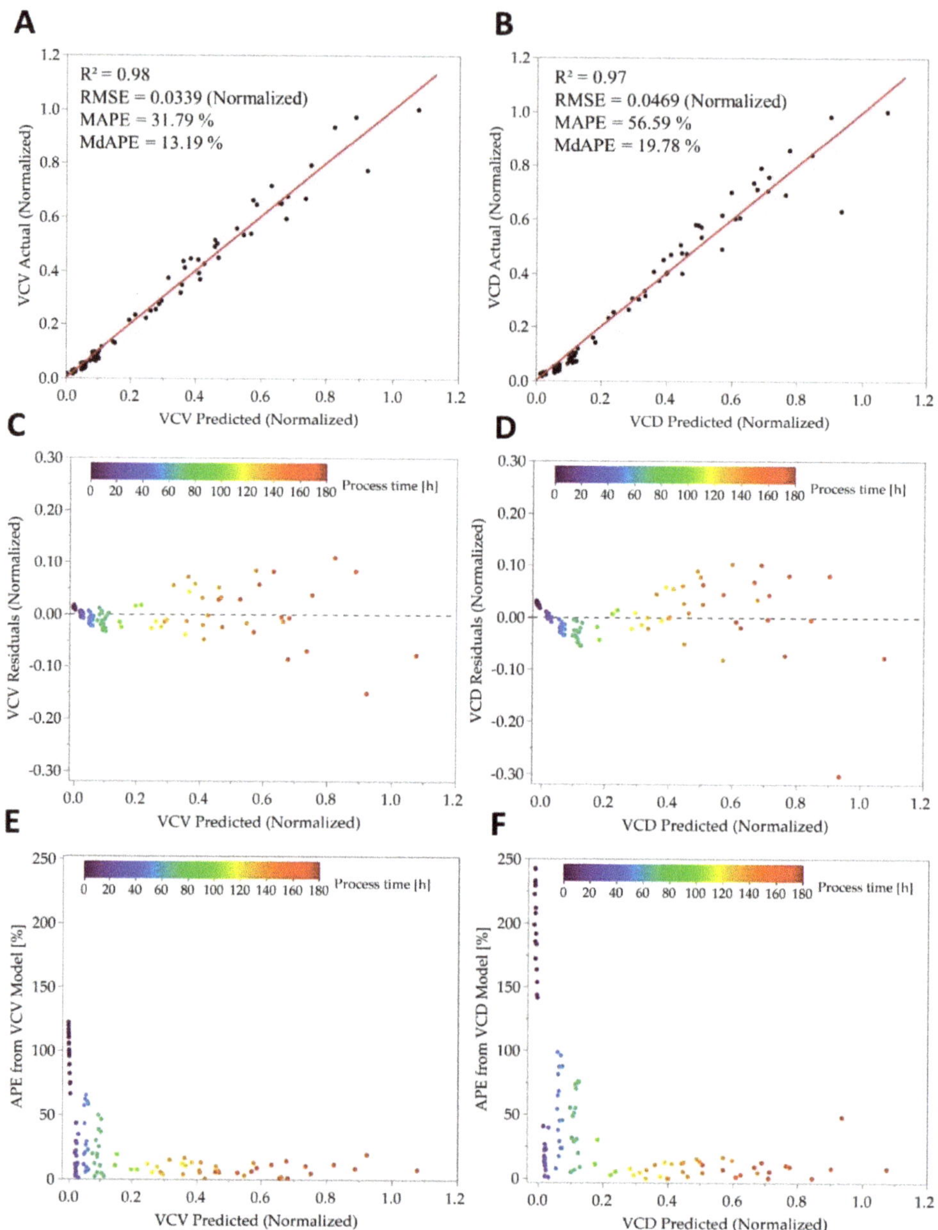

Figure 5. (**A**) Soft-sensor model for VCV prediction with model assessment RMSE and R^2. Black dots represent normalized values, red line describes the found model with prediction confidence $\alpha = 0.05$. (**B**) Soft-sensor model for VCD prediction with model assessment RMSE and R^2. Black dots represent normalized values, red line describes the found model with prediction confidence of $\alpha = 0.05$. (**C**) VCV normalized residuals plotted against normalized predicted values with process time indication. (**D**) VCD normalized residuals plotted against normalized predicted values with process time indication. (**E**) VCV-model-derived absolute percentage errors plotted against normalized predicted VCV values with process time indication. (**F**) VCD-model-derived absolute percentage errors plotted against normalized predicted VCD values with process time indication.

Even if a high cell concentration in a given sample might decrease the measurement error, the probability is increased when covering a more characteristic amount of cells in the analyzed sample because pre-dilution procedures are prone to cause unintended consequential errors [34,35]. Therefore, we consider the user-dependent and manually applied dilution step as the root cause for the observed residual increase during the course of the processes. The second input variable, process time, is further expected to represent an indirect measure of biomass growth rate.

The estimation functions are listed below:

$$VCV_{Predicted,Normalized} = -0.0907 + 0.425 \cdot OUR_{Normalized} + 0.00194 \cdot PT \tag{12}$$
$$+ (PT - 74.736) \cdot ((OUR_{Normalized} - 0.207) \cdot 0.00571)$$

$$VCD_{Predicted,Normalized} = -0.0994 + 0.362 \cdot OUR_{Normalized} + 0.00243 \cdot PT \tag{13}$$
$$+ (PT - 74.736) \cdot ((OUR_{Normalized} - 0.207) \cdot 0.00554)$$

3.3. Real-Time Prediction and Quality of Online OUR Monitoring

Using the identified models as a biomass soft sensor under real-time circumstances was considered as the chosen path of validation in this work. The estimator equations were implemented in SEEQ to perform online biomass prediction of dynamic state for the continuous processes (P-15, P-16 and P-17) with unknown cell lines (C-15, C-16 and C-17) to both models. The processes were executed in the same manner as described above, hence they are technical replicates, such as the processes P-01–P-14. In order to remove the signal noise from the calculated OUR, two signal-smoothing algorithms were applied in real time. As Bassey et al. [36] found the Savitzky–Golay (SG) filter algorithm to be well suited for gas-sensor-derived signal smoothing, we also applied the SG filter to remove signal distortions from the OUR signal. In addition, we tested a locally estimated scatterplot smoothing (LOESS)-based algorithm on the OUR signal to evaluate its influence on the final prediction quality. Both algorithms represent moving average functions that investigate a filter time window of 25 min with a permanent output frame of 30 s.

Using this approach, the SG applies a polynomial regression of first order, whereas the LOESS filter uses the best-fit line, which can either be a linear or a higher polynomial function. Due to the growth rate of animal cells of about 24 h, we consider the filter time delay to be negligible. Both soft-sensor models can predict the biomass in terms of VCV and VCD with good prediction accuracy regardless of whether the real-time OUR smoothing was done with the SG or LOESS algorithm (Figure 6A–D). Nevertheless, referring to model assessment parameters, the VCV model shows a significantly higher goodness of fit in each case (Tables 3 and 4). Calculated MAPE values are half the magnitude from the VCV model ($MAPE_{VCV,LOESS/SG} \approx 14\%$) compared to VCD-model-derived MAPE ($MAPE_{VCD,LOESS/SG} \approx 33\%$). Therefore, the VCV model is leading to predictions that are more precise on average. Beyond that, the difference between MAPE and MdAPE values is still noticeable in a comparable period after process start (Figure 6A,B) as the same root cause of a high signal-to-noise ratio creates a high local APE density. However, MdAPE values between both models are quite comparable and are in the range of $MdAPE_{VCD,LOESS/SG} \approx 8\%$ and, for the VCV model, $MdAPE_{VCV,LOESS} = 6.6\%$ and $MdAPE_{VCV,SG} = 8.3\%$. Half of the prediction errors are located above and below these values and, in reference to the calculated average prediction errors, the VCV model has the best prediction performance validated on the novel cell lines C-15, 16 and 17.

In contrast to offline-based measurements which usually consists of only one or a few measurement points per day, the prediction provides a continuous description of biomass during the processes, filling in the gaps between offline-derived measurements (Figure 6E,F). These real-time predictions can be further utilized to calculate other meaningful process variables in a soft-sensing manner, such as production or consumption rates (see Section 3.5). Additionally, high quality online biomass forecasts enable a verification of erroneous offline-based readings, revealing possible measurement errors.

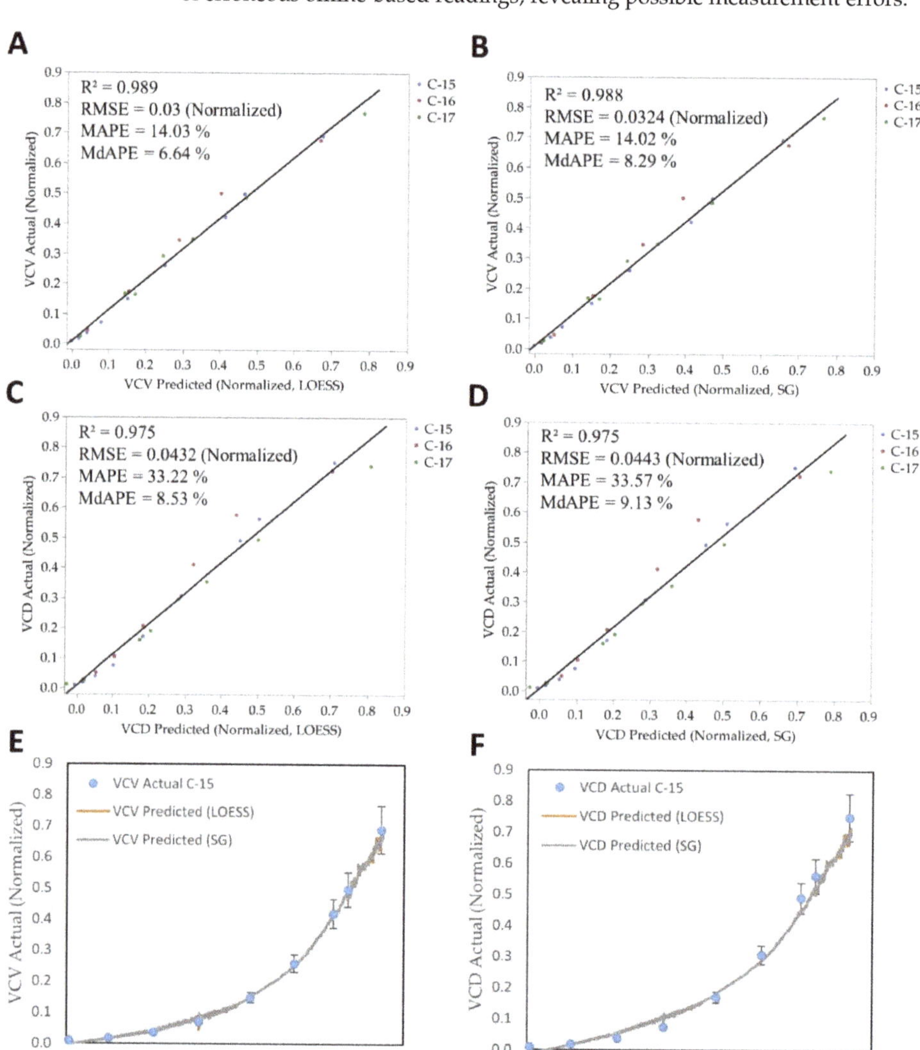

Figure 6. (**A**) Normalized measured VCV vs. real-time normalized predicted VCV data using the LOESS filter algorithm with model assessment for all three processes represented by colored dots. (**B**) Normalized measured VCV vs. real-time normalized predicted VCV data using the SG filter algorithm with model assessment for all three processes represented by colored dots. (**C**) Normalized measured VCD vs. real-time normalized predicted VCD data using the LOESS filter algorithm with model assessment for all three processes represented by colored dots. (**D**) Normalized measured VCD vs. real-time normalized predicted VCD data using the SG filter algorithm with model assessment for all three processes represented by colored dots. (**E**) Exemplary normalized predicted $VCV_{LOESS/SG}$ values (orange and gray lines) and actual normalized VCV values from process P-15 with clone C-15 represented by blue dots. Error bars describe an assumed 11% error for all VCV measurements. (**F**) Exemplary normalized predicted $VCD_{LOESS/SG}$ values (orange and gray lines) and actual normalized VCD values from process P-15 with clone C-15 represented by blue dots. Error bars describe an assumed 10% error for all VCD measurements. Biomass measures VCV and VCD are normalized to the maximum value in the training data set (P-01-P-14).

Table 3. Overview of measured VCV and real-time predicted VCV data using SG or LOESS algorithm.

Process	R^2		RMSE (Normalized)		MAPE [%]		MdAPE [%]	
	SG	LOESS	SG	LOESS	SG	LOESS	SG	LOESS
P-15	0.999	0.998	0.017	0.015	18.13	18.15	4.93	3.85
P-16	0.974	0.980	0.057	0.052	14.94	14.61	14.35	13.77
P-17	0.995	0.994	0.025	0.025	7.99	8.23	6.61	6.60

Table 4. Overview of measured VCD and real-time predicted VCD data using SG or LOESS algorithm.

Process	R^2		RMSE (Normalized)		MAPE [%]		MdAPE [%]	
	SG	LOESS	SG	LOESS	SG	LOESS	SG	LOESS
P-15	0.997	0.995	0.035	0.033	33.09	32.57	9.85	8.29
P-16	0.965	0.971	0.072	0.068	16.75	15.21	19.43	11.16
P-17	0.996	0.995	0.025	0.031	48.81	49.71	6.79	7.99

Since the increase in biomass is always accompanied by a growth in cell number and cell volume, we consider a description of the biomass solely by cell number in terms of VCD to be insufficient. VCD is a coarse measure of the viable biomass, because even small changes in mean cell diameter result in large differences in cell volume [37]. An analysis of cell size, especially its distribution during fermentation process time, can deliver valuable information that cannot be seen by only looking at cell numbers. Besides the fact that trypan blue-based automatic cell counting enables a differentiation in viable and nonviable cells, numerous publications can be found that highlight the advantages and also the necessity of cell size in terms of cell volume measurements [37–41]. Mammalian cell volume differs not only between cell lines but also during an ongoing process, which leads to changing biomass in terms of volume and cellular mass itself. In addition, the process mode, growth conditions and other parameters can influence cell size. For example, larger cells tend to consume more oxygen than smaller cells, and the rapid adaptability of cells to process conditions such as osmolality, where a rise results in cell size increase, underlines the advantages of having cell size measured [22]. All factors support our preference for more accurate correlations for a VCV-based biomass description. Furthermore, it has been demonstrated that packed cell volume measurements can reach errors below 5%, whereas standard trypan blue cell counting techniques still struggle with errors up to 15% [42].

3.4. Biomass-Specific Oxygen Demand and Key Metabolism Analysis

The metabolism of CHO cell lines during classical batch and fed-batch cultivation is highly dynamic, and metabolic steady-state descriptions can be used to analyze the coherences by mechanistic modeling approaches [43]. These significant metabolic changes originate from alterations in the dynamic cell environmental media matrix composition, such as substrate and cofactor consumption, (toxic) metabolite production and shifts in chemicophysical parameters, such as medium osmolality, buffer capacity and redox potential [44–47]. Perfusion cell cultivation processes can be used to overcome these media matrix variations by an optimized constant replacement of conditioned media with fresh media and by using parameters such as the cell-specific perfusion rate (CSPR). Nevertheless, the optimization and analysis of CSPR was not the goal of this study.

We analyzed the biomass-specific OUR and metabolism of key substrates and metabolites of tested CHO cell lines in SUB perfusion processes in more detail. As shown previously [14,48], the volumetric OUR of tested CHO cell lines followed the previously described cell density kinetics during cell cultivation (Figure 3B). The observed cell density formation consequently showed differences among each tested CHO cell line, and the final volumetric OUR of the perfusion processes showed a constant increase over process time. At the end of the cultivation, the volumetric OUR showed a broad variation between all tested CHO cell lines and, for C-05 and C-13, up to more than 100% more than the respective variance observed for viable cell densities (Figure 3B). The cell-specific OUR (qOUR), however, showed an initial slight increase followed by highly homogenous qOUR for all tested clones and plateaued on a stable level of approximately 41.7 amol cell^{-1} s^{-1} from day 5 until the end of the perfusion process at day 7 (Figure 3D). The observed level of qOUR fits well with previously reported qOURs for CHO suspension cells [24,27,48–50]. Plotting the viable cell volume of the process (VCV) vs. qOUR revealed very high qOUR and, subsequently, a fast decrease of qOUR in the beginning of the perfusion cultivation where low biomass was available, followed by a stable plateauing of qOUR (Figure 3F). Both observations, the initial increase in qOUR followed by a stabilization at a lower qOUR level at the later cell cultivation phases and the higher biomass levels, confirm previously reported trends for CHO cells in perfusion cultivations [27,51]. The early qOUR peaks were attributed to an initial metabolic acclimation phase when cells were seeded into an unconditioned media with high substrate concentrations and the cultivation conditions at start of cell culturing.

To understand the reason for this shift in early and late qOUR kinetics, we analyzed the concentration and consumption/production rates of glucose, lactate, glutamine and ammonia as key substrates in mammalian cell cultures. Significant changes in volumetric glucose and glutamine substrate availability, as well as lactate and ammonium byproduct levels, were observed by using the applied perfusion process strategy. Both glucose and glutamine levels dropped during the course of the perfusion process, with an earlier decline in glutamine, which may be due to additional abiotic degradation (Figure 7A). Both byproducts, lactate and ammonium showed an initial increase followed by an intermediate plateau phase between day 3 and 5 and a final metabolic inverse shift with a decreasing level of lactate and, subsequently, an increase in ammonium from day 5 until the end of perfusion fermentation at day 7 (Figure 7A). The analysis of the cell-specific rates of these substrates and metabolites emphasizes the metabolic shift at day 5 with a stagnation in low levels of cell-specific glutamine consumption qGln and lactate formation rates qLac (Figure 7B).

The yield coefficients $Y_{Lac/Glc}$ and $Y_{NH4/Gln}$ are characteristic bioprocess key parameters (KPI) for assessing the metabolic status of cellular systems and the utilized pathways for energy production. By applying these parameters, we temporally analyzed the yield coefficient $Y_{Lac/Glc}$ and $Y_{NH4/Gln}$ along the perfusion process time. Through our analysis, we identified three distinct metabolic phases: (i) from day 0 to day 3, a phase of high anaerobic lactate production and glutaminolysis-driven ammonium formation with a clone-dependent $Y_{Lac/Glc}$ of 1–2 mol/mol and $Y_{NH4/Gln}$ of 0.5–3.6 mol/mol, (ii) from day 3 to day 5, a metabolic transition phase switching to aerobic metabolism and low glutaminolysis activity and (iii) from day 5 to day 7, an almost complete aerobic phase with practically no lactate production and clone-dependent increasing glutaminolysis again ($Y_{Lac/Glc}$ of 0.03–0.64 mol/mol, $Y_{NH4/Gln}$ of 0.7–1.7 mol/mol) (Figure 7C,D).

The yield analysis by $Y_{Lac/Glc}$ and $Y_{NH4/Gln}$ suggests an alternative reason for the observed metabolic switch rather than substrate limitation since glucose and glutamine are available in the fermentation media matrix in high amounts during the whole perfusion process (Figure 7A). The limitation of pyruvate during the perfusion process was identified as a putative reason for the metabolic switch. The slight increase of cell-specific glucose consumption qGluc and of the glutaminolysis and ammonium formation qS$_{NH4}$ from day 5 onward correlates with the limitation of pyruvate (Figure 8A) and stagnation of cell-specific pyruvate consumption rate qPyr (Figure 8B). In general, pyruvate is an important alternative, energy-generating carbon source for fast proliferating mammalian cell lines and for reducing cell growth-inhibiting ammonium production in cell cultures [52]. Analysis of qPyr vs. the available global pyruvate concentrations in the culture suggests a concentration-dependent shift of qPyr at levels lower than 2 mM, which correlates with the increase of cell-specific ammonium formation with the drop in cell doubling time (data not shown).

In principle, the accumulations of cytostatic/toxic metabolic byproducts, other than lactate and ammonium, originating from the amino acid break-down metabolism in CHO fermentation processes are well characterized triggers which induce decreased biomass formation and increased cell doubling time [53]. In our study, however, we focused on the classical cell culture substrates and metabolites yet encouraged the analysis of these amino acid break-down products in the future to allow for optimized perfusion process designs with efficient depletion of known and unknown cytostatic/toxic metabolic byproducts.

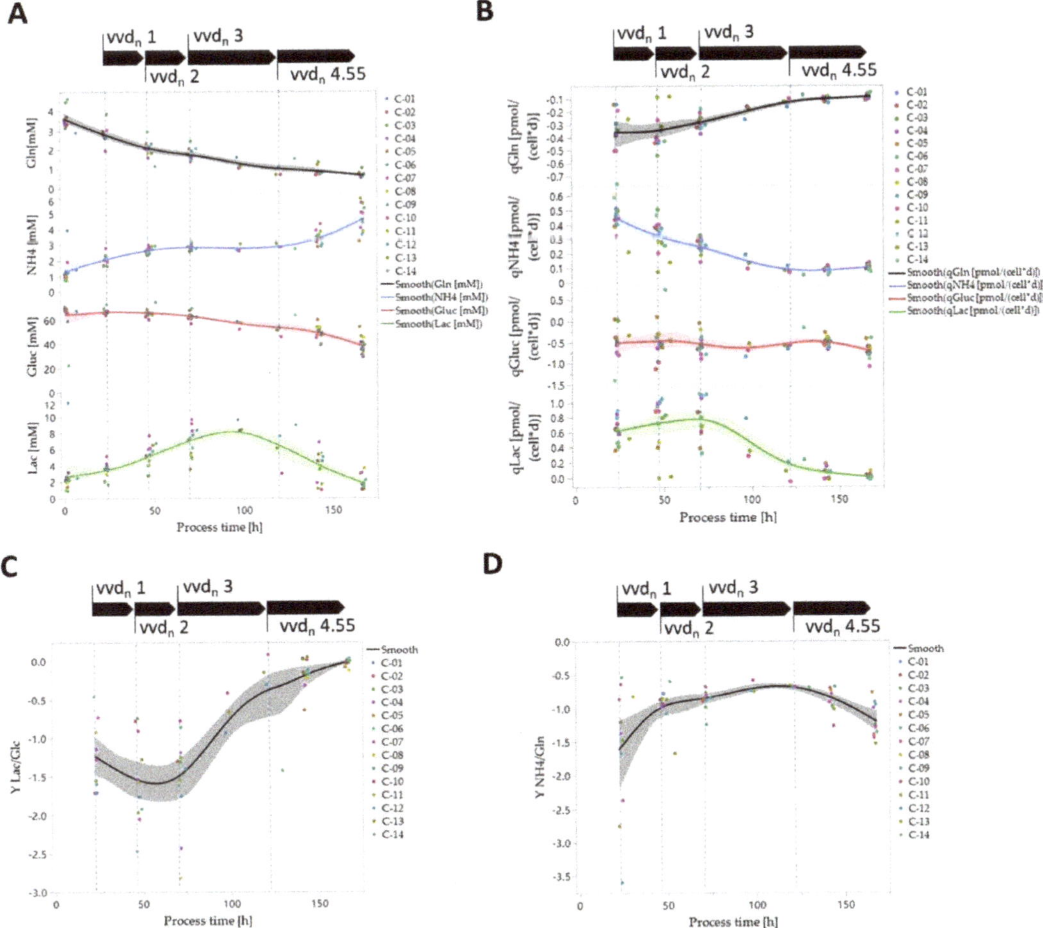

Figure 7. Metabolic analysis of the dynamic state of continuous CHO SUB processes. Kinetic of key substrates glucose and glutamine and metabolites lactate and ammonium concentrations (**A,B**) cell-specific rates. Time-resolved analysis of the cell-specific rate-based yield coefficients (**C**) $Y_{Lac/Glc}$ and (**D**) $Y_{NH4/Gln}$. The colored dots represent the tested 14 clones and the black, blue, red and green lines represent the fit of Gln concentration or cell-specific Gln consumption/production rate qGln, NH_4^+ concentration or cell-specific NH_4^+ consumption/production rate qNH4, glucose concentration or cell-specific glucose consumption rate and lactate concentration or cell-specific lactate consumption/production rate, respectively. The black, blue, red and green areas highlight the confidence of the fits with $\alpha = 0.05$. Black arrows and blue dotted lines show the perfusion rate protocol with respective normalized perfusion rate (in volume media per volume fermenter and day, vvd_n) and timing strategy.

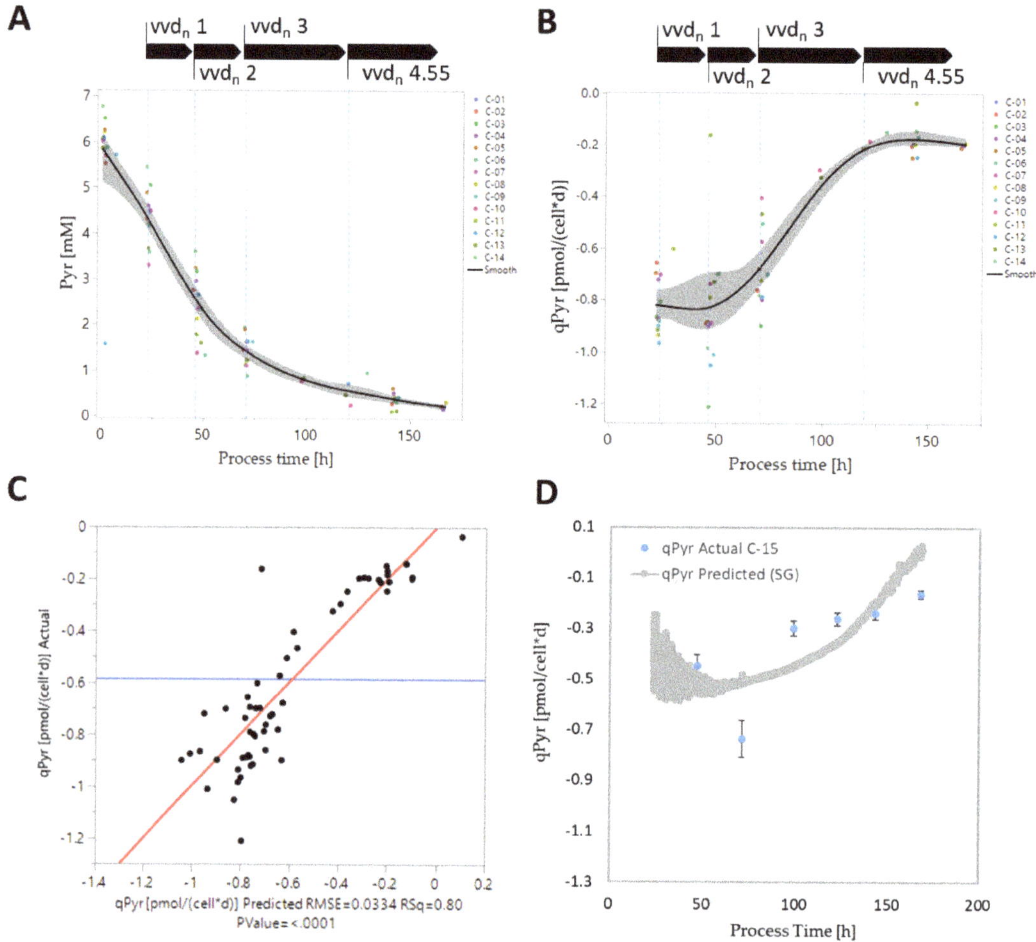

Figure 8. Real-time prediction of qPyr in dynamic state of continuous CHO SUB processes. (**A**) Analysis of pyruvate concentration over perfusion process time. The colored dots represent the tested 14 clones from the training data set and the black line represents the fit with $\alpha = 0.05$. (**B**) qPyr cell-specific consumption/production rates. The black area highlights the confidence of the fit with = 0.05. (**C**) Actual vs. predicted plot of a logistic regression model for qPyr for all tested clones from the training data set C-1 to C-14 (black dots) with regression model prediction (red line) and mean of all tested qPyr (blue line). (**D**) Online prediction of qPyr for a model validation perfusion process with C-15 (grey line) with qPyr actuals (blue dots). Error bars describe an assumed 10% error for actual qPyr values.

Mammalian amino acid metabolism is highly dependent upon the availability of bioavailable oxygen as an electron acceptor to allow for an indirect regeneration of redox equivalents NAD^+ and FAD in the tricarbon cycle (TCA), which are finally needed for the oxidative phosphorylation and energy production in cells [54]. Since there is no report that describes the correlation of specific amino acid and key metabolite consumption/production rates qS with qOUR of CHO cell lines in SUB continuous processes, we aimed to analyze this important investigation in our experimental set-up. Unexpectedly, cell-specific qGluc showed no correlation to qOUR (R^2: 0.007, RMSE: 0.31 pmol cell^{-1} d^{-1}) yet the following important metabolic rates of key substrates and metabolites revealed a sound correlation: qGln (R^2: 0.389, RMSE: 0.13 pmol cell^{-1} d^{-1}), qAla (R^2: 0.540, RMSE: 0.06 pmol cell^{-1} d^{-1}), qPyr (R^2: 0.521, RMSE: 0.23 pmol cell^{-1} d^{-1}), qLac (R^2: 0.324, RMSE:

0.32 pmol cell^{-1} d^{-1}) and qNH4 (R^2: 0.741, RMSE: 0.08 pmol cell^{-1} d^{-1}) (Figure 9A). In addition, the cell-specific product formation rate qP revealed no correlation to the cell biomass-specific OUR (R^2: 0.062, RMSE: 4.42 pg cell^{-1} d^{-1}) (Figure 9B).

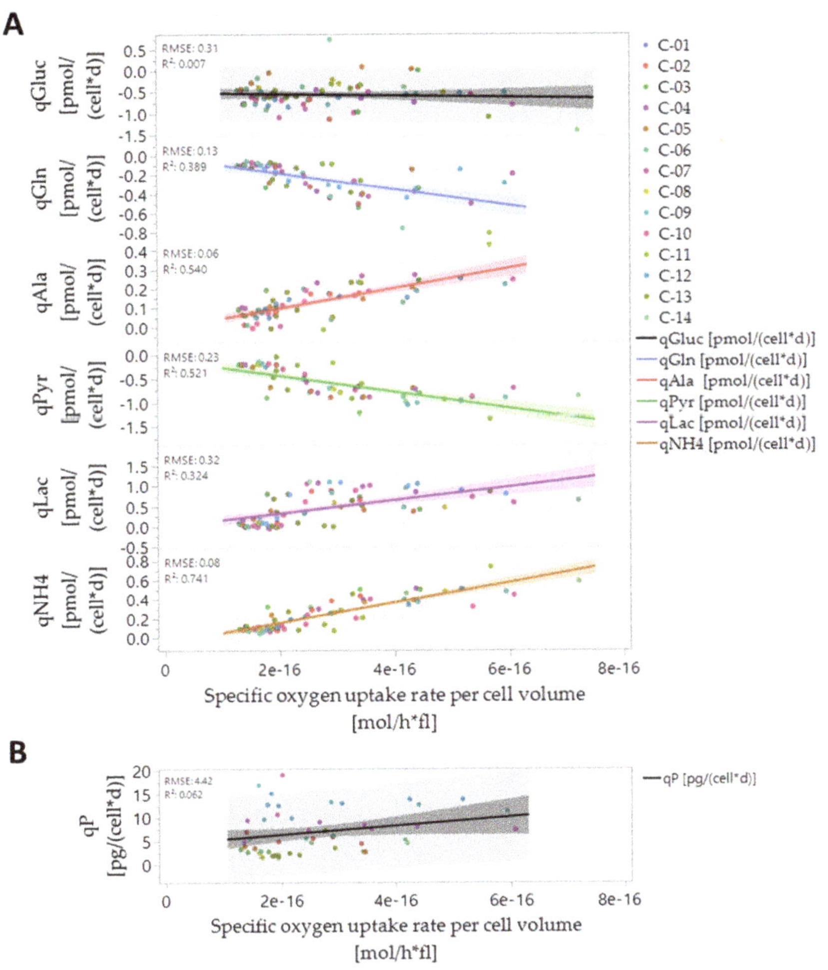

Figure 9. Correlation of (**A**) cell-specific substrate and metabolite formation/consumption rates and (**B**) product formation rate. The colored dots represent the tested 14 clones and for (**A**) the black, blue, red, green, violet and brown lines represent the fit of qGluc, qGln, qAla, qPyr, qLac and qNH4 cell-specific consumption/production rates and for (**B**) the black line represent the fit of qP. The dark black, blue, red, green, violet and brown areas highlight the confidence of the fits with α= 0.05 and light-colored areas the respective confidences of the predictions.

3.5. Online Prediction of Cellular Metabolic Rates

As shown in the previous section, the calculations and analyses of biomass-specific substrate consumption and metabolite production rates, qS, are mandatory to identify distinct cell metabolic phases, which can be preferably used to optimize perfusion media and rates for an efficient continuous cultivation of CHO cell lines. Solely monitoring global substrate and metabolite concentrations is not sufficient to allow for an equivalent characterization of cell cultivation processes, such as the described CHO perfusion process in SUBs.

As a proof of concept, we developed a soft-sensor-based real-time prediction of the cell-specific pyruvate consumption/production rate qPyr by using available real-time estimates for qOUR and biomass measures, as described before. The importance of an immediate estimation of the metabolic pyruvate flux into the cell is justified by its central role in the direct and indirect control of the cellular energy metabolism. Pyruvate is funneled into the TCA by the pyruvate dehydrogenase complex and/or by the anaplerotic reaction regulated by the pyruvate carboxylase [55,56]. Therefore, monitoring coupled with tailored control of qPyr is generally envisioned to improve the cellular energy state and avoid the lactate accumulation in cell culture fermentation processes.

qPyr correlated well with discrete cell-specific qOUR values (R^2 of 0.521, RMSE of 0.23) by using the discrete qPyr and real-time predicted qOUR data of the 14 different CHO cell lines and perfusion processes (Figure 9), suggesting the possibility to directly use this important information on the respiratory metabolism for a soft-sensing approach for real-time qPyr prediction. As a proof of concept, a suitable logistic multiregression model was generated for the generalized, sigmoid qPyr time course by simply using the available online OUR data and the predicted, SG-smoothed VCD, VCV and cell- and biomass-specific qOUR values (R^2 of 0.8, RMSE of 0.0334 pmol cell^{-1} d^{-1}) (Figure 8C). The used estimation functions for qPyr prediction are shown in following Equation (14):

$$qPyr_{Predicted} = \quad Logist\left(1.046 \cdot 10^{12} + 67.365 \cdot Logist(OUR) - 0.259 \right. \\ \cdot Logist(VCD_{Predicted}) + 1.87 \cdot 10^{14} \cdot qOUR_{cellvolumePredicted} \\ \left. -2.092 \cdot 10^{12} \cdot qOUR_{cellPredicted}\right) \tag{14}$$

We validated the prediction estimation model for qPyr by using a validation data set with CHO clone C-15 and perfusion process P-15 (Table 1). By this, the real-time prediction and discrete actuals for qPyr showed a technically relevant, good correlation in this validation data set (Figure 8D). The reason for the observed offset likely originates from the erroneous discrete qPyr measurement and respective error propagation by the calculation and/or by the cell-specific metabolic nature, often described for CHO cells with a high genetic plasticity [57]. The first 24 h of the prediction were not used due to the high noise in the OUR signal due to reasons described before. In general, more elaborated non-linear modeling approaches, such as decision trees and artificial neuronal nets, may also be used in the future for an increased precise estimation of cell-specific rates such as qPyr. Regardless, using these powerful modeling approaches requires large, annotated data sets that can be technically realized simply over a longer period of time.

4. Conclusions

In this work, we present, for the first time, an off-gas-based soft sensor for real-time biomass prediction in SUB continuous processes with CHO cell lines. The 14 different CHO cell lines that were used to build the soft-sensor models cover a variety of phenotypically different CHO cell lines. Given the diversity of our training data set, we expect the resulting models to be applicable to a broad range of CHO cell lines. This application is underlined by a high prediction accuracy achieved by the models on the bioprocesses of three novel CHO cell lines which were previously unknown to both models. The detailed analysis of both the model residuals as well as the absolute percentage errors disclosed some weaknesses that are primarily process related. The noisy OUR raw signal that was observed during the onset of all cell cultivation processes is caused by the pH controller response leading to very high prediction errors for up to 80 h after the processes were started. Optimization of pH controller settings and strategies or using more basic pH set points could overcome these technical challenges (data not shown). In addition, a split into different forecast models where altered pH controller interferences are present could lead to lower prediction errors. In addition, alternative yet computational and model-calibration-intensive forecast approaches such as Kalman filtering could significantly increase the prediction quality and should be considered for further, more elaborated closed-loop variable predictions and process control strategies [33].

Our data also demonstrated that higher model accuracy was established when VCV instead of VCD was used as biomass depiction. This strengthens our strong belief in a paradigm change regarding biomass description in modern bioprocesses. VCD should no longer be the leading, or the only, measurement looked at when it comes to biomass determination. The cell size or volume, its distribution over time and, of course, the VCV should be used by default to accurately describe the biomass and all derived metabolic variables, such as mAB, lactate production rate, or glucose/oxygen consumption rates. Conclusions, based only on cell density measurements, can lead to wrong assumptions, calculations or other unforeseen misinterpretations, generating a fragmented picture of the biomass [38,40]. As modern bioprocesses can be highly complex and dynamic, the biomass and cellular metabolism analysis should be as comprehensive as possible to generate a comparable and reproducible data basis. Furthermore, the utilization of an off-gas-based soft sensor is easy to implement in SUB systems as well as in common stainless steel plants. For this purpose, the installation of any hard-type probes inside the bioreactor is not necessary and does not increase handling or decrease safety and therefore prevents possible contamination risks. The fundamental correlation of biomass growth and increasing oxygen demand can be used, optimized and extended to generate profound real-time knowledge on diverse bioprocess variables such as the shown biomass and metabolic nutrient rate soft sensor. Moreover, off-gas analysis can be used to determine the true bioreactor pH without any sampling or as non-invasive method for online pCO_2 monitoring, which underlines the flexibility and outstanding character of having an off-gas analyzer implemented and running [58].

Author Contributions: T.W. and O.P. contributed equally to this work. All authors have read and agreed to the published version of the manuscript.

Funding: This work is exclusively funded by F. Hoffman-La Roche AG.

Institutional Review Board Statement: Not applicable.

Informed Consent Statement: Not applicable.

Data Availability Statement: Restrictions apply to the availability to these data; therefore, data sharing is not possible.

Acknowledgments: The authors thank the whole Bioprocess Research Team for their support. The authors thank Frederik Schröter, Ulrike Vollertsen and Katja Montan for the measurements of amino acids using the described HT-MS method. In addition, we would like to thank Tobias Grosskopf for critical reading. Furthermore, we thank Michalea Poth for the helpful discussion and Kathryn Perez for proofreading.

Conflicts of Interest: The authors declare no conflict of interest.

References

1. Aehle, M.; Kuprijanov, A.; Schaepe, S.; Simutis, R.; Lübbert, A. Increasing batch-to-batch reproducibility of CHO cultures by robust open-loop control. *Cytotechnology* **2011**, *63*, 41–47. [CrossRef] [PubMed]
2. FDA. *Guidance for Industry: PAT—A Framework for Innovative Pharmaceutical Development, Manufacturing, and Quality Assurance;* FDA: Silver Spring, MD, USA, 2004.
3. Lopes, A.G. Single-use in the biopharmaceutical industry: A review of current technology impact, challenges and limitations. *Food Bioprod. Process.* **2015**, *93*, 98–114. [CrossRef]
4. Löffelholz, C.; Husemann, U.; Greller, G.; Meusel, W.; Kauling, J.; Ay, P.; Kraume, M.; Eibl, R.; Eibl, D. Bioengineering Parameters for Single-Use Bioreactors: Overview and Evaluation of Suitable Methods. *Chem. Ing. Tech.* **2013**, *85*, 40–56. [CrossRef]
5. Junne, S.; Neubauer, P. How scalable and suitable are single-use bioreactors? *Curr. Opin. Biotechnol.* **2018**, *53*, 240–247. [CrossRef]
6. Mandenius, C.-F. Measurement Technologies for Upstream and Downstream Bioprocessing. *Processes* **2021**, *9*, 143. [CrossRef]
7. Gnoth, S.; Jenzsch, M.; Simutis, R.; Lübbert, A. Process Analytical Technology (PAT): Batch-to-batch reproducibility of fermentation processes by robust process operational design and control. *J. Biotechnol.* **2007**, *132*, 180–186. [CrossRef]
8. Justice, C.; Brix, A.; Freimark, D.; Kraume, M.; Pfromm, P.; Eichenmueller, B.; Czermak, P. Process control in cell culture technology using dielectric spectroscopy. *Biotechnol. Adv.* **2011**, *29*, 391–401. [CrossRef]
9. Busse, C.; Biechele, P.; De Vries, I.; Reardon, K.F.; Solle, D.; Scheper, T. Sensors for disposable bioreactors. *Eng. Life Sci.* **2017**, *17*, 940–952. [CrossRef]

10. Steinwandter, V.; Zahel, T.; Sagmeister, P.; Herwig, C. Propagation of measurement accuracy to biomass soft-sensor estimation and control quality. *Anal. Bioanal. Chem.* **2016**, *409*, 693–706. [CrossRef]

11. Kiviharju, K.; Salonen, K.; Moilanen, U.; Eerikäinen, T. Biomass measurement online: The performance of in situ measurements and software sensors. *J. Ind. Microbiol. Biotechnol.* **2008**, *35*, 657–665. [CrossRef]

12. Mandenius, C.-F.; Gustavsson, R. Mini-review: Soft sensors as means for PAT in the manufacture of bio-therapeutics. *J. Chem. Technol. Biotechnol.* **2015**, *90*, 215–227. [CrossRef]

13. Luttmann, R.; Bracewell, D.G.; Cornelissen, G.; Gernaey, K.V.; Glassey, J.; Hass, V.C.; Kaiser, C.; Preusse, C.; Striedner, G.; Mandenius, C.-F. Soft sensors in bioprocessing: A status report and recommendations. *Biotechnol. J.* **2012**, *7*, 1040–1048. [CrossRef]

14. Pappenreiter, M.; Sissolak, B.; Sommeregger, W.; Striedner, G. Oxygen Uptake Rate Soft-Sensing via Dynamic k L a Computation: Cell Volume and Metabolic Transition Prediction in Mammalian Bioprocesses. *Front. Bioeng. Biotechnol.* **2019**, *7*, 195. [CrossRef] [PubMed]

15. Metze, S.; Ruhl, S.; Greller, G.; Grimm, C.; Scholz, J. Monitoring online biomass with a capacitance sensor during scale-up of industrially relevant CHO cell culture fed-batch processes in single-use bioreactors. *Bioprocess Biosyst. Eng.* **2020**, *43*, 193–205. [CrossRef] [PubMed]

16. Aehle, M.; Kuprijanov, A.; Schaepe, S.; Simutis, R.; Lübbert, A. Simplified off-gas analyses in animal cell cultures for process monitoring and control purposes. *Biotechnol. Lett.* **2011**, *33*, 2103–2110. [CrossRef]

17. Zeng, A.-P.; Byun, T.-G.; Deckwer, W.-D. On-line estimation of viable biomass of a microaerobic culture using exit gas analysis. *Biotechnol. Tech.* **1991**, *5*, 247–250. [CrossRef]

18. Dorresteijn, R.C.; Numan, K.H.; De Gooijer, C.D.; Tramper, J.; Beuvery, E.C. On-line estimation of the biomass activity during animal-cell cultivations. *Biotechnol. Bioeng.* **1996**, *51*, 206–214. [CrossRef]

19. Goldrick, S.; Umprecht, A.; Tang, A.; Zakrzewski, R.; Cheeks, M.; Turner, R.; Charles, A.; Les, K.; Hulley, M.; Spencer, C.; et al. High-Throughput Raman Spectroscopy Combined with Innovate Data Analysis Workflow to Enhance Biopharmaceutical Process Development. *Processes* **2020**, *8*, 1179. [CrossRef]

20. Downey, B.J.; Graham, L.J.; Breit, J.F.; Glutting, N.K. A novel approach for using dielectric spectroscopy to predict viable cell volume (VCV) in early process development. *Biotechnol. Prog.* **2014**, *30*, 479–487. [CrossRef]

21. Yoon, S.-J.; Konstantinov, K.B. Continuous, real-time monitoring of the oxygen uptake rate (OUR) in animal cell bioreactors. *Biotechnol. Bioeng.* **1994**, *44*, 983–990. [CrossRef]

22. Wagner, B.A.; Venkataraman, S.; Buettner, G.R. The Rate of Oxygen Utilization by Cells. *Free Radic. Biol. Med.* **2011**, *51*, 700–712. [CrossRef]

23. Ruffieux, P.-A.; von Stockar, U.; Marison, I.W. Measurement of volumetric (OUR) and determination of specific (qO2) oxygen uptake rates in animal cell cultures. *J. Biotechnol.* **1998**, *63*, 85–95. [CrossRef]

24. Goudar, C.T.; Piret, J.M.; Konstantinov, K.B. Estimating cell specific oxygen uptake and carbon dioxide production rates for mammalian cells in perfusion culture. *Biotechnol. Prog.* **2011**, *27*, 1347–1357. [CrossRef]

25. Fleischaker, R.J.; Sinskey, A.J., Jr. Oxygen demand and supply in cell culture. *Appl. Microbiol. Biotechnol.* **1981**, *12*, 193–197. [CrossRef]

26. Pereira, S.; Kildegaard, H.F.; Andersen, M.R. Impact of CHO Metabolism on Cell Growth and Protein Production: An Overview of Toxic and Inhibiting Metabolites and Nutrients. *Biotechnol. J.* **2018**, *13*, e1700499. [CrossRef] [PubMed]

27. Heidemann, R.; Lütkemeyer, D.; Büntemeyer, H.; Lehmann, J. Effects of dissolved oxygen levels and the role of extra- and intracellular amino acid concentrations upon the metabolism of mammalian cell lines during batch and continuous cultures. *Cytotechnology* **1998**, *26*, 185–197. [CrossRef] [PubMed]

28. De Jonge, L.; Heijnen, J.; van Gulik, W. Reconstruction of the oxygen uptake and carbon dioxide evolution rates of microbial cultures at near-neutral pH during highly dynamic conditions. *Biochem. Eng. J.* **2014**, *83*, 42–54. [CrossRef]

29. Moreno, J.J.M.; Pol, A.P.; Abad, A.S. Using the R-MAPE index as a resistant measure of forecast accuracy. *Psicothema* **2013**, *25*, 500–506. [CrossRef]

30. Bausch, M.; Schultheiss, C.; Sieck, J.B. Recommendations for Comparison of Productivity Between Fed-Batch and Perfusion Processes. *Biotechnol. J.* **2019**, *14*, e1700721. [CrossRef]

31. Wu, L.; Lange, H.C.; Van Gulik, W.M.; Heijnen, J.J. Determination of in vivo oxygen uptake and carbon dioxide evolution rates from off-gas measurements under highly dynamic conditions. *Biotechnol. Bioeng.* **2002**, *81*, 448–458. [CrossRef]

32. Bloemen, H.H.J.; Wu, L.; van Gulik, W.M.; Heijnen, J.J.; Verhaegen, M.H.G. Reconstruction of the O_2 uptake rate and CO_2 evolution rate on a time scale of seconds. *AIChE J.* **2003**, *49*, 1895–1908. [CrossRef]

33. Tuveri, A.; Pérez-García, F.; Lira-Parada, P.A.; Imsland, L.; Bar, N. Sensor fusion based on Extended and Unscented Kalman Filter for bioprocess monitoring. *J. Process. Control* **2021**, *106*, 195–207. [CrossRef]

34. Sarkar, S.; Lund, S.P.; Vyzasatya, R.; Vanguri, P.; Elliott, J.T.; Plant, A.L.; Lin-Gibson, S. Evaluating the quality of a cell counting measurement process via a dilution series experimental design. *Cytotherapy* **2017**, *19*, 1509–1521. [CrossRef] [PubMed]

35. Cadena-Herrera, D.; Lara, J.E.E.-D.; Ramírez-Ibañez, N.D.; Lopez-Morales, C.A.; Pérez, N.O.; Flores-Ortiz, L.F.; Medina-Rivero, E. Validation of three viable-cell counting methods: Manual, semi-automated, and automated. *Biotechnol. Rep.* **2015**, *7*, 9–16. [CrossRef] [PubMed]

36. Bassey, E.; Whalley, J.; Sallis, P. An Evaluation of Smoothing Filters for Gas Sensor Signal Cleaning. In Proceedings of the Fourth International Conference on Advanced Communications and Computation, Paris, France, 8–9 February 2014; pp. 19–23.

37. Sonderhoff, S.A.; Kilburn, D.G.; Piret, J.M. Analysis of mammalian viable cell biomass based on cellular ATP. *Biotechnol. Bioeng.* **1992**, *39*, 859–864. [CrossRef]
38. Pan, X.; Dalm, C.; Wijffels, R.H.; Martens, D.E. Metabolic characterization of a CHO cell size increase phase in fed-batch cultures. *Appl. Microbiol. Biotechnol.* **2017**, *101*, 8101–8113. [CrossRef]
39. Nielsen, L.K.; Reid, S.; Greenfield, P.F. Cell cycle model to describe animal cell size variation and lag between cell number and biomass dynamics. *Biotechnol. Bioeng.* **1997**, *56*, 372–379. [CrossRef]
40. Milo, R. What is the total number of protein molecules per cell volume? A call to rethink some published values. *BioEssays* **2013**, *35*, 1050–1055. [CrossRef]
41. Frame, K.K.; Hu, W.-S. Cell volume measurement as an estimation of mammalian cell biomass. *Biotechnol. Bioeng.* **1990**, *36*, 191–197. [CrossRef]
42. Stettler, M.; Jaccard, N.; Hacker, D.; De Jesus, M.; Wurm, F.M.; Jordan, M. New disposable tubes for rapid and precise biomass assessment for suspension cultures of mammalian cells. *Biotechnol. Bioeng.* **2006**, *95*, 1228–1233. [CrossRef]
43. Popp, O.; Müller, D.; Didzus, K.; Paul, W.; Lipsmeier, F.; Kirchner, F.; Niklas, J.; Mauch, K.; Beaucamp, N. A hybrid approach identifies metabolic signatures of high-producers for chinese hamster ovary clone selection and process optimization. *Biotechnol. Bioeng.* **2016**, *113*, 2005–2019. [CrossRef]
44. Yuk, I.H.; Zhang, J.D.; Ebeling, M.; Berrera, M.; Gomez, N.; Werz, S.; Meiringer, C.; Shao, Z.; Swanberg, J.C.; Lee, K.H.; et al. Effects of copper on CHO cells: Insights from gene expression analyses. *Biotechnol. Prog.* **2014**, *30*, 429–442. [CrossRef]
45. Takagi, M.; Hayashi, H.; Yoshida, T. The effect of osmolarity on metabolism and morphology in adhesion and suspension chinese hamster ovary cells producing tissue plasminogen activator. *Cytotechnology* **2000**, *32*, 171–179. [CrossRef] [PubMed]
46. Park, J.H.; Noh, S.M.; Woo, J.R.; Kim, J.W.; Lee, G.M. Valeric acid induces cell cycle arrest at G1 phase in CHO cell cultures and improves recombinant antibody productivity. *Biotechnol. J.* **2016**, *11*, 487–496. [CrossRef] [PubMed]
47. Chevallier, V.; Andersen, M.R.; Malphettes, L. Oxidative stress-alleviating strategies to improve recombinant protein production in CHO cells. *Biotechnol. Bioeng.* **2020**, *117*, 1172–1186. [CrossRef]
48. Deshpande, R.R.; Heinzle, E. On-line oxygen uptake rate and culture viability measurement of animal cell culture using microplates with integrated oxygen sensors. *Biotechnol. Lett.* **2004**, *26*, 763–767. [CrossRef]
49. Jorjani, P.; Ozturk, S.S. Effects of cell density and temperature on oxygen consumption rate for different mammalian cell lines. *Biotechnol. Bioeng.* **1999**, *64*, 349–356. [CrossRef]
50. Gray, D.R.; Chen, S.; Howarth, W.; Inlow, D.; Maiorella, B.L. CO_2 in large-scale and high-density CHO cell perfusion culture. *Cytotechnology* **1996**, *22*, 65–78. [CrossRef] [PubMed]
51. Super, A.; Jaccard, N.; Marques, M.P.C.; Macown, R.J.; Griffin, L.D.; Veraitch, F.S.; Szita, N. Real-time monitoring of specific oxygen uptake rates of embryonic stem cells in a microfluidic cell culture device. *Biotechnol. J.* **2016**, *11*, 1179–1189. [CrossRef] [PubMed]
52. Genzel, Y.; Ritter, J.B.; König, S.; Alt, R.; Reichl, U. Substitution of Glutamine by Pyruvate to Reduce Ammonia Formation and Growth Inhibition of Mammalian Cells. *Biotechnol. Prog.* **2008**, *21*, 58–69. [CrossRef]
53. Mulukutla, B.C.; Mitchell, J.; Geoffroy, P.; Harrington, C.; Krishnan, M.; Kalomeris, T.; Morris, C.; Zhang, L.; Pegman, P.; Hiller, G.W. Metabolic engineering of Chinese hamster ovary cells towards reduced biosynthesis and accumulation of novel growth inhibitors in fed-batch cultures. *Metab. Eng.* **2019**, *54*, 54–68. [CrossRef] [PubMed]
54. Jeremy, M.; Berg, J.L.T. *The Citric Acid Cycle. Biochemistry*, 5th ed.; W H Freeman: New York, NY, USA, 2002.
55. Möller, J.; Bhat, K.; Guhl, L.; Pörtner, R.; Jandt, U.; Zeng, A. Regulation of pyruvate dehydrogenase complex related to lactate switch in CHO cells. *Eng. Life Sci.* **2021**, *21*, 100–114. [CrossRef] [PubMed]
56. Gupta, S.K.; Srivastava, S.K.; Sharma, A.; Nalage, V.H.H.; Salvi, D.; Kushwaha, H.; Chitnis, N.B.; Shukla, P. Metabolic engineering of CHO cells for the development of a robust protein production platform. *PLoS ONE* **2017**, *12*, e0181455. [CrossRef] [PubMed]
57. Dhiman, H.; Gerstl, M.P.; Ruckerbauer, D.; Hanscho, M.; Himmelbauer, H.; Clarke, C.; Barron, N.; Zanghellini, J.; Borth, N. Genetic and Epigenetic Variation across Genes Involved in Energy Metabolism and Mitochondria of Chinese Hamster Ovary Cell Lines. *Biotechnol. J.* **2019**, *14*, 1800681. [CrossRef] [PubMed]
58. Klinger, C.; Trinkaus, V.; Wallocha, T. Novel Carbon Dioxide-Based Method for Accurate Determination of pH and pCO2 in Mammalian Cell Culture Processes. *Processes* **2020**, *8*, 520. [CrossRef]

processes

Article

FPGA-Based Implementation of an Optimization Algorithm to Maximize the Productivity of a Microbial Electrolysis Cell

José de Jesús Colín-Robles [1], Ixbalank Torres-Zúñiga [1,*], Mario A. Ibarra-Manzano [1] and Víctor Alcaraz-González [2]

[1] C. A. Telemática, Departamento de Ingeniería Electrónica, Universidad de Guanajuato, Carr. Salamanca-Valle de Santiago km 3.5+1.8, Salamanca 36885, Mexico; jdj.colinrobles@ugto.mx (J.d.J.C.-R.); ibarram@ugto.mx (M.A.I.-M.)

[2] Centro Universitario de Ciencias Exactas e Ingeniería, Universidad de Guadalajara—CUCEI, M. García Barragán 1451, Guadalajara 44430, Mexico; victor.alcaraz@cucei.udg.mx

* Correspondence: ixbalank@ugto.mx; Tel.: +52-464-647-9940 (ext. 2406)

Abstract: In this work, the design of the hardware architecture to implement an algorithm for optimizing the Hydrogen Productivity Rate (HPR) in a Microbial Electrolysis Cell (MEC) is presented. The HPR in the MEC is maximized by the golden section search algorithm in conjunction with a super-twisting controller. The development of the digital architecture in the implementation step of the optimization algorithm was developed in the Very High Description Language (VHDL) and synthesized in a Field Programmable Gate Array (FPGA). Numerical simulations demonstrated the feasibility of the proposed optimization strategy embedded in an FPGA Cyclone II. Results showed that only 21% of the total logic elements, 5.19% of dedicated logic registers, and 64% of the total eight-bits multipliers of the FPGA were used. On the other hand, the estimated power consumption required by the FPGA-embedded optimization algorithm was only 146 mW.

Keywords: MEC; hydrogen production; online optimization; golden section search; super-twisting controller; FPGA

Citation: Colín-Robles, J.d.J.; Torres-Zúñiga, I.; Ibarra-Manzano, M.A.; Alcaraz-González, V. FPGA-Based Implementation of an Optimization Algorithm to Maximize the Productivity of a Microbial Electrolysis Cell. *Processes* **2021**, *9*, 1111. https://doi.org/10.3390/pr9071111

Academic Editors: Philippe Bogaerts and Alain Vande Wouwer

Received: 19 May 2021
Accepted: 15 June 2021
Published: 25 June 2021

Publisher's Note: MDPI stays neutral with regard to jurisdictional claims in published maps and institutional affiliations.

1. Introduction

Nowadays, biotechnological systems represent a very attractive option for hydrogen production. The degradation of organic matter through the use of bacteria has gained great interest in the scientific community because hydrogen can be produced in a clean way [1,2]. In contrast to current industrial methods, in which unfortunately 90% of the hydrogen produced requires the use of fossil fuels generating a large amount of CO_2 (10 tonnes of CO_2 per ton of H_2) [3], Microbial Electrolysis Cells (MEC) represent a great alternative to produce hydrogen because they require less energy compared to the classic techniques to produce hydrogen, such as the electrolysis of water [4,5].

A MEC is an electrochemical device which uses electroactive microorganisms as catalysts to convert the organic matter to hydrogen and provides a novel approach for producing economically viable hydrogen from a wide range of renewable biomass sources [6,7]. Furthermore, a waste biorefinery based on MECs to produce clean and renewable electrofuel and valuable chemical compounds holds the flexible potentials for pollutants removal and CO_2 capture [8]. Broadly speaking, unlike a Microbial Fuel Cell, a MEC requires the induction of a constant voltage supply generating a potential difference between the electrodes to produce a flow of hydrogen as a result of the degradation of the organic matter that is fed to the MEC.

Other widely biological approaches used for the production of hydrogen in a clean way include Dark Fermentation (DF) in which bioreactors are fed by wastewater with a high concentration of organic matter from domestic and industrial origin. However, its efficiency to produce hydrogen compared to a MEC is relatively low (40% or less) [9]. Generally a MEC is fed with a controlled flow of wastewater which is rich in Volatile Fatty

Acids (VFAs) that in turn might come from another Wastewater Treatment Plant (WWTP) like a DF bioreactor.

The production of hydrogen at the industrial scale through biotechnological systems is a challenge that has been dealt with from different approaches. For instance, in [10] an optimization scheme to maximize the hydrogen productivity of a DF is presented. In such study the optimization is achieved by a heuristic strategy with a nonlinear observer consisting in a Luenberger observer coupled to a super-twisting observer. Then, a super-twisting controller is used to lead the DF process to its maximum hydrogen productivity rate. In [11] the optimization is focused on the effect of the operating conditions such as pH, temperature, nutrient availability and substrate concentration. This involves mathematical modeling of a fermentation process in such a way that biohydrogen production can be predicted. On the other hand in [12] the hydrogen productivity was reported to increase from 0.13 to 0.82 m^3 [H_2] per m^3 per day improving the conductivity of the electrode in a MEC and increasing the population of bacteria in the cathode biofilm. Another work related to hydrogen optimization is presented in [13] where the authors demonstrated that the MEC efficiency can be improved through the reduction of the apparent resistance. The optimization strategy is integrated by both perturbation and observation algorithms designed to track the minimal apparent resistance and adjusting the applied voltage used as control input. Other works in literature are focused in MEC construction details, for example, in [14] an effective strategy to improve the productivity performance through an improved anode arrangement is presented. In such work, the anode is strategically located in such a way that the solution resistance, the biofilm and the whole physical system are reduced. The polarization of the MEC was considerably reduced, affecting directly 72–118% the rate of hydrogen production.

The possibility of being able to implement control algorithms using digital systems such as microcontrollers, Graphic Processing Units (GPUs) and Field Programmable Gate Arrays (FPGAs) has been of great interest due to its great processing capacity, resources optimization and low energy consumption. Besides, the parallelism in the execution of the algorithms has given to the FPGAs a great advantage over other digital systems based on microcontrollers and microprocessors. For example, in [15,16], an FPGA-based fuzzy-logic controller is implemented and analyzed, and it is concluded that this technology is a good choice for implementation. The parallelism offered by FPGAs is used in [17,18] to implement complex control algorithms for a AC-DC converter and a DC-DC converter, respectively. In these works the FPGA processing efficiency is highlighted. In [19] both, the optimization of 80% of the hardware and reduction of 40% of the power consumption of a distributed-arithmetic (DA)-based proportional-integral-derivative (PID) controller compared to a multiplier-based scheme is demonstrated for temperature control. The efficiency of the complete digital control system is demonstrated using a Xilinx Spartan-2E FPGA. More recently, in [20] the authors proposed a combination of a direct torque control, space vector modulation, input-output feedback linearisation, a second-order super-twisting speed controller, and sliding-mode-load torque and stator-flux observers with stator resistance estimation implemented in an FPGA. This control strategy demonstrated robustness in presence of stator resistance variations and uncertainties when it was applied to an induction motor drive. An interesting pipeline implementation of a super-twisting controller to control ground vehicles is presented in [21]. The super-twisting controller was used to control the lateral and yaw velocities in the vehicle dynamics that are described by a discrete time model. The resulting implementation required shorter sampling times and can be synthesized in a low-cost FPGA. A classical Proportional-Integral-Derivative (PID) controller implemented in FPGA is proposed in [22]. With the objective to accelerate the execution of the algorithm, to obtain great precision and to get highly commercial ability, the implementation was based on smooth motion interpolation. The results from numerical simulations and practical tests, demonstrated its correct performance. Nevertheless, to the best of the authors knowledge, there is not FPGA-based control implementations applied to bioprocesses.

In the present work the optimization problem of maximizing the Hydrogen Production Rate (HPR) in a MEC is addressed. The productivity function is approximated from the MEC model in steady state, for which, a point of maximum performance in a well-defined operating region is ensured. Using the golden section search optimization algorithm coupled to a robust super-twisting controller, the MEC is online brought to its maximum hydrogen production performance. The proposed optimization strategy is embedded in an FPGA throughout different digital architectures that are executed in parallel without hardware sharing. The resulting digital architecture has mainly two advantages, first, the portability to be synthesized in an FPGA card from any manufacturer, and second, the low power consumption compared to a personal computer. The implementation of the optimization algorithm in an FPGA has the great advantage of being described in hardware. This allows an easy adaptation in the use of communication protocols with external devices.

The rest of the paper is organized as follows: in Section 2 the mathematical model of the MEC is described, and the objective function as the HPR is presented. A description of the optimization problem is described in detail in Section 3. In Section 4 the optimization problem is addressed by using the Golden Section Search algorithm coupled to the discrete time super-twisting controller. In addition, the maximum HPR numerically computed is verified analytically. The FPGA-based implementation of the optimization algorithm is presented in Section 5 including numerical algorithms for the implementation of arithmetic operations like division, multiplication and square root. The results are presented in Section 6, where numerical simulations are carried out in an FPGA to verify the performance, including both the truncation error and the synthesis report of the digital architecture. Finally some conclusions are pointed out in Section 7.

2. Mathematical Model

One of the most used Microbial Electrolysis Cell (MEC) configurations currently consists mainly of two chambers that are separated by a cathode membrane (see Figure 1). In the anode chamber, the anode is covered by a biofilm where the existence of anodophilic and methanogenic bacteria is considered. The degradation of VFAs in the MEC takes place in the anode chamber, where hydrogen protons and electrons are produced. Protons pass through a ionic membrane to the cathodic chamber where the production of hydrogen occurs. A relatively small voltage is supplied to the system generating a potential difference between the two electrodes, which allows the electrons released in the anode by the anodophilic bacteria to circulate and pass to the cathode to combine with the hydrogen protons. In the degradation process there is a competition between two types of microorganisms, anodophilic and methanogenic, to decide who will consume the substrate.

This behavior is modeled by the following system of Ordinary Differential Equations (ODEs) [23]:

$$\dot{s} = (s_{in} - s)D_{in} - k_a \mu_a x_a - k_m \mu_m x_m \tag{1}$$

$$\dot{x}_a = \mu_a x_a - k_{d,a} x_a - \alpha_a D_{in} x_a \tag{2}$$

$$\dot{x}_m = \mu_m x_m - k_{d,m} x_m - \alpha_m D_{in} x_m, \tag{3}$$

where s is the acetate concentration (mg/L^{-1}), while x_a and x_m are the concentration of the anodophilic and acetoclastic methanogenic microorganisms, respectively (mg/L^{-1}); D_{in} is the dilution rate, $D_{in} = F_{in}/V_{reac}$ (d^{-1}), where F_{in} is the input flow rate (Ld^{-1}) and V_{reac} is the reactor volume (L); α_a and α_m are the dimensionless biofilm retention constants. μ_a and μ_m are the growth rates (d^{-1}) for anodophilic and acetoclastic methanogenic microorganisms, respectively, which are defined as follows:

$$\mu_a = \mu_{max,a} \frac{s}{k_{s,a} + s} \frac{1}{1 + e^{-\frac{F}{RT}\eta}} \tag{4}$$

$$\mu_m = \mu_{max,m} \frac{s}{k_{s,m} + s}, \tag{5}$$

where $\mu_{max,a}$ and $\mu_{max,m}$ are the maximum grown rates (d^{-1}), $k_{s,a}$ and $k_{s,m}$ are the half-rate Monod constants (mg (s) L^{-1}), F is the Faraday constant (C mol^{-1} e^{-1}), R is the ideal gas constant (J mol^{-1}K^{-1}), T is the temperature (K), $\eta = E_{anode} - E_{Ka}$ is the local potential, where E_{anode} is the anode potential (V) and E_{Ka} is the half-maximum-rate anodic Electron Aceptor (EA) potential (V) i.e., the potential that occurs when $S = k_{S,a}$ and the rate is half of the maximum rate [24].

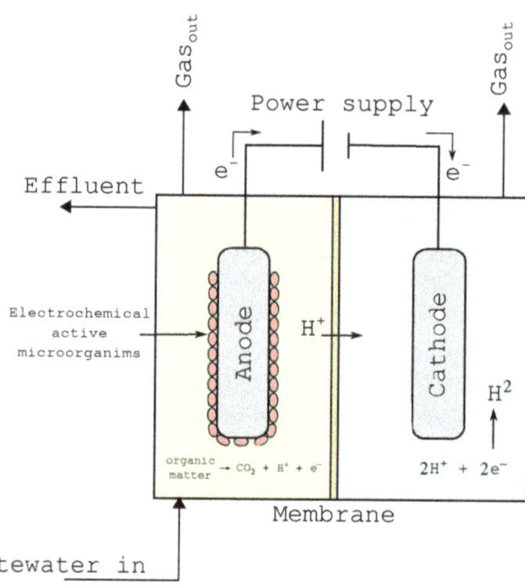

Figure 1. Schematic diagram of the MEC.

MEC Productivity

The hydrogen flow rate in the MEC is modeled by Equation (6), where it can be seen that the hydrogen produced is closely related to the current generated from the flow of electrons between the electrodes.

$$Q_{H_2} = Y_{H_2} A_a \frac{I_{MEC}}{mF} \frac{RT}{P}, \tag{6}$$

where Y_{H_2} is the dimensionless cathode efficiency, A_a is the anode area (m^2), m is the electrons per mol specie (mol e$^-$ mol^{-1} M^{-1}) and P is the pressure inside the cathodic chamber (atm). In the Equation (6) the methanogenic microorganisms consumption is neglected and it is considered that only anodophilic microorganisms are responsible for acetate degradation. The current in the MEC is modeled as:

$$I_{MEC} = \left(\gamma_s k_a \mu_a x_a L_f (1 - f_s^0) + \gamma_x b x_a L_f \right) A_a, \tag{7}$$

where γ_s and γ_x (mFM^{-1}W$_s^{-1}$) are the yield coefficients related to the number of coulombs that it is possible to obtain from W_s (g mol^{-1}) and W_x (g mol^{-1}), i.e., the substrate, and the biomass respectively; f_s^0 is the dimensionless fraction of electrons used for cell synthesis, b is the endogenous decay coefficient (d^{-1}) and L_f is the biofilm thickness (m).

The hydrogen production rate (HPR) $Q_{H_{2,p}}$ is defined as the hydrogen flow rate produced per volume of reactor ($L[H_2]$ L^{-1}d^{-1}):

$$Q_{H_{2,p}} = \frac{Q_{H_2}}{V_{reac}}, \tag{8}$$

where Q_{H_2} is the hydrogen flow rate defined by Equation (6).

3. Problem Statement

The HPR is function of both, the dilution rate D_{in} and the inlet acetate concentration s_{in}. D_{in} is the optimization variable, while s_{in} is considered as a disturbance. As it can be seen in Figure 2, the HPR presents a maximum hydrogen productivity point related to an optimal dilution rate $(Q_{H_2,p\,max}, D_{in,opt})$ within a range of concentrations for the inlet acetate s_{in} [2000, 6000] mL^{-1}. Therefore, the optimization problem consists in calculating the value of the optimal dilution rate $D_{in,opt}$ that ensures the maximum performance $Q_{H_2,p\,max}$ in the MEC.

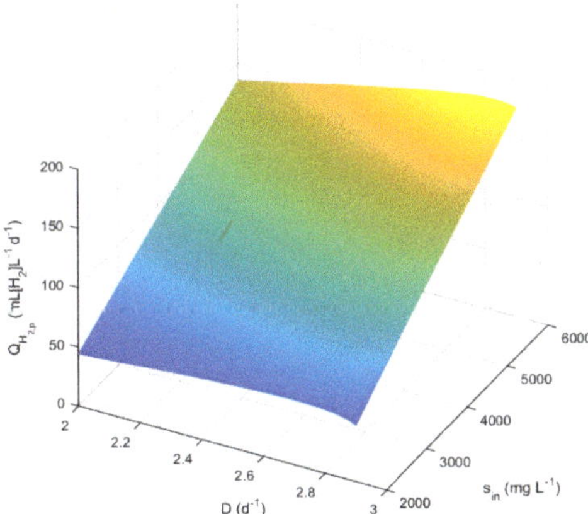

Figure 2. MEC hydrogen productivity rate in steady state on the operating region.

Maximizing the HPR in the MEC is possible if and only if a $D_{in,opt}$ of the productivity function $Q_{H_2,p}(D_{in}, s_{in})$ can be computed in an open neighborhood region (Γ) for each acetate concentration in the inlet s_{in}. Ensuring the existence of $D_{in,opt}$ implies the following assumptions [25]:

Assumption 1. *The function $Q_{H_2,p}$ is twice continuously differentiable in Γ with respect to D_{in} such that:*

$$\frac{\partial Q_{H_2,p}(D_{in,opt}, s_{in})}{\partial D_{in}} = 0$$

$$\frac{\partial^2 Q_{H_2,p}(D_{in}, s_{in})}{\partial D_{in}^2} < 0$$

(9)

Assumption 2. *The function $Q_{H_2,p}$ is convex, unimodal and any $D_{in,opt}$ is a global maximizer for each s_{in} in the operating region.*

The optimization problem to maximize the hydrogen production rate in the MEC is proposed as:

$$\max_{D_{in}} \; Q_{H_2,p}(D_{in}, s_{in})$$

such that: (10)

$$\dot{x}(t) = f(x, D_{in}, s_{in})$$
$$y(t) = Q_{H_2,p}(x),$$

where $x = [s, x_a, x_m]^T$ is the state vector, $f(x, D_{in}, s_{in})$ is defined by Equations (1)–(5) and the measured output $Q_{H_2,p}(x)$ is the hydrogen production rate defined by Equations (6)–(8).

As it is shown in Figure 2, only a maximum $Q_{H_2,pmax}$ can be observed for each maximizer $D_{in,opt}$ in the operating region.

The optimization problem is online solved by the GSS algorithm coupled to a super-twisting controller. The GSS algorithm calculates the value $Q_{H_2,pmax}$ using a hydrogen productivity function in relation to both, the dilution rate and the inlet acetate concentration of the MEC. The super-twisting controller uses $Q_{H_2,pmax}$ as a reference to track the MEC productivity to the maximum value. The optimization scheme described before is depicted in Figure 3.

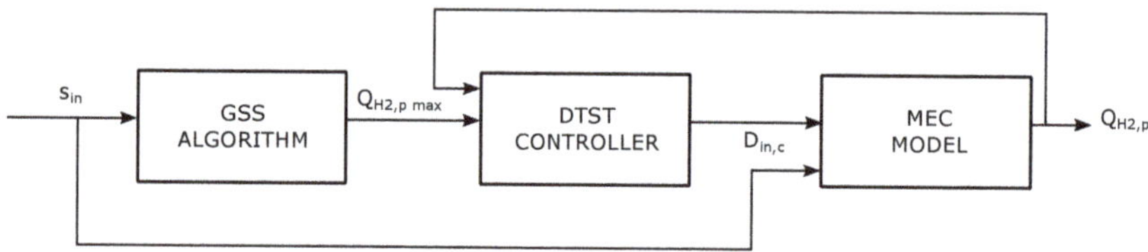

Figure 3. Optimization scheme of the MEC.

In order to optimize the hardware resources and to reduce the power consumption, the optimization strategy to maximize the HPR of the MEC is embedded in an FPGA. This way, the energy cost required to bring the MEC to its maximum HPR can be considerably reduced.

4. Optimization of the MEC Productivity

An optimum point $(Q_{H_2,pmax}, D_{in,opt})$ is possible if and only if the MEC achieves an steady state $[s^*, x_a^*, x_m^*]$. The operating point of the system (1)–(3) as function of s_{in} and D_{in} is given in steady state as:

$$s^* = \frac{k_{s,a}k_{d,a} + k_{s,a}\alpha_a D_{in}}{\frac{\mu_{max,a}}{\psi} - k_{d,a} - D_{in}\alpha_a} \tag{11}$$

$$x_a^* = \frac{(s_{in} - s^*)D_{in}}{k_a\mu_a} \tag{12}$$

$$x_m^* = 0, \tag{13}$$

with

$$\psi = 1 + e^{-\frac{F}{RT}\eta}. \tag{14}$$

The objective function $Q_{H_2,p}(D_{in}, s_{in})$, defining the input-output map in steady state, is therefore expressed as:

$$Q_{H_2,p}(D_{in}, s_{in}) = \frac{L_f A_{sur} Y_{H_2} A_a RT D_{in}}{mFPV_{reac}}(s_{in} - s^*)\left[\gamma_s(1 - f_s^0) + \frac{\gamma_x b\psi(k_{sa} + s^*)}{k_a\mu_{max,a}s^*}\right]. \tag{15}$$

In this work, the acetate concentration in the inlet s_{in} is assumed as known.

4.1. The Golden Section Search Algorithm

Golden ratio (φ) has been of a great interest to mathematicians, physicists, philosophers and artists. In antiquity, civilizations like Egyptians used the φ number as the main criterion for the construction of the Great Pyramids. The Parthenon in Greece was also built based on φ [26].

In relation to nature, the golden ratio is considered a natural constant that sets the standard in reproduction, growth patterns of living beings such as plants and animals. Their geometric relationship is described in Figure 4, where a line A–C is divided into two segments l and r by a point B where l is greater than r such that the ratio l/r is equal to the ratio $(l + r)/l$.

The Golden Section Search (GSS) algorithm is an iterative process suggested to optimize one-dimensional, unimodal and well behaved functions [27], taking into account that the optimum value must be into a search region defined by a lower bound (A) and an upper bound (C), as described in Figure 4.

$$\varphi = \frac{l}{r} = \frac{l+r}{l} = 1.618033988... \tag{16}$$

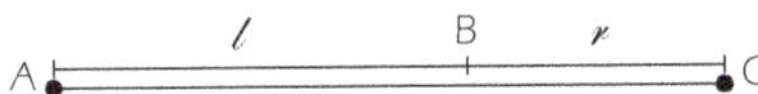

Figure 4. The golden ratio.

The optimization of the HPR in the MEC begins defining the search region of Equation (15). In this case, the search region is defined by $D_{in,A} = 1\,\mathrm{d}^{-1}$ and $D_{in,C} = 3\,\mathrm{d}^{-1}$. Then, two inner evaluation points $D_{in,1}$ and $D_{in,2}$ are selected as function of φ.

$$D_{in,1} = D_{in,A} + d \tag{17}$$

$$D_{in,2} = D_{in,C} - d, \tag{18}$$

with

$$d = (\varphi - 1)(D_{in,A} - D_{in,C}). \tag{19}$$

The error used by the GSS algorithm to stop its operation is defined as:

$$err = (\varphi - 1)\left| \frac{D_{in,C} - D_{in,A}}{D_{in,opt}} \right|. \tag{20}$$

The complete GSS algorithm to calculate the optimum point $(D_{in,opt}, Q_{H_2,p\,max})$ is presented in Algorithm 1.

4.2. GSS Validation

First, let us analyze the stability of the nonlinear system (1)–(3) by calculating the eigenvalues (λ_i) of its linear approximation. The indirect Lyapunov method establish conditions that allow us to obtain conclusions about the stability of the nonlinear system in an operating point.

Consider the nonlinear system (1)–(3) with the operating point $x^* = [s^*,\ x_a^*,\ x_m^*]$ that has the following linear approximation

$$\dot{\bar{x}} = A\bar{x} + B_u\bar{u} + B_w\bar{w}, \tag{21}$$

where $\bar{x} = x - x^*$, A, B_u and B_w are the Jacobian matrices of the system, $\bar{u} = D_{in} - D_{in}^*$ and $\bar{w} = s_{in} - s_{in}^*$ respectively.

The indirect Lyapunov method states that the nonlinear system (1)–(3) is asymptotically stable if and only if $Re(\lambda_i) < 0$ of the matrix A, $\forall \lambda_i$, $i = 1, 2, 3$, defined as:

$$A = \left.\frac{\partial f(x, D_{in}, s_{in})}{\partial x}\right|_{(x^*, D_{in}^*, s_{in}^*)}. \tag{22}$$

As it can be seen in Figure 5 the eigenvalues of the matrix A are Hurwitz in the operating region of the MEC. It must be pointed out that the closer the dilution rate is to the value $3\,\mathrm{d}^{-1}$, the more the eigenvalues λ_1 and λ_2 approach the origin.

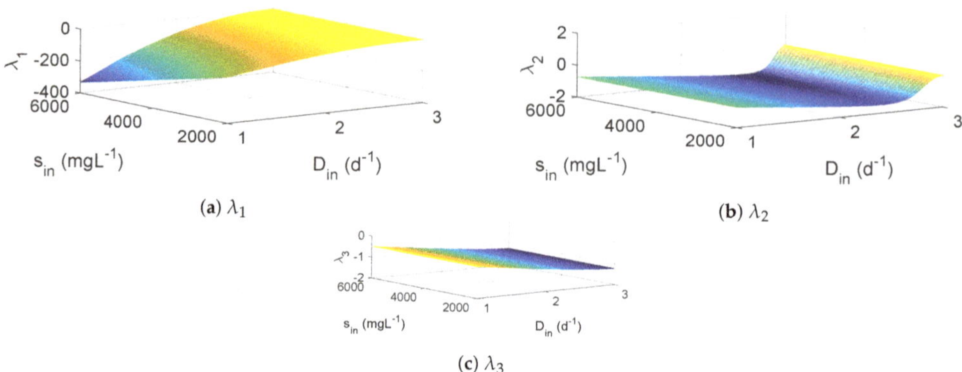

Figure 5. Eigenvalues of the MEC model (1)–(3) linearized in the operating region.

Algorithm 1: GSS algorithm description

Input: $(D_{in,A}, D_{in,B}, tolerance)$
Result: $(Q_{H_2,p}max, D_{in,opt})$

$f_1 = Q_{H_2,p}(D_{in,1})$;
$f_2 = Q_{H_2,p}(D_{in,2})$;
while $err > tolerance$ **do**
    if $(f_1 > f_2)$ **then**
        $D_{in,A} = D_{in,2}$;
        $D_{in,2} = D_{in,1}$;
        $D_{in,1} = D_{in,A} + d$;
        $f_2 = f_1$;
        $f_1 = Q_{H_2,p}(D_{in,1})$;
        $D_{in,opt} = D_{in,1}$;
        $Q_{H_2}max = f_1$
    else
        $D_{in,C} = D_{in,1}$;
        $D_{in,1} = D_{in,2}$;
        $D_{in,2} = D_{in,C} - d$;
        $f_1 = f_2$;
        $f_2 = Q_{H_2,p}(D_{in,2})$;
        $D_{in,opt} = D_{in,2}$;
        $Q_{H_2,p}max = f_2$;
    end
    $err = (\varphi - 1)\left|\dfrac{D_{in,C} - D_{in,A}}{D_{in,opt}}\right|$;
end

The optimum value $D_{in,opt}$ is then obtained by differentiating the objective function (15) with respect to D_{in} and equating the result to zero (first-order optimally condition), which leads to

$$\frac{\partial Q_{H_2,p}}{\partial D_{in}} = \left(\frac{Y_{H_2} A_a RT}{mFP V_{reac}}\right)\frac{\partial I^*_{MEC}}{\partial D_{in}} = 0, \tag{23}$$

where

$$\frac{\partial I^*_{MEC}}{\partial D_{in}} = L_f A_{sur} [D_{in}(s_{in} - s^*)(\frac{\psi \gamma_x b}{k_a \mu_{max,a}} \frac{\partial \rho}{\partial D_{in}}) + (\gamma_s(1 - f_s^0) + \frac{\gamma_x b}{k_a \mu_a})(s_{in} - s^* - D_{in}\frac{\partial s^*}{\partial D_{in}})] \tag{24}$$

$$\frac{\partial \rho}{\partial D_{in}} = \frac{\frac{\partial s^*}{\partial D_{in}}(s^* - (k_{sa} + s^*))}{s^{*2}} \tag{25}$$

$$\frac{\partial s^*}{\partial D_{in}} = \frac{k_{sa}\alpha_a\left(\frac{\mu_{max,a}}{\psi} - k_{d,a} - D_{in}\alpha_a\right) + \alpha_a(k_{s,a}k_{d,a} + k_{s,a}\alpha_a D_{in})}{\left(\frac{\mu_{max,a}}{\psi} - k_{d,a} - D_{in}\alpha_a\right)^2} \tag{26}$$

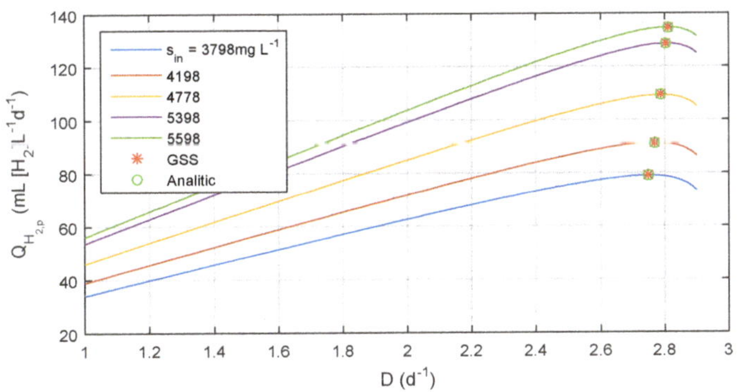

Figure 6. Hydrogen productivity for different s_{in}.

Figure 6 shows the $Q_{H_2,p max}$ value calculated both, by the GSS Algorithm 1 and by substituting $D_{in,opt}$, calculated by setting the Equation (23) equal to zero (see Figure 7), in Equation (15). As it can be seen, the results of the GSS algorithm match the results obtained analytically.

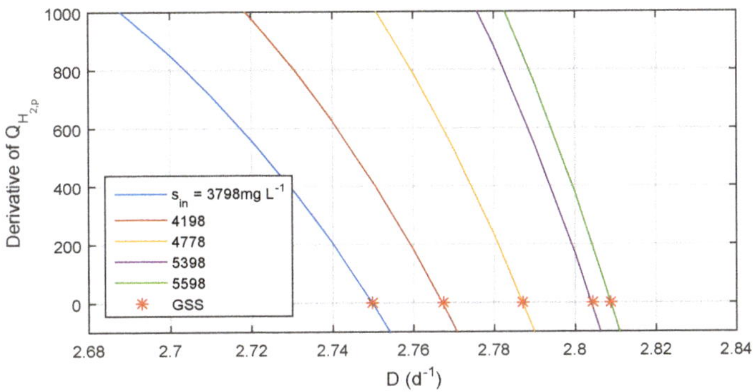

Figure 7. Derivative of $Q_{H_2,p}$ respect to D_{in}.

4.3. Super-Twisting Controller

The MEC model (1)–(3) can be rewritten as follows:

$$\dot{x} = \gamma(x) + g(x)D_{in} \tag{27}$$

$$y = Q_{H_{2,p}}(x), \tag{28}$$

where $\gamma(x)$ and $g(x)$ are vector functions defined as:

$$\gamma(x) = \begin{pmatrix} -k_a\mu_a x_a - k_m\mu_m x_m \\ (\mu_a - k_{d,a})x_a \\ (\mu_m - k_{d,m})x_m \end{pmatrix} \tag{29}$$

$$g(x) = \begin{pmatrix} s_{in} - s \\ -\alpha_a x_a \\ -\alpha_m x_m \end{pmatrix}. \tag{30}$$

The relative degree σ of System (27) and (28) is computed by differentiating the output with respect to time as [28]:

$$\dot{y} = \frac{\partial Q_{H_{2,p}}(x)}{\partial x}\dot{x} = \beta\gamma(x) + \beta g(x)D_{in}, \tag{31}$$

where $\beta = \left[\frac{\partial Q_{H_{2,p}}}{\partial s}, \frac{\partial Q_{H_{2,p}}}{\partial x_a}, \frac{\partial Q_{H_{2,p}}}{\partial x_m}\right]$. Hence, the relative degree of the system (27) and (28) is $\sigma = 1$.

In this work the super-twisting controller, Equations (32) and (33), is therefore considered to track the maximum hydrogen flow rate computed by the GSS algorithm with the sliding variable defined as the tracking error [29].

$$D_{in,c} = -\rho_1\sqrt{|\epsilon_c|}sign(\epsilon_c) + D_{nom} \tag{32}$$

$$\frac{dD_{nom}}{dt} = -\rho_2 sign(\epsilon_c) \tag{33}$$

In the super-twisting controller (32) and (33), the tracking error is defined as:

$$\epsilon_c = Q_{H_{2,p}max} - Q_{H_{2,p}}, \tag{34}$$

ρ_1 and ρ_2 are the controller gains that ensure the finite-time stability of the tracking error, while $D_{in,c}$ is the control input necessary to bring the MEC to the maximum value $Q_{H_{2,p}max}$.

For implementation purposes in an FPGA, the discrete time super-twisting controller (DT-STC) is considered. The representative numerical solution showed in the Equations (35) and (36) is obtained from Equations (32) and (33) using the Euler's method. The controller uses the value $Q_{H_{2,p}max}$ as a reference to carry the real productivity to its maximum value in finite time.

$$D_{in,c}[k] = -\rho_1\sqrt{|\epsilon_c|}sign(\epsilon_c) + D_{nom}[k] \tag{35}$$

$$D_{nom}[k+1] = D_{nom}[k] - \tau\rho_2 sign(\epsilon_c), \tag{36}$$

In Equation (36), τ (*d*) is the sampling time considered.

5. FPGA-Embedded Optimization Algorithm

The FPGA-based implementation of the optimization algorithm is depicted in Figures 8 and 9. Following the scheme presented in Figure 3, the implementation block diagram is integrated by the GSS algorithm digital architecture coupled to the DTSTC digital architecture. A finite state machine (FSM) and a down programmable counter are used to ensure the proper operation of the optimization algorithm embedded in the FPGA.

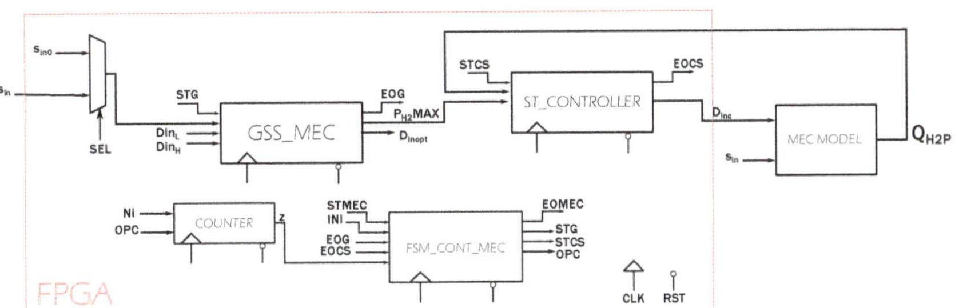

Figure 8. FPGA-based implementation of the hydrogen optimization algorithm.

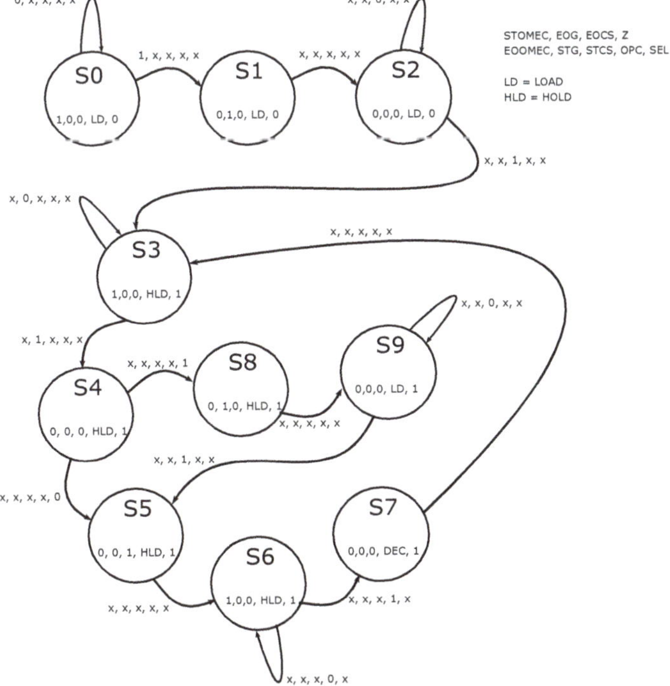

Figure 9. FSM_CONT_MEC module in the FPGA-based optimization algorithm.

The digital architecture of the optimization algorithm uses a fixed point format (16,24) to represent all the input-output signals and inner operations. The hardware description used to develop the digital architecture was VHDL and the target board used was the Cyclone II EP2C35F672C6 integrated in the ALTERA DE2 educational board with a clock frequency $f_{CLK} = 50$ MHz.

The modules GSS_MEC and ST_CONTROLLER were designed for an easy interaction with the FSM_CONT_MEC module and any other external device through the STG, EOG, STCS and EOCS signals. When the input signals STG and STCS are assigned to the logical value '1' by the FSM_VCONT_MEC module, they will produce a busy mode of their respective modules due to the latency time in the calculation of their final results. The busy mode is indicated by the output signals EOG ='0' and EOCS = '0'. On the other hand, when EOG = '1' and EOCS = '1', it means that the modules GSS_MEC and ST_CONTROLLER have finished and the results are ready to be read.

5.1. Operation of the FPGA-Embedded Optimization Algorithm

The FSM depicted in the Figure 9 is a great help for understanding the operation of the digital architecture. The FPGA execution can be divided in two steps, the initialization step, which is controlled by the states S0 to S2, and the normal operation, which is controlled by the remanding states of the FSM_CONT_MEC module. The initialization is executed when the FPGA is energized and the INI signal has a binary value '1'. Otherwise, the FPGA remains in standby mode until an external source changes the value of that signal. In such case, the initialization is started by a push-button (see the state S0). When INI = '1' the FSM changes to the state S1 where STG = '1' and SEL = '0' in the GSS_MEC module and the two-one multiplexer. This will start the calculation of $Q_{H_2,p,max}$ with the initial value $s_{in,0}$. In the next clock cycle, the EOMEC signal in the GSS_MEC module will change from logic '1' to logic '0' indicating that this module is in the process of calculating $Q_{H_2,p,max}$. At the same time, without any condition, a transition is made to the state S2 where the FSM is waiting by the logic value '1' in the EOMEC signal indicating that the result is ready. When $Q_{H_2,p,max}$ is ready to be used by the ST_CONTROLLER module, the FSM make a transition to the state S3 where the initialization step is done, and the system now is in the normal operation where SEL = '1' and it is waiting for an external device to set the value STOMEC = '1'. During the initialization step, the down counter is loaded with an initial value decreased by one every sampling period until reaching the optimization period.

In the normal operation, the ST_CONTROLLER module and the down counter are executed every sampling period with the aim of controlling the HPR in the MEC, and decreasing the initial value of the counter. When the down counter reaches the value zero, this means that the optimization period has expired and the GSS_MEC module is executed to generate a new $Q_{H_2,p,max}$, after that, the down counter is reloaded with the initial value.

The normal operation starts in the state S3 and the digital architecture reads s_{in} by SEL = '1' in the multiplexer. When the signal STOMEC = '1', the FPGA-based optimization algorithm generates the control input $D_{in,c}$ of the MEC after a latency time, otherwise, the system is in standby. The execution of the ST_CONTROLLER and the down counter are managed by the states S5 to S7 in the FSM every sampling period, while the states S8 and S9 manage the GSS_MEC MODULE and the reinitialization of the down counter when the optimization period has been reached. In order to know when the GSS_MEC module should be executed, the FSM reads the signal Z from the down counter in the state S4. When Z = '0' this means that the optimization period has not yet elapsed and the FSM is currently executing the ST_CONTROLLER module, otherwise, when Z = '1' the FSM executes one more time the GSS_MODULE and generates a new $Q_{H_2,p,max}$ in function of the current value s_{in}. The down counter is reinitialized as well.

The most used arithmetic operations in the optimization algorithm are product, addition, division and square root. The hardware description was developed using standard VHDL and therefore the designs presented in this work do not belong to any manufacturer.

5.2. GSS Implementation

The digital architecture of the GSS optimization strategy, described in Algorithm 1, is depicted in Figure 10. The digital architecture of such algorithm is made up of registers, full adders, 8-bit embedded multipliers, multiplexers and full comparators using the previously mentioned fixed point format. Notice that the objective function shown in Equation (15) was programmed in the block $Q_{H_2,p}$. Its implementation needed a simplified representation with the objective to calculate the hydrogen productivity with few hardware resources and small latency time. By precalculating constant parameters and making a separation by variables the following objective function is obtained:

$$Q_{H_2,p} = \beta_1 x_a^*(s_{in}, D_{in})(\beta_2 \mu_a(s^*(D_{in})) + \beta_3), \tag{37}$$

where the values of constants β_1, β_2 and β_3 are defined in Table 1.

Table 1. Constant parameters in $Q_{H_{2,p}}$.

Constant Parameter	Value
β_1	2.1906×10^{-8}
β_2	316.825
β_3	40

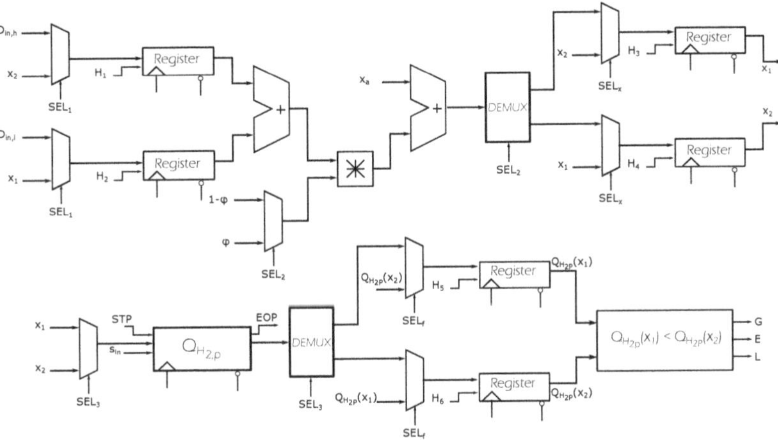

Figure 10. Digital architecture of the GSS algorithm.

The complete comparator that determines if $f_1 > f_2$, in Algorithm 1, was designed taking into account that the operation involves real numbers and therefore the classical definition of a complete comparator of binary numbers is not sufficient for this implementation.

5.3. DTSTC Implementation

The digital architecture of the DTSTC (see Figure 11) is simpler than that one of the GSS algorithm. Although only combinational elements are required, its response speed is quite fast to generate the control action compared to the speed of change to generate the reference computed by the GSS algorithm.

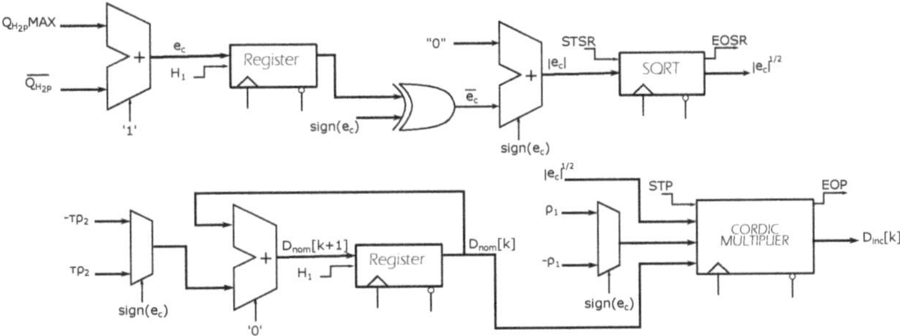

Figure 11. Digital architecture of the DTSTC.

The controller correction term $\sigma_1\sqrt{|\epsilon|}$ requires a digital circuit capable of computing the square root of the tracking error. Particularly in this work, the Pencil and Paper algorithm [30] proved to be very useful as a basis for the design of the SQRT arithmetic circuit.

The arithmetic circuit of the multiplier in the DTSTC architecture is based on the Coordinate Digital Computer Algorithm (CORDIC) with its rotating linear version (see Figure 12) [31], i.e.,:

$$
\begin{aligned}
x_{j+1} &= x_j, \\
y_{j+1} &= y_j + \sigma_j 2^{-j} x_j, \\
z_{j+1} &= z_j + \sigma_j 2^{-j},
\end{aligned}
\tag{38}
$$

with

$$
\sigma_j = \begin{cases} -1 & \text{if } z_j \geq 0 \\ +1 & \text{otherwise.} \end{cases}
\tag{39}
$$

The results obtained after a sequence of fixed micro-rotations are given in the following way:

$$
\begin{aligned}
x_f &= x_{in}, \\
y_f &= y_{in} + x_{in} z_{in} \\
z_f &= 0.
\end{aligned}
\tag{40}
$$

The resulting operation y_f in Equation (40) has the necessary shape to implement the DTSTC. As it can be seen in Equation (35), $D_{in,c}$ can be calculated from the final result y_f by these two arithmetic operations; i.e., the product and the addition. The CORDIC-based Multiplier Digital Circuit presented in the Figure 12 has the shape necessary to implement DTSTC without the need of using embedded multipliers in the FPGA and it has a short latency time.

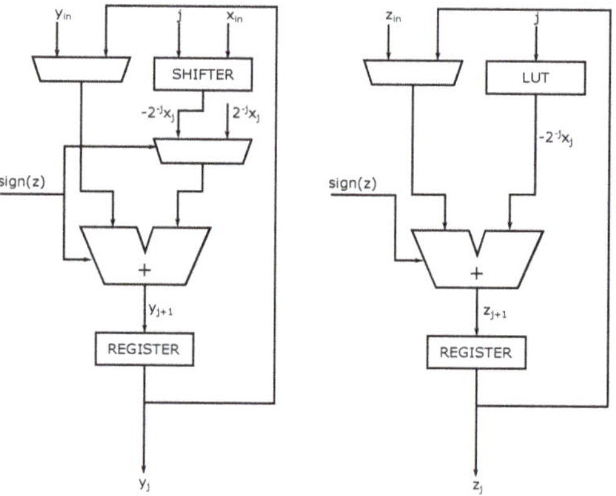

Figure 12. Digital architecture of the linear vectoring CORDIC.

6. Results

The feasibility of the FPGA-embedded optimization algorithm was demonstrated through numerical simulations. The MEC model (1)–(6) was simulated in Matlab, the ODEs were solved by the stiff solver ode15s. The parameters used in the numerical simulations are listed Table 2. In order to demonstrate the robustness of the optimization strategy proposed, modified parameters between $\pm30\%$ from their nominal value were considered. The hardware required for the verification test is depicted in the Figure 13. As it can be seen, a serial communication was used to communicate the FPGA with Matlab, which was executed in a personal computer with Windows 10, Intel Core i7 and memory RAM DDR3 of 32 GB. In these conditions, six hours were needed to perform the verification test of the optimization algorithm in a Cyclone II FPGA running at $f_{clock} = 50$ MHz and a reception-transmission data rate of 70 Mbps. The operation time of the MEC simulated in the computer was of 200 d with a sampling period $\tau = 0.004\,d$. The hardware resources in the target board are summarized in Table 3.

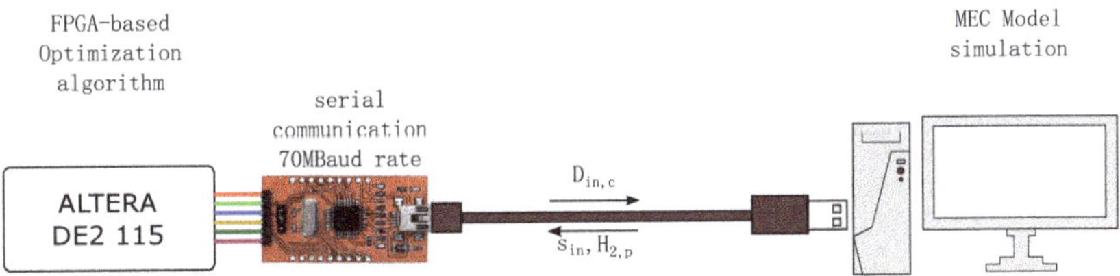

Figure 13. Implementation scheme for numerical simulation tests

Table 2. MEC Model parameter with uncertainties.

Description	Symbol	Value	Variation (%)
Gas ideal constant ($\mathrm{J\,mol^{-1}K^{-1}}$)	R	8.31	0.00
Faraday constant ($\mathrm{C\,mol^{-1}e^{-1}}$)	F	96,485	0.00
Temperature (K)	T	298.15	-20.00
Yield coefficient ($\mathrm{mg\,(s)\,mg^{-1}}\,(x_a)$)	k_a	0.667	$+15.00$
Yield coefficient ($\mathrm{mg\,(s)\,mg^{-1}}\,(x_m)$)	k_m	4.7067	-20.00
Microbial decay ($\mathrm{d^{-1}}$)	$k_{d,a}$	0.05 $\mu_{max,a}$	$+5.00$
Microbial decay ($\mathrm{d^{-1}}$)	$k_{d,m}$	0.05 $\mu_{max,m}$	$+2.00$
Biofilm retention constant of x_a	α_a	0.5	$+12.00$
Biofilm retention constant of x_m	α_m	0.5	$+5.00$
Maximum grown rate ($\mathrm{d^{-1}}$)	$\mu_{max,a}$	1.97	$+28.00$
Maximum grown rate ($\mathrm{d^{-1}}$)	$\mu_{max,a}$	0.30	$+14.00$
Half-rate constant ($\mathrm{mg\,(s)\,L^{-1}}$)	$k_{s,a}$	20	$+15.00$
Half-rate constant ($\mathrm{mg\,(s)\,L^{-1}}$)	$k_{s,m}$	80	-15.00
Local potential (V)	η	0.3	$+10.00$

Table 3. Specifications of the FPGA ALTERA DE2.

Device	Digital Elements	Total Resources
EP2C35F672C6	Logic Elements(L.E.)	33,216
	Registers	3967
	Number of pins	475
	Embedded Multipliers	70
FPGA	RAM bits (Kb)	4
	PLLs	4
	$f_{max,CLK}$	120 MHz
	RS-232 transceiver and 9-pin connector	120 Kbits/s
	Expansion Headers	two 40-pin
	Toggle switches	18
	Push button switches	4

The inlet acetate concentration s_{in} used to feed the MEC in the numerical simulations is depicted in Figure 14. The digital architecture verification test of the MEC optimization algorithm consists mainly in comparing the results obtained from the FPGA working with the fixed point format (16,24) with the results of the same algorithm executed in Matlab in a floating point representation format.

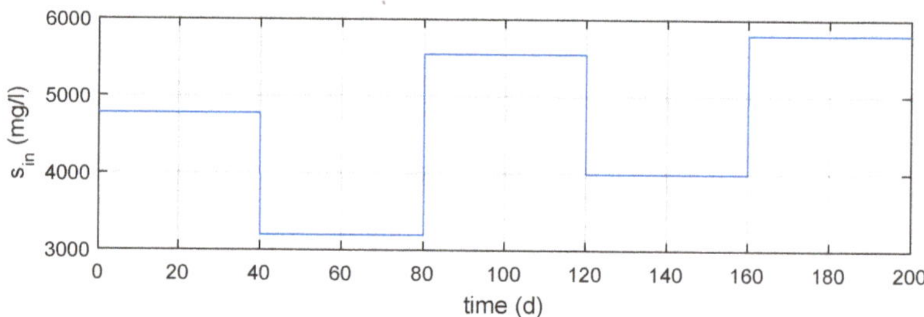

Figure 14. Inlet Acetate concentration (s_{in}).

The resulting HPR obtained by executing the optimization algorithm both in the FPGA and in Matlab is shown in Figure 15. The green dashed-line represents the HPR by the MEC model, the red line represents the maximum HPR computed in Matlab, while the blue dashed-line represents the maximum HPR computed by the FPGA.

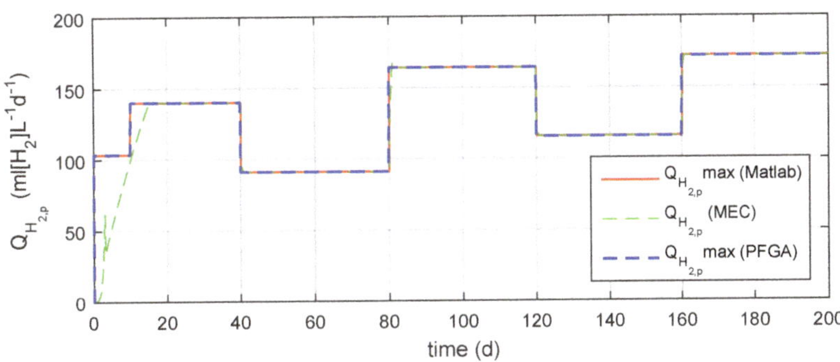

Figure 15. $Q_{H_{2,p}max}$ results in Matlab and FPGA.

On the other hand, the dilution rate computed by the optimization algorithm both in the FPGA and in Matlab is shown in Figure 16. The green dashed-line represents the optimum dilution rate computed by the GSS algorithm, the red line represents the dilution rate computed by the DTSTC in Matlab, while the blue dashed-line represents the dilution rate computed by the DTSTC in the FPGA. As it can be seen, the numerical representation format used to design the optimizer's digital architecture reduces properly the truncation error due to the finite number of bits.

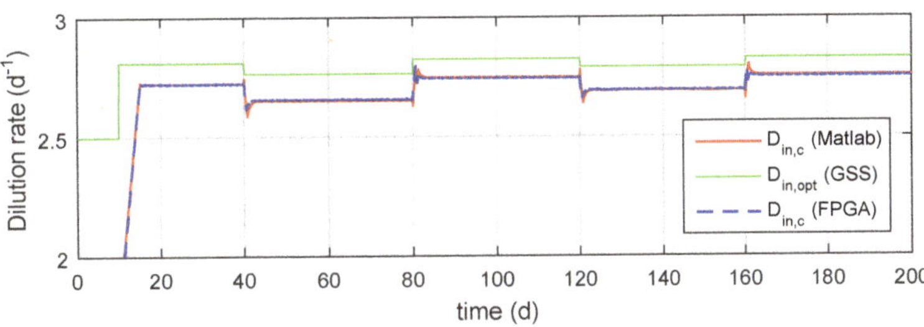

Figure 16. Dilution rate ($D_{in,c}$) generated by FPGA and Matlab.

Initially, the optimization algorithm requires eighteen days to reach the optimal point, as shown in Figure 15. The super-twisting controller requires this period for the control error to converge to zero using the gains specified in Table 4. In this transitory period, the GSS algorithm is initialized with $105\,\mathrm{mL}[H_2]\,\mathrm{mL}^{-1}\mathrm{d}^{-1}$ and this value was taken as the initial reference for the DTSTC.

Table 4. Discrete time super-twisting controller gains.

Gain	Value
ρ_1	0.09
ρ_2	0.19

Once the tracking error has converged to zero, the GSS algorithm reads the inlet acetate concentration value s_{in}, every optimization period equivalent to $D_{in,max}^{-1} = 0.33\,\mathrm{d}$ to update the maximum productivity value $Q_{H_{2,p}max}$ used as reference by the DTSTC.

The acetate concentration in the MEC is showed in the Figure 17. It is easy to see in Figures 18 and 19 that the most of the acetate used to feed the MEC is consumed by the anodophilic bacteria x_a because there is a inhibition process in the methanogenic bacteria growing x_m. As expected, the current between the MEC electrodes is closely related to the HPR (see Figure 20).

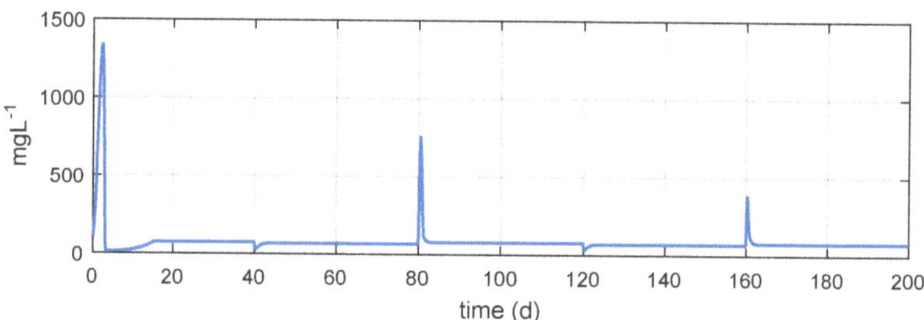

Figure 17. Acetate concentration s in the MEC.

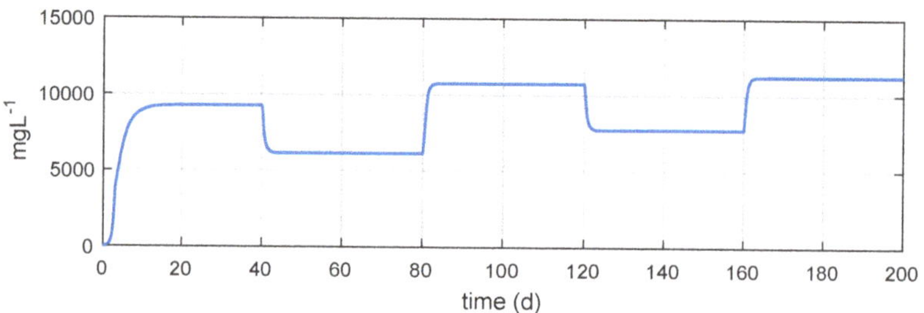

Figure 18. Anodophilic biomass concentration x_a in the MEC.

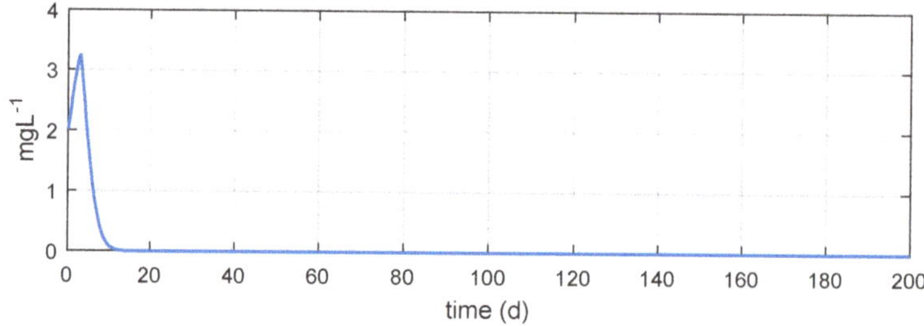

Figure 19. Methanogenic biomass concentration x_m in the MEC.

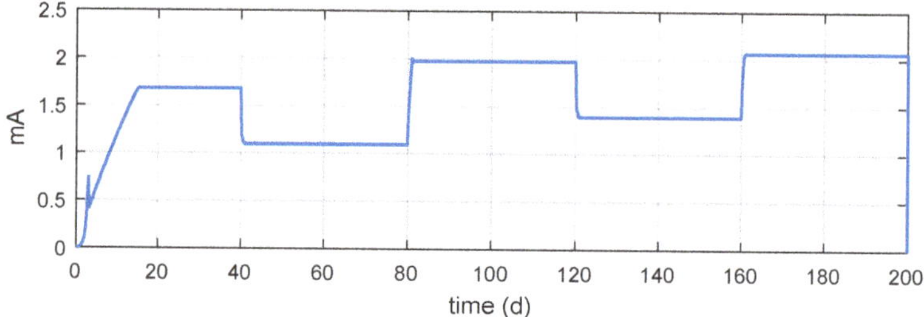

Figure 20. Current intensity in the MEC.

6.1. Error Analysis

The truncation error in the digital architecture of the optimization algorithm is mainly due to the bits fixed quantity in the representation format established in this work. If the resolution in the intermediate operations required to run the optimization algorithm on the FPGA is not sufficient, the truncation error will propagate in such a way that the results obtained are greatly affected.

Figures 21 and 22 show the behavior of the truncation error throughout the simulation process. It can be seen that the error is small enough to determine that the (16,24) format is sufficient to implement the optimization algorithm architecture in the FPGA.

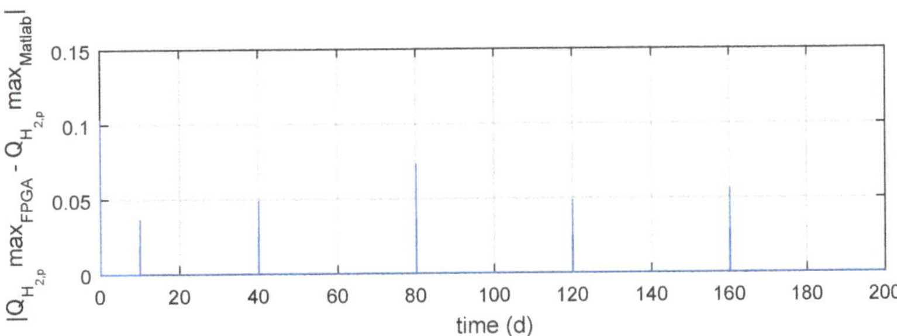

Figure 21. Truncation error in GSS algorithm implementation.

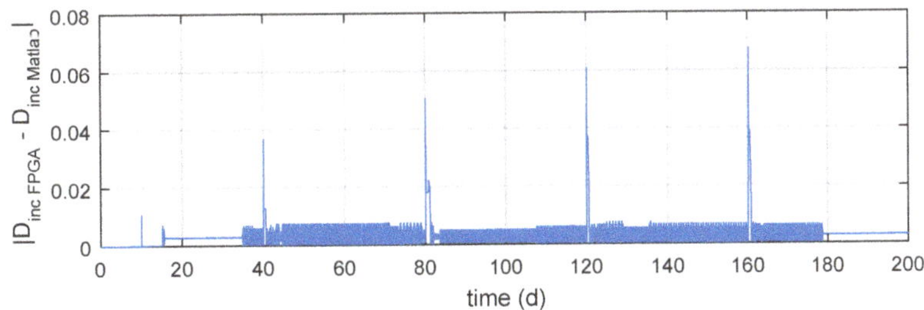

Figure 22. Truncation error in DTSTC implementation.

6.2. Hardware Report

The FPGA hardware resources needed for embedding the digital architecture of the optimization algorithm on Figure 8 are summarized in Table 5. Only a 21% of the total logic elements (L.E.), 5.19% of dedicated logic registers (D.L.R.) and 64% of total eight-bits multipliers (8b-Mult.) in the chip Cyclone II were used. The input to output delay in the implementation was of 150 µs. The estimated power consumption required by the EP2C35F672C6 device using the aforementioned hardware resources is 146 mW. This estimate was calculated by the PowerPlay Early Power Estimator spreadsheet for Cyclone II family v8.0 SP1.

Table 5. Hardware resources used by the optimization algorithm.

Digital Elements	Resources	Used	%
Total L.E.		7089	21.34%
Register only	33,216	291	0.87%
LUT/Register		1472	4.43%
D.L.R.		1724	5.19%
M4K	483,340	0	0.00%
8b-Mult.	70	45	64.00%
I-O delay (No. cycles)	50 MHz	7500	150 µs

The hardware resources used by the most important functional blocks in the optimization algorithm are summarized in Tables 6–8. As it can be seen in Tables 6 and 7, 64%

of the total 8-b multipliers in the FPGA are used in the GSS algorithm, where 48.57% is destined to the $Q_{H_{2,p}}$ functional block where the objective function defined by Equation (15) is processed. It should be pointed out that the $Q_{H_{2,p}}$ block is part of the GSS algorithm functional block (see Figure 10). The GSS algorithm needs at least 4 cycles in the worse of the cases to reach the tolerance error (err = 0.001) defined by Equation (20). Therefore, embedded multipliers most be used in the GSS algorithm digital architecture to have a short latency time.

Table 6. Hardware resources used by GSS algorithm.

Digital Elements	Resources	Used	%
Total L.E.		5849	17.60%
Register only	33,216	191	0.57%
LUT/Register		1044	3.14%
D.L.R.		1235	3.71%
8b-Mult.	70	45	64.00%
I-O delay (No. cycles)	50 MHz	7500	150 μs

Table 7. Hardware resources used by $Q_{H_{2,p}}$ block.

Digital Elements	Resources	Used	%
Total L.E.		5180	15.59%
Register only	33,216	183	0.55%
LUT/Register		727	2.18%
D.L.R.		910	2.74%
8b-Mult.	70	34	48.57%
I-O delay (No. cycles)	50 MHz	249	4.97 μs

Table 8. Hardware resources used by DTSTC algorithm.

Digital Elements	Resources	Used	%
Total L.E.		1165	3.51%
Register only	33,216	100	0.30%
LUT/Register		374	1.13%
D.L.R.		473	1.42%
8b-Mult.	70	0	0.00%
I-O delay (No. cycles)	50 MHz	200	4 μs

On the other hand, the hardware resources used in the DTSTC and its inner functional block, the CORDIC Multiplier, are summarized in Tables 8 and 9. Most DTSTC inner operations are implemented using a CORDIC-based multiplier that has a latency time of 1.48 μs in the worse of the cases, before the tracking error converges to zero. After that, the multiplier is executed faster than 1.48 μs. It should be noted that the CORDIC-based multiplier internally uses an 8-bit expansion in the fractional part to substantially improve the truncation error generated by the fixed-point format considered (see Figure 22).

Table 9. Hardware resources used by CORDIC multiplier.

Digital Elements	Resources	Used	%
Total L.E.		900	2.70%
Register only	33,216	99	0.30%
LUT/Register		219	0.66%
D.L.R.		300	0.90%
8b-Mult.	70	0	0.00%
I-O delay (No. cycles)	50 MHz	74	1.48 µs

Finally, the arithmetic operation $\sqrt{|\epsilon|}$ in the DTSTC is processed by the SQRT functional block, which is based on the Pencil and Paper algorithm. Its digital architecture is primarily based on bit additions and shifts. Table 10 shows the hardware resources needed.

Table 10. Hardware resources used by SQRT.

Digital Elements	Resources	Used	%
Total L.E.		153	0.46%
Register only	33,216	1	0.00%
LUT/Register		89	0.26%
D.L.R.		90	0.27%
8b-Mult.	70	0	0.00%
I-O delay (No. cycles)	50 MHz	74	1.3 µs

7. Conclusions

In this work an FPGA-embedded optimization algorithm to maximize the hydrogen production rate (HPR) of a microbial electrolysis cell (MEC) using the golden section search (GSS) algorithm coupled to a discrete-time super-twisting controller (DTSTC) was presented. The correct performance of the GSS algorithm was analyzed analytically. Furthermore, it was proven that the relative degree of the MEC model is one, a necessary condition to use the DTSTC to bring the HPR to its maximum performance point in finite time.

To reduce the power consumption required to bring the MEC to its maximum performance, a digital architecture of the optimization algorithm was designed and embedded in an FPGA. Although the FPGA used in this work was the Cyclone II of ALTERA, the digital architectures presented in this work were designed to be implemented in any FPGA, regardless of the manufacturer.

The results of the FPGA-embedded optimization algorithm showed a correct performance with low hardware resources and low power consumption compared with a personal computer. Besides, the truncation error generated by the fixed point format used in this work was practically negligible.

Such results allow us to conclude that the implementation of control and optimization algorithms in FPGAs represents an excellent alternative to replace personal computers. Particularly, as demonstrated in the previous section, the FPGA-embedded optimization algorithm proposed to maximize the HPR in the MEC, represents a lower cost alternative in terms of consumed power and resources.

Author Contributions: Conceptualization, I.T.-Z.; methodology, I.T.-Z.; software, J.d.J.C.-R.; validation, I.T.-Z., M.A.I.-M. and V.A.-G.; formal analysis, J.d.J.C.-R., I.T.-Z., M.A.I.-M. and V.A.-G.; investigation, J.d.J.C.-R.; resources, I.T.-Z. and M.A.I.-M.; writing—original draft preparation, J.d.J.C.-R.; writing—review and editing, I.T.-Z., M.A.I.-M. and V.A.-G.; supervision, I.T.-Z. and V.A.-G.; project administration, I.T.-Z.; funding acquisition, I.T.-Z. All authors have read and agreed to the published version of the manuscript.

Funding: This research was funded by the Mexican Council of Science and Technology (CONACyT) under the PhD Grant 614412/327289.

Institutional Review Board Statement: Not applicable.

Informed Consent Statement: Not applicable.

Acknowledgments: This study was funded by the University of Guanajuato and University of Guadalajara. It has been partially supported by the Mexican Council of Science and Technology (CONACyT) under Grant 614412/327289 and the National System of Researches (SNI) program.

Conflicts of Interest: The authors declare no conflict of interest.

References

1. Luo, G.; Xie, L.; Zou, Z.; Wang, W.; Zhou, Q. Exploring optimal conditions for thermophilic fermentative hydrogen production from cassava stillage. *Int. J. Hydrogen Energy* **2010**, *35*, 6161–6169. [CrossRef]
2. Rupprecht, J. From systems biology to fuel—Chlamydomonas reinhardtii as a model for a systems biology approach to improve biohydrogen production. *J. Biotechnol.* **2009**, *142*, 10–20. [CrossRef] [PubMed]
3. Maddy, J.; Cherryman, S.; Hawkes, F.; Hawkes, D.; Dinsdale, R.; Guwy, A.; Premier, G.; Cole, S. *Hydrogen 2003: Report Number 1: ERDF Part-Funded Project Entitled: A Sustainable Energy Supply for Wales: Towards the Hydrogen Economy*; University of Glamorgan: Wales, UK, 2003.
4. Liu, H.; Grot, S.; Logan, B.E. Electrochemically assisted microbial production of hydrogen from acetate. *Environ. Sci. Technol.* **2005**, *39*, 4317–4320. [CrossRef] [PubMed]
5. Sangeetha, T.; Muthukumar, M. Catholyte performance as an influencing factor on electricity production in a dual-chambered microbial fuel cell employing food processing wastewater. *Energy Sources Part A Recover. Util. Environ. Eff.* **2011**, *33*, 1514–1522. [CrossRef]
6. Xing, D.; Yang, Y.; Li, Z.; Cui, H.; Ma, D.; Cai, X.; Gu, J. Hydrogen Production from Waste Stream with Microbial Electrolysis Cells. In *Bioelectrosynthesis: Principles and Technologies for Value-Added Products*; Wiley-VCH Verlag GmbH & Co.: Weinheim, Germany, 2020; pp. 39–70.
7. Flores-Estrella, R.A.; de Jesús Garza-Rubalcava, U.; Haarstrick, A.; Alcaraz-González, V. A Dynamic Biofilm Model for a Microbial Electrolysis Cell. *Processes* **2019**, *7*, 183. [CrossRef]
8. Zhen, G.; Lu, X.; Kumar, G.; Bakonyi, P.; Xu, K.; Zhao, Y. Microbial electrolysis cell platform for simultaneous waste biorefinery and clean electrofuels generation: Current situation, challenges and future perspectives. *Prog. Energy Combust. Sci.* **2017**, *63*, 119–145. [CrossRef]
9. Nath, K.; Das, D. Improvement of fermentative hydrogen production: Various approaches. *Appl. Microbiol. Biotechnol.* **2004**, *65*, 520–529. [CrossRef] [PubMed]
10. Torres Zúñiga, I.; Villa-Leyva, A.; Vargas, A.; Buitrón, G. Experimental validation of online monitoring and optimization strategies applied to a biohydrogen production dark fermenter. *Chem. Eng. Sci.* **2018**, *190*, 48–59. [CrossRef]
11. Nath, K.; Das, D. Modeling and optimization of fermentative hydrogen production. *Bioresour. Technol.* **2011**, *102*, 8569–8581. [CrossRef]
12. Verea, L.; Savadogo, O.; Verde, A.; Campos, J.; Ginez, F.; Sebastian, P. Performance of a microbial electrolysis cell (MEC) for hydrogen production with a new process for the biofilm formation. *Int. J. Hydrogen Energy* **2014**, *39*, 8938–8946. [CrossRef]
13. Tartakovsky, B.; Mehta, P.; Santoyo, G.; Guiot, S. Maximizing hydrogen production in a microbial electrolysis cell by real-time optimization of applied voltage. *Int. J. Hydrogen Energy* **2011**, *36*, 10557–10564. [CrossRef]
14. Liang, D.W.; Peng, S.K.; Lu, S.F.; Liu, Y.Y.; Lan, F.; Xiang, Y. Enhancement of hydrogen production in a single chamber microbial electrolysis cell through anode arrangement optimization. *Bioresour. Technol.* **2011**, *102*, 10881–10885. [CrossRef] [PubMed]
15. Masmoudi, N.; Hachicha, M.; Kamoun, L. Hardware design of programmable fuzzy controller on fpga. In Proceedings of the FUZZ-IEEE'99, 1999 IEEE International Fuzzy Systems, Conference Proceedings (Cat. No. 99CH36315), Seoul, Korea, 22–25 August 1999; Volume 3, pp. 1675–1679.
16. Obaid, Z.A.; Sulaiman, N.; Marhaban, M.; Hamidon, M. Analysis and performance evaluation of PD-like fuzzy logic controller design based on MATLAB and FPGA. *Common Knowl.* **2010**, *10*, 11.
17. Zumel, P.; De Castro, A.; Garcia, O.; Riesgo, T.; Uceda, J. Concurrent and simple digital controller of an AC/DC converter with power factor correction. In Proceedings of the APEC, Seventeenth Annual IEEE Applied Power Electronics Conference and Exposition (Cat. No. 02CH37335), Dallas, TX, USA, 10–14 March 2002; Volume 1, pp. 469–475.

18. Charaabi, L.; Monmasson, E.; Slama-Belkhodja, I. Presentation of an efficient design methodology for FPGA implementation of control systems. Application to the design of an antiwindup PI controller. In Proceedings of the IEEE 2002 28th Annual Conference of the Industrial Electronics Society, IECON 02, Seville, Spain, 5–8 November 2002; Volume 3, pp. 1942–1947.
19. Chan, Y.F.; Moallem, M.; Wang, W. Design and implementation of modular FPGA-based PID controllers. *IEEE Trans. Ind. Electron.* **2007**, *54*, 1898–1906. [CrossRef]
20. Krim, S.; Gdaim, S.; Mtibaa, A.; Faouzi Mimouni, M. FPGA-based real-time implementation of a direct torque control with second-order sliding mode control and input–output feedback linearisation for an induction motor drive. *IET Electr. Power Appl.* **2020**, *14*, 480–491. [CrossRef]
21. Lúa, C.A.; Di Gennaro, S.; Guzman, A.N.; Ortega-Cisneros, S.; Domínguez, J.R. Digital implementation via FPGA of controllers for active control of ground vehicles. *IEEE Trans. Ind. Inform.* **2019**, *15*, 2253–2264. [CrossRef]
22. Ngo, H.Q.T.; Nguyen, H.D.; Truong, Q.V. A Design of PID Controller Using FPGA-Realization for Motion Control Systems. In Proceedings of the 2020 International Conference on Advanced Computing and Applications (ACOMP), Quy Nhon, Vietnam, 25–27 November 2020; pp. 150–154.
23. Flores-Estrella, R.; Rodríguez-Valenzuela, G.; Ramírez-Landeros, J.; Alcaraz-González, V.; González-Álvarez, V. A simple microbial electrochemical cell model and dynamic analysis towards control design. *Chem. Eng. Commun.* **2020**, *207*, 493–505. [CrossRef]
24. Kato Marcus, A.; Torres, C.I.; Rittmann, B.E. Conduction-based modeling of the biofilm anode of a microbial fuel cell. *Biotechnol. Bioeng.* **2007**, *98*, 1171–1182. [CrossRef] [PubMed]
25. Nocedal, J.; Wright, S. *Numerical Optimization*; Springer Science & Business Media: New York, NY, USA, 2006.
26. Akhtaruzzaman, M.; Shafie, A.A. Geometrical substantiation of Phi, the golden ratio and the baroque of nature, architecture, design and engineering. *Int. J. Arts* **2011**, *1*, 1–22. [CrossRef]
27. Koupaei, J.A.; Hosseini, S.M.M.; Ghaini, F.M. A new optimization algorithm based on chaotic maps and golden section search method. *Eng. Appl. Artif. Intell.* **2016**, *50*, 201–214. [CrossRef]
28. Khalil, H.K.; Grizzle, J.W. *Nonlinear Systems*; Prentice Hall: Upper Saddle River, NJ, USA, 2002; Volume 3.
29. Shtessel, Y.; Edwards, C.; Fridman, L.; Levant, A. *Sliding Mode Control and Observation*; Springer: New York, NY, USA, 2014.
30. Behrooz, P. *Computer Arithmetic: Algorithms and Hardware Designs*; Oxford University Press: Oxford, UK, 2000; Volume 19, p. 512583.
31. Ercegovac, M.D.; Lang, T. *Digital Arithmetic*; Elsevier: Amsterdam, The Netherlands, 2004.

Article

Simple Gain-Scheduled Control System for Dissolved Oxygen Control in Bioreactors

Mantas Butkus, Donatas Levišauskas and Vytautas Galvanauskas *

Department of Automation, Kaunas University of Technology, LT-51367 Kaunas, Lithuania;
mantas.butkus@ktu.lt (M.B.); donatas.levisauskas@ktu.lt (D.L.)
* Correspondence: vytautas.galvanauskas@ktu.lt; Tel.: +370-37-300-291

Abstract: An adaptive control system for the set-point control and disturbance rejection of biotechnological-process parameters is presented. The gain scheduling of PID (PI) controller parameters is based on only controller input/output signals and does not require additional measurement of process variables for controller-parameter adaptation. Realization of the proposed system does not depend on the instrumentation-level of the bioreactor and is, therefore, attractive for practical application. A simple gain-scheduling algorithm is developed, using tendency models of the controlled process. Dissolved oxygen concentration was controlled using the developed control system. The biotechnological process was simulated in fed-batch operating mode, under extreme operating conditions (the oxygen uptake-rate's rapidly and widely varying, feeding and aeration rate disturbances). In the simulation experiments, the gain-scheduled controller demonstrated robust behavior and outperformed the compared conventional PI controller with fixed parameters.

Keywords: PID (PI) control; gain-scheduling; mathematical model; biotechnological cultivation process; dissolved oxygen concentration

Citation: Butkus, M.; Levišauskas, D.; Galvanauskas, V. Simple Gain-Scheduled Control System for Dissolved Oxygen Control in Bioreactors. *Processes* **2021**, *9*, 1493. https://doi.org/10.3390/pr9091493

Academic Editors: Philippe Bogaerts and Alain Vande Wouwer

Received: 2 July 2021
Accepted: 21 August 2021
Published: 25 August 2021

Publisher's Note: MDPI stays neutral with regard to jurisdictional claims in published maps and institutional affiliations.

1. Introduction

Intense global competition, business strategies that are mainly based on profit, promptly developing social and economic conditions, high interest in better-quality control, increased safety concerns, and stringent environmental norms are prompting many process industries to automate their operations using accurate, robust, reliable, efficient, optimal, adaptive and intelligent advanced control systems [1,2]. Control-system design is greatly influenced by the number of nonlinearities present within the process. Classical controllers, such as proportional–integral–derivative (PID) or proportional–integral (PI) are adequate if the nonlinearity encountered is very mild. In presence of significant number of nonlinearities, however, such linear models are ineffective, since even small disturbances can force the process away from the operating point [3]. Control quality is influenced by the controller's ability to provide a stable performance while dealing with process variability and disturbances [1,2,4]. Accurate control of technological parameters during microorganism cultivation processes is necessary for retaining currency with desired technological regimes and reproducibility of processes. However, the dynamical parameters of batch and fed-batch cultivation processes vary widely over the cultivation cycle. Therefore, conventional control systems with fixed-gain controllers are not able to provide the required performance [5]. Temperature, pH, dissolved oxygen concentration, and other basic process variables are usually controlled in these systems [6].

Adaptive control systems of various complexity have been developed for the automatic control of cultivation process parameters under time-varying operation conditions. The system, based on process tendency models and online measurements of process variables [4,7,8], provides high-quality control under extreme operating conditions (oxygen uptake rate rapidly changing within a wide range, feeding and aeration rate disturbances). However, development of a model-based control algorithm is a time-consuming task, and, in addition, online measurements of the process variables require advanced instrumentation of the controlled process. Expert, knowledge-driven adaptive fuzzy systems are effective; however, they require deep process knowledge [3,6,9]. An approach of development for the control systems of dissolved oxygen concentration (DOC) and pH based on artificial neural network (ANN) models is presented in [10,11]. A. Mészáros et al. present ANNs that are trained off-line to predict the nonlinear dynamics of controlled processes and the inverse ANNs are used in the control systems as feedback controllers [10]. Du, Xianjun, et al. developed a radial basis function neural network based adaptive PID controller for DOC control [11]. Such development of ANN model-based control systems requires a sufficient amount of informative process data and time expenses for training the ANNs. For to these reasons, application of complex control systems is not attractive in industrial bioprocess-control engineering practice.

Model–reference adaptive control (MRAC) uses a reference model of the process that defines how the process output should respond to a command signal [12]. Although MRAC is a good alternative to PID it must be tuned for each particular process, and the tuning depends on the presence of lag, delay and other factors. For non-well-known processes, the controller must be tuned experimentally, and it could be a disadvantage from a commercial or business point of view [3].

Several gain-scheduling approach-based control systems have been developed for adaptive control of batch bioreactors. In the control systems, the oxygen uptake rate (*OUR*) [13,14] and the carbon dioxide evolution rate (*CER*) [15] are used as gain-scheduling variables. In the control systems, the *OUR* and *CER* are estimated from the online analysis of an exhaust gas. A requirement for practical realization of the above systems is that the bioreactor system is equipped with the exhaust gas analyzer.

DOC control systems have been also developed [16,17], in which the PID (PI) controller adaptation does not require additional measurements of process variables. The controller adaptation is based on the online statistical analysis of controller input and output data. Computer simulations of the control systems performance show the working capacity of the adaptation algorithms. However, optimal values of the algorithm tuning parameters are determined by a "trial and error" approach that is time-consuming. Various other PID controller-tuning approaches are presented in [18–21]. A feedforward–feedback controller was proposed, in [22], in which processes that evolve exponentially were controlled.

In order to simplify controller adaptation algorithms and practical realization of the adaptive control systems for cultivation process control, in this contribution the authors propose the gain scheduling approach, which is based on controller input/output signals only and does not require additional online measurements of cultivation process variables for adaptation of controller parameters.

2. Materials and Methods

2.1. Development of Adaptation Algorithm for DOC Control

Dynamics of the dissolved oxygen concentration (DOC) in culture medium can be represented by a simple tendency model based on the mass balance for DOC:

$$\frac{dc}{dt} = K_L a(c_{\text{sat}} - c) - OUR, \tag{1}$$

where $K_L a$ is oxygen transfer coefficient:

$$K_L a = \alpha u^\beta q^\gamma, \tag{2}$$

c is DOC, c_{sat} is saturation value of DOC, *OUR* is oxygen uptake rate, u is stirring speed (control variable), q is air supply rate, α, β and γ are parameters, and t is time.

Linearization of eq. (1) around the process state point at time t_k with respect to the state (c) and the control (u) variables represents the DOC dynamics equation at time t_k:

$$\frac{d\Delta c}{dt} = -\left[\alpha u^\beta q^\gamma\right]_{t\,=\,t_k}\Delta c + \left[\alpha\beta u^{\beta-1}q^\gamma(c_{sat} - c)\right]_{t\,=\,t_k}\Delta u. \tag{3}$$

From Equation (3), the DOC dynamics can be represented by a first-order transfer function model:

$$G_{\Delta c/\Delta u}(s) = \frac{\Delta c(s)}{\Delta u(s)} = \frac{K_{pr}(t_k)}{T_{pr}(t_k)s + 1}, \tag{4}$$

$$\text{where } K_{pr}(t_k) = \left[\frac{\beta(c_{sat} - c)}{u}\right]_{t\,=\,t_k}, \tag{5}$$

$$T_{pr}(t_k) = \left[\frac{1}{\alpha u^\beta q^\gamma}\right]_{t\,=\,t_k}. \tag{6}$$

$K_{pr}(t_k)$ and $T_{pr}(t_k)$ are process controller gain and integration time constant at time point t_k, respectively, s is Laplace operator.

The resultant dynamics of controlled process in the DOC control system also depends on the stable dynamical parameters of the motor–stirrer system and the DOC electrode. As the time constants of the above control system elements are significantly smaller, compared with the time constant $T_{pr}(t_k)$, their influence on controlled-process dynamics is taken into account by adding some time delay to the transfer function model (4).

Therefore, dynamics of the DOC control process can be roughly represented by the first-order-plus-time delay (FOPTD) model:

$$G_{\Delta c/\Delta u}(s) = \frac{K_{pr}(t_k)}{T_{pr}(t_k)s + 1}e^{-\tau}, \tag{7}$$

where τ is time delay representing influence of the control system elements dynamics.

According to PI controller tuning rules (Ziegler–Nichols, internal model control (IMC), etc. [23]), the controller gain K_c is proportional to the ratio $T_{pr}/K_{pr}/\tau$ and the integration constant T_i is proportional to the resultant time constant T_{pr}. Taking into account the functional relationships (5), (6), and assuming that the controlled value of the DOC during cultivation process is close to the set-point value ($c \cong c_{set}$), a character of relationships between the controller tuning parameters and the controller output and the set-point signals can be estimated:

$$K_c \sim T_{pr}/K_{pr},/\tau = \frac{1}{\alpha u^\beta q^\gamma}\frac{u}{\beta(c_{sat} - c)}. \tag{8}$$

Based on relationship (8), the gain scheduling algorithm for controller gain adaptation takes the following form:

$$K_c(t_k) = \frac{K_{Kc}}{(u(t_k))^{\beta-1}(c_{sat} - c_{set}(t_k))}, \tag{9}$$

where u and c_{set} are the gain scheduling variables; K_{Kc} is coefficient for tuning the controller to obtain desired performance of the control system (approximate values of the coefficient can be taken from the desired controller tuning rules). The power β of stirring speed in the oxygen transfer rate estimation Equation (6) is typically $\beta \cong 2$ [24] and the formula (9) for scheduling the controller gain coefficient can be reduced to:

$$K_c(t_k) = \frac{K_{Kc}}{u(t_k)(c_{sat} - c_{set}(t_k))}. \tag{10}$$

A character of relationships between the controller integration constant and the controller output is the following:

$$T_i \sim T_{pr} = \frac{1}{\alpha u^\beta q^\gamma}.$$ (11)

Based on relationship (11), the gain scheduling algorithm for controller integration time constant adaptation takes the following form:

$$T_i(t_k) = \frac{K_{Ti}}{(u(t_k))^2},$$ (12)

$$K_{Ti} = k_{Ti}\frac{1}{\alpha q^\gamma}.$$ (13)

where k_{Ti} is coefficient for tuning the controller to obtain desired performance of the control system (approximate value of the coefficient can be taken from the desired controller tuning rules). DOC model parameter values and initial conditions of the state variables are given in Table 1.

Table 1. DOC model parameter values and initial conditions of the state variables.

Model Parameters		
$H = 0.7906$ L mmol^{-1}	$\varepsilon = 0.15$	$T_{el1} = 10$ s
$T_{el2} = 2$ s	$T_q = 2$ s	$T_u = 1$ s
$\alpha = 0.8 \cdot 10^{-7}$	$\beta = 2$	$\gamma = 0.2$
	$v_{mol} = 0.0224$ l mmol^{-1}	
Initial Conditions		
$c_{el}(0) = 10\%$	$q(0) = 2$ s^{-1}	$u(0) = 2.5$ s^{-1}
$c_a(0) = 0.0266$ mmol L^{-1}	$y_{O2}(0) = 0.2099$	$a_{el}(0) = 10\%$

2.2. Mathematical Model of the Biotechnological Process

To simulate the biotechnological process, a mathematical model of an *E.coli* fed-batch process similar to the one presented in [25] was used:

$$\frac{dx}{dt} = \mu x - \frac{F_s + F_{pH}}{V}x,$$ (14)

$$\frac{ds}{dt} = -q_s x + \frac{F_s S_0}{V} - \frac{(F_s + F_{pH})s}{V}$$ (15)

$$\frac{dV}{dt} = F_s + F_{pH} - F_{smp}$$ (16)

$$\mu = \mu_{max}\frac{s}{K_s + s}\frac{K_i}{K_i + s}\frac{c_a}{c_a + k_c}$$ (17)

$$q_s = \frac{\mu}{Y_{xs}} - m,$$ (18)

$$F_s = \frac{\mu_{set} x V}{Y_{xs}(S_0 - s)},$$ (19)

where x—biomass concentration in the cultivation medium, gl^{-1}; μ—biomass specific growth rate, lh^{-1}; V—cultivation medium volume, l; S_0—substrate concentration in feed, gl^{-1}; F_{smp}—sampling rate- Y_{xs}—biomass/substrate yield coefficient, gg^{-1}; c_a is DOC in absolute units, mmol L^{-1}; k_c is parameter, mmol L^{-1}. The Luedeking–Piret model was used to calculate the oxygen uptake rate [26]:

$$OUR = \mu Y x V + m x V$$ (20)

Values of the model parameters are given in Table 2.

Table 2. Biotechnological model parameter values and initial values of the state variables.

Model Parameters		
$Y = 0.8646$ gg^{-1}	$m = 0.018$ gg^{-1} h^{-1}	$Y_{xs} = 0.52$ gg^{-1}
$\mu_{max} = 0.737\ 1$ h^{-1}	$K_i = 93.8$ gl^{-1}	$S_0 = 450$ gl^{-1}
$K_s = 0.02$ gl^{-1}	$k_c = 0.00265$ mmol L^{-1}	$F_{smp} = 0.025$ lh^{-1}
Initial Conditions		
$V(0) = 45$ L	$x(0) = 0.25$ gl^{-1}	$s(0) = 0.5$ gl^{-1}

A set of equations is used to model and simulate the controlled process [8]:

$$\frac{dq}{dt} = \frac{1}{T_q}(q_{set} - q), \tag{21}$$

$$\frac{du}{dt} = \frac{1}{T_u}(u_{set} - u), \tag{22}$$

$$\frac{dc_a}{dt} = -OUR_v + \alpha u^\beta q^\gamma \left(\frac{y_{O_2}}{H} - c_a\right), \tag{23}$$

$$\frac{dy_{O_2}}{dt} = \frac{q}{V}\left(\frac{1}{\varepsilon} - 1\right)(0.21 - y_{O_2}) - \alpha u^\beta q^\gamma \left(\frac{1}{\varepsilon} - 1\right)\left(\frac{y_{O_2}}{H} - c_a\right)v_{mol}, \tag{24}$$

$$\frac{da_{el}}{dt} = \frac{1}{T_{el1}}\left(100\frac{c_a H}{0.21} - a_{el}\right), \tag{25}$$

$$\frac{dc_{el}}{dt} = \frac{1}{T_{el2}}(a_{el} - c_{el}), \tag{26}$$

where q_{set} is set value of air supply rate, lh^{-1}; u_{set} is set value of stirring speed (control variable), h^{-1}; y_{O2} is portion of oxygen in exhaust gas, -; OUR_V is volumetric oxygen uptake rate, mmol L^{-1} h^{-1}, a_{el} is auxiliary variable, %; c_{el} is signal from dissolved oxygen (DO) electrode, %; H is Henry's constant, L mmol^{-1}; V is volume of cultural liquid, l; v_{mol} is volume of mmol of gas, L mmol^{-1}; T_q, T_u, T_{el1}, T_{el2} are time constants of air supply system, motor-stirrer system, and DOC electrode, respectively, s; ε is gas holdup in the gas-liquid dispersion. The dynamics of air supply and stirring systems is modelled by Equations (21) and (22). Equations (23) and (24) represent mass balances on oxygen in liquid and gaseous phases. Equations (25) and (26) are used to model the second-order dynamics of DOC electrode. A scheme of the DOC control system is depicted in Figure 1.

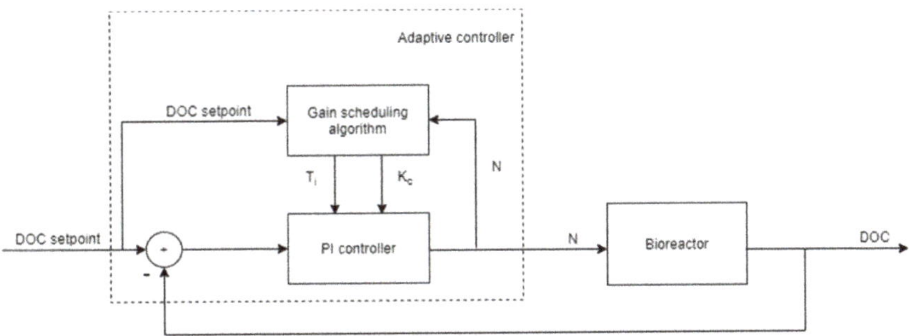

Figure 1. Block diagram of the dissolved oxygen concentration (DOC) control system.

As shown in Figure 1, the DOC adaptive control system uses only controller input/output signals for the gain scheduling algorithms.

The DOC measurements were simulated by adding Gaussian noise:

$$c_{\mathrm{el\_m}}(t_k) = c_{\mathrm{el}}(t_k) + \sigma \, Randn, \tag{27}$$

where c_{el_m} is measured value of DOC; σ is standard deviation estimated from real measurements ($\sigma \cong 0.2\%$), *Randn* is a sequence of normalized Gaussian random numbers.

In the gain-scheduling and PI-control algorithms the time discretization step $\Delta t = 0.18$ s was used throughout the simulation experiments. The simulations were carried out in Matlab/Simulink environment.

Performance of the DOC adaptive control system was investigated for set-point tracking and disturbance rejection. The developed system was compared with the standard control approach with fixed PI controller parameters presented in Table 3.

Table 3. Fixed PI controller parameters for standard control.

	Fixed Parameter Values	
	K_c	T_i
DOC control	$50\%^{-1}\,\mathrm{h}^{-1}$	$3.6 \times 10^{-4}\,\mathrm{h}$

The performance of the developed control algorithm was evaluated by calculating the mean absolute error (MAE) and comparing it to the MAE of the conventional system with fixed controller parameters.

3. Results and Discussion

3.1. DOC Control System Performance

3.1.1. DOC Set-Point Tracking Performance

The bioprocess was simulated by numerically solving the Equations (14)–(26) and by applying the controller parameter adaptation rules defined by the Equations (12) and (13) for DOC control. Typical trajectories of the bioprocess variables are presented in Figure 2 for the case when the DOC set-point tracking quality was investigated.

After inoculation, the biomass x (Figure 2a) grows in batch mode (until 1 h) consuming a small initial amount of substrate s (Figure 2b). Culture broth volume V in the bioreactor (Figure 2c) changes due to the feeding flow of the substrate Fs (Figure 2d), which is initiated at the end of the batch phase (~1 h). The biomass specific growth rate depends on the actual substrate concentration and DOC level (Equation (17), Figure 2e). Substrate oxidation and subsequent biomass growth result in oxygen consumption, which is reflected by the oxygen uptake rate OUR (Figure 2f).

During the cultivation process, DOC level is controlled by a PI controller. Both standard and gain-scheduled PI control systems were investigated and compared for the DOC control.

To reduce the large number of the presented figures, the plots with trajectories of the process variables (x, s, V, F_s, OUR) in the investigation of DOC disturbance rejection will be omitted. Only the plots for the controlled variable (DOC), manipulated variable (N), disturbance (q), and controller parameters (K_c, T_i) will be presented and discussed.

First, performance of the DOC adaptive control system was investigated for tracking set-point. In the simulation experiments, time profile of the DOC set-point change, depicted in Figure 3a was selected for the simulation fo close-to-realistic operating conditions in fed-batch cultivation process.

Performance of the gain-scheduled controller for step changes of the set-point at 5 and 7.5 process hours is presented in Figure 4. The investigated adaptive control algorithm yields in lower tracking error and shorter rise time.

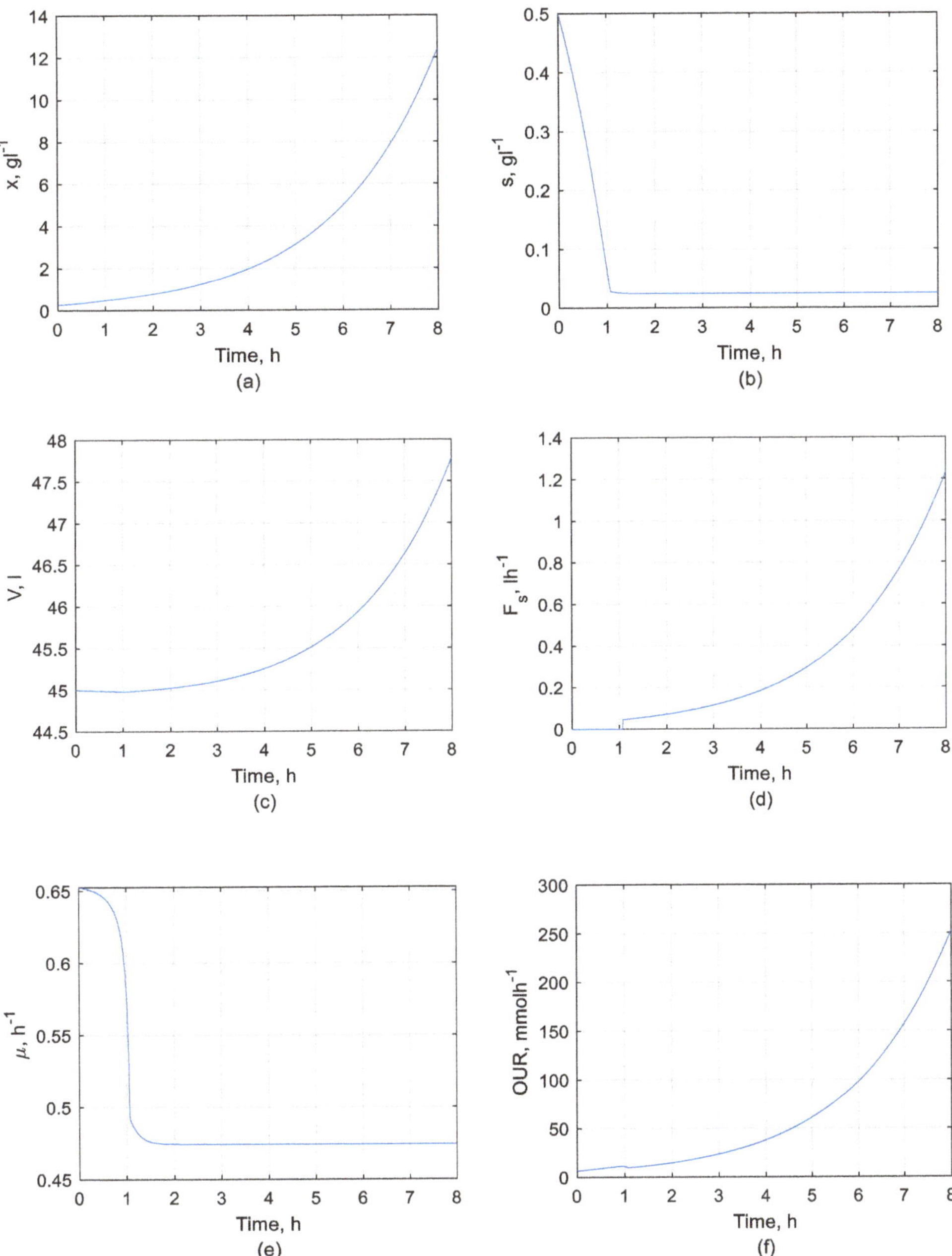

Figure 2. Trajectories of x (**a**), s (**b**), V (**c**), Fs (**d**), μ (**e**), and oxygen uptake rate (OUR) (**f**) during a DOC set-point tracking simulation run.

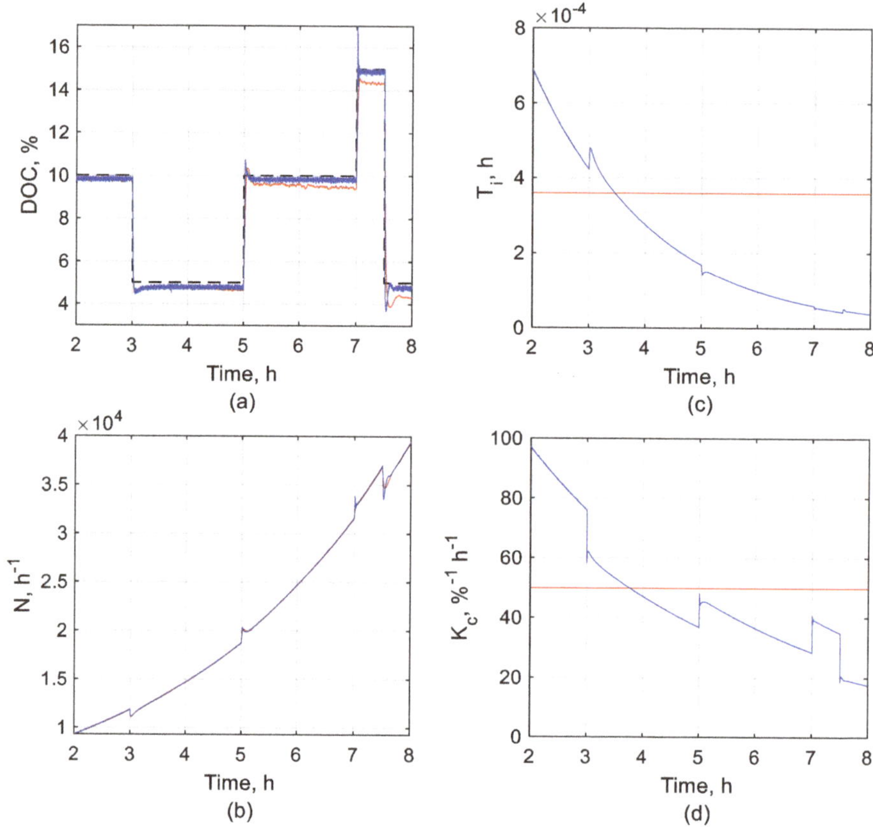

Figure 3. Trajectories of DOC (**a**), controller tuning parameter T_i (**b**), stirring speed N (**c**), and controller tuning parameter K_c (**d**). Set-point change: PI controller with fixed parameters (red), adaptive PI controller with Gain Scheduling (blue); DOC set-point (black).

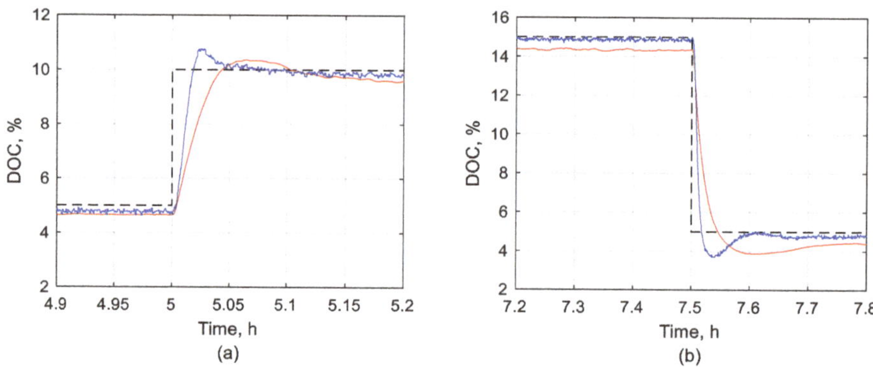

Figure 4. DOC responses (**a**) and (**b**) to set-point change: PI controller with fixed parameters (red), adaptive PI controller with Gain Scheduling (blue), DOC set-point (black).

3.1.2. DOC Disturbance Rejection Performance

To evaluate the performance of disturbance rejection, the system was simulated at a constant set-point of 10%. Air supply rate change was selected to simulate the disturbance. The change of the air supply rate is depicted in Figure 5.

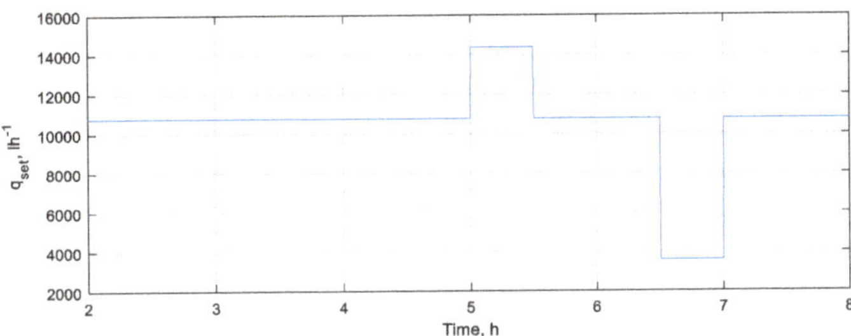

Figure 5. Air supply rate change during the simulation.

The system response and control performance are depicted in Figure 6a. The trajectory of the manipulated stirring speed N is presented in Figure 6b. Figure 6c,d highlight the adaptation of the controller tuning parameters during the simulation run.

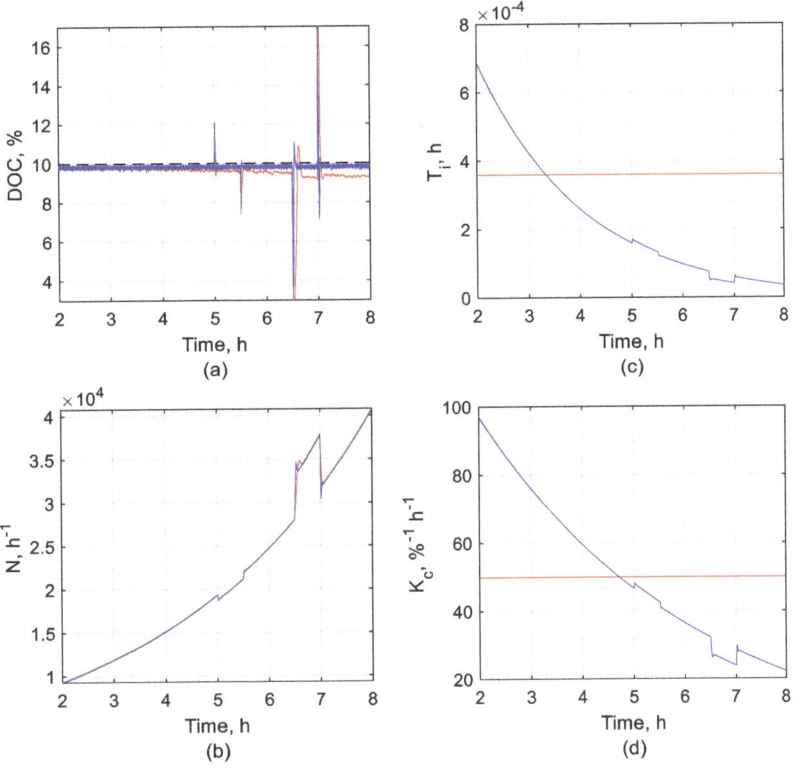

Figure 6. Trajectories of DOC (**a**), controller tuning parameter T_i (**b**), stirring speed N (**c**), and controller tuning parameter K_c (**d**). Disturbance rejection when using: PI controller with fixed parameters (red), adaptive PI controller with Gain Scheduling (blue); DOC set-point (black).

Performance of the gain-scheduled controller for disturbance compensation (the air supply rate step change from 10,800 lh^{-1} to 14,400 lh^{-1} occurred at t = 5.5 h, and from 14,400 to 3600 lh^{-1} occurred at t = 7 h) is presented in Figure 7. The adaptive control system yields lower tracking error, as well reduces the overshoot.

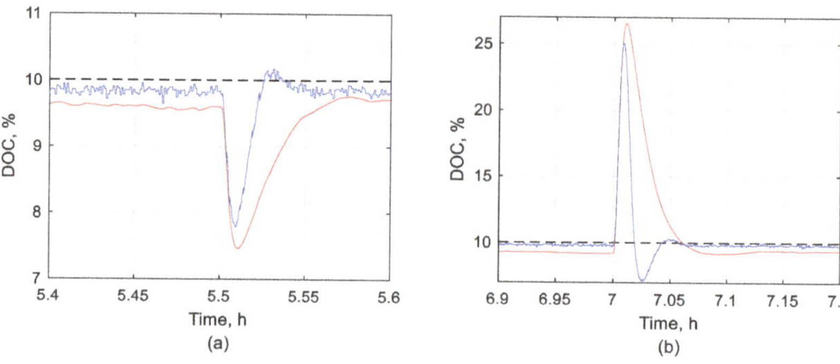

Figure 7. DOC disturbance compensation cases (**a**) and (**b**): PI controller with fixed parameters (red), adaptive PI controller with Gain Scheduling (blue), DOC set-point (black).

Simulation results show that the gain scheduled PI controller ensures good control quality of DOC under extreme operating conditions and evidently outperforms the conventional PI controller. The integration time constant T_i and the controller gain K_c changed in a wide range, therefore reflecting the significantly varying dynamics of the process. Analysis of the simulation results shows that the adaptive system has reduced the mean absolute error more than 2 times for the investigated control schemes. The rise time of the transient processes caused by the set-point change was approx. 2 times shorter for the adaptive system (see Figure 4a,b). However, both investigated systems yielded similar rise times in case of disturbance rejection (see Figure 7a,b). The control performance of the investigated systems is summarized in Table 4. The adaptive control algorithm outperforms the standard system approx. 2 times in terms of mean absolute error.

Table 4. Tuning parameters and MAE values for the investigated DOC control systems.

Control Type	Tuning Parameters	Mean Absolute Error	
		Disturbance Rejection	Set-Point Tracking
Standard DOC	$K_c = 50\%^{-1}\,h^{-1}$, $T_i = 3.6 \times 10^{-4}\,h$	0.166	0.071
Adaptive DOC	$K_{Ti} = 0.6 \times 10^5$, $K_{Kc} = 1.5 \times 10^5$	0.063	0.028

4. Conclusions

In this paper, a simple adaptive control system for the set-point control and disturbance rejection of dissolved oxygen concentration is proposed, in which gain scheduling of PID (PI) controller is based on the controller input/output signals only and, therefore, does not require online measurements of process variables for development of gain scheduling algorithms. Realization of the proposed system does not depend on the instrumentation level of the bioreactor and is attractive for practical application.

The controller input/output-based gain scheduling algorithms were developed for set-point tracking and disturbance rejection during DOC control for bioreactor operating both in batch and fed-batch mode. Performance of the gain-scheduled PI controller under extreme operating conditions was investigated by computer simulation. The results demonstrate obvious advantage of the proposed control system compared to conventional PI control systems.

In future work, the authors are planning to perform further experimental investigation by testing the system under real conditions.

Author Contributions: Conceptualization, M.B., D.L. and V.G.; methodology, M.B., D.L. and V.G.; software, M.B. and V.G.; writing—original draft preparation, M.B.; writing—review and editing, V.G.; supervision, V.G.; project administration, V.G.; funding acquisition, M.B., D.L. and V.G. All authors have read and agreed to the published version of the manuscript.

Funding: This research was funded by the European Regional Development Fund according to the supported activity "Research Projects Implemented by World-class Researcher Groups" under Measure No. 01.2.2-LMT-K-718.

Institutional Review Board Statement: Not applicable.

Informed Consent Statement: Not applicable.

Conflicts of Interest: The authors declare no conflict of interests.

References

1. Boudreau, M.A.; McMillan, G.K. *New Directions in Bioprocess Modelling and Control: Maximizing Process Analytical Technology Benefits*; ISA: Durham, NC, USA, 2007.
2. Dochain, D. *Bioprocess Control*; ISTE: London, UK, 2008.
3. Singh, P.K.; Bhanot, S.; Mohanta, H.K.; Bansal, V. Design and implementation of adaptive fuzzy knowledge based control of pH for strong acid-strong base neutralization process. *J. Engl. Res.* **2020**, *8*.
4. US Food and Drug Administration: Guidance for Industry PAT—A Framework for Innovative Pharma-Ceutical Development, Manufacturing and Quality Assurance. 2004. Available online: https://www.fda.gov/regulatory-information/search-fda-guidance-documents/pat-framework-innovative-pharmaceutical-development-manufacturing-and-quality-assurance (accessed on 27 May 2021).
5. Simutis, R.; Lübbert, A. Bioreactor control improves bioprocess performance. *Biotechnol. J.* **2015**, *10*, 1115–1130. [CrossRef]
6. Butkus, M.; Repšytė, J.; Galvanauskas, V. Fuzzy logic-based adaptive control of specific growth rate in fed-batch biotechnological processes: A simulation study. *Appl. Scien.* **2020**, *10*, 6818. [CrossRef]
7. Levišauskas, D. An algorithm for adaptive control of dissolved oxygen concentration in batch culture. *Biotechnol. Tech.* **1995**, *9*, 85–90. [CrossRef]
8. Levišauskas, D.; Simutis, R.; Galvanauskas, V. Adaptive set-point control system for microbial cultivation processes. *Non. An. Mod. Cont.* **2016**, *21*, 153–165. [CrossRef]
9. Babuška, R.; Damen, M.R.; Hellinga, C.; Maarleveld, H. Intelligent adaptive control of bioreactors. *J. Intell. Manuf.* **2003**, *14*, 255–265. [CrossRef]
10. Mészáros, A.; Andrášik, A.; Mizsey, P.; Fonyo, Z.; Illeová, V. Computer control of pH and DO in a laboratory fermenter using a neural network technique. *Bioprocess. Biosyst. Eng.* **2004**, *26*, 331–340. [CrossRef]
11. Du, X.; Wang, J.; Jegatheesan, V.; Shi, G. Dissolved oxygen control in activated sludge process using a neural network-based adaptive PID algorithm. *Appl. Scien.* **2018**, *8*, 261. [CrossRef]
12. Palancar, M.C.; Aragón, J.M.; Miguéns, A.J.A.; Torrecilla, J.S. Application of a model reference adaptive control system to pH control. effects of lag and delay time. *Ind. Eng. Chem.* **1996**, *35*, 4100–4110. [CrossRef]
13. Cardello, R.J.; San, K.-Y. The design of controllers for batch bioreactors. *Biotechnol. Bioeng.* **1988**, *32*, 519–526. [CrossRef]
14. Kuprijanov, A.; Gnoth, S.; Simutis, R.; Lübbert, A. Advanced control of dissolved oxygen concentration in fed batch cultures during recombinant protein production. *Appl. Microbiol. Biotechnol.* **2009**, *82*, 221–229. [CrossRef]
15. Gnoth, S.; Kuprijanov, A.; Simutis, R.; Lübbert, A. Simple adaptive pH control in bioreactors using gain-scheduling methods. *Appl. Microbiol. Biotechnol.* **2010**, *85*, 955–964. [CrossRef] [PubMed]
16. Hwang, Y.B.; Lee, S.C.; Chang, H.N.; Chang, Y.K. Dissolved oxygen concentration regulation using auto-tuning proportional-integral-derivative controller in fermentation process. *Biotechnol. Tech.* **1991**, *5*, 85–90. [CrossRef]
17. Levisauskas, D.; Simutis, R.; Galvanauskas, V.; Urniezius, R. Simple control systems for set-point control of dissolved oxygen concentration in batch fermentation processes. *Chem. Eng. Trans.* **2019**, *74*, 127–132.
18. Smets, I.Y.; Claes, J.; November, E.J.; Bastin, G.P.; Van Impe, J.F. Optimal adaptive control of (bio) chemical reactors: Past, present and future. *J. Process. Control.* **2004**, *14*, 795–805. [CrossRef]
19. Bastin, G.; Impe, J.F. Nonlinear and adaptive control in biotechnology: A tutorial. *Eur. J. Control* **1995**, *1*, 37–53. [CrossRef]
20. Butkus, M.; Simutis, R.; Galvanauskas, V. Unified structure of adaptive system for control of basic process variables in biotechnological cultivation processes: Ph control system case study. *Chem. Eng. Trans.* **2021**, *86*, 985–990.
21. Galvanauskas, V.; Simutis, R.; Vaitkus, V. Adaptive control of biomass specific growth rate in fed-batch biotechnological processes. A comparative study. *Process* **2019**, *7*, 810. [CrossRef]
22. Brignoli, Y.; Freeland, B.; Cunningham, D.; Dabros, M. Control of specific growth rate in fed-batch bioprocesses: Novel controller design for improved noise management. *Process* **2020**, *8*, 679. [CrossRef]

23. Levine, W.S. (Ed.) *The Control. Handbook*; IEEE/CRC Press: Boca Raton, FL, USA, 2011.
24. Villadsen, J.; Nielsen, J.; Liden, G. *Bioreaction Engineering Principles*; Springer: Berlin/Heidelberg, Germany, 2011.
25. Levisauskas, D.; Galvanauskas, V.; Henrich, S.; Wilhelm, K.; Volk, N.; Lübbert, A. Model-based optimization of viral capsid protein production in fed-batch culture of recombinant Escherichia coli. *Bioproc. Biosyst. Eng.* **2003**, *25*, 255–262. [CrossRef]
26. Luedeking, R.; Piret, E.L. A kinetic study of the lactic acid fermentation. Batch process at controlled pH. *Biotechnol. Bioeng.* **2000**, *67*, 636–644. [CrossRef]

MDPI
St. Alban-Anlage 66
4052 Basel
Switzerland
Tel. +41 61 683 77 34
Fax +41 61 302 89 18
www.mdpi.com

Processes Editorial Office
E-mail: processes@mdpi.com
www.mdpi.com/journal/processes